AF687061

The Smart Campus Manifesto

Transforming IT and OT
for the AI-Enabled Sustainable Smart Campus

Dr. Steven Burrell

Point Peter Press
St. Marys, Georgia

Copyright © 2026 by **Steven C Burrell**

All rights reserved. No part of this publication may be reproduced, distributed, or transmitted in any form or by any means without prior written permission.

This book is an independent publication and is not affiliated with, authorized, sponsored, or endorsed by any private or corporate entity.

Steven Burrell / Point Peter Press
St. Marys, Georgia

Book Layout © 2017 BookDesignTemplates.com

The Smart Campus Manifesto:
Transforming People, Technology, and Processes for the AI-Enabled Sustainable Smart Campus / Steven Burrell. -- 1st ed.

April 19, 2026

ISBN 979-8-9952263-0-7

THE AUTHOR

Dr. Steven C. Burrell is an executive leader, author, strategist, and futurist with over 40 years of experience driving higher education forward through intentional, human-centered technology transformation. Recognized by EdTech Magazine in 2020 as one of the top 30 higher education IT influencers, he has shaped his career by linking digital innovation to institutional mission, culture, and student success.

Dr. Burrell has served as Vice President and Chief Information Officer at five institutions, including doctoral research universities and community colleges, most recently as CIO at Northern Arizona University. Throughout his career, he has led complex statewide and institutional strategies, multi-million-dollar budgets, and large-scale digital infrastructure initiatives focused on access, resilience, and innovation. A committed mentor and educator, Dr. Burrell has played a key role in developing the next generation of higher education IT leaders through programs such as the Next Leaders Fellowship, EDUCAUSE leadership initiatives, Internet2, and national community leadership efforts.

Beyond campus leadership, he is a trusted advisor in the public, private, and nonprofit sectors. He has contributed to statewide AI policies, Indigenous nation technology planning, and national research and education networks. He has served on international CIO advisory boards for Microsoft© and Salesforce©. Additionally, he has served on public boards throughout the U.S. and was part of Arizona Governor Katie Hobbs' AI Advisory Taskforce. His work has been instrumental in bridging the digital divide in rural and tribal communities across Southeastern states. Currently, he works as an independent consultant and podcast host, guiding institutional leaders in the future of education and smart infrastructure.

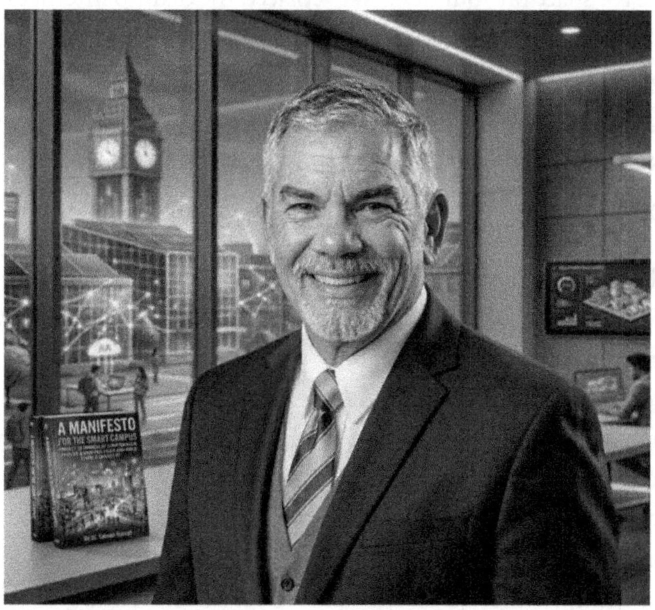

Forward

The quest for the smart campus has been decades in the making. Throughout my 40+ year career, I worked closely with colleagues in Facilities Services and institutional leaders to develop the smart campus. In the early 80s, it was about integrating projectors and computers into classrooms and laboratories. We saw great opportunities to improve learning through these new "personal" computing and emerging networking technologies. I grappled with the VHS vs. BetaMax decision and faced challenges from most faculty in keeping the overhead projector in these new "smart" classrooms. Facilities "intelligence" was limited to analog standalone controllers for HVAC and lighting, and IT was focused on mainframe data processing and reporting (on paper). We ditched the card catalog for mainframe computer-based "smart" capabilities in our libraries. Then, in 1986, we watched Apple's *The Knowledge Navigator* video with awe and wonder, dreaming of the day when we could talk to our computer as if it were a graduate assistant.[1]

By the early 90s, the Internet arrived on campus. We expanded our dial-up modem pools to connect people to campus, and we searched for space and raceways to install data wiring closets. We also explained to the space-planning committee why it was a bad idea to collocate our network gear in the janitor's closet. By the mid-90s, we were experimenting with UBC's open-source WebCT learning management system and were captivated by Mosaic and Netscape Navigator. We built more computer labs and connected additional classrooms, offices, and residence hall rooms to the network. We still made money by charging .39 cents per minute for long-distance voice services on campus. The Chair of the English Department was furious at me for installing spell and grammar check on computers in the writing lab.

On the first day of the 2000s, we dodged Y2K Armageddon, but soon after, the "dot-com" market bubble burst, and we pumped the technology brakes. We shifted our focus to establishing IT standards and staying neutral in the Apple© vs. Microsoft© battle for space on your desk. We replaced the dial-up internet modem pool with impressive T1 lines that offered 1.544 Mbps of bandwidth. Many of us fell in love with Blackberry (2002) and later the Apple© iPhone (2007). By the end of the decade, Google's Chrome browser dominated Windows™, and Safari owned the Mac OS desktop. Campus Ethernet networks became commonplace, and various wireless networks were emerging everywhere. Everyone was craving ubiquitous access to technology and information - can you hear me now?

When the 2010's rolled around, technology was accelerating again at a dizzying pace and finding its way into nearly every aspect of our lives at home and at work. Hundreds of web applications and smartphone apps orchestrated our day. We deployed electronic door locks and biometric devices for facility access and monitored them with "smart" surveillance tools. Students were taking their LMS-web-hosted course entirely on their smartphones. We stopped checking out resources from the library in favor of the now-commonplace digital variants, which we could read while in the coffee shop, where the circulation desk was when Dad went to school. All the information we needed was right there on our phone. Like the apps on our phones, the world was conforming to our "favorites," and we loved the confirmation bias in our media feeds. And then, COVID hit. So, we dropped a pin on a map, experienced our first robot-delivered coffee and bagel, and created the "smart campus at home." I suppose necessity is the mother of all invention.

If the notion of the smart campus evolved slowly, almost predictably over the past four decades, the last five years have been a rapid escalation to a crescendo - a confluence of technological capabilities. All coming together to open up doors to "smart." This confluence is challenging us to rethink not only *what* we do with technology, but *why* we do it in the first place. This timing couldn't be better.

If necessity truly is the mother of all invention, then Higher Education Institutions are on the brink of an incredible cycle of re-invention. And we must adapt. We are increasingly challenged to respond to the current crisis in a rapidly changing world, with new generational demographics and cultures, shifting socio-economic paradigms, confusing government policies, unpredictable inflationary economics, declining trust and public funding, and AI-driven workforce needs that evolve week to week.

Ultimately, this convergence opens opportunities for institutions to revisit traditional viewpoints, reevaluate priorities, and pursue innovation to transform the higher education experience around the concept of a smart campus. The Smart Campus shifts our strategic focus from merely riding the wave of change, initiating localized IT projects, pursuing sustainability goals, or managing decommissioned and deferred maintenance of building assets to an intentionally designed, dynamic, and smart campus driven by data, made intelligent by AI, and resilient to disruptions.

The outcomes are as significant as they are measurable: institutions are achieving 15–25% reductions in energy costs, extending the lifespan of vital infrastructure, and using real-time analytics to boost energy efficiency and sustainability. More importantly, these technologies directly influence the "human edge" of education by enabling seamless experiences—from automated administrative tasks to proactive wellness alerts—that substantially enhance student retention and success. By connecting operational efficiency with quality of life, the smart campus has become a living lab for the future, anticipating the needs of every student, researcher, and staff member within a responsive, adaptive, and intelligent ecosystem.

That old saying, “if you’ve seen one university, you’ve seen one university,” rings especially true in the context of the Smart Campus. Each institution will approach opportunities, develop strategies, and implement tactics that are meaningful to its unique mission and situation. It's impossible to create a one-size-fits-all manual for such a broad topic, given the differences among institutions. My goal in writing this “Manifesto” about smart campuses isn't to provide a step-by-step recipe, but rather to offer a “culinary” concept that sparks thinking and discussion around innovation and discovery. This approach can help expand our shared knowledge, enabling institutional leaders to advance their campuses. After all, even the most elite institutions must be smart to evolve and change or otherwise die.

Best wishes, and enjoy the journey toward a smarter campus.

Steve Burrell

CONTENTS

Part V Executing Vision, Strategy, and Plans

Appendices

Part I

Introduction

Executive Summary

The AI-enabled, sustainable smart campus is not a technology program, a facilities upgrade, or a sustainability initiative. It is an institutional operating model.

Higher education is in an era defined by persistent volatility. Climate disruption, aging infrastructure, workforce scarcity, declining public trust, constrained capital, and the accelerating impact of artificial intelligence are no longer episodic challenges—they are the operating conditions of the institution. In this environment, campuses designed on decades-old assumptions and operating norms are becoming structurally fragile.

A modern campus is a cyber-physical system—a living sentient ecosystem where buildings, utilities, safety systems, data platforms, and people continuously interact. Decisions about the physical campus are no longer separate from questions of cybersecurity, financial resilience, student retention and success, and stakeholder trust. Treating these domains as isolated silos introduces hidden risk, increases costs, and undermines long-term viability.

The future of higher education smart campuses will not be characterized by those who deploy the most advanced technology, the most sensors, or the most siloed pilot projects—rather, it will be shaped by leaders who act collaboratively with purpose, govern effectively, and integrate people, platforms, data, and strategy to create a sustainable and responsive campus that is a living system that speaks to us

Emerging and Converging Technologies Create New Opportunities

Smart campus technologies are essential tools, not just gadgets needing justification. They serve as leverage points—mechanisms enabling institutions to maximize value, resilience, and learning from the same physical space, workforce, and limited capital. When implemented as an integrated system rather than isolated pilots, these technologies transform campus operations, capabilities, and adaptability. They excel by simultaneously delivering four main benefits: creating efficiencies, enabling new economic opportunities, improving human experience, and increasing institutional effectiveness. Evolving IT and OT capabilities require new leadership approaches, shared governance, and purpose-built architecture that automagically turns sensors into insights, platforms into systems, and ambition into durable institutional capabilities to adapt, adopt, and grow through improved decision-making and services.

Convergence of Leadership

This Manifesto posits that institutions prepared for the future should purposefully integrate Information Technology (IT), Operational Technology (OT), facilities management, sustainability initiatives, and governance within a cohesive architectural and decision-making system. Through the use of governed data estates, interoperable platforms, and strong leadership, the campus network becomes more scalable, easier to control, and better positioned for continuous enhancement. Effective smart campus leadership starts when silos are eliminated, establishing integrated CTO–CFaO perspectives, collaboration, and governance that foster operational clarity and trust, creating aligned priorities, language, culture, and decision-making across the institution.

Beyond Resilience Towards Antifragility

This transformation of leadership and practice centers on moving from resilience to antifragility. Resilience helps institutions withstand disruptions and recover, while antifragility enables them to learn, adapt, and grow through disruptions. Given the challenges of climate instability, cyber-physical risks, supply chain volatility, and rapid AI advances, resilience alone is inadequate. Institutions need to develop campuses that become smarter and more efficient under stress, rather than more fragile. Resilience helps universities survive disruption. Antifragility helps them evolve.

Sustainability as an Economic and Operational Imperative

In this model, sustainability is more than a stand-alone function tasked with meeting targets or producing reports; it is a core economic and operational principle. Through circular campus design, comprehensive lifecycle visibility, material intelligence, and feedback-based capital planning, sustainability shifts from standalone initiatives to a lasting institutional strength—helping to cut waste, manage costs, and enhance long-term resilience. A sustainable smart campus isn't just about achieving targets; it's about maintaining long-term viability.

Governance at the Foundation

The Manifesto underscores that governance is the foundational framework of the smart campus. Governance provides clarity on decision rights, standards, and guardrails that prevent fragmentation and ensure modernization advances safely and strategically toward antifragility. Without well-defined decision rights, cybersecurity measures, data stewardship, and ethical AI oversight, modernization risks devolving into disjointed pilots, vendor lock-in, and superficial innovation. High-functioning governance enables institutions to expand intelligence securely while maintaining trust, accountability, and alignment with their mission.

Redefining Operational Efficiency and Anticipating Change

Traditional campus operations rely on lagging indicators such as alarms after failures, budgets after overruns, and maintenance after equipment degradation. In contrast, smart campuses use AI-driven fault detection and predictive maintenance to identify equipment issues, inefficiencies, and impending failures weeks or months in advance. This approach reduces emergency work orders, prolongs asset life, and stabilizes campus operations. Condition-aware building automation dynamically adjusts HVAC, lighting, and ventilation based on occupancy, weather, and energy data, minimizing waste while maintaining comfort. Additionally, integrated operational dashboards replace scattered reports with real-time data, allowing quicker and more informed decisions across IT, Facilities, Safety, and Sustainability teams. These are not incremental gains. Institutions routinely achieve double-digit reductions in energy, maintenance, and downtime costs—not by cutting services, but by eliminating invisibility.

Creating New Economies that Turn the Campus into a Value Engine

The smart campus introduces new economic models to environments that have historically been treated as sunk costs. AI-enabled Digital Twins transform buildings and infrastructure into financial decision assets—enabling scenario modeling for capital planning, climate adaptation, and space consolidation before capital is committed. **Grid-**interactive energy systems—microgrids, battery storage, and demand-response automation—turn campuses from passive energy consumers into active market participants, reducing exposure to price volatility while funding modernization. Lifecycle and material intelligence (asset analytics, material passports, circular procurement) convert waste, early replacement, and deferred maintenance into reclaimable value. In this model, cost savings are not incidental. They become self-funding mechanisms for continuous reinvestment—especially critical in an era of declining enrollment margins and capital scarcity.

Elevating the Campus Experience Through Human-Centered Design

Smart technologies matter most where they become *invisible*—quietly improving daily life. Adaptive learning and working environments automatically tune light, temperature, acoustics, and air quality to improve cognitive performance, health, and inclusion. Seamless digital–physical services—from wayfinding and space booking to safety alerts and service requests—replace friction with predictability. Ambient interfaces and smart campus apps personalize experience without surveillance, guided by strong governance and privacy-first design.

For students, faculty, and staff, the result is not "more technology"—it is less friction, fewer disruptions, and a campus that responds intelligently rather than bureaucratically.

Improving Institutional Effectiveness and Agility

A key but often unseen benefit of a smart campus is faster institutional decision-making. By creating shared data estates and semantic models, the campus synchronizes Technology, Facilities, Sustainability, Safety, and Academic planning around a unified, trusted operational core. This synchronization reduces decision delays: leaders no longer argue over data accuracy but instead focus on what actions to pursue. Robust governance ensures that this increased speed does not compromise security, privacy, ethics, or lifecycle accountability, enabling safe scaling of innovation. Performance metrics now focus on outcomes—linking energy consumption, space efficiency, safety, and human experience directly to mission goals. As a result, strategic planning becomes a continual, iterative process rather than a one-time event, giving the institution the agility to adapt confidently and coherently—transforming insights into action rapidly enough to meet the challenges of a volatile world.

The Strategic Case

Institutions do not invest in smart campuses because they want more technology. They invest because the alternative is fragility. In a world of climate volatility, cyber-physical risk, enrollment uncertainty, labor shortages, and constrained public funding, campuses optimized for yesterday's assumptions will fail quietly—and at great expense.

A smart campus strategy offers a unique opportunity to integrate cost efficiency, carbon savings, risk management, and an enhanced user experience within a single operational framework. It enables institutions to maximize the value of existing assets rather than delay or replace them. This boosts resilience without increasing complexity, creating opportunities for funding sustainability through operational intelligence rather than relying on sporadic grants. It also supports safe, predictable, and ethical scaling of innovation. This approach isn't about being "cutting edge," but about ensuring viability, governability, and adaptability for the coming decades.

The Bottom Line

At the core of this work is a call for IT and OT convergence—not just at the network level, but across governance, architecture, data, and leadership. The convergence of IT and Facilities is not a technical footnote; it is a leadership discipline. Technology and Facilities leaders operate under different risk models, time horizons, and success metrics, all of which are valid. When unmanaged, these differences create fragility. When aligned through a shared architecture, interoperable data estates, and clear decision rights, they create durable institutional value that neither domain can achieve on its own. A smart campus is not a collection of disparate technology systems. It is an integrated capability stack—one that turns physical infrastructure into an active partner in mission delivery. Institutions that build leadership and invest intentionally will operate more efficiently. They will learn faster, spend smarter, innovate more deeply, respond better, and endure longer.

The Rise of the Sentient Campus

The sentient campus is the moment infrastructure begins listening, learning, analyzing, and informing decision-making towards an environment that is ultimately responsive, anticipatory, and humane.

For many years, campus management was divided into two separate areas: facilities teams handled the physical infrastructure, like bricks, mortar, chillers, and boilers, while IT teams managed the digital infrastructure, including networks, servers, and data. These areas seldom overlapped, except maybe when cooling a server room. That separation is now a thing of the past.

Today, a "Smart Building" is more than just a structure with energy-efficient windows; it functions as a complex data source within a broader institutional network. We need to shift from managing isolated static assets to coordinating an integrated Digital-Physical Ecosystem that is consistently deployed across the campus. This approach expands the concept of a stand-alone smart building to include technology-enabled pedestrian ways, digital signage, wayfinding, parking, transportation, and personalized applications and services, forming a broader set of "smart" features that, together, create a **Sentient Campus**.

Higher education is nearing a pivotal moment. Institutions face rising operational costs, demographic changes, and shifting enrollment trends, all amid growing political instability that could alter funding structures and academic agendas. Concurrently, demands for equity, accessibility, and digital advances are intensifying, and environmental concerns have transformed sustainability from a mere goal into a core institutional duty.

These mounting pressures demand more than just careful management and budget cuts during a difficult year; they call for a fundamental shift in approach. Small, standalone upgrades or isolated projects won't suffice. Universities that succeed will be those that undertake bold transformations to create campuses that are integrated, intelligent, and sustainable—spaces designed to operate dependably, adapt quickly, and demonstrate transparent, measurable stewardship over time.

In this changing environment, depending on old methods is more than inefficient—it risks making the campus irrelevant. A forward-looking campus combines smart technologies with sustainable design to boost building and utility efficiency, improve resilience, and enhance everyday learning and living experiences. This approach is not just a technological trend but a strategic response to ongoing structural challenges.

The development of an **AI-enabled Sustainable Smart Campus** is no longer aspirational—it is strategic. It is not merely about installing sensors or improving utility dashboards. It is about using digital transformation to reimagine the student experience, support faculty innovation, and build agile, efficient, and environmentally responsible operational systems.

When done well, smart campus initiatives become catalysts for cross-functional collaboration—aligning academic, administrative, and auxiliary units around shared outcomes and common platforms. This work goes beyond institutional limits. Universities can collaborate with industry partners and local communities to accelerate research, test new solutions, and share their progress—enhancing regional influence and bolstering their leadership in sustainability and innovation.

The AI-enabled Sustainable Smart Campus exceeds resilience, showcasing anti-fragility and serving as a catalyst for enterprise transformation. It connects infrastructure and services through secure digital platforms, allowing the institution to operate with greater insights, measurable results, and fostering a culture of collaboration and agility. This enables the institution to adapt quickly to changing markets, expectations, and responsibilities.[2]

The Emergence of Operational Technology

Transforming a university into a cognitive, sustainable, smart campus involves a significant shift for the CTO. They must move beyond traditional digital data boundaries, crossing from IT infrastructure into physical systems. Today's IT portfolio is not limited to server racks or access points; it increasingly encompasses boilers, chillers, security systems, and specialized research labs that form the institution's **Operational Technology (OT)** refers to the hardware and software that detect or cause a change through the direct monitoring and/or control of physical devices, processes, and events within the enterprise. In the campus context, OT is the nervous system of the physical plant.

This environment consists of three closely linked but separate spheres: the essential "Core OT Domains" that support everyday operations, the "**New Edge**" technologies that drive strategic growth, and Intelligent systems that drive new efficiencies and capabilities - all drivers of the university's future.

Core OT Domains

- **Building Automation Systems (BAS)**
  Acting as the interconnected digital brain of the campus, these systems leverage sensor networks to control critical infrastructure such as HVAC and lighting, necessitating strict governance to balance occupant comfort and energy efficiency against cybersecurity risks.
- **Utilities and Energy**
  This infrastructure forms the strategic foundation of campus resilience and decarbonization, requiring robust OT security to ensure the uninterrupted delivery of power, heating, and cooling necessary for instruction and research.
- **Physical Security**
  Evolving into complex cyber-physical platforms that integrate access control and surveillance, these systems demand rigorous configuration and network segmentation to protect digital integrity and fulfill the institution's duty of care.
- **Specialized Research and Operational Support**
  Highly sensitive environments like clean rooms and cold storage require the strictest safety and uptime standards, as poorly timed digital changes or network disruptions can cause catastrophic research or compliance failures.

The "New Edge"

As campuses modernize, a new frontier emerges where traditional Operational Technology evolves beyond its mechanical roots and becomes a core driver of institutional digital strategy. This "New Edge" represents the moment when rapidly expanding sensor grids and cloud-connected systems shift OT from isolated building controls to a strategic platform that shapes innovation, resilience, and long-term competitiveness. Three examples of this new edge are:

- **Sensor Grids**
  While providing the foundational, real-time visibility needed for predictive intelligence, these rapidly scaling sensor networks demand disciplined lifecycle management and network segmentation to prevent the dangerous proliferation of unmanaged "**Shadow OT**."
- **Digital Twins**
  By fusing spatial records, asset metadata, and real-time telemetry, digital twins create a continuously updated, predictive model of the campus that drives intelligent maintenance and carbon-aware load management.

- **EV Charging Networks**
  Operating at the intersection of grid interaction, payment processing, and vendor cloud operations, electric vehicle (EV) infrastructure requires strict institutional standardization to mitigate overlapping physical, financial, and cyber vulnerabilities.

The Emergence of Intelligent OT

As Operational Technology becomes a strategic digital platform, a new layer of intelligence is emerging across the campus—one driven by Artificial Intelligence and the expanding power of Digital Twins. These capabilities transform raw operational signals into prediction, automation, and insight, enabling the campus to think, respond, and adapt in ways previously impossible.

- **Artificial Intelligence (AI)**
  AI systems analyze real-time data from energy systems, HVAC, space use, and campus services to optimize performance, predict failures, personalize learning, and automate routine operations. This evolution shifts the campus from reactive management to proactive, continuously improving decision-making.
- **Digital Twins**
  Digital Twins create dynamic, data-driven virtual replicas of buildings, infrastructure, and campus environments. By integrating real-time telemetry with spatial and asset data, they enable simulation, scenario planning, predictive maintenance, and energy optimization—turning the campus into a living, learning system by fusing spatial records, asset metadata, and real-time telemetry. Digital twins create a continuously updated, predictive model of the campus that drives intelligent maintenance and carbon-aware load management.

To effectively oversee this complex cyber-physical environment—balancing rapid innovation with safety and uptime—the CTO must implement a new operational approach. This manifesto aims to provide the strategic clarity required to view these systems not as isolated utilities but as interconnected assets that serve the institution's mission.

The Cognitive Engine of the Smart Campus

In higher education's changing landscape, Artificial Intelligence (AI) functions as more than a simple tool—it's the core cognitive force powering the smart campus. By allowing machines to mimic human intelligence—learning from data, recognizing patterns, forecasting results, and automating decisions—AI is revolutionizing institutional operations, teaching methods, and innovation. Key to this change is Machine Learning (ML), a vital subset of AI that consistently improves performance by adjusting to real-world data without needing frequent reprogramming.

A modern AI-enabled smart campus utilizes these technologies to gain valuable insights, provide tailored experiences, optimize operations, and promote sustainability. Its extensive network of IoT sensors, digital infrastructure, and interconnected systems continuously produces data. AI processes this data to enable institutions to respond more quickly, operate more efficiently, and develop sustainable, engaging environments. The main ways AI is transforming the smart campus of the future include the following key areas.

Intelligent Energy Management

AI systems continuously analyze real-time data from heating, cooling, lighting, and power systems to optimize energy use across campus buildings. Rather than relying on static schedules, AI dynamically adjusts environmental controls based on occupancy, weather patterns, time of day, and historical usage. The result: reduced energy waste, lower costs, and a significantly smaller carbon footprint—all without sacrificing comfort or functionality.

Condition-Based Predictive Maintenance

The days of reactive repairs and costly downtime are over. AI-driven predictive maintenance platforms monitor essential campus systems—from HVAC units to elevators—and anticipate failures before they occur. This allows maintenance teams to tackle issues proactively, extend the lifespan of infrastructure, minimize disruptions, and decrease resource-intensive emergency interventions.

Personalized Learning and Academic Success

AI is revolutionizing how students learn by creating tailored educational pathways. Machine learning algorithms identify individual learning styles, track progress, and pinpoint areas of struggle—offering real-time feedback, personalized content, and adaptive assessments. This precision not only enhances outcomes and engagement but also reduces reliance on one-size-fits-all instruction, contributing to a more sustainable and resource-efficient learning model.

Smart Classrooms and Adaptive Learning Environments

Learning spaces now operate in real time, with AI-powered smart classrooms that automatically adjust lighting, temperature, and AV systems based on occupancy, time, or teaching requirements. These dynamic environments improve comfort and efficiency, reducing resource consumption—making classrooms, laboratories, and libraries as smart as the content they host.

Sustainable Research and Innovation

Faculty researchers use AI to model climate systems, enhance renewable energy production, improve urban planning, and accelerate progress in clean technology. These efforts not only deepen academic knowledge but also contribute to a sustainable planet and establish the university as a leader in environmental efforts.

Automated Campus Operations

AI-powered chatbots and automation tools are optimizing a range of administrative tasks, including enrollment, financial aid, IT support, and scheduling. These digital assistants provide immediate assistance, reduce staff workloads, and ensure services are available 24/7. This enables human staff to concentrate on more strategic, high-value activities, fostering a more responsive and student-focused environment.

Optimized Waste and Transportation Systems

AI boosts operational sustainability in less-obvious yet vital areas. Waste management systems utilize AI to track bin levels and improve collection routes. Additionally, campus transportation—including shuttles, bikes, and autonomous vehicles—relies on AI for route optimization, real-time tracking, and emissions reduction, thereby enhancing efficiency and reducing the environmental footprint.

Strategic Data-Driven Decision Making

AI enables faculty and administrators across departments to analyze complex datasets and extract actionable insights. It helps optimize classroom usage, compare sustainability achievements, and identify patterns in student success. By transforming raw data into institutional intelligence, AI supports more informed planning and policy-making.

Why Resilience Is No Longer Enough

For decades, higher education has measured operational success through the lens of **reliability and resilience.** A resilient campus withstands disruption and returns to normal: systems recover after a storm, classes resume after an outage, and buildings reopen after a crisis. Resilience has been necessary—and for a long time sufficient. Today, it is no longer enough.

Universities now operate in a world defined not by isolated shocks, but by **persistent volatility**. Financial pressure from demographic shifts, accelerating climate risk, escalating cyber threats, workforce scarcity, regulatory scrutiny, and rapid advances in artificial intelligence are not episodic challenges. They are structural conditions. In this environment, merely surviving disruption and returning to baseline leaves institutions progressively weaker, more expensive to operate, and less capable of fulfilling their mission. What higher education now requires is **antifragility**.

> ***Resilience helps universities survive disruption.***
> ***Antifragility helps them evolve.***

From Resilience to Antifragility

Resilience focuses on **absorption and recovery**. Antifragility goes further: it describes systems that **improve through stress, learning from disruption,** and emerging stronger, more adaptive, and more predictable over time.

An antifragile campus does not simply endure failures—it **harvests intelligence from them**. It converts variability, uncertainty, and shock into inputs for learning, redesign, and improvement. Instead of asking, *"How quickly can we return to normal?"* an antifragile institution asks, *"How do we become better prepared, more efficient, and more capable because this happened?"*

This distinction matters profoundly for higher education. Universities cannot pause operations, reboot infrastructure, or suspend their social obligations when conditions deteriorate. They must remain open, safe, and trustworthy—often precisely when stress is highest. Resilience helps them survive. Antifragility helps them **evolve.**

Why Antifragility Matters Now for Higher Education

Modern campuses are no longer static collections of buildings supported by back-office systems. They are **living cyber-physical ecosystems:** networks of digital services, operational technology, data platforms, people, policies, and physical infrastructure. Decisions in one domain increasingly cascade across others. A cyber incident becomes a life-safety concern. A climate event becomes a financial and reputational risk. A staffing shortage becomes an operational bottleneck amplified by deferred maintenance and fragmented systems.

In this context, fragility accumulates quietly. Deferred maintenance, siloed technology decisions, opaque vendor dependencies, unmanaged data, and cultural separation between IT and Facilities create environments that appear stable—until they are not. When disruption arrives, these systems fail abruptly, expensively, and publicly.

Antifragility offers a different path.

An antifragile institution designs its campus so that:

- Failures are observable, not hidden.
- Systems degrade gracefully, not catastrophically.
- Decisions are informed by real-time operational truth, not lagging reports.
- Learning is continuous, not episodic.
- Governance, data, and architecture reduce risk *before* automation or scale is introduced.

Rather than optimizing solely for uptime, antifragile campuses optimize for **learning velocity**—how quickly they detect weak signals, test responses, validate outcomes, and institutionalize improvements.

Antifragility as a Leadership Discipline

Antifragility is not a technological feature. It is an institutional operating philosophy expressed through governance, architecture, culture, and metrics.

Technology enables antifragility—but leadership determines whether it emerges.

For Technology and Facilities leaders, antifragility reframes success. The goal is no longer to eliminate all variability or to design perfect systems. Instead, it is to create environments where:

- Operational stress produces insight rather than blame.
- Data reveals emerging risk before it becomes a crisis.
- Cross-functional teams respond with clarity and shared authority.
- Systems become more predictable, not more brittle, as complexity increases

This requires a deliberate shift away from project-centric modernization toward platform-centric, learning-oriented **design**. It demands shared governance between IT and Facilities, disciplined architecture, ethical and transparent data practices, and continuous feedback loops that connect sensing to decision-making.

The Antifragile Campus

An antifragile smart campus is not defined by the number of sensors, dashboards, or AI models it deploys. It is defined by what happens *after* disruption.

Over time, an antifragile campus becomes:

- More energy-efficient because anomalies are detected early and corrected permanently.
- More secure because near-misses and attempted intrusions improve segmentation and controls.
- More sustainable because operations adapt dynamically to climate, occupancy, and grid conditions.
- More humane because staff are supported by systems that reduce cognitive load rather than increase it.
- More trustworthy because decisions are transparent, measurable, and aligned with institutional values.

Antifragility—not resilience—is the true end state of smart campus transformation. The chapters that follow outline how Technology and Facilities leaders can intentionally design governance, architecture, data estates, and human systems that allow universities not merely to survive the decade ahead, but to become stronger because of it.

In an era of constant disruption, the most important question for higher education *It is no longer a question of whether the campus can withstand stress.* *It is whether the campus can learn from it and improve as a result.*

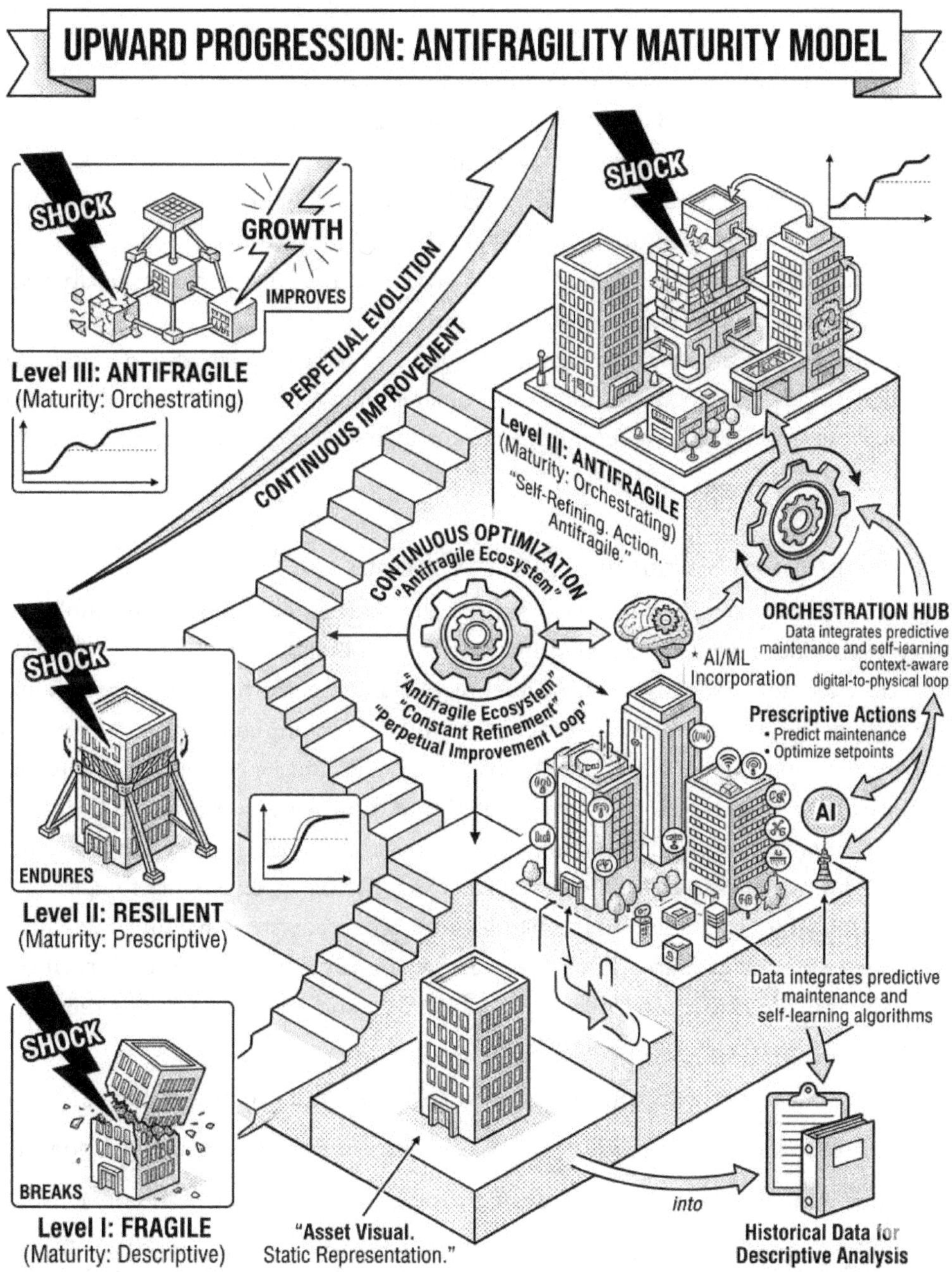

Figure 1. The Antifragile Maturity Model

Why the Campus Must Evolve Now

The traditional campus operations approach—characterized by static buildings, isolated systems, and reactive maintenance—fails to meet the growing demands for responsiveness, resilience, and responsible management. Small, incremental improvements can help, but they do not address the fundamental issue: universities need to overhaul their campuses as an integrated, well-governed system rather than a patchwork of separate projects.

The physical campus is no longer merely a backdrop for education. It is a visible expression of institutional identity, priorities, and future readiness. From the moment a prospective student arrives, the built environment communicates what the university stands for—through its quality, functionality, and care. Design and condition are not simply aesthetic concerns; they influence recruitment, retention, and reputation. As competition intensifies, institutions must confront a strategic question: **"Does the campus experience reflect the excellence the university claims to deliver?"**

At the same time, "campus experience" now extends far beyond classrooms and grounds. It is shaped by the reliability and comfort of learning and living spaces; the safety and accessibility of movement across campus; and the consistency of digital services that support everyday life.

Also, sustainability performance is now equally visible and consequential—energy, carbon, water, and waste outcomes have become part of the institution's public credibility. In this context, modernization is not an optional technology agenda. It is an institutional imperative: to deliver a campus that performs reliably, serves people well, and demonstrates measurable progress toward stated commitments.

Finally, ***the AI-enabled sustainable smart campus is not a technology initiative — it is an institutional strategy*** that unifies the digital and physical campus into a coordinated operating model. Its goal is to strengthen mission delivery, operational excellence, and long-term resilience through governed, interoperable, data-driven modernization.

Defining the AI-Enabled Sustainable Smart Campus

An AI-enabled sustainable smart campus is far more than a collection of modern technologies—it is a new institutional operating model that unifies digital infrastructure, physical systems, data governance, and mission outcomes into a cohesive, adaptive ecosystem. At its core, this model transforms buildings, utilities, mobility, and campus services into measurable, manageable, and continuously improving components of a larger whole. Through secure connectivity, interoperable platforms, intelligent automation, and a governed data estate, the campus evolves from reactive to predictive, from fragmented to integrated, and from static to responsive. Defining this paradigm is essential because it frames smart-campus modernization not as a technology upgrade but as a strategic commitment to resilience, sustainability, operational excellence, and an elevated campus experience for all.

An AI-driven Sustainable Smart Campus is a digitally connected ecosystem that integrates intelligent infrastructure, sustainability efforts, and advanced IT/OT systems to improve institutional performance and the campus experience. Its main characteristic is coordination, in which data and systems work together smoothly through a shared architecture and governance. In this model:

- Buildings and infrastructure become measurable and manageable in near real time.
- Digital services become integrated and human-centered rather than fragmented by department.
- AI turns operational and experience data into predictions, optimizations, and proactive interventions.
- Sustainability becomes operationalized through continuous measurement and improvement.

Technologies like AI, IoT, digital twins, modern data architectures, and cloud/edge computing are not useful on their own. They only become strategic assets when combined into a unified platform approach.

The Internet of Things (IoT) provides real-time data on energy consumption, indoor climate, occupancy, and equipment health. AI further enhances this data by transforming it into actionable operational insights—identifying anomalies, forecasting future requirements, and optimizing systems.

Digital twins are dynamic, data-driven virtual models of buildings or campuses that enable monitoring, simulation, scenario analysis, and predictive maintenance.

AI integrated into the digital twin offers insights into operational parameters and knowledge that enhance equipment and personnel performance. This leads to higher efficiencies, new operational economies, and improved reliability, safety, and investment strategies.

When these capabilities are intentionally designed, integrated, and managed, they turn the campus from simply reactive to adaptive, supporting ongoing performance enhancements rather than single solutions. Additionally, collaborative leadership between the Chief Technology Officer (CTO) and the Chief Facilities Officer (CFaO) fosters a smart campus strategy that is regularly assessed based on outcomes rather than features alone.

The Core Strategy

An AI-driven Sustainable Smart Campus combines technology, operations, governance, and personnel into a unified system that enhances current efficiency and prepares for a rapidly changing future. It converts the campus into a responsive, resilient, data-focused ecosystem that continually learns, adapts, and provides tangible benefits over time. Although each institution will identify its priorities, goals, and strategies, the primary purpose of the AI-enabled Sustainable Smart Campus is to align mission objectives, operational excellence, and sustainability within a cohesive strategic framework.

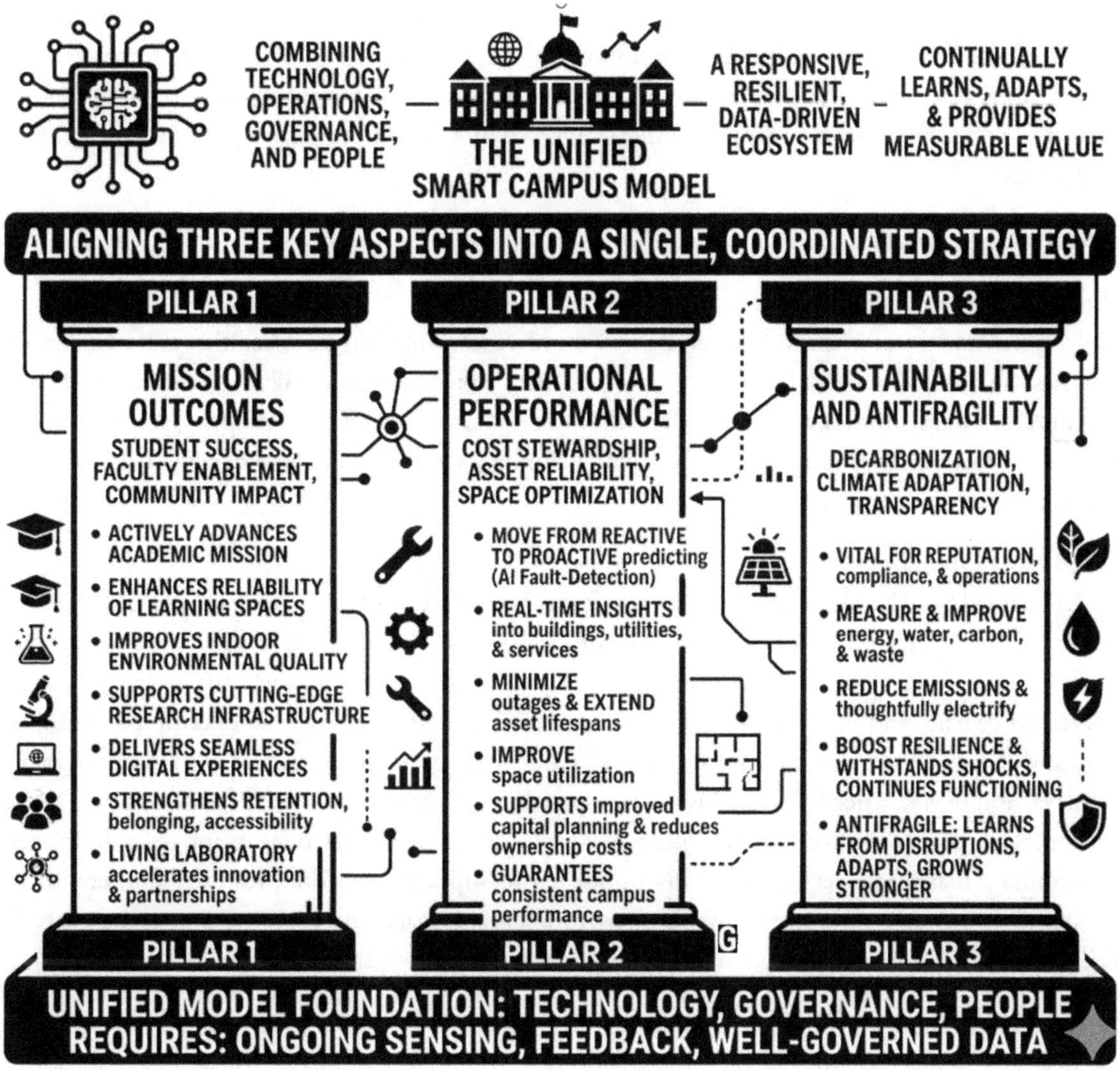

Figure 2. Pillars of the Smart Campus Strategy

Mission Outcomes

Student Success and Faculty Enablement

A modern campus must actively advance the academic mission. An AI-enabled environment enhances the reliability of learning spaces, improves indoor environmental quality, supports cutting-edge research infrastructure, and delivers seamless digital experiences. It strengthens retention, a sense of belonging, accessibility, and institutional competitiveness. The smart campus becomes a living laboratory that accelerates innovation, research, partnerships, and community engagement.

Community Impact

An AI-enabled Sustainable Smart Campus extends its value well beyond institutional boundaries, acting as a force multiplier for the surrounding community. By optimizing energy use, reducing emissions, and integrating with local grids and transportation systems, the campus strengthens regional sustainability and resilience while lowering shared environmental impact. Its secure, data-driven infrastructure enhances public safety coordination, emergency response, and climate preparedness, benefiting both the campus and the community alike. At the same time, the smart campus functions as a living laboratory—supporting workforce development, local innovation, research partnerships, and pilot programs that municipalities and industry can replicate. Through transparent governance, ethical data practices, and community-oriented design, the campus becomes not just a place of learning but a trusted civic partner that advances economic vitality, environmental stewardship, and social well-being for the region it serves.

Operational Performance

Cost Stewardship, and Asset Reliability

Siloed systems and reactive maintenance are now outdated. Combining IT, Facilities, and sustainability operations allows the campus to move from reacting to problems to proactively predicting issues using data. Real-time insights into buildings, utilities, and services help minimize outages, extend asset lifespans, and improve space utilization. This approach supports improved capital planning, reduces overall ownership costs, and guarantees consistent campus performance.

Space and Asset Optimization

An AI-enabled Sustainable Smart Campus transforms how spaces and assets are managed from a static planning process to an ongoing, data-driven approach. By incorporating real-time occupancy data, scheduling tools, building automation telemetry, and digital twins, institutions gain accurate insights into the actual usage of classrooms, labs, offices, residence halls, and shared areas—beyond assumptions. This visibility enables campuses to adjust HVAC, lighting, and cleaning services based on true demand, reduce energy and maintenance costs in underused areas, and enhance availability where needed most. Over time, spatial analytics guide strategic decisions such as consolidation, renovation, repurposing, or divestment, helping avoid unnecessary new construction and debt. In this framework, space becomes a managed asset—dynamically optimized to support education, research, and community life—while extending the facility's lifespan, lowering environmental impact, and ensuring that each square foot aligns with the institution's mission through measurable efficiency and responsible management.

Sustainability and Antifragility

Decarbonization, Climate Adaptation, and Transparency

Environmental performance is now vital for reputation, compliance, and operations. AI and digital tools facilitate the measurement and improvement of energy, water, carbon, and waste management. The campus can reduce emissions, manage demand, electrify thoughtfully, and boost resilience to climate and grid challenges. Transparent reporting builds trust with students, regulators, donors, and accreditors — turning sustainability goals into concrete actions.

Evolving from Disaster Recovery to Antifragility

For decades, universities primarily considered risk and continuity in terms of disaster recovery. Recently, this perspective has shifted toward resilience, meaning a resilient campus can withstand shocks and continue functioning under stress. Currently, higher education faces climate volatility, cyber threats, financial issues, workforce shortages, regulatory pressures, and rapid AI innovations. This necessitates a more advanced strategy: antifragility. An antifragile campus not only survives disruptions but also learns from them, adapts, and grows stronger. Achieving antifragility involves more than redundancy; it requires ongoing sensing, well-governed data, and feedback mechanisms across both physical and digital campuses.

Obstacles to the Smart Campus

The primary challenge in creating a smart campus often relates to cultural and governance issues, known as the "Bridge Problem." This problem arises because IT and Facilities leaders typically have different goals and varying degrees of risk tolerance.

The Divergence of Priorities

While both the CTO/CFaO are dedicated to ensuring a safe, reliable, and high-performing campus, they pursue this goal using fundamentally different risk models, time horizons, and operational practices. These differences are not mere philosophical preferences but are rooted in structural realities formed by years of distinct professional experiences.

The CTO focuses on safeguarding data, identities, and digital resilience in an environment marked by rapid updates and persistent, constantly changing cyber threats. As a result, the CTO's focus is guided by the Confidentiality, Integrity, and Availability (CIA triad) of data and systems. From this perspective, any vulnerable system is considered a risk and must be addressed immediately to prevent a "zero-day" attack.

In contrast, the CFaO oversees life-safety systems with lifetimes spanning multiple decades, making uptime and safe failure modes crucial. The Facilities leader emphasizes the Safety, Availability, and Sustainability (SAS) triad, trusting that even a security patch that could lead to a system shutdown could threaten human safety or interfere with research. Rebooting a hospital wing or heating plant during active operations is simply not an option.

CFaOs and CTOs also work on different time horizons: ***CTOs plan refresh cycles over months or years and measure obsolescence rapidly, while CFaOs focus on asset lifespan over decades with outage windows linked to seasons and occupancy.***

CTOs operate under the premise that least privilege, session logging, and ephemeral access are table stakes. The CFaO relies on longstanding vendor relationships, where its "always-on" access to physical infrastructure is convenient for ensuring uptime. If Facilities approach IT as just another utility, it will underestimate cyber risk, data governance obligations, and the operational dependencies created by shared networks, identity access systems, and cloud services.

Additionally, CTOs view data as an institutional asset—curated, governed, and reusable. Meanwhile, CFaO data is more operational—valuable for quick diagnosis and dispatch, even if it is messy.

CTOs justify their expenses by mitigating risk and enhancing enterprise capabilities, such as identity, segmentation, and data estate. In contrast, CFaOs justify their spending by relating investments to improved reliability, reduced downtime, increased comfort, and energy efficiency.

These examples illustrate how CFaOs' and CTOs' operational priorities have been shaped by their professional roles. When these priorities conflict without proper governance, it can lead to friction, delays, and vulnerabilities within the institution. Recognizing these differences is crucial for developing a convergence strategy that balances both perspectives, ensuring smart-campus modernization is sustainable, safe, scalable, and resilient.

Shared Outcomes: The Case for Alignment

Why make the effort to bridge this gap? Because managing these systems separately is no longer financially or operationally sustainable. When IT and Facilities work together, we achieve five transformative results:

- **Energy and Decarbonization**
  Smart metering and AI-driven BAS optimization are the only viable pathways to meeting aggressive Net Zero goals without prohibitive capital expenditure.
- **Indoor Air Quality (IAQ) and Health**
  post-pandemic, the ability to monitor and report on air quality is a baseline expectation for student and employee safety.
- **Institutional Resilience**
  Converged systems allow for predictive maintenance—fixing a critical pump before it fails—rather than reactive triage that disrupts campus operations.
- **Occupant Experience**
  The "user" of the campus (student, researcher, employee) expects a seamless experience. Navigation, room booking, and comfort control should be as intuitive as a smartphone app.
- **Lifecycle Cost Optimization**
  Data-driven decisions extend the life of physical assets and reduce the Total Cost of Ownership (TCO).

By pinpointing and addressing these friction points within campus governance and planning, a previously contentious culture war transforms into a strategic design approach. This leads to fewer cascading failures, quicker recovery, increased trust in vendor work, and the creation of a data infrastructure that truly supports optimization and twin-driven operations, rather than just producing more dashboards and leading to operational drift.

Leveraging the Technology and Facilities Leaders

Most likely, initial efforts to integrate IT and OT are already in progress; this is not a future trend but the current state of our infrastructure. Leaders need to decide whether to guide this convergence deliberately or let it unfold chaotically. A common but often overlooked source of conflict between IT and Facilities is not their values or goals—it's time.

The strategic opportunity is to treat the lifecycle gap as a design constraint—not a cultural barrier. When both leaders internalize each other's time horizons, they can build a modernization approach that is:

- Phased, not disruptive (aligned to capital cycles and outage windows)
- Secure by design (aligned to modern threat realities)
- Interoperable and standards-driven (to avoid lock-in across decades)
- Operationally adoptable (so insight becomes action, not dashboards)

In short, the CTO learns to architect for longevity and safety-critical operations, while the CFaO learns to operate in a world of continuous digital risk and rapid technology change. That shared understanding is what turns "smart campus" from a collection of projects into a resilient operating model.

Learning the Other Perspective

The modern campus is no longer a set of independent buildings supported by a separate IT utility. It is a unified, cyber-physical enterprise in which operational technology, digital infrastructure, and institutional trust are inseparable. In that environment, the CTO and the Chief Facilities Officer become co-architects of institutional performance.

For the CTO, understanding the reality of facilities is a strategic necessity. Facilities is not just another system to manage; it represents shared responsibility for enhancing the campus experience and optimizing the built environment—where uptime is a public commitment, safety is unconditional, and asset lifecycles span decades.

The realization that OT is not managed like laptops and servers, and that the campus cannot be rebooted, is crucial. When CTOs grasp controls modernization, commissioning, deferred maintenance, and building-level constraints, they can design cybersecurity, data architecture, and network services that safeguard the institution without interrupting its mission. Additionally, they view the data estate—and the eventual digital twins—as practical operational tools instead of just aspirational dashboards.

For CFaOs, understanding the IT perspective is equally essential. As buildings become equipped with more sensors and connect to the cloud, facility systems face the same risks as enterprise systems, including privacy issues, cyber threats, vendor vulnerabilities, and compliance requirements.

Exhibit 1. CTO and CFaO Cross Domain Knowledge Map

What the CTO Needs to Know About Facilities and OT	What the Facilities Leader Needs to Know About IT
Facilities mission and "always-on" obligations: life safety, compliance, research continuity, and the reality that many buildings must operate 24/7 with minimal tolerance for downtime.	IT mission and risk model: confidentiality, integrity, and availability—plus how cybersecurity risk, privacy obligations, and audit requirements shape architecture choices.
Deferred maintenance as a technology risk multiplier: aging equipment, "patchwork" upgrades, and failing controls increase failure rates, expand cyber-attack surface, and complicate integration.	Cybersecurity fundamentals for cyber-physical environments: segmentation, least privilege, endpoint hardening, logging, and incident response—applied to OT without breaking uptime.
Core OT systems and how they interrelate: BMS/BAS, EMS, SCADA (where applicable), lighting controls, metering/submetering, access control, video, district energy plants, and CMMS/work orders.	Identity and access management (IAM): authentication, RBAC, MFA, privileged access, vendor access models, and service accounts—especially for contractors and integrators.
Key building standards and protocols: BACnet (esp. BACnet/IP), Modbus (RTU/TCP), LonWorks, KNX, DALI, plus IoT messaging like MQTT and typical integration patterns (gateways, drivers, brokers).	Networks as a shared utility: routing, VLANs, firewalls, DNS/DHCP, Wi-Fi design, private cellular/CBRS, quality of service, and why small network changes can destabilize building controls.
Controls modernization pathways: controller refresh vs overlays, point mapping, tagging strategy, sequence-of-operations governance, commissioning/retro-commissioning.	Data estate architecture: normalization, metadata/tagging, time-series storage, APIs, event streams, and integration patterns that enable analytics and digital twins.
Facilities success metrics: IAQ, comfort, energy intensity, service calls, uptime, asset life, maintenance backlog, and customer experience.	Service management discipline: change control, release management, SLAs, outage communications, and operational handoffs from IT to facilities workflows.
Vendor and integrator ecosystem: mechanical/electrical contractors, controls vendors, commissioning agents, security vendors, ESCOs—each with different incentives and maturity.	Vendor risk management: contract language, data ownership, patch responsibilities, third-party access controls, and avoiding brittle one-off integrations.
Safety and regulatory context: code compliance, permitted work, life-safety boundaries, and why some systems cannot be "experimented" on in production.	Privacy and compliance: occupancy, access, and video data implications; retention; minimization; and governance that preserves campus trust.

What the CTO Needs to Know About Facilities and OT	What the Facilities Leader Needs to Know About IT
Digital twin readiness: accurate spatial/static data (BIM/as-builts, GIS), reliable live telemetry, and operational workflows that convert insight into action.	Operational analytics mindset: fault detection and predictive maintenance only matter if they drive dispatch, prioritization, verification, and continuous improvement.

Achieving modern facilities goals—such as energy efficiency, indoor air quality, predictive maintenance, and resilience—relies increasingly on integration, identity management, data governance, and service discipline. When facilities leaders grasp networks, IAM, segmentation, and data architecture, they enhance safety and trust, allowing them to modernize operations while safeguarding the community's confidence.

Integrated deployment strategies avoid the notion of traditional "pilot projects" and "innovation theater" to practical tests of reusable technology and process patterns that inform planning and coordinated roadmaps.

Most importantly, when leaders understand each other's perspectives, the institution benefits beyond just technical coordination: it develops a shared decision-making framework. Budgeting moves from a focus on 'IT projects versus facilities projects" to investments that emphasize lifecycle management, risk mitigation, and mission success. Integrated deployment strategies avoid the notion of traditional "pilot projects" and "innovation theater" to practical tests of reusable patterns of technology and process that inform planning and coordinated roadmaps. Ultimately, the smart campus evolves from a series of individual pilots into a sustainable, enduring operational model—where the built environment acts as a strategic foundation for sustainability, resilience, learning, and enhancing human experience.

Mastering this cross-domain knowledge map enables the CTO/CFaO to go beyond their usual silos, creating a common architectural language and mutual understanding needed to handle the challenges of IT/OT convergence. However, developing this shared vocabulary is only the first step in the transformation.

The Digital Twin: Convergence of Technology and Intelligence

A great deal has been written about human and artificial intelligence. Never before has a technological innovation advanced so quickly, affecting higher education and nearly every other industry globally. An intelligent campus features physical infrastructure that not only exists but also actively interacts. It continually learns, adapts, and evolves. Central to this evolution is a significant innovation in campus management: Digital Twins. These digital replicas merge human and artificial intelligence. Although twins are still in their early stages, they are expected to become standard platforms capable of incredible functionality in the future. This is why, when designing a smart campus, their role is crucial, which is why it is discussed in this introduction.

A Digital Twin is a real-time virtual replica of a physical asset, system, or campus. It functions as a dynamic, data-driven simulation rather than a static model, displaying current conditions, behaviors, and performance. By integrating data from IoT sensors, building systems, and campus operations, digital twins provide a centralized platform for monitoring, predicting, and improving areas such as space utilization and energy efficiency.

In a modern smart campus initiative, digital twins serve as the bridge between the physical and digital worlds—allowing institutions to visualize, simulate, and plan more efficiently than before.

Predictive Maintenance and Infrastructure Intelligence

Digital twins allow early detection of issues such as wear and tear or failures in critical systems like HVAC, electrical, and mechanical equipment before they occur. By modeling real-time performance and environmental factors, they enable facilities teams to plan maintenance proactively, reduce energy waste, and avoid costly, unexpected outages. This approach not only prolongs the lifespan of assets but also maintains a consistent, comfortable environment for students and staff.

Space Utilization and Environmental Optimization

By continuously analyzing occupancy data and behavioral trends, digital twins help institutions reevaluate their use of space. This visibility highlights underused classrooms, overcrowded labs, and energy-intensive buildings, enabling targeted upgrades. Systems such as lighting, heating, and ventilation can be automatically adjusted in real time based on actual usage, minimizing waste and creating optimal settings for learning and collaboration.

Integrated Operations and Seamless Systems Management

Digital twins serve as control centers for campus management by integrating data from multiple systems, such as lighting, irrigation, water consumption, and access control. This unified interface offers a comprehensive overview of the campus environment, facilitating smarter, faster decisions. For example, a sudden increase in energy consumption can trigger automated conservation measures, and weather forecasts can be used to adjust irrigation plans proactively.

Scenario Planning and Emergency Readiness

Digital twins extend beyond simple management by providing advanced simulation functions. They enable users to explore questions such as what happens if a building loses power during finals, how to control foot traffic during campus events, or what the most sustainable energy plan is in a heatwave. These tools help leaders create multiple "what-if" scenarios—testing assumptions, confirming plans, and readying for emergencies with accuracy.

Sustainability by Design

Most importantly, digital twins help campuses reach their climate and sustainability goals. They enable ongoing improvements by monitoring energy consumption, emissions, and system inefficiencies in real time. This leads to smarter resource management, less waste, lower carbon footprints, and progress toward net-zero targets, all based on accurate data and practical insights.

The Digital Twin transforms the campus from a set of isolated systems into a unified, data-driven entity that can assess its own performance and respond accurately. By integrating spatial records, real-time telemetry, asset data, and analytics into a cohesive operational model, the twin serves as the institution's intelligence layer—facilitating predictive maintenance, adaptable environments, and fact-based planning. Its true value lies not just in visualization but in the disciplined architecture, data governance, and cross-functional practices that ensure the model remains accurate, reliable, and actionable. In this form, the Digital Twin becomes a vital institutional asset: a continuously evolving guide for decision-making, investments, and operational excellence.

Why the Smart Campus Matters

Leaders at our institution increasingly expect smart-campus initiatives to deliver outcomes that are visible, measurable, and aligned with their mission, rather than just showcasing technical capabilities. As your manifesto highlights, the true value of an AI-enabled sustainable smart campus depends on how data, infrastructure, and governance work together to enhance operational reliability, decision-making, sustainability efforts, campus experience, and the institution's role as a living laboratory for learning and research. These results are particularly meaningful at the executive level because they connect modernization with key institutional priorities—such as cost management, resilience, student success, faculty support, and progress on sustainability goals—transforming digital initiatives from mere technology projects into comprehensive enterprise strategies with wide-ranging, organization-wide benefits.

Enhanced Operational Efficiency and Reliability

A digital or smart campus revolutionizes campus life by transforming buildings and infrastructure into measurable, manageable, and optimizable systems. By integrating and monitoring data from facilities, utilities, and services in real-time, the institution shifts from a reactive approach to emergencies to a proactive one—identifying issues early, intelligently prioritizing repairs, and reducing disruptions in classrooms, labs, and residence halls. This strategy not only cuts costs and waste but also enhances campus dependability, ensuring consistent temperature control, lighting, air quality, and essential services. As a result, the daily experience of learning and working becomes more stable, safe, and less frustrating.

Data-Informed Decision-Making

A digital or smart campus enhances lives by empowering leaders to base decisions on data rather than guesses. A robust data system helps CTOs, Facilities managers, and other leaders spot trends in usage, energy, maintenance, and student services. They can run simulations of different scenarios before making investments. This method moves decision-making from opinion-driven debates to virtual testing, outcome prediction, and clear investment priorities. It reduces risk, accelerates consensus, and ensures that spending supports the institution's mission, experience, and resilience.

Sustainability Performance with Transparency

A smart campus is crucial to transforming lives by making sustainability tangible—both visible and improvable—while maintaining accountability. AI-driven monitoring of energy, carbon, water, and waste converts climate objectives into active management, not just annual targets. Using real-time data, institutions can identify inefficiencies, verify savings, and prioritize high-impact areas, providing trustworthy reports to students, boards, regulators, and communities. Essentially, the campus exemplifies responsible resource management—reducing emissions and resource consumption in measurable, repeatable, and dependable ways.

Elevated Campus Experiences

A digital or smart campus significantly impacts lives by enhancing the daily experiences of students, staff, and faculty, making them more seamless and responsive. Integrating services such as space management, scheduling, navigation, safety alerts, and support systems enables the campus to anticipate needs proactively rather than merely reacting to problems. When deployed ethically and transparently, analytics enable better space utilization, address bottlenecks, and improve safety and well-being, particularly during disruptions. This creates a campus environment that feels more intuitive, accessible, and predictable, and fosters a sense of belonging and achievement.

A living laboratory for instruction and research

A digital or smart campus transforms its community by serving as a platform for learning and discovery. It leverages a well-documented, accessible data infrastructure, sensor networks, digital twins, and governed experimentation to support interdisciplinary research and hands-on learning at an extraordinary scale. Students can work on real operational challenges, faculty can test new approaches and assess outcomes, and collaborations with cities, utilities, and industries foster sustainability and resilience. Over time, the campus evolves beyond its traditional role, becoming a continuous source of knowledge, workforce development, and societal advancement.

Trusted governance

Reliable governance with defined decision rights, accountability, and transparency forms the institutional backbone that enables a smart campus strategy to grow safely and credibly. Governance is the coordination mechanism that prevents pilot sprawl, enforces standards, and guarantees consistent decision-making across IT, Facilities, sustainability, finance, and academic sectors. It clarifies who holds authority over architecture, cybersecurity, data stewardship, and deployment order, ensuring that complex cyber-physical systems develop within a stable framework rather than through informal exceptions.

Clear decision rights and a structured escalation process—embodied in entities such as the Smart Campus Steering Committee (SCSC)—establish accountability by defining roles, responsibilities, and required deliverables, including shared scorecards, RACI models, and time-limited exceptions.

This openness fosters trust among stakeholders, allowing leaders to communicate what data is gathered, why systems are interconnected, how risks are mitigated, and how progress is evaluated. Ultimately, trustworthy governance turns modernization from isolated projects into a unified enterprise strategy that leaders can endorse, fund, and sustain over the long term.

Exhibit 2. Smart Campus Outcomes Summary

Outcome	Summary
Financial stewardship backed by measurable returns	Smart-campus modernization reduces waste, improves reliability, and shifts operations from reactive to proactive, producing measurable, recurring savings in energy, maintenance, and lifecycle costs.
Risk reduction across cyber, physical, operational, and climate domains	Converged IT/OT systems improve safety, reduce outages, strengthen cybersecurity posture, and enhance institutional resilience against climate and grid disruptions.
Trusted governance with clear decision rights, accountability, and transparency	Governance structures—such as a joint steering committee, shared scorecards, and cross-silo decision rights—ensure consistent, transparent, and accountable modernization across IT, Facilities, and institutional leadership.
Institutional competitiveness through a high-performing, future-ready campus	A modern, reliable, and responsive campus enhances recruitment, retention, faculty satisfaction, research quality, and the institution's public reputation.
Mission alignment that strengthens teaching, research, and community value	Smart campus capabilities support academic excellence, research continuity, and community partnerships by enhancing reliability, transparency, and responsiveness.
Mission outcomes: student success, faculty enablement, community impact	Improved indoor environments, seamless services, resilient infrastructure, and accessible experiences directly support student learning, faculty productivity, and community engagement.
Operational performance: cost stewardship, asset reliability, space optimization	Integrated data and modernized systems reduce downtime, extend asset life, and optimize space utilization, enabling better capital planning and daily operational stability.
Sustainability and resilience: decarbonization, climate adaptation, transparency	Real-time visibility, continuous optimization, and data-driven strategies help the institution reduce emissions, conserve resources, and build climate-resilient operations, with transparent reporting.

Why This Manifesto Is Worth Your Time

This Manifesto is not aspirational—it is operational. The time you invest in reading it returns practical value: a leadership-ready operating model maturity roadmap for moving beyond fragmented IT projects and siloed facilities operations toward an architected, intelligent campus system designed for resilience, sustainability, and mission impact. This is not a vision of possible futures; it is a guide to decisions, structures, and actions leaders can take now.

As you read, you'll be encouraged to consider your current organizational landscape—focusing on how Technology and Facilities leaders interact. Do your CTO, Facilities leaders, architects, and operations managers have a common platform for strategy discussions? Do they align on risk, priorities, and outcomes? If not, the Manifesto highlights a significant opportunity. Building this connection isn't just about structure—it's a leadership necessity.

The Manifesto is designed to spark informed dialogue. Technology and Facilities leaders are encouraged to read it together and use it as a catalyst for structured, collegial conversations surrounding governance, architecture, risk, sustainability, and human aspects systems that underpin transformation. These conversations are not abstract; they are essential to breaking the cycle of disconnected initiatives and short-lived pilot projects.

The return on time improves even more when planning and service teams work together on the Manifesto, guided by the CTO and CFaO. Reading and discussing it together helps uncover assumptions, identify friction points, and foster a common understanding of each domain's constraints, responsibilities, and success metrics. This approach shifts from isolated efforts to a unified purpose, transforming coordination challenges into clearer architectural insights.

In summary, this Manifesto values the reader's time by providing not only ideas but also a practical, repeatable approach to thinking and acting. This ensures that every hour invested in modernizing the campus builds resilience, fosters learning, and adds long-term institutional value.

Part II

The Technology Leaders' Point of View

Chapter 1
The Chief Technology Officer's Vision

The modern Chief Technology Officer must evolve from a digital services leader into the integrative architect of the institution's cyber-physical future.

The IT Operating Context

The modern Higher Education Chief Technology Officer (CTO) operates in a challenging environment marked by the polycrisis in higher education.[3] The polycrisis involves overlapping challenges such as demographic changes, financial instability, socio-political issues, stricter compliance requirements, rapid advancements and complexities in AI, and growing expectations from campus stakeholders. The CTO's role now extends beyond maintaining email service and network security, as the obstacles to success are structural, mission-related, and existential in nature.

Most CTOs are currently managing three interrelated issues that contribute to this ongoing crisis:

- **The Demographic and Financial "Vice Grip."**
  The Condition
  The "Enrollment Cliff" is no longer a forecast; it is an operational reality. The pool of traditional college-aged students is shrinking, destabilizing the tuition-dependent revenue model. Simultaneously, inflation and talent costs are rising.

  The CTO's Reality
  The CTO is trapped in a paradox of ***"Do More with Less."*** The institution demands "Level 4" strategic innovation (AI, personalization, digital transformation) but provides "Level 1" funding. The traditional IT levers for efficiency—cloud migration, vendor consolidation, and licensing negotiation—have been pulled. They offer diminishing returns. The CTO is managing a structural deficit where the cost of maintaining legacy "Utility" services consumes the entire budget, leaving no capital for the innovation the Cabinet demands.
- **The Asymmetric Cyber-Physical War**
  The Condition
  The threat landscape has evolved. Universities are prime targets for ransomware attacks not only because of their data but also because of their open, vulnerable networks. The attack surface has grown from the data center to the physical campus, including lab equipment, smart building controllers, and unsecured IoT devices.

  The CTO's Reality
  While the CTO is responsible for institutional security, they lack oversight of the rapidly expanding attack surface: Shadow OT. Vulnerabilities in vendor-managed HVAC controllers or departmental "smart"

refrigerators enable attackers to bypass the enterprise firewall. The CTO is battling this security challenge with a critical blind spot: the new 'perimeter' now encompasses thousands of physical devices overseen by Facilities or academic departments outside IT control.

- **The "Experience Expectation" Gap**
  The Condition
  Students, faculty, and staff are shaped by the seamless, highly personalized experiences of the consumer internet, like Amazon and Netflix. They do not see a difference between "digital" and "physical" experiences and anticipate a smooth, frictionless journey across campus.

  The CTO's Reality
  The CTO finds it challenging to deliver a seamless experience because the digital layer ends at the screen. While the student app (IT) functions, issues such as unreliable lecture hall Wi-Fi, a freezing room, and static wayfinding (Facilities) hinder the experience. Although the CTO oversees "Student Success," they are structurally isolated from the physical environment where that success takes place.

> ***The cybersecurity perimeter no longer ends at the firewall.***
> ***It now includes things like boilers, chillers, fume hoods, and life-safety systems.***

In response, the CTO is not only leading transformational projects but also acting as an "Integrative Architect," crossing traditional boundaries as a polymath to build collaboration and partnerships that leverage technology and reshape services. Therefore, the CTO *must* adopt operational technology. The shift to OT is not "scope creep"; it is a strategic move essential to tackling the core issues that endanger the growth of education.

Solving the Financial Crisis: IT as a Value Generator

The CTO cannot achieve prosperity solely by cutting IT budgets. The Integrative Architect understands that the significant inefficiencies on campus are not in the server room, but in the boiler room.

- **The Pivot**
  By applying IT practices like data analytics, automation, and predictive modeling to the physical plant, the CTO enables cost savings worth millions in energy and helps avoid future maintenance costs.
- **The Result**
  The CTO shifts from being a "cost center" that requests budget to a "value generator" that returns capital to the institution. This financial return supports the innovation needed to address the enrollment crisis.

Closing the Security Gap: The CISO of Everything

The CTO, serving as the "Integrative Architect," understands that risk does not respect departmental org charts.

- **The Pivot**
  The CTO manages the Cyber-Physical perimeter governance. By applying Zero Trust Architecture and network segmentation to OT, the CTO eliminates vulnerabilities associated with Shadow OT.
- **The Result**
  The institution becomes more antifragile as the CTO shifts focus from simply "defending the network" to actively "protecting the mission," securing the continuous operation of services, teaching, research, and maintaining campus safety.

Defining the Future: The Integrative Mandate

The CTO's pinnacle role is to serve as the **Integrative Architect** and leader who fuses systems, data, and culture to create new institutional capabilities and drive antifragility.

- **The Pivot**
  The integration of IT and OT represents the culmination of this fusion. It combines the digital nervous system with the physical skeletal structure, forming a "Cognitive Campus" that learns, adapts, and responds to its users.

- **The Result**
  The CTO provides the seamless, "Amazon-like' experience students expect—featuring intelligent wayfinding, personalized settings, and responsive services—filling the gap between the app and the user campus.

The shift toward Operational Technology is the critical momentum needed to escape the "Perfect Storm." It enables the CTO to move beyond managing infrastructure challenges and become the Integrative Ecosystem Architect—the sole leader capable of unifying the fragmented digital and physical worlds of today's university.

The Strategic Discipline of Technology Operations

For many institutional leaders, "IT Operations" is often equated with the Help Desk—the team you contact when email stops working. Yet, for today's enterprise, Technology Operations (Tech Ops) represents a strategic, high-stakes area. It functions as the university's digital nervous system, vital for keeping infrastructure and applications supporting research, teaching, and administration secure, resilient, and flexible.

This framework describes how Technology Operations combines architectural planning, compliance automation, and thorough engineering to support the institution's digital goals.

The Core Mission: Purpose and Scope

The scope of Technology Operations goes beyond just "keeping the lights on." Its main goal is to ensure the institution’s digital environment is secure, dependable, and supports academic objectives. This scope is divided into two vital areas:

- **Infrastructure Services**
  The foundational "digital concrete" of the campus includes the cloud and on-premise compute power, the network transport layer, and the sprawling mesh of edge devices and IoT sensors.

- **Application Services**
  The stewardship of the software that powers the business, from the Enterprise ERP and Student Information Systems to identity management and data integration layers.

Key Operational Activities

To maintain this digital estate, the Tech Ops organization manages a continuous lifecycle of health and hygiene activities:

- **System Observability and Health**
  Moving beyond simple monitoring (is it on?) to deep observability (is it healthy?), tracking network latency, server loads, and application performance in real-time.

- **Data Hygiene and Governance**
  The "janitorial" work of the digital age—purging obsolete records, archiving research data, and ensuring database integrity to reduce risk and storage costs.
- **Cybersecurity and Compliance**
  The relentless rhythm of patch management, vulnerability scanning, and regulatory alignment (FERPA, GDPR, NIST) to harden the attack surface.
- **Asset Management (ITAM)**
  A fiduciary discipline tracking hardware lifecycles and software license compliance to optimize millions in annual technology spend.

The Four Pillars of Tech Ops Strategy

A mature Tech Ops organization maintains four key operational postures. The main objective is to transition focus from reactive firefighting to proactive engineering.

- **Preventive Operations (Planned)**
  Scheduled, routine tasks designed to avoid outages, such as patching windows, load balancing tests, and backup verifications.
- **Predictive Operations (AIOps)**
  The transformative use of AI in operations leverages machine learning to identify subtle anomalies, such as slow memory leaks or abnormal traffic patterns, and addresses them proactively before users notice any slowdown.
- **Corrective/Reactive Operations (The War Room)**
  Incident response following an outage is unavoidable; however, such costly disruptions can damage trust and should be minimized through resilience engineering.
- **Technical Debt Management (Strategic Deferral)**
  This represents the digital form of deferred maintenance—a deliberate choice to delay refactoring code or upgrading hardware to meet a deadline. This decision acknowledges that "interest" will accrue later as increased complexity and risk.

People and Structure: The Hybrid Engineering Team

High-performing Technology Operations teams have shifted from isolated roles like "server admins" and "network admins" to a unified, converged structure.

- **In-House Engineering**
  A blend of DevOps engineers, System Architects, and Service Desk analysts who bridge the gap between software development and operational stability.
- **Strategic Partners**
  A reliance on Managed Service Providers (MSPs) for 24/7 "eyes on glass" monitoring and specialized cybersecurity firms for penetration testing, allowing internal teams to focus on strategy.

The Technology Backbone

Just as facilities require tools to repair buildings, Tech Ops relies on software to oversee its operations. The contemporary operational stack relies on:

- **IT Service Management (ITSM)**
  Platforms like ServiceNow that serve as the "command center" for incident management, change control, and user support.

- **Observability and AIOps Platforms**
  Tools like Splunk or Datadog that provide a real-time MRI of the digital environment, correlating vast amounts of data to pinpoint root causes.
- **Orchestration and Automation**
  The use of "Infrastructure as Code" (tools like Ansible or Terraform) to automate repetitive tasks, ensuring that a server built today is identical to one built tomorrow, eliminating human error.

Measuring Success: The Digital Scorecard

Success in Tech Ops is no longer measured solely by "uptime." It is measured by velocity, efficiency, and experience:

- **DORA Metrics**
  The industry standard for speed and stability, tracking Deployment Frequency, Lead Time for Changes, Mean Time to Restore (MTTR), and Change Failure Rate.
- **FinOps and Cost Efficiency**
  The capability to track cloud spend per student or per service, ensuring that digital resources are optimized and not wasted.
- **Experience Level Agreements (XLAs)**
  Moving beyond Service Level Agreements (SLAs) that measure availability to measuring the actual quality of the user experience—latency, friction, and satisfaction.

Strategic Distinctions

Ultimately, executive leadership needs to differentiate between Technology Operations, which focuses on the tactical execution of reliability, and Technology Strategy, which involves designing the future architecture. A strong operational budget, based on Zero Trust security principles and Disaster Recovery (DR) preparedness, should not be seen as overhead; instead, it acts as an insurance policy enabling the organization to innovate safely.

Redefining the CTO's Vision

For decades, the university maintained a division of labor that was convenient but ultimately fragile: Information Technology handled the data center, wireless network, and screens, while Operational Technology (Facilities) was responsible for the boiler room, chillers, and physical locks. These two areas rarely communicate, each working on different schedules, budgets, and risk levels.

Today, viewing these functions as separate is no longer just outdated; it poses an existential threat to the institution's financial stability and sustainability objectives. Leaders must move beyond letting the boiler room and data center function in isolation. The driving force for a transformative campus leadership vision is the Art of the Possible—the understanding that the university's physical infrastructure is becoming more integrated and responsive.

The Catalyst: The Campus as a Cognitive Organism

The "Art of the Possible" is achieved when we cease seeing buildings as fixed containers for education and instead see them as dynamic, data-producing nodes within the network.

As inexpensive IoT sensors, dense connectivity, and edge computing become widespread, each physical asset—ranging from HVAC air handlers to smart waste bins and -80°C research freezers—now possesses a digital signature. When these operational technology (OT) assets are integrated with AI-powered analytics, the campus transitions from a mere collection of buildings to a Cognitive Organism.

- **Predictive, not reactive**
  Imagine a campus that no longer waits for a chiller to fail during a July orientation. Instead, AI algorithms analyze vibration and acoustic telemetry in real time, predicting failure weeks in advance and automatically generating a preventive work order.
- **Context-aware operations**
  Imagine an HVAC system that doesn't run on a rigid 8-to-5 schedule, but dynamically adjusts based on real-time Wi-Fi occupancy data, weather forecasts, and the registrar's class schedule.
- **Spatial intelligence**
  Imagine utilizing anonymized spatial data to redesign the physical flow of the campus, identifying underutilized real estate to save millions in deferred maintenance or new construction.

The Modified Vision: The Cyber-Physical Architect

This realization necessitates a significant shift in the CTO's vision. You can't just be a digital services provider anymore; you need to become the Architect of Cyber-Physical Fusion.

A new approach is needed because simply attaching "smart" sensors to existing infrastructure without a unified governance strategy leads to disorder. For example, if Parking invests in a smart garage system, Housing purchases smart locks, and Sustainability acquires smart meters—without a shared data ontology or security framework—you haven't created a smart campus. Instead, you've just assembled a costly, unmanageable collection of "shadow OT."

The new vision requires radical interoperability. It places the CTO as the key enforcer of a unified integration platform, delivering enterprise-grade Identity Management, Zero Trust cybersecurity, and semantic data standards to the physical plant.

Articulating The Value of the Smart Campus

How does the CTO use this updated vision to help the President, CFAO, and Board see the investment's worth? By framing technical integration as essential for the institution's survival. The benefits of the AI-Enabled Sustainable Smart Campus are highlighted through three strategic pillars.

Financial and Ecological Sustainability (The ROI)

Achieving net-zero carbon emissions relies heavily on integrating AI and OT. Using digital twins and AI-powered microgrids enables the university to optimize energy use in real time. This is more than just a sustainability marketing tactic; it's about saving millions on utility expenses that can instead support the university's primary educational goals.

Institutional Resilience and Safety

In a converged environment, physical safety and cybersecurity become the same. Integrating IT and OT provides a unified view for incident response. A combined dispatch system can link a swipe-card access anomaly with a sudden decrease in building pressure, enabling campus safety to respond immediately to a physical threat or hazardous lab leak with full situational awareness.

The Frictionless Human Experience

Ultimately, technology should prioritize serving people. A unified campus design responds to student needs instead of requiring students to conform to bureaucratic systems. It fosters an environment where wayfinding apps easily lead students to open study areas, smart lighting enhances safety during evening strolls, and the physical setting naturally supports their well-being.

By illustrating the "Art of the Possible," the CTO changes the dialogue. Instead of requesting the Cabinet to fund a technology project, you're inviting them to invest in the institution's sustainable, resilient, and truly human future.

Crafting the CTO's Vision Statement

Higher education institutions often see vision statements as simply aspirational poetry. Yet, within an AI-Enabled Sustainable Smart Campus, the vision should go beyond inspiration; it must act as an operational guide. It provides a decision-making framework that helps leaders clarify what to standardize, pilot, fund, and, importantly, what to reject.

Without a common, mission-focused direction, campuses often suffer from scattered investments, vendor-dependent systems, and inconsistent governance. A clear and effective vision simplifies this complexity by breaking down departmental barriers and providing a unified story that emphasizes the institution's core needs.

A well-designed smart campus vision effectively accomplishes three key strategic goals:

- **It Grounds the Work in the Mission**
  It clearly connects technological modernization with the university's primary goals—supporting student achievement, empowering faculty, involving the community, and securing long-term institutional strength.
- **It Paints the Future in Human Terms**
  A vision should clearly explain how the daily experience will be improved for students, faculty, staff, and visitors as the campus becomes more measurable, manageable, and more effectively responsive.
- **It establishes Inflexible Guardrails**
  It clearly outlines the non-negotiable essentials. A campus that sacrifices individual privacy, increases cyber risks, or depends on fragile, proprietary systems is not genuinely "future-ready," regardless of how advanced the technology seems.

Because it establishes essential boundaries, the highest university leadership must approve the vision. Cabinet approval is more than just symbolic; it is a binding commitment. It demonstrates to the entire institution that, when faced with tough decisions, the organization will stick to the unified strategy rather than fall back on departmental independence.

Sample Vision Statements From the CTO's Point of View

Ultimately, a compelling vision underpins responsible modernization. The sample vision statements below convert these principles into clear, actionable directives, equipping leaders with the language to align teams, steer investments, and anchor the transformation toward a trustworthy, sustainable campus.

"We will build *the secure, interoperable digital foundation that connects buildings, infrastructure, and campus services—so the university can reduce cost and carbon, improve reliability, and measurably elevate the student and faculty experience."*

"My vision is a future-ready university where digital capability is as essential as power and water: a governed data estate, identity fabric, and integration platform that anticipates needs, optimizes operations, and accelerates teaching, research, and responsible stewardship."

"We will orchestrate IT/OT convergence through clear standards, cybersecurity baselines, and shared governance—so real-time insight and predictive operations can scale safely across the campus portfolio."

"We will establish a trusted institutional data estate—secure, privacy-forward, and decision-grade—so leaders can plan with evidence, decarbonize with confidence, and deliver seamless, equitable services across learning, living, and working."

"We will design the smart campus as an open, secure platform—built on common integration patterns and lifecycle governance—so innovation can move faster than silos, and adoption can scale without increasing risk."

Later in the manifesto, we will examine the Facilities Leader's Vision and then present a shared joint vision for the AI-Enabled Sustainable Smart Campus.

Chapter Summary

Chapter 1 redefines the role of the Chief Technology Officer to suit a new institutional landscape where digital and physical campuses are linked. As buildings, utilities, security, and research environments become software-defined and data-centric, the CTO's scope now includes managing the cyber-physical ecosystem alongside traditional IT functions. The chapter presents the CTO as the Integrative Architect, responsible for merging information technology and operational technology through secure infrastructure, shared governance, and a unified data platform that promotes reliability, sustainability, and a cohesive campus experience.

The CTO is no longer simply a digital services provider.
They are the integrative architect of the cyber-physical institution.

This chapter argues that this change is more than mere scope creep; it's a strategic decision driven by financial challenges, rising cyber-physical risks, and increasing expectations from students, faculty, and staff. The CTO employs enterprise principles such as identity, segmentation, interoperability, telemetry, and change management within the built environment to achieve clear benefits: reduced energy consumption and emissions, enhanced operational resilience, and more human-focused spaces. Essentially, Chapter 1 envisions a future where the AI-powered, sustainable smart campus is not just a network of connected devices but a governed, adaptive learning ecosystem that ensures reliable operation today and enables intelligent responses to future uncertainties.

Chapter 2
The Chief Technology Officer's Strategy

The CTO secures, governs, and integrates OT to deliver operational excellence, measurable sustainability, and a seamless campus experience.

Transforming a modern campus into an AI-enabled, sustainable smart environment positions the Chief Technology Officer (CTO) at a pivotal point of leadership in the institution's history. What was once considered peripheral—such as OT HVAC controllers, metering systems, security, and utility infrastructure—has become central to a smart campus, closely linked with the university's digital core, risk management, and core mission objectives. As Operational Technology (OT) systems transition to IP networks, leverage cloud platforms, and connect with enterprise data, the CTO's responsibilities expand beyond managing digital services alone. Now, the CTO is tasked with shaping the cyber-physical campus, ensuring the physical environments and digital systems operate together reliably, securely, and with purpose.

This is not just an infrastructure side quest or an optional expansion of the traditional IT domain. Instead, it serves as the foundational architecture that enables three strategic outcomes, crucial for the institution's ability to operate, compete, and achieve its mission amidst an era of volatility, limited resources, increasing expectations, and advanced technological opportunities.

Initially, ***the CTO must achieve operational excellence.*** Universities can no longer sustain a reactive approach where buildings silently fail, energy inefficiencies go unnoticed, and vital systems degrade without detection. Modernizing operational technology—through identity, segmentation, telemetry, and integration—shifts the focus from reactive responses to proactive planning. Predictive maintenance becomes feasible when equipment can inform operators of its status. Energy efficiency becomes automatic as the campus continually optimizes its operations. Disruptions diminish when systems are observable, interconnected, and managed with disciplined change processes. For the CTO, operational excellence is more than a technical goal; it's essential for the reliable functioning of classrooms, residence halls, labs, and campus life overall.

Secondly, ***the CTO facilitates measuring sustainability performance.*** Decarbonization targets are now public commitments or just marketing slogans. Institutions' plans and progress are scrutinized by students, regulators, donors, and accreditors. Reliable improvement, verification, and trust in sustainability metrics rely on the data infrastructure that the CTO establishes. Key elements such as carbon visibility, transparent reporting, and resilient operations depend on governed telemetry from sensors, meters, building systems, and distributed energy resources, integrated into the institution's data ecosystem. Achieving this requires integration approaches that prioritize safety, privacy, cybersecurity, and data accuracy. When implemented effectively, the CTO enables the institution to shift from merely estimating its environmental impact to actively managing it in real time. This transformation makes sustainability a core operational capability rather than just a compliance statement.

Third, ***the CTO influences the overall campus experience.*** The campus now combines physical and digital touchpoints—covering comfort, air quality, room readiness, mobility, accessibility, safety, and responsiveness.

These experiences depend on the underlying systems operating unseen. A healthy, adaptable, and transparent campus environment is not only a result of sensors but also relies on governance, data architecture, and integration practices that give sensors purpose. When these systems are connected, trustworthy, and understandable, campus services become more seamless, predictable, and human-centered. The CTO facilitates this transformation by designing a cyber-physical framework that promotes well-being, minimizes friction, and builds community trust through transparency and consistent operations.

These three outcomes collectively redefine the CTO's role. This chapter of the manifesto focuses on outlining the essential strategy that supports them: clear decision rights regarding identity, segmentation, remote access, and data governance; a secure, interoperable OT architecture; a well-managed institutional data estate; disciplined change management; and a partnership model with Facilities that ensures uptime while fostering innovation. The CTO's OT Strategy extends beyond technology; it aims to foster reliability, sustainability, and trust across the institution. It involves designing a campus capable of ongoing adaptation, response, and improvement. Ultimately, it transforms the built environment into a strategic asset that consistently advances mission outcomes.

The Strategic Domains of the CTO

Historically, the Chief Technology Officer's scope was limited to the data center and the edges of the wireless network. Yet, the modern smart campus has expanded this boundary, integrating the CTO into the realm of Operational Technology (OT). This shift—from overseeing digital assets to managing cyber-physical systems—significantly increases both complexity and risk.

Unlike a traditional IT server failure that halts digital services, an OT system breach can cause tangible consequences, such as the loss of climate-sensitive research samples, heat loss in residence halls, or security breaches. To build an intelligent university, the CTO must transcend traditional IT boundaries and oversee a broader range of physical and digital systems. These systems, once isolated and managed solely by facilities staff, are now interconnected digital platforms demanding strict IT governance, identity control, and cybersecurity.

These domains form the essential components of this new operational landscape, encompassing both the vital systems that sustain the physical campus and the innovative "new edge" technologies shaping its strategic trajectory. Ultimately, elevating the CTO from a conventional technology manager to the Integrative Ecosystem Architect hinges on mastering these core OT and 'new edge' areas. The goal extends beyond merely linking Building Automation Systems to the enterprise or integrating IoT sensors into cloud dashboards; it aims for radical interoperability secured by strict protocols. By implementing unified identity management, Zero Trust segmentation, and semantic data standards across these physical systems, the Architect bridges the long-standing, risky gap between IT and Facilities. This approach transforms a fragmented collection of vulnerable, siloed infrastructure into a unified cognitive entity—one that continuously enhances energy efficiency, safety, and research continuity, making the built environment a strategic driver rather than just a passive space.

Transforming a traditional university into an AI-enabled, sustainable, smart campus requires more than just isolated technology acquisitions. It needs to be managed as a comprehensive institutional overhaul. Combining Operational Technology (OT) with Information Technology (IT) demands a structured framework to prevent the campus from becoming overwhelmed by its own complexity. These five domains form the core architecture of this implementation, offering the essential structure to govern, secure, connect, optimize, and ultimately adopt the technologies shaping the institution's future.

Technical Governance and Enterprise Alignment

Establishing cross-functional governance, decision-making authority, technical standards, and funding discipline for all smart campus projects is essential. It involves formally aligning IT, Facilities, Sustainability, Campus Safety, and Data Leadership. The biggest risk to a smart campus isn't technological gaps but poor coordination. Without enterprise-wide alignment, institutions risk falling into "Shadow OT," where departments like Parking, Housing, and Utilities purchase smart systems that don't communicate with one another, such as a smart garage, smart locks, and smart meters.

Digital Governance serves as the solution to the fragmented, vendor-driven proliferation. It creates an institutional "Supreme Court" for overseeing technology procurement and deployment. The CTO and essential cabinet members establish clear decision rights in this domain, such as determining who owns the data from a smart building, who funds the continuous maintenance of an IoT sensor network, and who has the authority to approve new physical security systems.

This governance structure promotes adoption by enforcing portfolio discipline and clear standards. Vendors must follow the institution's guidelines rather than setting their own, ensuring that modernization spending aligns with mission goals. It makes expenditures auditable, supports sustainability and resilience, and prevents the campus from becoming a collection of isolated pilot projects.

Securing IT/OT and Managing Risk

The definition and strict enforcement of security and interoperability standards specifically tailored for connected OT systems. This involves Identity and Access Management (IAM), Multi-Factor Authentication (MFA), strict network segmentation with "zones and conduits," remote access controls, ongoing monitoring, and vendor accountability.

When a university links its physical infrastructure to the digital network, its attack surface grows significantly, and the focus shifts from data breaches to physical safety. Compromising an air handler or a lab's negative pressure system directly endangers the campus community's safety.

This domain compels the institution to maintain a "Secure-by-Design" approach. It requires OT networks to be heavily segmented from the main IT networks, preventing malware on a student's laptop from spreading to critical systems like the central chiller plant. Additionally, it tackles the significant risk posed by vendor remote access. Facilities vendors frequently need access to oversee their installed systems; this architecture ensures that such access is tightly restricted, continuously monitored, and based on the principle of "least privilege."

This strict architectural approach significantly minimizes cyber-physical risks and contains the "blast radius" of potential breaches. It enables the institution to rapidly expand its IoT and smart-building connectivity without jeopardizing essential operations. Ultimately, it assures the Board of Trustees that developing a modern, intelligent campus will not threaten the safety or ongoing continuity of the institution.

Institutional Data Estate and Interoperability Platform

The development of a central campus data and integration platform encompasses an API and event-streaming backbone, a unified semantic layer (campus ontology), and rigorous data cataloging and quality controls. These elements ensure that both OT and enterprise data are highly reusable across all domains. A smart campus produces vast amounts of data, but raw data is meaningless if it uses multiple terminologies. For example, the HVAC system might identify a room as "Bldg4_Rm101," whereas the Registrar refers to it as "Science Lecture A."

This domain addresses the ***Babel Problem*** by establishing a central integration platform that dismantles proprietary vendor silos. Its API framework facilitates seamless communication among different systems, while the semantic layer establishes a common dictionary—ensuring that occupancy sensors, energy meters, and academic scheduling tools interpret data consistently. This approach turns data, which was once just a byproduct of building operations, into a valuable, well-organized asset for the institution.

This interoperability drives 'decision-grade' insight, significantly reducing the time and cost of integrating new systems. More critically, it is essential for scaling Artificial Intelligence and Digital Twins. Without this unified data infrastructure, AI cannot operate, and transparent, precise ESG (Environmental, Social, and Governance) sustainability reporting becomes unfeasible.

AI Enablement and OT/Digital Twin Capability Development

The responsible development of AI capabilities—including model governance, algorithmic explainability, and performance monitoring—culminates in the deployment of OT and Digital Twins as a central "control tower" for the campus ecosystem. Once the sensor telemetry is established and the data estate is organized and secure, the institution can shift from merely monitoring the campus to actively optimizing it. Various applications, from remote classroom support to a Digital Twin, a real-time virtual replica of the physical campus, are powered by a continuous flow of OT telemetry and other data comprising the Data Estate.

Mastering this domain represents a maturity stage in which digital skills and AI agents and models analyze large data streams to identify inefficiencies, forecast equipment failures, and simulate intricate energy scenarios. It requires strict governance and stewardship to ensure that AI is not acting independently on campus. Instead, it operates within "Human-in-the-Loop" systems, providing highly accurate suggestions that human operators review and approve.

> *The essential ROI of an AI-enabled smart campus is*
> *The ability to transform passive observation into active, prioritized action management.*

This is where the ROI of an AI-enabled smart campus truly comes into play. It transforms passive observation into active management. Facilities teams use OT systems and the Digital Twin to simulate "what-if" scenarios for severe weather, power outages, and major infrastructure component failures, greatly enhancing emergency preparedness and resilience to campus shocks. This shifts the campus from a reactive system fueled by dashboard alarms to a constantly evolving, predictive entity that reduces carbon emissions and operational expense. Experience, Adoption, and Change Enablement

The intentional coordination of the human aspect in modernization involves role-specific training, integrating new digital insights into everyday physical workflows, and implementing strict service management practices across IT and OT operations. Even the most advanced AI, OT, and precise Digital Twin are pointless if maintenance technicians refuse to utilize them. The cultural gap between IT (the keyboard) and Facilities (the wrench) remains significant.

This domain acknowledges that implementing technology is fundamentally about understanding human psychology and establishing a new locus of trust. It emphasizes Product Management and User Experience (UX) for operational staff. When an AI system predicts a pump failure, it doesn't merely send a cryptic code to a database. Instead, it creates a clear, prioritized work order directly on the technician's mobile device, including the physical location and necessary parts.

By actively managing change and prioritizing staff experience as a key result, this domain minimizes friction and prevents technology abandonment. It ensures modernization is seen not as an IT requirement that complicates jobs, but as an essential set of tools that empower users. This is the crucial final step that transforms a multi-million-dollar installation into a sustainable, highly effective institutional capability.

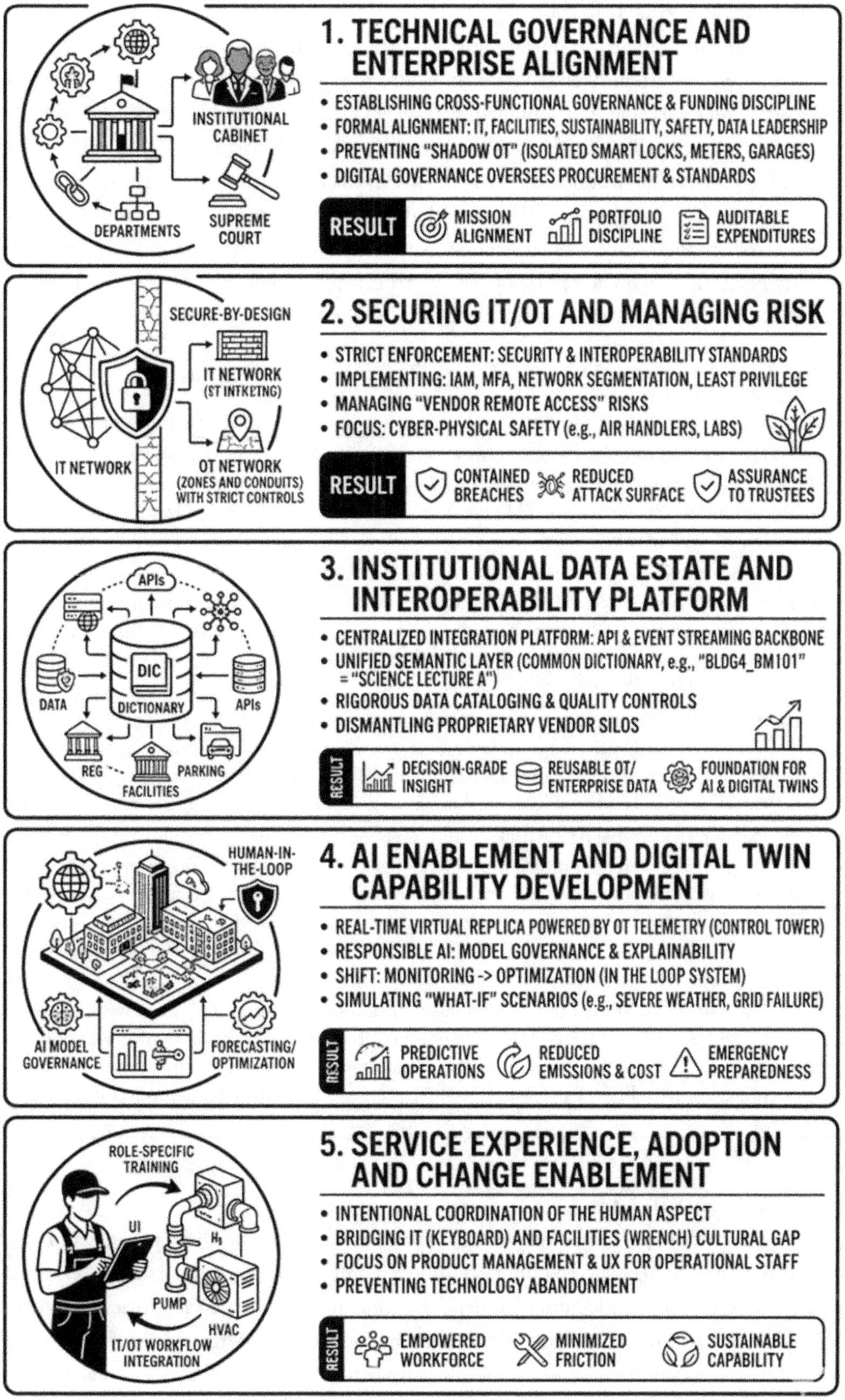

Figure 3. The Strategic Domains of the CTO

The CTO as Integrative Architect

Ultimately, implementing these five pillars transforms the CTO from simply managing digital utilities into the institution's Integrative Ecosystem Architect. Through strong governance, securing the cyber-physical boundary, unifying data assets, and deploying AI thoughtfully, the CTO creates a highly resilient technological foundation. The crowning achievement of this architecture, however, is found in the last pillar—the human experience.

When the CTO advocates adoption and reduces obstacles to change, they embrace their role as the Chief Motivator. They offer crucial support and intuitive tools that enable facilities operators, campus security officers, and network engineers to turn their daily, often unseen work into tangible progress toward the university's top academic and sustainability goals. In this integrated vision, the smart campus evolves from a mere network of connected devices into a cohesive, intelligent system that actively promotes the institution's mission.

Exhibit 3. Summary of the Strategic Domains of the CTO for the Smart Campus

Domain	What It Means	Value of a Smart Campus
Technical Governance and Enterprise Alignment	Establish cross-functional governance, decision rights, standards, and funding/portfolio discipline for smart campus initiatives—aligning IT, Facilities, sustainability, safety, and data leadership.	Prevents fragmented pilots and vendor-driven sprawl; accelerates adoption through clarity; ensures modernization remains mission-aligned, auditable, and trust-preserving.
Secure IT/OT Architecture and Cyber-Physical Risk Management	Define and enforce the security and interoperability guardrails for connected OT: IAM/MFA/least-privilege, segmentation ("zones and conduits"), remote-access controls, monitoring/response, and vendor accountability.	Reduces cyber-physical risk and blast radius; protects safety and continuity; enables scalable connectivity without destabilizing operations; strengthens board-level confidence in modernization.
Institutional Data Estate and Interoperability Platform	Build the campus data and integration platform (API/event backbone, semantic layer, data catalog/quality, identity, and governance) that makes OT and enterprise data reusable across domains.	Creates decision-grade, cross-domain insight; enables AI and digital twins to scale; reduces integration costs; supports transparent sustainability reporting and consistent user experiences.
AI Enablement and Digital Twin Capability Development	Develop responsible AI capabilities (model governance, explainability, monitoring) and treat digital twins as a capstone "control tower" that becomes powerful once data maturity and governance are in place.	Converts visibility into optimization and scenario planning; supports resilience and emergency readiness; enables selective automation with human oversight; makes the campus continuously improving rather than dashboard-driven.
Service Experience, Adoption, and Change Enablement	Orchestrate the human and operational adoption side: role-based enablement, workflow integration, product management for platforms, and service management practices across IT/OT operations.	Ensures tools improve operator and user experience; reduces friction and non-adoption; sustains outcomes over time; turns modernization from an installation project into an institutional capability.

CTO Decision Rights in OT

Most campuses have seen operational technology expand through decentralized projects and vendor implementations. While this approach may succeed on a building-by-building basis, it falters at the enterprise level due to inconsistent security standards, undocumented dependencies, and fragmented identities, networks, and data systems. The CTO's role is not to oversee Facilities operations but to set enterprise-level decision rights that ensure secure connectivity, lasting governance, and reusable integrations—while still allowing Facilities to manage daily operations, safety measures, and priorities.

A helpful way to frame this is: Facilities is responsible for operating the building, while the CTO oversees the governance, security, and integration of connected systems to ensure safe campus operations at scale. These CTO decision rights should be clearly defined, enforceable, and considered non-negotiable in procurement, architecture, and operational procedures.

Identity and Access Management for OT

In a converged campus, identity serves as the foundation for accountability and risk management, not just a technical preference. OT environments often gather shared vendor accounts, local admin credentials, and ad hoc remote access pathways that lack visibility to enterprise oversight. This leads to two key issues: the institution cannot identify who accessed critical systems, and it cannot consistently prevent over-privileged access from causing cyber-physical incidents.

> ***There is to be no privileged OT access.***
> ***without enterprise identity, segmentation, and auditability.***

The CTO's right to decide is to mandate that privileged access—whether by staff or vendors—adheres to an enterprise-standard identity approach whenever possible: individual authentication, least privilege, and traceable authorization. This approach reflects modern "zero trust" principles: access is granted based on verified identity and role, not network proximity or past convenience.

Non-negotiable expectation:

- No privileged OT access without enterprise controls.

What the CTO must require (minimum posture):

- Enterprise identity integration where feasible (SSO / centralized accounts)
- MFA and least privilege for privileged functions
- Role-based access tied to job function and vendor role.
- "Break-glass" emergency procedures that are time-bound, logged, and reviewed.

This isn't about making Facilities less responsive. It's about guaranteeing that access remains predictable, auditable, and revocable—allowing the campus to modernize without increasing hidden risks.

Segmentation and "Zones and Conduits" Logic

Segmentation is the most effective method to minimize cyber-physical risk by limiting the blast radius. An unsegmented, flat network can turn a single compromised device into a vulnerability that affects the entire campus; in contrast, structured segmentation blocks lateral movement and reduces the risk of an IT incident affecting building operations. ISA/IEC 62443 formalizes this strategy with "zones and conduits"—grouping assets with similar security needs into zones and strictly controlling communication between them through designated conduits.[4]

The CTO's authority includes establishing the campus segmentation model and implementing it as a design standard. However, this does not entail that Facilities lose control. It ensures that control networks stay stable and protected, even when data is shared externally for analytics.

Non-negotiable expectation:

- OT is segmented from IT by default; high-consequence OT domains are further restricted.

What the CTO must require (campus-ready model):

- OT separated from enterprise IT and public networks by default.

- Additional restricted zones for sensitive domains (e.g., BAS core, utilities plant, physical security, research-critical infrastructure)
- Conduits that are explicit, minimal, monitored, and governed (what can talk to what, and why)
- A documented "exception" process with risk acceptance and compensating controls

The National Institute of Standards and Technology's (NIST) OT guidance highlights that securing operational technology must consider specific performance, safety, and reliability needs—hence the importance of segmentation and customized controls.[5]

Remote Access for Vendors and Staff

Remote access is vital for modern campus operations, but it is often misused, leading to security incidents. The campus should see remote access as a controlled feature rather than just a casual tool. Unregulated 'always-on' tunnels, shared vendor credentials, and direct remote connections to controllers are like leaving service doors wide open. While these may simplify immediate support, they compromise long-term security and accountability.

The Cybersecurity and Infrastructure Security Agency (CISA) offers guidance tailored for OT environments on configuring and managing remote access, highlighting the importance of secure architectures, consistent policy enforcement, and controls that minimize exposure.[6]

Non-negotiable expectation:

- Remote access is centrally governed, time-bound, logged, and monitored.

What the CTO must require (minimum posture):

- No unmanaged "always-on" vendor tunnels
- Approved access methods only (e.g., jump hosts, privileged access paths, controlled VPN/ZTNA patterns)
- MFA, least privilege, and time-bound access windows
- Logging and monitoring of sessions; session recording or equivalent auditing for privileged work where feasible
- A formal exception process with explicit risk acceptance and documented compensating controls.

This decision also safeguards Facilities by minimizing "mystery changes," enhancing forensic clarity, and reducing recovery time when issues arise.

Data Governance for OT Telemetry and Control Data

As operational technology (OT) data integrates into the campus data architecture, the institution moves beyond merely gathering building telemetry. Instead, it is developing a facilities data estate capable of influencing decisions on safety, space, energy, and human behavior. This development carries responsibilities, including safeguarding privacy, establishing proper access controls, ensuring ethical use, and maintaining transparency. Therefore, the CTO must consider OT data governance as an integral part of enterprise data governance, assigning the same level of importance to data streams that influence campus safety, experience, and institutional trust as to student and research data.

This aligns with the governance emphasis in NIST Cyber Security Framework (CSF) 2.0, which elevates "GOVERN" as a core function—establishing, communicating, and monitoring cybersecurity risk management strategy, expectations, and policy in alignment with mission and stakeholder expectations.[7]

Non-negotiable expectation:

- OT data is governed as an institutional asset with clear purpose, controls, and accountability.

What the CTO must require (decision-right scope):

- Clear rules for what data is collected, retained, and for how long.
- Where it is stored, under what security controls, and with what auditability.
- Who can access which datasets, for what purpose (especially for AI and digital twins)
- How data is shared with vendors, researchers, and partners—with explicit agreements and constraints
- A "no mission creep" posture: operational purpose first, privacy-forward defaults, and transparent community norms

Done well, this approach enables scalability. Done poorly, it creates suspicion and political blowback that can stall modernization.

What Facilities Should Own—and Why It Matters

A mature model preserves Facilities' operational authority because it is essential to safety, uptime, and user experience. Facilities should remain accountable for:

- Operational performance and service levels (comfort, uptime, response times)
- Safety constraints and commissioning standards (what "safe state" means; how systems fail safely)
- Maintenance workflows and field adoption (how work actually gets done)
- Asset lifecycle decisions (repair/replace strategy, capital sequencing, vendor performance)

These tasks are not merely "IT-adjacent;" they are central to campus operations. The CTO's responsibility is to guarantee that as these systems connect—allowing data to flow outward and access inward—the institution maintains robust governance, security, and interoperability, ensuring Facilities can operate confidently.

The CFaO owns how the building runs. *The CTO owns how connected systems are governed, secured, and integrated.* *So the building can run safely.*

Exhibit 4. CTO Decision Rights in OT

Decision Right	Non-Negotiable Expectation	What the CTO Must Require (Minimum Posture)	Why It Matters (Board-Ready Value)
Identity and Access Management (IAM) for OT	No privileged OT access without enterprise controls.	Enterprise identity integration where feasible (SSO/centralized accounts); MFA; least privilege; role-based access; "break-glass" emergency procedures that are time-bound, logged, and reviewed.	Creates predictable, auditable, revocable access; reduces "mystery changes," improves accountability, and limits cyber-physical exposure.
Segmentation (ISA/IEC 62443 "Zones and Conduits")	OT is segmented from IT by default; high-consequence OT domains are further restricted.	OT separated from enterprise IT/public networks; restricted zones for BAS, utilities plant, physical security, and research-critical environments; conduits are explicit, minimal, monitored, and governed; documented exception process.	Limits blast radius and prevents lateral movement; reduces the likelihood that an IT incident becomes an operations/safety incident; enables safe scaling of connectivity.
Remote Access (Vendors and Staff)	Centrally governed, time-bound, logged, and monitored remote access.	No unmanaged "always-on" tunnels; approved access methods only (jump hosts, controlled VPN/ZTNA); MFA and least privilege; session logging/recording for privileged work; formal exception process.	Reduces a common exploit path; improves forensic clarity and speeds recovery; protects continuity while preserving necessary vendor support.
Data Governance for OT Telemetry and Control Data	OT data is governed as an institutional asset with clear purpose, controls, and accountability.	Define what data is collected/retained and where it is stored; security controls and auditability; role-based access for defined purposes (esp. AI/twins); rules for vendor/research sharing; privacy-forward defaults and "no mission creep."	Protects trust, privacy, and institutional legitimacy; enables reusable, decision-grade data for analytics, digital twins, and transparent reporting.
Scope Clarifier: What Facilities Owns	Facilities own how the building runs (operational performance, safety constraints, maintenance workflows, asset lifecycle); CTO owns governance of how connected systems are secured and integrated.	Clear split of authorities institutionalized in governance, procurement, and operating practice.	Preserves uptime and safety while ensuring connectivity is safe at scale; avoids role confusion and prevents fragmented, vendor-driven sprawl.

Translating OT Modernization into Board-Level Narratives

Governing Boards frequently provide funding to lower risk exposure and enhance institutional capacity and capability. The CTO's role is to frame OT modernization as a resilience program that manages risk while delivering measurable benefits.

Resilience and continuity (board language)

OT modernization is best understood as **mission continuity infrastructure**. It reduces the likelihood that a localized failure (a controller fault, vendor tunnel issue, misconfigured change, or brittle integration) cascades into a broader disruption that impacts instruction, research, housing, and safety operations. From a board perspective, the key point is not "more connectivity," but **more controlled, observable, and recoverable operations**—with clear fallback modes, rehearsed procedures, and fewer single points of failure. NIST's OT security guidance reinforces that OT environments have distinct performance, reliability, and safety requirements and therefore require controls tailored to those constraints—precisely because disruption can be immediate and consequential.

Operational translation (what changes in practice)

- Fewer cascading failures from unmanaged changes (formal change enablement + rollback plans)
- Faster recovery through known-safe operating modes and tested playbooks
- Reduced dependence on "tribal knowledge" by institutionalizing documentation, monitoring, and shared ownership

Board-level statement (sample)

" *We are funding OT modernization as a resilience and continuity program: it will reduce the likelihood and impact of disruptions that affect instruction, research continuity, housing, and safety operations, and it will measurably shorten recovery time when incidents occur.*"

Safety and duty of care (board language)

On a connected campus, OT takes on added importance for safety. The board doesn't require alarmist rhetoric—just clear governance framing: when building and utility systems affect ventilation, access/egress, lab conditions, alarms, and emergency procedures, cyber vulnerabilities can lead to operational disruptions and, in certain situations, physical safety risks. Therefore, OT/ICS guidance considers the reliability and safety contexts significantly different from standard IT settings and advocates security methods that accommodate operational limitations.

Operational translation (what changes in practice):

- Stronger protection of systems that affect life-safety and critical environments (e.g., ventilation, access control, specialized research spaces)
- OT-aware monitoring and incident response (joint table-top exercises with Facilities + Safety + IT)
- Clear decision rights so response is fast, proportionate, and accountable (who can authorize what action, and when)

Board-level statement (sample)

"OT modernization strengthens our duty of care by improving the security and reliability of systems that influence safety-relevant outcomes—such as ventilation, access/egress, and critical research environments—while ensuring response procedures and decision rights are clear and rehearsed."

Avoided downtime and cost containment (a CFAO-friendly bridge)

Modern OT minimizes unnecessary downtime and emergency-response expenses by shifting from reactive firefighting to predictive, condition-based maintenance. The board can view this as disciplined cost management with clear benefits: reduced outage times for critical systems, fewer after-hours emergencies, lower overtime costs, and earlier prevention of asset damage. It also safeguards "hidden value" specific to higher education, such as research continuity and the reputational risk posed by visible failures in housing, classrooms, and safety systems. This approach underpins the financial reasoning behind Phase 1 and Phase 2 maturity—focusing on visibility and fault detection first, followed by consistent optimization.

Operational translation (what changes in practice):

- Reduced reactive work orders and overtime through earlier detection and prioritization.
- Avoid equipment damage by catching drift and faults before failure.
- Protected continuity for labs and specialized environments (fewer preventable interruptions)

Board-level statement (sample):

"This investment reduces avoidable downtime and emergency costs by improving visibility, fault detection, and preventive action—yielding measurable reductions in outage hours, reactive work orders, overtime, and avoidable asset damage while protecting research continuity."

Trust (privacy, ethics, and institutional credibility)

The scalability of smart campus capabilities depends on the institution's governance, which must safeguard privacy, enhance cybersecurity, and maintain community trust. For boards, this issue concerns institutional legitimacy: widespread sensing and integrated analytics might seem like surveillance unless the university clearly defines its purpose, boundaries, and accountability. The board must ensure that governance is considered a fundamental design principle, not just a supplementary policy. According to NIST CSF 2.0, 'Govern' is now a core function, highlighting that cybersecurity risk strategies, expectations, and policies must be clearly established, communicated, and continuously monitored. Moreover, governance should embed cybersecurity into the overall enterprise risk management framework.

Operational translation (what changes in practice):

- Transparent policies on what is sensed, why, and what is explicitly *not* collected.
- Role-based access controls, strong identity, and monitoring/auditability
- Vendor accountability and data ownership clarity (including contract requirements and offboarding)

Board-level statement (sample):

"We will scale smart campus capabilities only with explicit governance—clear privacy boundaries, role-based access controls, auditable accountability, and vendor obligations—so modernization strengthens institutional trust and credibility rather than eroding it."

Overarching Sample Board Statement

"OT modernization is a resilience and safety initiative aimed at minimizing disruptions through clear decision rights for identity, segmentation, and remote access, backed by a governed data foundation. After verifying the baseline—covering inventory, segmentation, monitoring, and tested responses—we will expand the use of analytics, AI, and digital twins to decrease downtime, enhance sustainability, and improve the campus experience. This approach embodies disciplined modernization: begin with small, proven steps, demonstrate value, and scale confidently."

Chapter Summary

Chapter 2 positions the Chief Technology Officer as a key institutional strategist responsible for more than just digital systems—also shaping the entire cyber-physical campus. As Operational Technology integrates with IT, the CTO must ensure that building systems, utilities, sensors, and digital services function within a unified framework centered on identity, segmentation, secure integration, and governed data. This chapter outlines the CTO's OT strategy as the basis for three key enterprise outcomes: operational excellence, measurable sustainability, and a significantly enhanced campus experience.

To achieve these outcomes, the CTO must establish clear decision rights, develop a secure, interoperable operational technology (OT) architecture, oversee a reliable institutional data estate, and enable AI and digital twin technologies to convert raw signals into predictive insights. The chapter emphasizes that modernization is not merely a technology initiative; it involves a fundamental shift in governance and operational models, requiring close collaboration with Facilities and a disciplined focus on risk management, architecture, and change processes. Essentially, Chapter 2 presents the CTO as the key architect of a safer, more resilient, and smarter campus environment.

Chapter 3
The IT Architecture

IT architecture is the invisible force that turns sensors into insight, platforms into systems, and ambition into durable institutional capability.

Operational Technology (OT) data represents the campus's "physical truth"—a continuous flow of signals that show building operations, energy consumption, space utilization, and changing conditions in real time. For a sustainable smart campus, this data must be securely captured, integrated across different systems, standardized within a unified Data Estate, and transformed into insights that support decision-making.

For the CTO, this isn't just an infrastructure task; it's the foundational architecture that enables three key strategic outcomes.

1. **Operational excellence**
   (predictive maintenance, energy optimization, and less disruption)
2. **Sustainability performance**
   (carbon visibility, accountable reporting, resilient operations)
3. **Institutional experience**
   (healthy spaces, responsive services, transparent stewardship)

This chapter outlines a CTO's architecture strategy that links OT systems and IoT sensing to a campus Digital Twin, avoiding a fragile, proprietary "science project" that cannot scale.

Digital Governance and Enterprise Alignment

Effective institutions implement a strategic, phased plan rooted in governance by clearly outlining roles and policies that connect IT, facilities, and academic sectors. They encourage collaboration across different functions and ensure vendors are accountable for interoperability and long-term consistency. Their core approach starts with small projects, showcasing value before scaling up to standards rather than relying on isolated solutions.

In creating a sustainable smart campus, governance is central—serving as the fundamental framework that supports all other aspects. The AI-driven Sustainable Smart Campus (SSC) forms a network of interconnected systems, individuals, data, and principles. Governance guarantees that this intricate system operates smoothly. It harmonizes the institution's mission with practical implementation, ethical considerations with new technologies, and campus-wide projects with clear accountability measures.

While discussions about smart campuses often emphasize technological advances and environmental goals, governance is key to guiding their priorities, funding, security, and overall success. Developing intelligent campuses relies on effective governance, which involves establishing adaptable, transparent, and inclusive frameworks that mirror the interdisciplinary and cross-functional aspects of this work. Therefore, the goal is to create management systems that are transparent, secure, and stakeholder-driven.

Strategic Functions of Smart Campus Governance

Policy and Standards Infrastructure

At the core of governance is the institutional policy framework. Universities need to establish and periodically revise policies related to IT procurement, data management, AI deployment, accessibility, privacy, and cybersecurity. These policies should conform to national and international regulations such as FERPA, GDPR, and CCPA, while also tackling emerging ethical issues specific to AI and pervasive sensing. Planning for this involves identifying existing governance bodies that can be utilized and deciding where new structures or committees are necessary. Institutions must also commit to regular policy review cycles and ensure policies are communicated clearly to all campus constituents.

Ethical and Accountable AI Governance

AI brings about significant new responsibilities. Ethical guidelines should guide the development, implementation, and supervision of algorithms that affect student services, facility automation, learning analytics, and public safety. Essential principles encompass fairness, transparency, explainability, and accountability. Governance teams should establish AI ethics committees or integrate these duties into current data governance bodies. Planning must encompass frameworks for AI auditing, bias detection, redress mechanisms, and opt-out options for sensitive systems. Additionally, institutions need to educate staff and students on ethical data practices and AI literacy.

Data Stewardship

An AI-enabled SSC handles large volumes of sensitive data, which requires effective data governance to prevent privacy breaches, operational issues, and reputational damage. Data stewardship involves establishing data ownership, quality standards, access permissions, usage rights, lifecycle supervision, and compliance measures. Campus IT and data governance leaders should collaboratively create a strategic data governance charter, backed by federated governance frameworks involving academic and operational units. This approach must be strengthened with comprehensive cybersecurity measures, such as zero-trust architecture, real-time threat monitoring, and cybersecurity awareness programs.

Risk Management and Institutional Resilience

Smart campus initiatives bring new risks, including technical, financial, reputational, and ethical concerns. Governance bodies need to be prepared to identify, monitor, and address these risks proactively. This involves conducting vendor risk assessments, implementing change management procedures, overseeing AI model governance, and coordinating incident response plans across departments. Planning involves creating comprehensive risk registers for SSC initiatives, ensuring risk management aligns with strategic planning, and performing scenario modeling to stress-test campus systems.

Stakeholder Engagement and Decision Rights

Effective governance requires inclusivity, involving students, faculty, facilities leaders, IT teams, academic affairs, legal, and external partners in planning and oversight of smart campus systems. Establishing formal shared governance models that specify decision rights, advisory roles, and communication channels is crucial. Planners ought to establish structured forums—like smart campus councils or innovation committees—to promote collaboration, transparency, and trust. Decision-making authority must be clearly identified and aligned with the institution's priorities.

Secure IT/OT Architecture and Risk Management

The modern smart campus signifies a major shift: the traditional separation, or "air gap," between Information Technology (IT) and Operational Technology (OT) has disappeared. Today, enterprise networks and user devices are closely connected with the physical infrastructure that supports the campus, including HVAC systems, lighting, water management, and access controls.

While this convergence drives institutional efficiency and innovation, it also significantly broadens the university's attack surface. Thousands of newly connected IoT devices, real-time data streams, and vendor cloud services turn previously isolated, mechanical assets into interconnected digital vulnerabilities.

Security as a Foundational Enabler, Not an Afterthought

In this integrated ecosystem, security isn't just an add-on perimeter; it must be embedded into the core architecture from the start. A "Secure-by-Design" strategy requires layered protections, comprehensive Identity and Access Management (IAM), consistent privacy policies, and extensive device governance. Previously handled in isolated operational units, security now needs to be managed within a cohesive, campus-wide digital environment.

From Data Protection to Kinetic Resilience

As higher education's digital presence grows, the risks of cyber threats extend beyond digital issues to include physical and kinetic dangers. Achieving sustainability goals—such as digital twins and energy management systems—relies heavily on secure, real-time data transmission and seamless system operation. A security breach here could expose more than sensitive information; it could disrupt essential operations, endanger physical safety, and significantly undermine stakeholder trust.

Therefore, comprehensive cybersecurity has become more than just a technical necessity; it is a key factor in fostering institutional resilience. It guarantees that the university's technological infrastructure—vital for a sustainable future—remains flexible, protected, and aligned with the institution's strategic risk management priorities.

Architectural Tactics for the Cyber-Physical Frontier

Securing a converged IT/OT campus demands the CTO move beyond the outdated "castle and moat" strategy. As the perimeter now includes thousands of smart devices, autonomous mowers, and remote diagnostics, the traditional moat is effectively breached. To protect this new cyber-physical landscape, the Integrative Architect needs to implement modern, targeted architectural strategies that assume the network is already under threat. Below is the tactical playbook for securing the AI-enabled smart environment campus.

Zero Trust Architecture for the Physical World

In a traditional IT environment, a device inside the firewall is generally trusted. In a smart campus, "trust but verify" is replaced with "never trust, always verify." The CTO must extend **Zero Trust** principles to operational technology. Every device—whether it is a student's smartphone or an 80°C freezer—must cryptographically prove its identity before communicating. Access is granted based strictly on the principle of "least privilege." A smart thermostat should communicate only with the specific building management server it reports to; it has no business querying the student information system or the wider internet.

Micro-Segmentation and "Zones and Conduits."

If a vulnerability in a networked vending machine enables a hacker to access the university's central financial database, it indicates a failure in the architecture. To prevent such breaches, the network should be aggressively segmented using micro-segmentation, often based on the ISA/IEC 62443 "Zones and Conduits" standard. This approach is comparable to watertight bulkheads on a submarine. The CTO is responsible for setting strict boundaries that separate enterprise IT, student Wi-Fi, and critical OT infrastructure. Communication between these zones is highly restricted and must go through secure, monitored conduits. If one zone, such as a smart parking sensor, is compromised, the bulkhead prevents lateral movement into critical systems, such as life-safety or research networks.

Privileged Access Management (PAM) for Vendor Ecosystems

One of the most critical yet often overlooked vulnerabilities on a modern campus is the vendor supply chain. Facilities teams frequently rely on external contractors to remotely oversee and service complex OT systems, such as elevators or chiller plants. Using broad-access VPNs is no longer acceptable for this purpose. Instead, the CTO must deploy OT-specific Privileged Access Management (PAM). When a vendor needs to troubleshoot an HVAC controller, they access a secure, session-recorded portal with strict monitoring. They are limited to that specific controller and only for the scheduled maintenance window. After completing their task, the access rights are automatically revoked.

Passive Monitoring and OT-Specific Anomaly Detection

IT security primarily depends on active scanning—sending pings to devices to identify vulnerabilities. However, OT devices, such as older Programmable Logic Controllers, are very sensitive; performing active scans can overload their limited processors and physically damage the equipment. Therefore, the CTO should implement passive continuous monitoring for the OT environment. This involves techniques such as port mirroring, in which the security system quietly observes traffic on the industrial network. AI-based anomaly detection models create a baseline of normal activity, such as "Pump A typically sends a 2-kilobyte status update every five minutes." If the pump suddenly tries to download a large file or connect to an external IP address, the system detects the anomaly immediately, all without risking harm to the delicate hardware.

Unified Incident Response

When a cyber incident happens in an OT environment, a solely digital response is inadequate. For example, shutting down a compromised network switch might eliminate the digital threat but could also disrupt ventilation in a hazardous chemistry lab. Therefore, the CTO must develop a converged Incident Response (IR) playbook that involves cross-training the IT Security Operations Center (SOC) alongside Facilities and Campus Police, dispatchers. Conducting tabletop exercises that simulate cyber-physical incidents is essential to ensure that, when an anomaly occurs, responses effectively balance digital containment, human safety, and physical system continuity.

Risk Domains Unique to Smart Campuses

As institutions update their operations, the traditional boundaries fade away. The campus no longer functions as a fortress with one primary entrance; instead, it becomes a permeable, highly interconnected network of physical and digital interactions. Within this setting, multiple risk areas demand concentrated attention from leadership.

1. Shadow IoT: The Invisible Attack Surface

In the legacy model, IT handled all technology procurement and installation. In the smart campus, technology becomes accessible to everyone. For instance, a biology researcher orders a Wi-Fi-enabled freezer monitor from Amazon; a facilities manager sets up a "smart" coffee machine in the breakroom; a student connects a crypto-mining rig in their dorm.

These unmanaged and unauthorized devices constitute Shadow IoT. They pose risks not because they are complicated, but because they are often hidden and lack basic security features, such as patching and strong passwords. Threat actors don't breach a smart vending machine just to steal soda; their goal is to gain network access. Once there, they can move laterally to access vital assets such as the financial data center or the research IP repository. Additionally, many of these devices use non-standard protocols such as Zigbee, LoRaWAN, or Bluetooth Low Energy, which traditional enterprise firewalls might not inspect or log, creating security blind spots.

2. Mobile App Security: The "Super App" Vulnerability

The modern "Campus Smart App" consolidates a wide range of features—such as door access, dining payments, class schedules, and transit tracking—into one unified interface. Although this represents a significant UX achievement, it also creates a substantial concentration of risk.

To quickly develop and launch these apps, developers often rely on third-party Software Development Kits (SDKs) for features such as maps, analytics, and payments. These SDKs function as "black boxes" code with permissions equal to those of the main app. If a third-party mapping SDK is compromised, it can threaten the security of the entire campus app.

Apps frequently ask for extensive permissions, such as location, contacts, and camera access, often beyond what they actually need. For example, a transit app only requires location data when the bus map is open. Still, many apps default to "Always Allow," generating detailed behavioral data that could be leaked or subpoenaed. Additionally, in the rush to release, developers sometimes embed API keys or credentials directly into the app binary. Skilled attackers can reverse-engineer the app to extract these secrets, allowing them to impersonate the institution and access backend servers.

3. Cloud and SaaS Environments: The Responsibility Mirage

Universities are rapidly migrating to the cloud, often assuming that "moving to the cloud" means "outsourcing the risk." This is the **Responsibility Mirage**. In every cloud contract (IaaS, PaaS, SaaS), the vendor secures the *infrastructure*, but the university must secure the *configuration*. For example, Amazon Web Services (AWS) will ensure that its physical data centers are secure. Still, if a university researcher leaves an S3 storage bucket configured to "Public," the data breach is the university's fault, not Amazon's.

A new emerging area of Shadow IT is known as "Shadow AI." Faculty and staff frequently input sensitive institutional information—such as grant proposals, student essays, and tenure reviews—into publicly accessible, unregulated Generative AI tools to enhance productivity. In the absence of an enterprise agreement, this data is frequently used as training material for public AI models, which can result in the unintended leakage of institutional intellectual property to the broader public.

Finally, Manual audits are impossible in a dynamic cloud environment where virtual servers spin up and down in seconds. Institutions must deploy Cloud Security Posture Management (CSPM) tools that continuously scan the cloud environment against policy, automatically flagging (and even auto-remediating) misconfigurations like unencrypted databases or overly permissive IAM (Identity and Access Management) roles.

The Institutional Data Estate and Interoperability Platform

The Data Estate is a cohesive, strategically organized collection of data sources, ranging from enterprise applications to detailed IoT telemetry. It has the potential to serve as a "single source of truth," boosting ROI, ongoing improvements, and improved campus outcomes.

The credibility of Digital Twin relies heavily on the quality of its Data Estate. An impressive-looking twin can be unreliable if the data is incomplete, untrusted, or poorly managed. Conversely, even a basic twin interface gains value when supported by a strong, well-structured data foundation.

This is why the CTO's role in architecture is crucial. Facilities can set up systems and focus on sustainability to define results. However, only a CTO who collaborates closely with facilities and risk leaders can effectively implement integration patterns, security standards, data governance, and reusable platforms across the organization.

The Campus Sensorium and Its Integration Challenge

A smart campus relies on networks and IoT sensor grids that continuously collect data on air quality, space utilization, HVAC, lighting, and water. When integrated with analytics and automation, these data points become actionable insights—supporting predictive maintenance, resource efficiency, and reducing carbon footprints.

The challenge is not that campuses lack data. The challenge is that campuses often have *too many incompatible streams*:

- OT controller and submeters
- occupancy sensors and environmental sensors
- security and access systems
- vendor clouds, gateways, and proprietary APIs

The CTO's architecture must therefore answer a practical question: How do we convert fragmented, high-volume telemetry into trusted, reusable insights—without compromising safety, reliability, or ***vendor supportability***?

The most scalable approach treats OT data as a structured product pipeline with clear layers and responsibilities. The CTO manages the integration fabric and governance; Facilities handles operational context and system performance; and Sustainability and Academic stakeholders focus on defining impact, outcomes, and use cases.

Layer 1: Signal Capture and Edge Mediation

OT systems and IoT devices produce a stream of raw data signals. The edge layer processes these signals to make them suitable for enterprise use without affecting real-time control. This is where you determine what should stay within OT-only loops and what can be safely shared for analytics.

Many occupancy sensor setups publish timestamps using MQTT, which are routed through Telegraf, stored in InfluxDB, and displayed in Grafana. The key takeaway is not the particular tools, but the pattern: **lightweight publish/subscribe → time-series database → visualization** to extend the sensor board's applications. MQTT is a popular lightweight publish/subscribe protocol, especially suited for constrained IoT environments, offering efficiency when needed.[8]

Layer 2: Connectivity and Wireless Fabric (Right Network for the Right Work)

Digital networks serve as the backbone of the smart campus, providing secure, resilient connectivity that enables real-time data exchange across systems and supports large-scale coordinated decision-making. Wireless technology is not a single, uniform solution; instead, it represents a strategic portfolio choice.

Wi-Fi 6 / Wi-Fi 6E - Dense, Human-Centric Connectivity

Wi-Fi 6 and 6E remain the backbone of high-density indoor connectivity across the smart campus. Designed to handle thousands of concurrent devices with improved spectral efficiency, Wi-Fi 6 excels in classrooms, residence halls, offices, libraries, and public spaces where human mobility and device diversity dominate. Features such as OFDMA and Target Wake Time improve performance and energy efficiency, while 6E's extension into the 6 GHz band reduces congestion for latency-sensitive applications.

From an OT perspective, Wi-Fi is best suited for systems near humans, such as occupant experience apps, way-finding, smart classroom tech, mobile maintenance tools, and edge analytics gateways. While it isn't ideal for safety-critical control loops or ultra-low-power sensing, it significantly helps connect people, devices, and applications within the larger digital environment.

Private 5G / CBRS - Deterministic, Controlled, and Resilient Connectivity

Private 5G, often deployed via CBRS in the U.S., brings carrier-grade reliability and policy management to campus settings. Unlike Wi-Fi, private 5G offers consistent performance, predictable latency, and robust device-level identification, making it ideal for mission-critical and mobility-heavy applications.

On a smart campus, private 5G performs exceptionally well outdoors, in utility corridors, around distributed assets, and for autonomous vehicles, robotics, and public safety. It is especially useful in areas with poor Wi-Fi coverage or when operational reliability is critical. Additionally, it provides secure connectivity for OT assets that cannot afford signal drops or uncontrolled roaming. Strategically, private 5G serves as a control and resilience framework rather than a substitute for Wi-Fi. Its benefits are realized when managed as shared infrastructure rather than used as a specialized solution experiment.

LoRaWAN - Ultra-Low-Power, Long-Range Sensing

LoRaWAN is specifically designed to connect numerous simple sensors over large areas while maintaining low power consumption. Its extended range and energy efficiency make it well-suited for applications such as meters, environmental sensors, leak detection, waste management, parking, irrigation, and other campus-wide telemetry, where small data packets and infrequent updates are enough.

The tradeoff involves latency and bandwidth limitations. LoRaWAN isn't suitable for real-time control or transmitting complex data streams. Instead, it offers broad visibility into conditions that are typically unmeasured. When incorporated into the institutional data infrastructure, LoRaWAN plays a vital role in sustainability analytics, maintenance forecasting, and digital twin data inputs.

LoRa's strategic advantage lies in its scalability and cost-effectiveness, allowing sensing across campuses or cities that would be difficult with higher-power networks.

Zigbee Dense, Building-Scale Mesh Networking

Zigbee fills a specific gap between Wi-Fi and LoRaWAN. It is a low-power, short-range protocol designed for dense, indoor mesh networks. It is often used in lighting controls, room sensors, switches, and building automation systems. Zigbee enables dependable, low-latency communication among numerous devices within a single building or floor, without requiring continuous connectivity to IP networks. Yet its advantages come with risks: unmanaged mesh networks can become unclear, unreliable, and hard to oversee as they scale up.

In the context of a smart campus, Zigbee should be regarded as local OT infrastructure rather than an enterprise network. Its data and control signals must always pass through gateways that enforce identity, security, and data normalization before connecting to campus platforms.

Exhibit 5. Comparison of Wireless Network Standards used on the Smart Campus

Dimension	Wi-Fi 6 / 6E	Private 5G / CBRS	LoRaWAN	Zigbee
Primary Scope	Indoor, human-centric	Campus-wide, mission-critical	Campus / regional sensing	Building / room-level
Typical Range	30–50 m	Hundreds of meters to kilometers	Several kilometers	10–100 m
Power Consumption	Medium	Medium	Very low	Low
Latency	Low	Very low, deterministic	High	Low
Bandwidth	High	High	Very low	Low
Topology	Star	Cellular	Star	Mesh
Best Use Cases	Classrooms, offices, mobile apps	Utilities, mobility, safety, robotics	Meters, environment, parking	Lighting, room sensors, BAS subsystems
Real-Time Control	Limited	Strong	No	Local only
Battery Operation	Limited	Limited	Excellent	Good
Identity & Security	Enterprise IT controls	Carrier-grade identity	Device-level keys	Often weak without gateways
Governance Risk	Moderate	Low (with policy)	Low	High if unmanaged
Strategic Role	Experience fabric	Resilience fabric	Visibility fabric	Local control fabric

The CTO's Role in Managing Campus Spectrum

No single wireless protocol can make a campus truly "smart." A smart campus is achieved when each network is intentionally designed for its specific role, and all telemetry—no matter how it is transmitted—is integrated into a governed data ecosystem that enables analytics, AI, and digital twins. For the CTO, understanding this spectrum is no longer just about "the Wi-Fi"; it is about protecting the institution's mission-critical operations.

Zigbee, LoRa, Wi-Fi, and private 5G are not rivals. Instead, they are different infrastructure dialects—each valuable only when used within a common architectural language. The CTO's architecture approach is straightforward but essential: align the network with the risk level and use case, and thoroughly document this alignment to ensure consistent implementation over time.

The University Spectrum Landscape

Navigating the radio frequency (RF) landscape of a modern university is like managing a microscopic, high-speed metropolitan transit system. On a "Smart Campus," every square inch is saturated with signals—from the basement sensors to the drones flying over the stadium. The campus spectrum is divided into **Controlled** (Licensed/Managed) and **Open** (Unlicensed) bands. Controlled spectrum provides "lanes" with guaranteed performance, while Open spectrum is a "public highway" where anyone can drive, leading to potential gridlock.

Exhibit 6. Radio Frequency Spectrum Across the Smart Campus

FREQUENCY	CATEGORY	BANDWIDTH	RANGE	PRIMARY USE CASES
800 - 900 MHZ	Mixed (Licensed/ISM)	Low	12.5 kHz – 1 MHz	Public Safety and IoT: Campus police radios (800MHz) and LoRaWAN/Smart Meters (900MHz).

1.9 - 2.5 GHZ	Controlled (DECT/EBS)	Moderate	5 MHz – 20 MHz	Voice & Private LTE: Secure wireless intercoms for theaters and "Educational Broadband Service" for private cellular.
2.4 GHZ	Open (ISM)	Moderate	20 MHz – 40 MHz	Legacy Wi-Fi: Highly congested; used by older student laptops and basic Bluetooth peripherals.
3.5 GHZ (CBRS)	Shared	High	10 MHz – 100 MHz	Private 5G: The "Innovation Band" for research, robotics, and high-security administrative data.
5 GHZ	Open (U-NII)	High	20 MHz – 160 MHz	Enterprise Wi-Fi: The primary high-speed data lane for students and faculty.
6 GHZ	Open	Ultra-High	80 MHz – 320 MHz	Wi-Fi 6E/7: The "Clean Air" band for AR/VR, 4K streaming, and dense lecture halls.
11 - 23 GHZ	Licensed PtP	High	40 MHz – 220 MHz	Backhaul: Mission-critical building-to-building links (the "Wireless Backbone").
60 - 80 GHZ	PtP mmWave	Extreme	2 GHz – 5 GHz+	Wireless Fiber: 10Gbps+ links connecting high-traffic research hubs and data centers.

The "Wild West" era of deploying wireless access points and hoping for the best is over. A CTO must transition to a **Spectrum Management** mindset for three critical reasons:

1. **The "Invisible" Interference**
   In an open spectrum (such as 2.4 GHz or 5 GHz), the university does not own the airwaves. Students bringing their own routers (ROGUE APs), Bluetooth-heavy research labs, and even poorly shielded microwaves in dorms can degrade the network. Without active management, your \$10M network investment can be crippled by a \$50 consumer device.

2. **Mission-Critical Reliability**
   Smart campuses rely on "connected" safety. If the spectrum used for campus security cameras or emergency "blue light" phones is congested, latency increases and reliability drops.

3. **Monetization and Asset Protection**
   Many universities hold **EBS (Educational Broadband Service)** licenses. These are valuable assets. In 2019, the FCC changed the rules to allow universities to sell or lease these to commercial carriers. A CTO must decide whether to keep this spectrum for a **Private 5G network** (to support autonomous campus shuttles or remote research) or monetize it to fund other IT initiatives.

The significance of managed spectrum on a Smart Campus involves capacity planning for potentially tens or hundreds of thousands of devices connecting to wireless spectrum. Intentionally designing the spectrum means shifting devices to a managed band, such as CBRS or 900 MHz, which keeps Wi-Fi bands free for students' academic activities. Allocating 2.4 MHz Wi-Fi bands for IoT and device controllers from external vendors, and moving users' laptops to higher-frequency Wi-Fi, help reduce congestion and improve performance for all users.

Using a managed-spectrum approach is also a sound cybersecurity principle. For instance, leveraging Private LTE/5G through CBRS enables SIM-based authentication, offering much higher security than typical Wi-Fi passwords. Additionally, for robotics or remote surgery research, the CTO can design a "Sliver" of spectrum with no interference—something that can't be achieved in the open 5 GHz band.

Layer 3: Data Ingestion into Platforms (The Data Transference Layer)

Campus OT uses a combination of open and legacy protocols, including BACnet, Modbus, LonWorks, and KNX, which are still common. More often, these systems now operate over IP networks to enable richer data, remote management, and scalability. As a result, the decision to ingest these systems involves both governance considerations and technical factors.

- What protocols are allowed where?
- What gateways are sanctioned?
- Which vendors must support open integrations?
- What data must be active versus confined to operational networks?

The CTO's guiding principle is achieving interoperability that remains robust, emphasizing the use of open, well-supported protocols and steering clear of custom integrations that might break due to staff changes, vendor shifts, or platform updates.

Layer 4: Normalization and Semantics (The "Meaning Layer" of the Campus)

This is the point at which most digital twin programs succeed or fail quietly. Normalization involves converting raw tags and vendor-specific names into a uniform campus language, such as "room," "air handler," "supply temp," "occupancy state," "kWh," "emissions factor," "work order," and their relationships.

Certain digital twin platforms, such as those built with a Digital Twins Definition Language (DTDL), focus on modeling entities and their relationships to form a graph of how real-world assets connect. Regardless of whether you use Microsoft©, the core architectural principle remains: an ontology—a shared model of concepts and relationships—is essential for creating digital twins that are both durable and scalable.

This is also where the term "spatial, static, and live data" becomes practical: a twin is more than "just BIM." It combines spatial context, static records, and real-time telemetry into a unified understanding system.

Layer 5: Storage and the Data Estate (Time-Series + Context + Governance)

OT data is typically high-frequency and time-indexed, including temperatures, vibrations, occupancy events, and meter readings. While it benefits from time-series storage formats, it also requires proper context.

- asset history
- space ownership and scheduling relationships
- sustainability reporting structures
- policy constraints about privacy and acceptable use

This is where the Data Estate serves as the institution's intelligence layer: it is curated, integrated, secure, and governed to reliably support operations, decision-making, and ongoing innovation in a 'living lab" environment.

Layer 6: Analytics, AI, and Automation (From Insight to Action)

The Data Estate is not just a destination; it serves as the fuel for Layer 6. This layer converts raw data into actionable insights, supporting operational efficiency, predictive maintenance, evidence-based decisions, and measurable sustainability progress. It operates across two levels of maturity:

Decision Support AI (The Copilot)

At this stage, the system examines complex variables to offer recommendations, run "what-if" simulations, and set priorities. For instance, an AI might suggest pre-cooling a building at 4:00 AM by shifting the campus cooling load, based on weather forecasts and utility rates. Still, the final decision rests with a human operator.

Closed-Loop Control (The Autopilot)

This represents the transition from predictive to autonomous systems. In a closed-loop setup, the Digital Twin identifies an issue —such as a CO_2 increase in a lecture hall —determines the best action—like boosting outside-air ventilation—and directly executes that command to the building system, all without human input. Note: Although closed-loop control greatly improves efficiency, it also comes with considerable risks. Achieving true autonomy demands advanced cybersecurity, strict governance, and fail-safe procedures. Automation is only reliable if it can be secured.

Layer 7: The Digital Twin Experience Layer (Workflows, Not Wow-Factor)

Layer 7 is where the digital twin finally becomes *useful at scale.* It is the point at which intelligence leaves dashboards and enters daily work, transforming the twin from a passive source of insight into an **active operating system** for the campus. At this layer, the measure of success is not visual sophistication, but whether the right action occurs for the right person at the right moment—often without requiring that person to ever "open the twin" at all.

> ***The goal is no longer just to see the campus, but to orchestrate it.***

This evolution depends on a fundamental shift: the introduction of Operational AI agents as the primary means by which digital twin intelligence is delivered and enacted.

Operational AI Agents: Intelligence That Works *For* People

Operational AI agents differ from general chatbots or basic analytics tools. They are role-aware, policy-guided, task-focused digital coworkers that work within the digital twin's environment, data, and governance rules. Their goal is straightforward yet impactful: to free staff from boring, repetitive, and mentally demanding tasks, while supporting precise, human-in-the-loop workflows that enhance expertise instead of replacing it. Each agent is created for a particular operational goal and designed for a specific human role—such as technician, sustainability officer, scheduler, safety lead, or researcher—rather than for generic automation. At Layer 7, the digital twin acts as the shared intelligence platform, with Operational AI agents serving as the virtual hands that advance work.

Facilities: From Alerts to Agent-Driven Action

For facilities teams, the biggest failure of early "smart building" programs was alert fatigue. Systems could identify faults, but they still required humans to interpret signals, triage priorities, gather context, and decide next steps—often under time pressure. In a predictive or condition-based approach:

- **Operational AI agents continuously monitor asset telemetry**, historical performance, warranties, and maintenance history.
- When design drift or risks emerge, the agent does not simply raise an alarm. It assembles *decision-ready context*: probable root cause, confidence level, recommended intervention window, required parts, safety constraints, and estimated impact if deferred.
- The agent can automatically **generate, prioritize, and route a work order**, pre-populated with location data, lock-out/tag-out notes, and relevant diagrams—leaving the technician to focus on judgment, craftsmanship, and safety rather than paperwork.

Over time, this replaces informal "tribal knowledge" with institutionalized operational memory, dramatically reducing wasted time, unnecessary site visits, and reactive firefighting. The technician remains the authority. The agent removes friction.

Sustainability: From Reporting Burden to Continuous Stewardship

Sustainability teams are often overwhelmed not by a lack of data but by the labor required to collect, normalize, verify, and explain it. Manual spreadsheet assembly and retroactive reporting consume time that should be spent on strategy and improvement. Operational AI agents change this dynamic:

- Sustainability agents continuously reconcile energy, water, waste, and emissions data against institutional baselines, utility factors, and regulatory schemas.
- They flag anomalies, explain variance drivers, and **prepare audit-grade disclosures in real time**, rather than at the end of the fiscal year.
- When leaders ask, *"What happens if we change this?,"* agents can simulate impacts across cost, carbon, comfort, and risk—producing narratives and evidence tailored to executive, board, or public audiences.

Here, agents act as **stewards of transparency**, ensuring that sustainability is operationalized daily—measured, verified, and improved—not merely reported.

Academic Mission: Agent-Enabled Living Laboratories

When operational data is governed, contextualized, and abstracted through a digital twin, it can be safely exposed to the academic mission. Operational AI agents play a critical role in enabling this without burdening facilities or risking operations. Examples include:

- **Research mediation agents** that provide students and faculty with curated, read-only digital twin views appropriate to their use case, protecting safety-critical systems while enabling experimentation.
- **Simulation agents** that allow classes to test algorithms against real campus conditions—energy markets, occupancy shifts, climate scenarios—without touching production systems.
- **Learning workflow agents** that translate operational challenges (e.g., fault detection, space optimization, microgrid control) into structured problem sets or capstone projects.

In this model, the campus becomes a "Living Lab" by default, not by exception, because agents manage the boundary between learning and operations.

Orchestration, Not Automation Theater

The power of Operational AI agents lies not in autonomy for its own sake but in orchestration. At Layer 7, agents coordinate workflows across systems and roles while adhering to governance, safety boundaries, and institutional values. Key principles apply:

- **Human-in-the-loop by design**: Agents recommend, prepare, and execute low-risk actions automatically, while escalating high-consequence decisions to humans.
- **Policy-bound execution**: Every agent operates within explicit decision rights, auditability, and ethical constraints.
- **Role-specific experience**: A technician, sustainability officer, and researcher experience the same twin through entirely different agent interactions tuned to their workflows.

This ensures that intelligence scales *with trust*, not ahead of it. When Operational AI agents are embedded into the Digital Twin Experience Layer:

- Work shifts from reactive to anticipatory

- Staff time shifts from administrative overhead to expert judgment
- Knowledge shifts from individual memory to institutional capacity
- The twin shifts from something you look at to something that works quietly on your behalf

At this point, the campus is no longer merely "smart." It becomes operationally intelligent—a coordinated, resilient, and human-centered system that learns continuously while keeping people firmly in control.

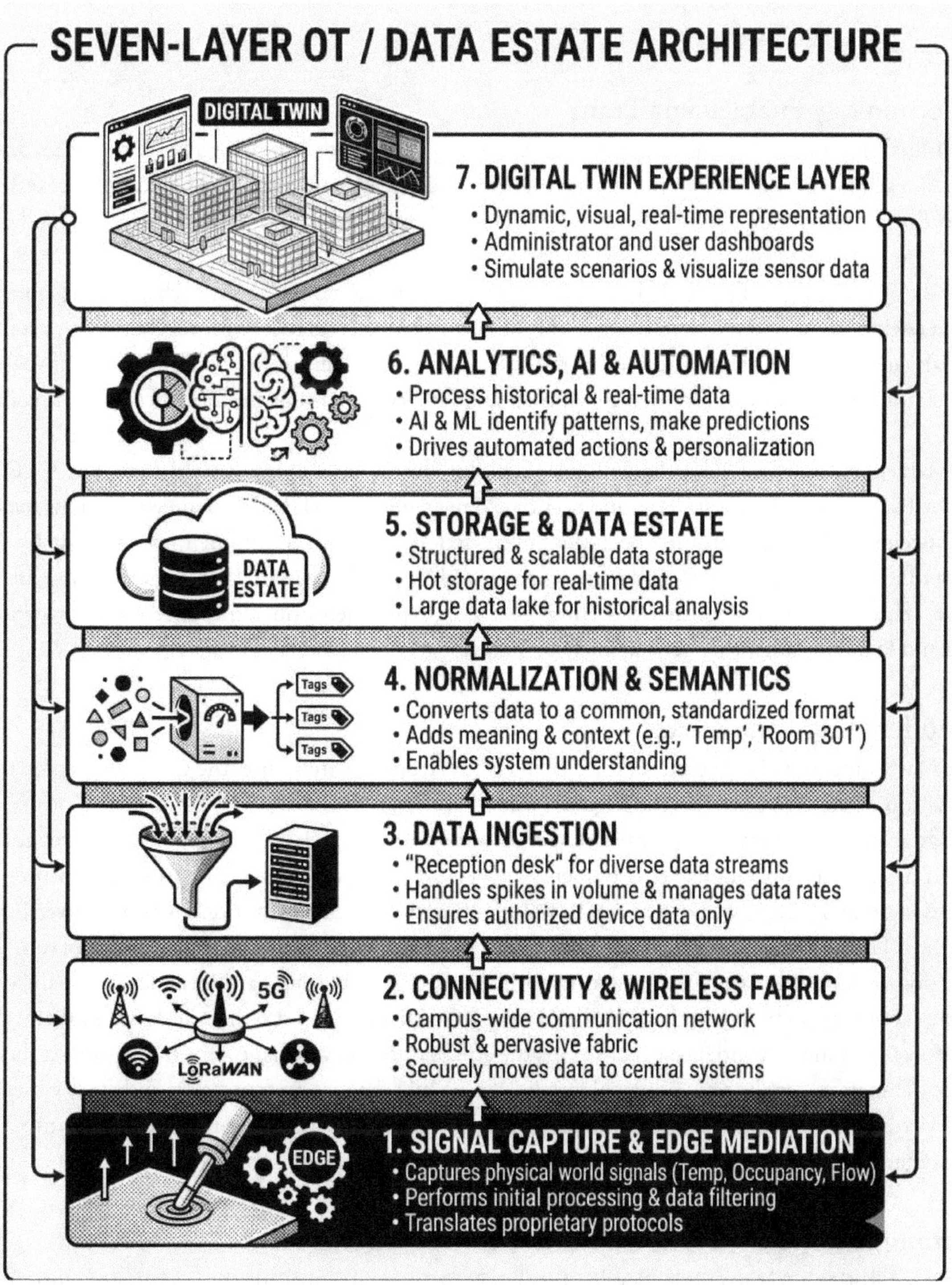

Figure 4. The Seven Layer OT/Data Estate Architecture

AI and Digital Twin Architecture

For the Integrative Ecosystem Architect, deploying Artificial Intelligence is more than just another software implementation. AI serves as the cognitive core of the sustainable smart campus, but like an engine, it requires a well-designed chassis and precise steering. The Chief Technology Officer's goal is to shift the institution from fragmented, departmental "shadow AI" toward a resilient, enterprise-grade infrastructure. This is accomplished through two main strategies: strong AI Enablement and the deployment of sophisticated Digital Twin technologies.

The Technology Enablement Team

The initial step in turning an AI-empowered university vision into reality is to build the necessary structural capacity. To achieve this, the CTO should create a dedicated Technology Enablement Team. This team functions as the core of the institution's AI development, balancing technical execution, strategic planning, and ethical oversight.

This is not a traditional IT helpdesk. The CTO must recruit and develop roles such as hybrid cloud orchestrators, data engineers, AI operations analysts, and cybersecurity specialists focused specifically on AI vulnerabilities. Their mandate is to construct a scalable infrastructure capable of handling intensive AI workloads—balancing centralized cloud computing with localized edge processing to ensure real-time performance without compromising security.

AI Enablement involves embedding ethical principles directly into the system's design. The CTO's role is to ensure the team applies AI Trust, Risk, and Security Management (AI TRiSM) frameworks. This includes establishing unchangeable audit trails that document every action, decision, and human review made by AI systems. Incorporating data minimization, model-provenance tracking, and adversarial-robustness testing into the infrastructure helps the CTO ensure that the university's AI environment inherently meets key regulations such as FERPA and HIPAA, rather than relying solely on policies.

The Student Capstone Project

Once the foundational data estate and AI enablement structures are secure, the CTO can deploy the ultimate manifestation of the smart campus: the Digital Twin.

A Digital Twin is not just a 3D visualization of campus buildings; it is an evolving, semantic replica of the university's physical and operational environment. Powered by real-time telemetry from IoT sensors, Building Automation Systems (BAS), and enterprise data silos, it functions as the institution's "control tower."

For the CTO, building this capability is transformative, shifting from passive data observation to active, predictive decision-making. Instead of only viewing dashboards that alert staff to chiller failures or spikes in library occupancy after they occur, AI models embedded within the Digital Twin run thousands of simulations to forecast these events in advance. Facilities teams can assess the effects of severe weather on the microgrid. At the same time, academic planners can analyze the potential impact of schedule changes on foot traffic and energy use—all within a secure virtual environment. This advancement transforms the campus from a reactive system into a predictive, resilient organism.

Governing Autonomy: From Generative to Operational AI

As the institution's AI capabilities develop further, its architecture must adapt to a crucial evolution: transitioning from Generative AI—which creates content and answers questions—to **Operational AI**, comprising Large Action Models that can independently carry out complex, multi-step tasks. Industry forecasts suggest that agentic

AI will soon be integrated into about one-third of all enterprise software, autonomously managing many routine operational decisions.

Before Operational AI can be implemented, institutions require three key foundations: integrated operational data (avoiding siloed systems), a Digital Twin acting as the AI's "brain" or 'memory," and mature cloud infrastructure capable of handling real-time building data at scale. Without these, AI remains limited to simple document interactions rather than managing physical systems assets.

The CTO's architectural pillar must now include developing Operational AI infrastructure and, in partnership with the CFaO, establishing the foundation for autonomy. This involves setting up strict operational guardrails and "Human-in-the-Loop" checkpoints. The architecture should clearly specify the actions an agentic system can take. For example, an AI agent might be permitted to automatically update HVAC settings based on the current weather. However, any intervention related to a student's academic standing or a significant financial purchase should trigger a mandatory approval process requiring human authorization.

The Architectural Value

Ultimately, AI Enablement and Digital Twin Capability Development turn insights into active processes. By establishing a strong Technology Enablement Team, embedding ethical governance into data systems, and employing Digital Twins to model and enhance reality, the CTO guarantees that the university not only survives the AI age but also leverages it as a cognitive engine. This approach aims for ongoing improvements in the human experience, environmental sustainability, and the university's academic mission.

Service Experience, Adoption, and Change Enablement

Creating an AI-enabled, sustainable smart campus depends heavily on the active involvement of the campus community. Even the most sophisticated digital twin or microgrid is ineffective without user engagement. For the Integrative Ecosystem Architect, implementing technology mainly involves understanding human psychology. The Service Experience, Adoption, and Change Enablement pillar highlights that human infrastructure is as crucial as digital and physical systems. This pillar oversees the human side of modernization, ensuring that new technologies are not only installed but also adopted and used effectively to support the institution's goals.

The Architectural Components

Constructing this pillar involves creating specific, durable structures that connect complex cyber-physical systems with the everyday activities of students, faculty, and operational staff.

The User Enablement Engine is Human Infrastructure

This component advances beyond traditional IT helpdesks by creating a proactive User Enablement Team that leads in promoting institutional fluency and cultural integration. Positioned at the nexus of people and technology, this team is responsible for developing role-specific learning paths, providing tailored support, and integrating AI literacy into key processes such as onboarding and faculty development. To expand this initiative, a formal network of AI Champions—trusted early adopters within academic and operational units—must be established. These champions serve as peer mentors, troubleshoot locally, and act as vital channels for grassroots feedback. Collectively, they build an active Community of Practice that fosters cross-functional collaboration and knowledge sharing across the organization.

Product Management and Contextual UX Design

In a converged IT/OT setting, the user experience (UX) should be designed to eliminate obstacles. For a maintenance technician, an AI-driven predictive alert about a failing pump needs to be clear and actionable, not just a cryptic database log. It should be delivered through a mobile-optimized workflow that includes the exact location, necessary parts, and safety procedures. This approach considers internal campus services as formal "products," demanding careful UX design that accounts for the specific context and operational needs of the end-user.

Formalized Change Management as "Operational Code"

Transformational change depends on creating new habits and dismantling old ones. Consequently, this architectural element integrates formal change management frameworks—like PROSCI ADKAR or Kotter's 8-Step Process—into the deployment lifecycle. Change management is viewed not merely as post-launch communication but as a fundamental structure that tackles the emotional, behavioral, and cultural aspects of technological transitions. It promotes psychological safety, open communication, and empathy-driven messaging to reduce the anxieties that often accompany AI and automation.

Cultivation of "Citizen Developers"

To ensure equal access to innovation and prevent central IT bottlenecks, the architecture should support citizen developers. Providing secure, governed access to low-code or no-code platforms enables non-technical faculty, staff, and students to participate rather than consume technology actively. The User Enablement Team assists by creating sandboxed environments for safe experimentation and promoting localized, domain-specific innovation that aligns with the institution's goals. Without these sandboxed environments, encouraging citizen developers to experiment with campus OT infrastructure risks them creating "Shadow OT."

Sentiment Telemetry and Continuous Feedback Loops

Similar to how the IT team monitors network latency, this pillar focuses on measuring human adoption and sentiment. By using continuous behavioral analytics and sentiment analysis tools, the institution can actively track user confidence, psychological safety, and engagement. This establishes a feedback loop where real-world usage data helps determine training requirements, interface improvements, and strategic changes.

The Architectural Importance of the Pillar

The Service Experience, Adoption, and Change Enablement pillar is crucial for successful technological progress. It serves as the main safeguard against superficial efforts such as "innovation theater" and the abandonment of technology. By actively managing change and emphasizing the service experience, this pillar helps avoid cultural clashes that often hinder complex IT/OT convergence projects. When tools are designed to be intuitive, and users are well-supported, resistance shifts to active participation.

Secondly, this pillar aims to empower the workforce strongly. When the technology leader promotes this human-centric approach, they fulfill their role as the institution's Chief Motivator. By easing administrative burdens and automating routine tasks, the university frees up staff mental capacity, enabling faculty, researchers, and facilities managers to focus on more strategic, high-value activities.

Ultimately, this architectural pillar ensures that campus modernization is not seen as an invasive technological order but as a fundamental institutional capability that empowers the community. It serves as the essential link that converts a set of advanced algorithms and physical sensors into a unified, intelligent system capable of continuous antifragility and achieving mission excellence.

The CTO's Failure Traps

Shifting from a traditional university to an AI-enabled, sustainable smart campus is not just a series of IT upgrades; it is a crucial paradigm shift with tangible physical impacts. For the Chief Technology Officer, serving as the Integrative Ecosystem Architect, this journey carries distinctive risks. Failures in smart campus projects usually stem not from technological flaws but from leadership's falling into known structural pitfalls that undermine governance, culture, or security. To successfully orchestrate this cyber-physical convergence, the CTO must anticipate and neutralize these five critical failure traps.

Trap 1: The "Shiny Object" Syndrome

Vendor-Driven Sprawl

The institution succumbs to "innovation theater," purchasing isolated smart solutions—a smart parking app here, a proprietary connected lighting system there—based on slick vendor demos rather than an institutional architecture. This creates a chaotic landscape of "Shadow OT." The campus becomes plagued by fragmented data silos that cannot communicate. The institution ends up paying for redundant cloud subscriptions and is effectively held hostage by vendor lock-in, making a unified Digital Twin or enterprise-wide AI optimization impossible.

The CTO must enforce a firm **API-First Mandate and Unified Governance.** No physical or digital asset may connect to the campus ecosystem unless it complies with the institution's open-standard interoperability requirements. The CTO must shift the institutional mindset from "buying smart products" to "investing in a resilient platform."

Trap 2: The IT/OT Cultural Chasm

Culture Eats Strategy

The CTO tries to impose traditional, rapid IT methods on the Facilities (OT) team without acknowledging the team's big cultural differences. IT views Facilities as resistant to change, while Facilities sees IT as reckless disruptors, unaware of the physics or safety of a 30-year-old chiller plant. These turf battles hinder deployment, with Facilities teams bypassing IT security to keep buildings operational, and IT sometimes pushing patches that accidentally disrupt critical systems, such as negative pressure in research labs.

The CTO should serve as the Chief Diplomat and Bridge Builder. To prevent avoidance, it's important to collocate IT network engineers with operational facility managers, set shared KPIs—such as evaluating success not only by network uptime but also by energy savings and zero safety incidents—and implement empathy-driven change management that acknowledges the "safety-first" operational realities of the physical plant.

Trap 3: The "Set and Forget" AI Fallacy

Semantic Rot

Believing that Artificial Intelligence and Digital Twins are autonomous and require no ongoing support, the institution overlooks funding for maintaining the data catalog and continuously monitoring AI models. As the physical campus changes—rooms are reconfigured, sensors degrade, and HVAC settings are adjusted—digital models are not updated, leading to 'semantic rot.' As a result, AI begins making decisions based on outdated or inaccurate data, leading to unnecessary energy consumption, false alarms, and diminishing trust among staff, who may eventually stop relying on the system.

The CTO should establish an AI TRiSM (Trust, Risk, and Security Management) Framework and set up a specialized data grooming team. Think of the Digital Twin as a dynamic system that requires continuous calibration. Implement rigorous model governance to identify algorithmic drift, ensuring AI decisions remain accurate, transparent, and aligned with real-world conditions.

Trap 4: The Perimeter Illusion

Security as an Afterthought

Relying on traditional IT security measures, such as the network firewall, to safeguard the integrated cyber-physical system is flawed. The institution believes that securing the core data center is sufficient, overlooking the thousands of new environmental sensors and vendor diagnostic ports that are now connected. This vastly increases the attack surface, creating critical vulnerabilities. A hacker could compromise an unsecured smart thermostat in a residence hall and use it as a backdoor to access the university's financial systems or even hold the campus heating system to ransom during winter storms.

The CTO needs to design a Zero Trust, micro-segmented environment, considering the network as already hostile. Enforce strict separation through "zones and conduits" between enterprise IT and vital OT infrastructure. Implement strong Privileged Access Management (PAM) for all external vendors and use passive anomaly detection to protect sensitive operational equipment without disrupting operations.

Trap 5: The Adoption Deficit

Ignoring Human Infrastructure

Investing heavily in the technology stack but neglecting the personnel who manage it is a common mistake. The CTO may believe that a better user interface can replace thorough training and smooth workflow integration. For example, an institution might spend millions on advanced predictive-maintenance AI. Yet frontline technicians ignore the alerts because they lack an understanding of the algorithm, or because the new digital workflow complicates their daily tasks. As a result, the technology ends up as a costly paperweight.

The CTO must thoroughly implement the Service Experience, Adoption, and Change Enablement architectural pillar. Develop user experiences tailored to the technician's specific field needs. Foster a network of embedded "AI Champions" for peer support and utilize ongoing sentiment telemetry to refine training programs. The technology should clearly make staff's jobs easier, safer, and more meaningful; otherwise, it will fail.

Chapter Summary

Chapter 3 lays the architectural foundation for transforming Operational Technology (OT) from fragmented, building-level systems into governed, reusable institutional intelligence that powers the AI-enabled sustainable smart campus. At its core, the chapter reframes OT not as a mechanical domain but as the primary source of "physical truth"—a continuous stream of telemetry that, when properly captured and governed, drives predictive maintenance, energy optimization, resilience, and strategic decision-making.

Operational Technology data is the institution's physical truth.

The chapter presents a multi-layered, enterprise-grade architecture starting with secure signal capture at the edge. It extends through connectivity, ingestion pipelines, semantic normalization, the institutional data estate, and finally, the analytics and digital-twin layers that transform data into insights and actions. This architectural approach—rather than any single vendor product—defines scalability, long-term value, and cybersecurity safety as intelligence spreads across the cyber-physical environment.

Architecture, not tools, determines whether intelligence scales safely.

Governance is emphasized, with key aspects such as identity, segmentation, remote access, data ownership, and AI oversight treated as non-negotiable architectural guardrails. The chapter clarifies that digital twins are not merely visualization projects; instead, they are operational layers that gain trustworthiness through clean, consistent data and are supported by disciplined commissioning, metadata standards, and human-in-the-loop controls.

Finally, the chapter presents a phased maturity roadmap: Stabilize → Integrate → Normalize → Optimize → (Selectively) Automate. This operational maturity roadmap guides institutions to avoid rushing into analytics or automation before establishing reliability, safety, and semantic coherence. By emphasizing architecture over tools, Chapter 3 portrays the CTO as the Integrative Architect responsible for transforming OT data into a resilient, cross-campus asset that enhances mission success, sustainability, and the daily human experience.

Part III

The Facilities Leader's Point of View

Chapter 4
The Chief Facilities Officer's Vision

The Chief Facilities Officer transforms the built environment into a predictive, sustainable, and human-centric platform that actively supports mission delivery.

Traditionally, a university's facilities department was judged by its silence and reliability. If the lecture halls remained warm in winter and cool in summer, and the lights operated smoothly, the CFaO was deemed effective. The campus was seen as a passive shell—an unchanging structure of brick, concrete, glass, and steel meant solely to support the academic purpose. Its operations were mostly reactive: a chiller broke down, a work order was issued, and repairs were made. The department was viewed as a necessary expense, overseeing the slow and inevitable decline of physical assets.

That era is decisively over.

Faced with heavy deferred maintenance, strict decarbonization rules, and swift digital transformation, the "invisible" facilities team needs to take a central role in institutional planning. The rise of AI-driven, sustainable smart campuses transforms the CFaO's responsibilities. The campus is no longer just a passive space; it is becoming proactive.

When the CFaO observes the modern university, they no longer see buildings and utility infrastructures. Instead, they perceive a dynamic, data-producing ecosystem—like a Cognitive Organism. Each element, such as air handlers, smart locks, occupancy sensors, and solar inverters, acts as a node within an extensive cyber-physical network. If the Chief Information Officer oversees the institution's digital nervous system, then the Chief Facilities Officer oversees its skeletal, muscular, and circulatory systems.

This calls for a significant visionary change. CFaOs must move beyond managing the slow decline of physical assets (deferred maintenance) to orchestrating a "Cognitive Organism" where every mechanical component is a node on a digital network. This shift revolves around the **Three Pillars of the Smart Campus Vision:**

- **Predictive Resilience:**
  utilizing IoT telemetry and AI-driven fault detection to identify equipment failures before they occur, moving away from a "run-to-failure" model.
- **Radical Sustainability**
  Shifting from passive measures like LED upgrades to creating a "Grid-Interactive Virtual Power Plant" that autonomously optimizes energy based on real-time carbon intensity and pricing.
- **Frictionless Occupant Experience**
  Designing invisible infrastructure that anticipates human needs, such as lighting that aligns with circadian rhythms and indoor air quality sensors that adapt ventilation minute-by-minute.

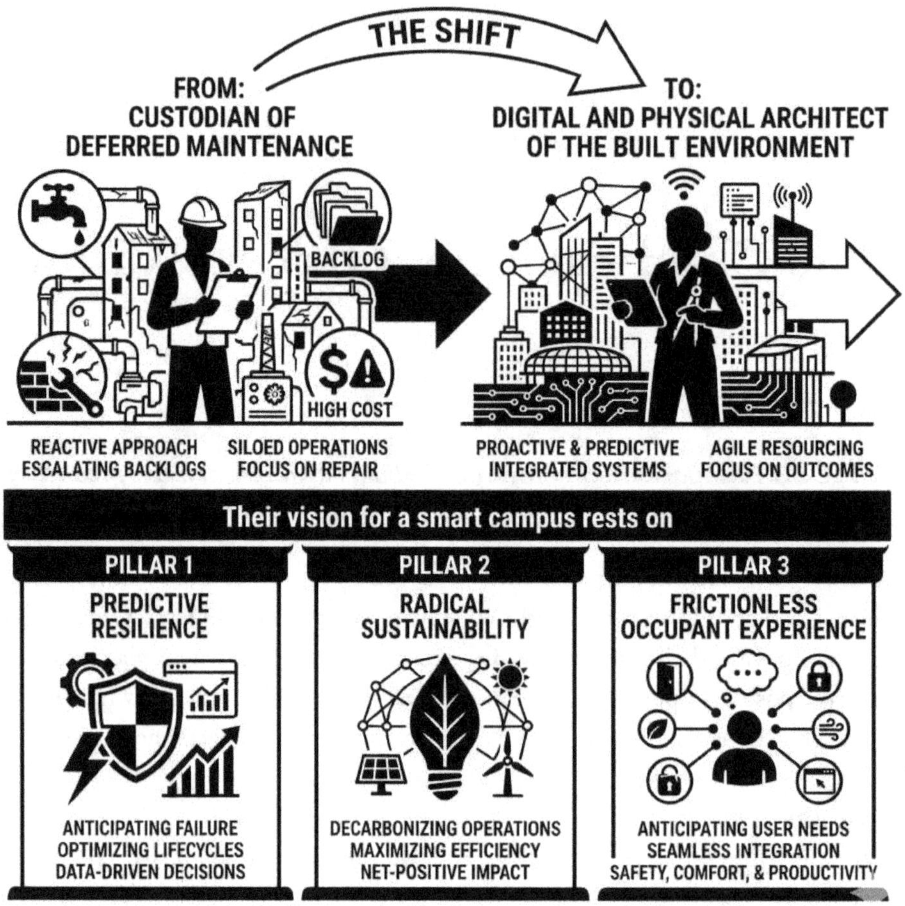

Figure 5. The Chief Facilities Officer Vision Shift

The CFaO's expertise must now encompass expertise in **Integrated Master Planning**, where digital requirements are hard-wired into architectural blueprints, and **Data-Driven Facilities Ops**, which treats facilities data as a managed institutional asset.

The Facilities Operating Context: A City That Must Stay Open

Facilities organizations in higher education oversee operations that resemble managing a small city rather than just a collection of buildings. Daily, they support teaching, host events, facilitate research with precise environmental needs, maintain livable residence halls, run dining and retail services, and ensure public spaces are safe. Unlike many commercial environments, universities are culturally open and physically accessible—visitors, contractors, deliveries, and community programs are integral to their identity. This openness adds complexity and heightens expectations for dependability.

This environment is especially challenging due to the disconnect between the slow pace of capital renewal and the rapid occurrence of operational failures. Capital cycles are purposefully slow, driven by reasons such as managing public and donor funds, adhering to procurement and permitting processes, and coordinating construction to avoid interfering with academic activities. However, building systems don't operate on these long timelines. A pump can fail suddenly, a controller can drift out of calibration as conditions change, and a valve can stick during the coldest day of the year. The facilities leader exists in that gap—a space between long-term renewal schedules and short-term disruptions.

This is why vendor ecosystems are crucial. Facilities uptime often relies on a mix of in-house staff and service agreements—covering controls, elevator maintenance, chiller services, safety testing, and energy performance partners. Outsourcing isn't a sign of weakness; it's a practical approach in a world of specialized systems. However, it creates dependencies that must be managed as institutional risks: remote access protocols, response time commitments, parts availability, and the vendor's capacity are not mere procurement concerns—they directly affect uptime.

The Strategic Discipline of Facility Maintenance

For executive leaders, it's common to see facility maintenance as merely a cost center—a routine task of replacing lightbulbs and repairing leaks. Nevertheless, contemporary facility maintenance should be viewed as a strategic, proactive discipline. It involves aligning the physical environment with the organization's long-term objectives, ensuring that buildings do more than exist—they actively support safety, operational efficiency, and the organization's reputation.

This framework explains how facilities maintenance combines planning, compliance, technology, and skilled labor to ensure long-term asset performance and occupant health.

The Core Mission: Purpose and Scope

Facility maintenance goes beyond just repairs; it primarily aims to protect the organization's productivity and capital. This broad responsibility is divided into two key areas:

- **Hard Services**
  The technical stewardship of the building's physical infrastructure, including HVAC, electrical grids, plumbing, and the building envelope (roofs, walls, and windows).
- **Soft Services**
  The management of the human experience and brand reputation encompasses janitorial services, sanitation, and groundskeeping.

Key Operational Activities

To accomplish this mission, the maintenance function must implement a range of activities to keep the campus resilient. Instead of handling tasks randomly, these activities are managed as ongoing processes:

- **System Inspections and Upkeep:** Routine auditing of mechanical health to prevent degradation.
- **Janitorial and Hygiene Services:** Ensuring a baseline of health and safety for all occupants.
- **Safety and Regulatory Compliance:** Managing the legal and physical risks associated with occupancy (e.g., fire codes, ADA compliance).
- **Asset Management and Lifecycle Planning:** The financial modeling of when equipment should be repaired versus replaced to optimize capital spend.

The Four Pillars of Maintenance Strategy

A mature maintenance organization does not treat all work equally. It balances four distinct strategies to optimize budget and reliability. The goal of executive leadership should be to move the organization away from reactive chaos and toward predictive precision.

- **Preventive Maintenance (Planned)**
  These are scheduled, recurring tasks (such as filter changes or coil cleaning) designed to prevent wear and tear. This is the baseline of a healthy system.
- **Predictive Maintenance (Optimized)**
  The gold standard of modern operations. This approach uses sensors and data analytics to anticipate issues before they occur, enabling intervention at the exact moment of need.
- **Corrective/Reactive Maintenance (The Cost of Failure)**
  Repairs are made after a system has already failed. This is the most expensive and disruptive form of maintenance and should be minimized through better planning.
- **Deferred Maintenance (Risk Management)**
  The strategic decision to postpone non-critical work due to resource constraints. While sometimes necessary, this creates a "shadow debt" that grows over time.

People and Structure: The Hybrid Workforce

Top maintenance operations usually adopt a hybrid approach, combining in-house and outsourced efforts to harness the advantages of both.

- **In-House Teams**
  Comprised of facility managers, general technicians, and custodial staff who hold institutional knowledge and provide immediate response.
- **Specialized Vendor Ecosystem**
  A network of external partners is utilized for high-complexity or highly regulated systems, such as elevator mechanics, fire safety engineers, and specialized HVAC technicians.

The Technology Backbone

Modern maintenance operates as a data-driven enterprise, rendering paper logs and whiteboard schedules outdated. Today, the department depends on advanced technology to keep situational awareness.

- **Computerized Maintenance Management Systems (CMMS)**
  The central "system of record" for all work orders, asset tracking, scheduling, and labor analytics.
- **Building Automation Systems (BAS)**
  The building's digital brain, the BAS, provides real-time control of HVAC, lighting, and energy consumption.
- **IoT Sensor Networks**
  The nervous system that enables predictive maintenance by continuously monitoring vibration, temperature, and usage patterns.

Measuring Success: The Scorecard

Effectiveness isn't about staff activity; it's about asset reliability and cost efficiency. So, how can we determine if the maintenance strategy is effective? Success is shown through:

- **Reliability Metrics**
  Key Performance Indicators (KPIs) such as **Mean Time Between Failures (MTBF)** (how reliable the equipment is) and **Mean Time to Repair (MTTR)** (how resilient the team is).
- **Cost Efficiency**
  Granular tracking of costs across labor, materials, and vendor contracts to identify waste.

- **Return on Investment (ROI)**
  Demonstrating value through reduced downtime, lower energy bills, and the extended lifecycle of expensive capital assets.

Strategic Distinctions

Lastly, leadership must differentiate between Facility Maintenance—focused on the tactical execution of duties—and Facility Management, which aligns the real estate portfolio strategically with the organization's future. Adequate maintenance funding and rigorous safety standards, including indoor air quality and fire system inspections, are essential foundational investments to support this broader strategic goal.

Key Issues Facing Facilities Leadership in Higher Education

Facilities leaders today are at the forefront of one of the most challenging periods in higher education operations. The physical campus, once seen as a stable, slowly changing asset, now faces mounting pressures from aging infrastructure, higher sustainability expectations, stricter regulations, and the swift digital transformation of building systems. These factors reveal existing vulnerabilities and introduce new strategic responsibilities for those managing the built environment.

This section highlights the key issues influencing modern facilities leadership, presenting them not just as isolated operational challenges but as interconnected, institution-defining problems that must be solved to ensure reliability, resilience, and long-term strategic value.

Deferred Maintenance: The Silent Compounding Risk

Deferred maintenance is commonly viewed as just a backlog figure on a balance sheet—an unchanging number referring to "old buildings." However, for facilities leaders, it represents more than a figure; it is an escalating problem that slowly erodes strategic options. The true risk lies not in its definition but in its effects: deferred maintenance builds up financially, operationally, and environmentally.

Emergency repairs are significantly more expensive than scheduled renewals. Unforeseen failures interfere with preventive maintenance, creating a cycle where neglect today causes crises tomorrow. Environmentally, worn-out envelopes and drifting controls silently raise energy consumption, making inefficiency seem normal. Most critically, it raises risk, as aging systems lead operators to rely on workarounds, temporary fixes, and undocumented exceptions.

Deferred maintenance is not a facilities problem— *It is a strategic risk multiplier.*

This compounding risk is particularly critical within the controls environment, which acts as the central nervous system of building performance. Delays or inconsistencies in control upgrades lead to the entire facility losing operational stability. Operators need to relearn the control logic of each building, troubleshooting slows down, and security becomes harder to enforce. Over time, the campus resembles a patchwork of isolated solutions. This fragmentation weakens resilience and hinders digital objectives: a fragile physical infrastructure makes digital systems more costly and less reliable. While sensors might be added to a failing system, their data cannot be fully trusted, expanded, or protected.

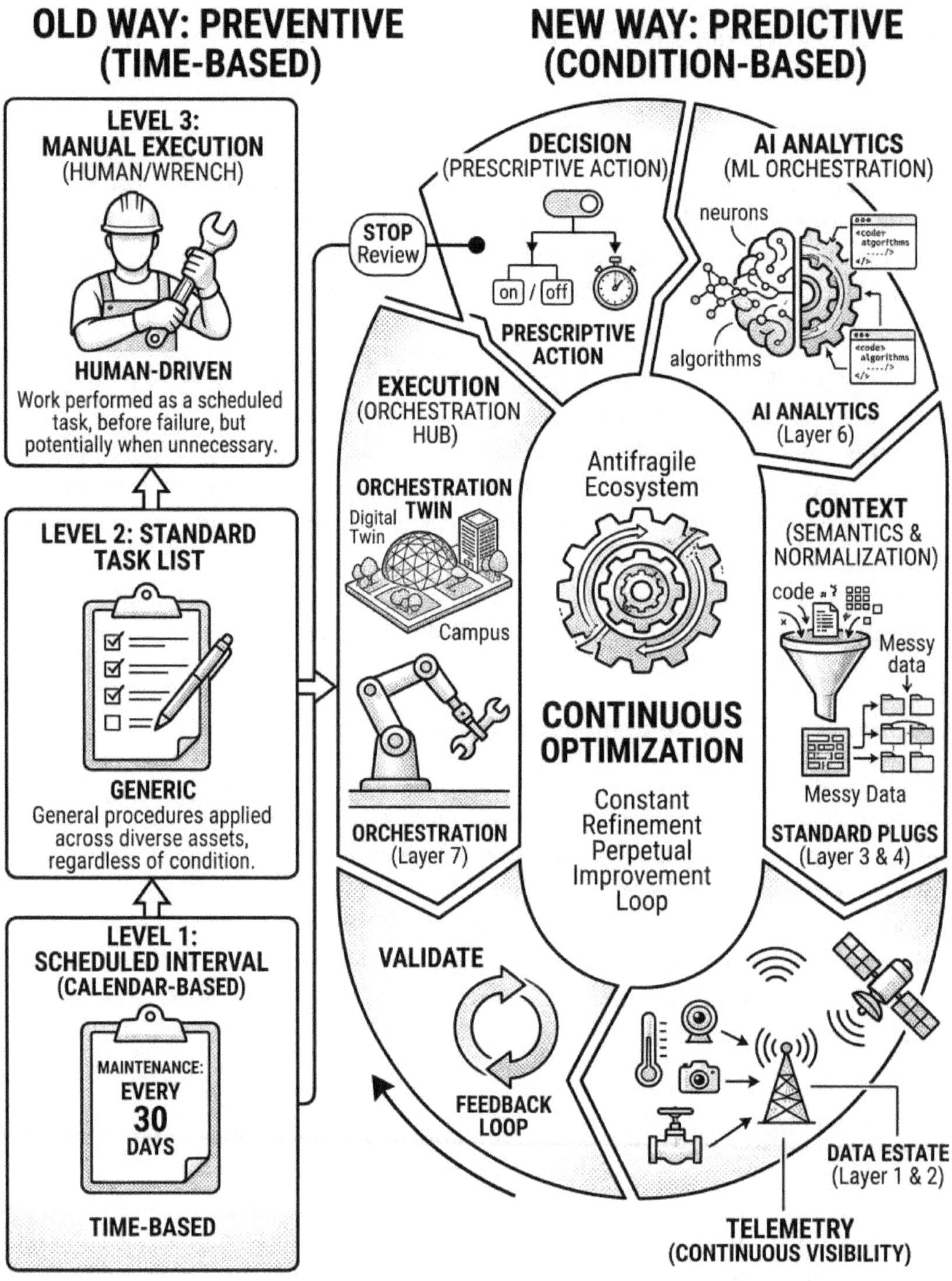

Figure 6. Preventive vs. Condition-Based Maintenance Approach

Breaking From the "Run-to-Failure" Paradigm

For many years, higher education has followed a "Run-to-Failure" approach. Although easy to manage administratively, this method leads to costly failures, such as a belt snapping or a boiler leaking. These incidents cause maximum operational downtime, high emergency repair costs, and reduced equipment lifespan. Shifting from this reactive approach to a proactive, Condition-Based Maintenance (CBM) strategy is among the most significant challenges a Chief Facilities Officer must tackle.

The challenge involves both cultural inertia and resource limitations. Facilities teams are often recognized for heroic firefighting—saving the day at 2:00 AM—rather than for the quiet, unseen work of prevention. Transitioning to a predictive approach means changing this mindset to favor data over gut feeling. Economically, institutions are caught in a competitive, declining-resource cycle; overwhelmed by urgent emergency repairs, they lack the time or funds to adopt technologies that could stop it. They are too busy bailing water to patch the hole in the boat.

Leveraging OT for Predictive Precision

The exit from this cycle is modernizing Operational Technology (OT). Condition-Based Maintenance escapes the trap by replacing scheduling with real-condition assessments. Unlike traditional Preventive maintenance, where a technician changes a filter every 90 days regardless of its condition—often leading to waste—this approach focuses on the actual state of equipment.

In a Condition-Based model, the filter is replaced only when a differential pressure sensor shows that airflow has decreased. The system forecasts failure before it happens. For example, a vibration sensor on a chiller pump doesn't just confirm operation; it detects tiny shaft misalignments weeks before bearing failure. This insight enables the team to plan maintenance, replacing a $50 seal rather than a $5,000 pump.

This feature relies on IoT Sensor Meshes and Edge Computing. Inexpensive, battery-powered sensors create a dense network within the physical plant, attaching to existing equipment to monitor vibration, temperature, and sound irregularities. Using Edge Computing, these sensors analyze data on-site—learning the motor's typical operating conditions and notifying the server only when deviations occur.

By deploying these sensor networks, the CFaO effectively leverages technology as a force multiplier. This enables an overextended team to oversee thousands of assets at once, reducing unnecessary calendar-driven tasks and directing limited labor hours to critical areas—aiming to prevent future emergencies before they occur.

Addressing Complexity through Modernization

Controls modernization is often presented as a step forward: new controllers, a modern BAS interface, improved analytics, and increased visibility. However, facility leaders focus on a simpler concern: will this improve operational ease and reliability, or will it add a lasting layer of complexity?

> ***You cannot automate what you cannot operate predictably.***

Modernization becomes complex when each building is treated as a unique technology deployment rather than a standard operation. The issue isn't the natural diversity of campuses, but this diversity leads to inconsistent naming, alarm logic, integration approaches, and documentation. In such scenarios, a "smart" campus is only understood by a few experts, making it fragile rather than truly intelligent.

Facilities-led modernization aims not merely for "more features," but for a repeatable operating system for buildings. In this system, systems behave consistently, data remains reliable, alarms are meaningful, and operators can transition between buildings without re-learning basic principles. This consistent foundation is essential for enabling advanced functions such as predictive maintenance, optimization, and digital twins.

Dealing with Workforce Constraints

Even top-tier architectures can falter if they overlook workforce dynamics. Facilities organizations face ongoing talent shortages, including a scarcity of skilled trades, retirement trends, and a growing dependence on specialized vendors. Additionally, the nature of the work is evolving. Modernizing controls and operational technology demands hybrid expertise—individuals who grasp physical systems, safety protocols, networks, telemetry, and troubleshooting through digital interfaces. This is why operator experience (UX) and training are not "nice to have." They are the delivery mechanism of value.

Operator experience is the delivery mechanism of value.

When interfaces are confusing, operators tend to find workarounds. Unmanaged alarm volumes turn alarms into noise, causing real signals to be overlooked. Without dashboards that facilitate work prioritization and clear actions, analytics lose their effectiveness and become mere show. If training happens sporadically rather than continuously, knowledge fades more quickly than systems evolve, especially as vendors change and software updates modify workflows.

Facilities modernization is effective when technology reduces cognitive load rather than adding to it. In practice, systems should guide staff by indicating the next best step: what to inspect first, what is malfunctioning, what can be deferred, and what demands urgent attention. The "smart campus" isn't just about screens; it creates a more confident and capable environment for those responsible for maintaining uptime.

Embodied Carbon and the Circular Campus

While many smart campus initiatives primarily aim to lower "Operational Carbon," which includes energy use from lighting, HVAC, and plug loads, the latest advances in sustainable architecture now emphasize a comprehensive approach to Embodied Carbon. For genuine institutional sustainability, the Integrative Architect needs to look beyond immediate metrics and oversee the campus's entire lifecycle, adopting a Circular Economy perspective.

Decoding Embodied Carbon

Embodied carbon refers to the CO_2 emissions linked to the complete lifecycle of building materials, including extraction, manufacturing, transportation, and demolition. In many contemporary high-efficiency buildings, embodied carbon may account for as much as 50% of total lifecycle emissions.

The "Carbon Lock-in"

Once a building is constructed, its embodied carbon is "spent." A smart campus strategy must prioritize the reuse of existing structures and the selection of low-carbon materials (such as mass timber or carbon-sequestering concrete) over new, carbon-intensive construction.

Intelligent Procurement

The Data Estate must expand to include Environmental Product Declarations (EPDs), allowing the institution to track the "carbon debt" of every new wing or renovation project.

The Material Passport: Assets as Future Resources

A fundamental element of the circular smart campus is the Material Passport. This digital dataset—embedded within the Building Information Model (BIM) and Digital Twin—details the value and properties of building materials to support their recovery and reuse.

Tracking the Inventory

Every steel beam, copper pipe, and glass panel is tagged and logged. When a building reaches the end of its useful life or undergoes a major renovation, these components are not viewed as waste but as a "material bank."

Predictive Deconstruction

Using the Digital Twin, campus planners can simulate the deconstruction process years in advance to identify which components can be sold, donated, or repurposed for new campus projects.

Closing the Loop: The Circular Economy Framework

The Circular Economy moves away from the conventional "Take-Make-Waste" linear model towards a regenerative system. For an AI-powered campus, this means:

- **Product-as-a-Service (PaaS)**
  Instead of purchasing thousands of LED fixtures or HVAC chillers, the campus enters into service contracts where the vendor retains ownership. The vendor is then incentivized to provide the most durable, repairable equipment possible, as they are responsible for the material at the end of its life.
- **AI-Driven Resource Matching**
  A sentient campus can use AI to match "waste" from one department with "needs" from another. For example, excess heat from a data center can be redirected to a greenhouse, or decommissioned IT hardware can be harvested for components to be used in student labs.

By incorporating Embodied Carbon tracking and Material Passports, the university turns its physical footprint from a liability into a regenerative asset. This change signifies that 'Sustainable' in the Smart Campus Manifesto means more than just reduced electricity costs; it represents a core commitment to conserving the planet's limited resources.

The "New Edge"
Where OT is Digital Strategy

The "new edge" marks the transition of Operational Technology from isolated building infrastructure to a crucial element of the institution's digital strategy. As campuses adopt digital twins, electric vehicle (EV) charging networks, and expanding IoT sensor arrays, these systems now operate at the intersection of physical assets, data, identities, and cloud platforms.

The smart campus network and sensor fabric serve primarily as operational infrastructure. Well-designed digital networks and IoT systems establish feedback loops that enable facilities to shift from reactive to anticipatory operations—identifying drift, detecting anomalies, and optimizing performance preemptively to prevent failures. Conversely, poorly designed systems add operational vulnerabilities: they increase the number of endpoints to oversee, generate more alerts to prioritize, introduce more integrations that may fail, and involve additional vendor interactions, thus enlarging the institution's exposure.

These edge domains are not merely side projects; they serve as accelerators. However, they pose institutional risks unless the foundational elements—such as identity, segmentation, remote access, monitoring, and governance—are established first. Therefore, it is essential to implement governance, identity management, segmentation, and secure data pathways to support responsible scaling of innovation at the edge.

Internet of Things Sensor Grids

A "new edge" digital strategy relies on data collected from various devices. Internet of Things (IOT) sensors began being integrated into building HVAC systems in the late 2000s, with widespread adoption accelerating between 2011 and 2015. Although building automation systems (BAS) had been around for a while, the move to IoT-enabled, connected, and smart sensors signified a shift toward real-time monitoring and, ultimately, predictive maintenance.

Today, intelligent facility devices are widespread. IoT sensor networks can quickly expand across buildings and open areas, providing continuous data on occupancy, air quality, energy consumption, environmental conditions, and equipment status.

Without proper onboarding, clear ownership, lifecycle oversight, and network segmentation, these sensors risk becoming "shadow OT"—unmanaged devices that compromise security and operational integrity. When properly integrated, sensor grids serve as the essential visibility layer, supporting predictive maintenance, optimization, and digital-twin-based intelligence.

Digital Twins

Digital twins integrate spatial data (including BIM and GIS), static asset details (such as CMMS/EAM and equipment metadata), and real-time operational telemetry to form a managed, continuously updated campus model. More than mere visualizations, these twins serve as predictive and diagnostic tools—facilitating anomaly detection, commissioning, maintenance, cost reduction, and carbon-aware load management. Their success relies on reliable data, clear access controls, and robust integration governance.

·EV Charging Networks

EV charging infrastructure functions as a hybrid system involving utilities, payment mechanisms, identity verification, and vendor-managed cloud services. This sector brings new demands for uptime, grid interaction, safety, and remote access, all of which the institution must regulate carefully. Without standardization and proper segmentation, EV charging may introduce vulnerabilities across physical, financial, and cyber risk areas.

The New Energy Edge

In the legacy campus model, energy infrastructure was massive, passive, and strictly hierarchical. The central plant burned fossil fuels, and the grid delivered electricity one-way. In the smart campus, Operational Technology (OT) transforms these assets into intelligent, communicative nodes in a localized energy internet.

This "New Edge" moves beyond the notion that "demand is the new supply" with energy. It redefines three critical physical assets, moving the campus from a fragile consumer to a dispatchable financial asset.

The Power Station: The Virtual Power Plant (VPP)

The central plant is no longer a passive "boiler room"; it is an orchestration engine. Using edge-compute controllers, the VPP aggregates thousands of distributed energy resources and presents them to the regional grid as a single, controllable asset. It doesn't just generate power; it trades it—automatically islanding the campus during grid instability or selling excess capacity back to the utility when spot prices spike.

Battery Storage (BESS): The Kinetic Stabilizer

Batteries have evolved from static emergency backups to the microgrid's "fast-twitch muscles." Because renewable energy is volatile, the Battery Energy Storage System (BESS) acts as a kinetic buffer. Its edge controllers perform sub-second frequency regulation to stabilize the local grid and execute "peak shaving"—discharging power instantly to flatten demand curves and eliminate costly utility demand charges.

Diversified Renewable Generation: The Green Baseload

Decarbonization requires a portfolio, not a single silver bullet. The smart campus integrates a diverse mix of generation assets into a unified "Green Baseload."

- **Solar and Wind Fields:** These harvest variable renewable electrons, driving the bulk of daytime electrical load.
- **Geothermal Exchange:** Turning the earth into a "thermal battery," these loops store waste heat underground in summer and extract it in winter, electrifying the heating/cooling load with massive efficiency.
- **Biomass Generation:** Acting as a dispatchable renewable engine, biomass plants convert organic campus waste into firm, carbon-neutral power that—unlike solar or wind—can be ramped up on demand, regardless of the weather.

Operational AI

In the 1980s, spreadsheets replaced ledgers, and word processing eliminated the typing pool. The 1990s saw email take over secretarial roles; the 2000s brought CAD replacing drafting tables; and in the 2010s, cloud computing supplanted data centers. Nonetheless, facilities management transitioned from paper-based CMMS to digital CMMS with minimal process changes. Today, **Generative AI** introduces new efficiencies and enhances human skills. **Operational AI**—focused on managing physical assets in real-time—differs significantly from Generative AI. While Generative AI improves productivity for knowledge workers, Operational AI interacts with chillers, predicts equipment failures, and autonomously adjusts HVAC systems based on occupancy and weather, evolving from a suggestion dashboard to a trusted autonomous partner in building management.

The built environment, especially higher education campuses, can benefit significantly from Operational AI integration due to universities' unique combination of size, complexity, and focus on sustainability. Campuses with centralized facilities management, aging infrastructure, ambitious carbon-neutral goals, and budget constraints serve as ideal testing grounds for Operational AI, delivering rapid returns on investment.

The Synthesis: The Autonomous Campus Energy Operating System

The true strength of the "New Edge" comes from the algorithmic integration of these energy systems into a new operating system. In this architecture, Solar and Wind supply affordable bulk power, while the BESS maintains stability during periods of cloud cover or light wind. Biomass and Geothermal serve as reliable, dispatchable sources to keep the system dependable when sunlight isn't available. All of this is managed by the Operational AI overseeing the Virtual Power Plant, transforming the campus into an antifragile, self-sufficient energy hub that can withstand grid failures and convert its physical infrastructure into a versatile financial asset.

What the Chief Facilities Officer Requires

A poorly planned smart campus strategy that ignores the actual facilities situation often fails. It leads to unsustainable pilot projects, an overload of tools that burden operators, and prevents verification of savings claims because the essential baselines and commissioning practices were not established.

The reality of facilities management requires a different approach. Prioritizing reliability is essential because automation is impossible without predictable operation. Deferred maintenance should be regarded as a strategic risk factor, as it compromises sustainability goals and occupant experience. Additionally, modernization should be approached as a change management process rather than just a procurement task, since new systems impact workflows, expectations, and accountability.

The traditional role of the CFaO centered on quiet dependability: ensuring lights stayed on, lawns remained manicured, and boilers functioned. It was a reactive responsibility, assessed by how quickly issues with assets were addressed and how well inevitable deterioration was managed.

Today, the traditional model is collapsing under modern institutional challenges. Transitioning to an AI-driven, sustainable smart campus is no longer a luxury or mere technological upgrade for the CFaO; it is a crucial response to increasing financial, environmental, and operational pressures. Below are the key factors compelling the CFaO to fundamentally change its outlook, along with six ways to communicate this new vision to the university community.

1. The Deferred Maintenance Avalanche and Financial Constraint

Higher education institutions are experiencing a historic crisis due to deferred maintenance. Relying on traditional capital planning, which involves replacing assets only when they age or fail catastrophically, is no longer financially sustainable. CFaOs are now required to achieve more with fewer resources. This situation drives a move toward AI-powered predictive maintenance, utilizing vibration and acoustic sensors to precisely identify where capital investments are necessary, thereby extending asset lifespans and preventing costly emergency replacements.

2. Aggressive Decarbonization and Climate Mandates

Institutions are now making binding commitments to reach net-zero carbon emissions. CFaOs can no longer rely solely on passive measures such as LED upgrades or improved insulation to meet these goals. Instead, they need to redefine their approach by adopting Grid-Interactive Efficient Buildings (GEB) and AI-powered microgrids. The physical infrastructure must evolve into an active participant in energy markets—anticipating weather conditions, predicting campus energy loads, and making autonomous decisions about when to store, use, or shed power.

3. The "Silver Tsunami" and Institutional Knowledge Loss

The facilities workforce is aging quickly. As master electricians and boiler operators retire, they take with them years of undocumented, intuitive knowledge about the campus's unique quirks. The CFaO needs to implement Digital Twins and AI knowledge bases to capture this operational expertise before it is lost, shifting the campus's reliance from individual specialists to a unified, intelligent system.

4. Post-Pandemic Space Utilization Realities

The hybrid model of modern learning and remote work has significantly changed how physical space is used. Running large HVAC systems in mostly empty buildings wastes resources. The CFaO should adopt IoT occupancy sensors and spatial analytics to adjust energy use in real time based on actual human presence. This approach helps the organization streamline operations and consider selling unused properties.

5. The Convergence of IT and Cyber-Physical Risk

Facilities systems (OT) are rapidly connecting to the digital network, often without involving the IT department. As a result, the CFaO can no longer overlook cybersecurity. Recognizing that a compromised building automation system could cause pipes to freeze or a lab to shut down, the CFaO should collaborate with the CTO. The strategy should now encompass a secure, governed architecture for the cyber-physical perimeter.

6. The Demand for a Frictionless, Health-Promoting Experience

Students and faculty now anticipate that the physical campus offers the seamless experience of their digital lives. They seek environments tailored to enhance cognitive function and well-being. The CFaO should imagine a campus where lighting automatically aligns with circadian rhythms, indoor air quality (IAQ) is continuously monitored to reduce the risk of airborne illnesses, and AI-driven wayfinding systems ensure the campus is accessible to everyone.

Sample Versions of the New CFaO's Vision

Driven by these factors, the CFaO's vision undergoes a significant transformation. It evolves from simply managing decline to actively orchestrating intelligence. The revised perspective sees buildings not as static spaces for academics but as dynamic, data-producing partners. The CFaO moves from overseeing a "cost center" to managing a "strategic asset." This vision is inherently cyber-physical and highly collaborative with IT and Academic leaders. It sharply focuses on leveraging predictive AI to ensure financial resilience, achieve sustainability, and create a seamless human experience.

To secure funding and ensure institutional alignment, the CFaO needs to communicate this new reality clearly. Here are six different versions of the vision, each tailored to emphasize different strategies, which can be shared with the President, the Board of Trustees, and the university community.

Vision 1: The "Cognitive Ecosystem"

"Our vision is to transform the physical campus from a collection of static buildings into a unified, cognitive ecosystem. By fusing artificial intelligence with our physical infrastructure, we will create an environment that learns from its occupants, predicts its own maintenance needs, and autonomously optimizes energy—ensuring our physical spaces actively propel the university's academic mission and financial resilience."

Vision 2: The "Net-Positive Engine" (Focus on Sustainability and Climate Leadership)

"We envision a campus that gives back more to the environment than it consumes. By leveraging AI-driven microgrids, digital twins, and predictive load management, Facilities Operations will transition the university from passive energy consumers into a dynamic, net-zero virtual power plant, setting the global standard for climate leadership in higher education."

Vision 3: The "Frictionless and Human-Centric Campus" (Focus on Student/Faculty Experience)

"Our vision is an invisible, intuitive physical infrastructure that anticipates the needs of our community. Through smart spatial analytics and responsive environmental controls, we will deliver a frictionless campus experience—where air quality, lighting, and accessibility automatically adapt to promote the health, safety, and cognitive well-being of every student and researcher."

Vision 4: The "Predictive Portfolio" (Focus on Financial Stewardship and ROI for the Board)

"We will redefine institutional stewardship by eliminating the era of reactive maintenance. Through the deployment of IoT telemetry and AI-driven fault detection, we envision a predictive portfolio of facilities. This intelligence will drastically extend the lifecycle of our capital assets, crush the deferred maintenance backlog, and convert operational waste into strategic funding for the university's future."

Vision 5: The "Living Laboratory" (Focus on Academic Co-Creation and Research)

"Our vision is to erase the boundary between campus operations and academic discovery. We will open our smart infrastructure, microgrids, and digital twins to our students and faculty, transforming the built environment into a continuous 'Living Laboratory' where the next generation of engineers, data scientists, and sustainability leaders can test and refine the technologies of tomorrow."

Vision 6: The "Resilient Cyber-Physical Foundation" (Focus on Risk Mitigation and Continuity)

"In an era of increasing physical and digital threats, our vision is to architect an unbreakable cyber-physical foundation. By deeply integrating Operational Technology with enterprise security and AI intelligence, we will create a

campus environment capable of predicting disruptions, autonomously isolating threats, and ensuring the absolute safety and continuity of the university's research and residential life."

Chapter Summary

Chapter 4 redefines the CFaO's role from simply managing aging infrastructure to serving as a strategic planner for the modern cyber-physical campus. Facilities operations, which were once characterized by quiet reliability and reactive repairs, now play a crucial role in achieving institutional resilience, sustainability, risk mitigation, and enhancing the student experience. The chapter highlights that the physical campus is no longer a passive environment but an active, data-driven ecosystem. Its performance directly impacts mission achievement, financial management, research continuity, and public reputation.

The CFaO's operating environment resembles a small city that must remain open at all times. It is influenced by long asset lifecycles, growing deferred maintenance, increasingly complex building systems, higher standards for indoor environmental quality, and more regulatory demands. As controls, utilities, security, and environmental systems become networked and cloud-connected, Facilities leaders must balance Operational Technology (OT) with IT governance, cybersecurity, data architecture, and identity management. The chapter highlights the CFaO as a co-leader, alongside the CTO, of the organization's integrated cyber-physical strategy.

Chapter 4 outlines **three pillars** of the CFaO's modern vision:

- Predictive **Resilience**, enabled by IoT telemetry, AI-driven fault detection, and digital twins.
- **Radical Sustainability**, requiring grid-interactive operations, continuous optimization, and verifiable decarbonization; and
- **A Frictionless Occupant Experience**, where environments dynamically adapt to promote health, accessibility, comfort, and cognitive performance.

To realize this vision, the CFaO needs to transform the organization from reactive repairs to proactive, data-driven operations. This shift should be supported by a modernized controls environment, a robust vendor ecosystem, and a workforce trained to operate at the nexus of mechanical systems and digital intelligence. Additionally, leadership must regard facilities data as a managed institutional asset—ensuring it is accurate, calibrated, interoperable, and aligned with the enterprise's overall goals.

Facilities is no longer managing static assets.
It is orchestrating a cognitive physical ecosystem.

Ultimately, Chapter 4 portrays the CFaO not just as a manager of decline but as a key strategist shaping the institution's future. By integrating planning, modernizing OT, ensuring lifecycle sustainability, and partnering closely with IT, the CFaO turns the built environment into a platform that advances the mission—protecting research, lowering carbon footprint, boosting financial resilience, improving daily experiences, and preparing the university for long-term competitiveness.

Chapter 5
The Chief Facilities Officer's Strategy

The physical campus is no longer managed as a cost center.
It must be reimagined as an intelligent, data-rich strategic asset
that drives antifragility, sustainable practices, and new operational efficiencies.

A vision, no matter how compelling, is merely a destination. Strategy is the physics of the journey. For the Chief Facilities Officer (CFaO), translating the ambitious vision of an AI-enabled, sustainable, smart campus into kinetic reality requires confronting some of the most entrenched financial, operational, and cultural headwinds in higher education. The strategy cannot simply be a procurement plan for new sensors; it must be a comprehensive operational redesign. It is the calculated deployment of capital, cyber-physical technology, and human infrastructure designed to transform the university's physical plant from a reactive cost center into a predictive, value-generating asset.

The CFaO strategy is not born in a vacuum; it is forged in the crucible of four urgent institutional realities that make the status quo untenable.

1. The Mathematics of Deferred Maintenance and Capital Scarcity

The traditional strategy of "run-to-failure" and age-based replacement has resulted in an avalanche of deferred maintenance that far exceeds available capital. Institutions can no longer afford to replace massive infrastructure simply because a spreadsheet indicates it is thirty years old, nor can they afford the catastrophic costs of emergency failures.

- *The Strategic Response*
  The strategy must pivot aggressively from reactive repair to **AI-driven predictive asset management**. By instrumenting the campus with IoT telemetry and utilizing machine learning for fault detection, the strategy aims to flatten the deferred maintenance curve—extracting maximum life from existing assets while directing scarce capital only toward systems that exhibit quantifiable, algorithmic signs of imminent failure.

2. The Imperative of the Climate Mandate

Universities have made binding commitments to aggressive decarbonization and net-zero targets. The era of achieving these goals through superficial energy conservation measures—like upgrading to LED bulbs or adding thicker insulation—has passed. The remaining carbon footprint is deeply embedded in the complex mechanics of central heating and cooling, as well as in reliance on the grid.

- *The Strategic Response*
  The strategy must redefine the campus as a **Grid-Interactive Virtual Power Plant**. This requires a strategy that implements intelligent microgrids, dynamic thermal storage, and automated demand-response

systems. The CFaO must architect a physical plant capable of interacting with the external energy market—autonomously shedding load, storing renewables, and optimizing consumption based on real-time pricing and carbon-intensity data.

3. The Redefinition of Academic Space

The post-pandemic shift toward hybrid learning, remote administration, and highly collaborative research has fundamentally decoupled the institution from traditional space utilization models. Massive lecture halls and rigid administrative blocks often sit empty, yet the energy required to condition these spaces remains constant.

- *The Strategic Response*
  The strategy must leverage **Dynamic Spatial Intelligence.** Using anonymized Wi-Fi telemetry, computer vision, and IoT occupancy sensors, the CFaO strategy shifts from conditioning spaces based on fixed schedules to conditioning them based on real-time human presence. This data-driven approach allows the institution to consolidate operations, redesign underutilized spaces, and potentially divest from unnecessary real estate.

4. The Convergence of Cyber-Physical Vulnerability

As operational technology (OT) is rapidly digitized to enable smart capabilities, the physical campus is exposed to the digital threat landscape. A compromised building automation system is no longer just an IT headache; it is a life-safety issue that can freeze pipes, disrupt negative-pressure research labs, or lock down residence halls.

- *The Strategic Response*
  The strategy must mandate **Secure IT/OT Convergence.** The CFaO can no longer operate a shadow network. The strategy demands deep, structural collaboration with the Chief Information Officer to enforce Zero Trust architecture, network micro-segmentation, and rigorous vendor access controls across the entire physical plant.

The Strategic Synthesis: A New Operational Reality

Driven by these key issues, the CFaO strategy becomes an exercise in radical integration. It requires abandoning the siloed procurement of closed, proprietary building systems in favor of an open, API-driven architecture.

The strategy dictates that every capital project, every equipment replacement, and every vendor contract must support the ingestion of operational data into a central Digital Twin. By aligning the physical realities of deferred maintenance and energy consumption with the digital capabilities of Artificial Intelligence, the CFaO strategy creates the indispensable cyber-physical foundation needed to realize the ultimate vision: a cognitive, sustainable campus that actively learns, adapts, and propels the institution's enduring mission.

The Strategic Domains of the Chief Facilities Officer

If the Chief Technology Officer manages the digital nervous system of the smart campus, then the CFaO oversees its skeletal, muscular, and circulatory systems. Traditionally, the Facilities leader was mainly seen as the guardian of deferred maintenance—ensuring the grass is mowed, boilers function, and roofs stay leak-free.

In the era of the AI-Enabled Sustainable Smart Campus, this role is dramatically elevated. The CFaO is no longer merely managing tangible assets such as concrete and steel; they now serve as the Architect of the Built Environment, tasked with transforming static physical infrastructure into active, data-driven platforms that support the institution's academic and sustainability goals. To succeed, the CFaO needs to develop expertise across five key strategic areas.

Integrated Master Planning

Traditional campus master planning was a static, ten-year effort that produced glossy architectural renderings. In contrast, modern CFaOs see Integrated Master Planning as a dynamic, multi-dimensional process. It involves directly incorporating digital requirements into architectural blueprints. A new science building is now assessed not only by its size or aesthetic appeal but also by its embodied carbon, its integration with the campus microgrid, and its readiness for inclusion into the central Digital Twin. The CFaO requires that no new construction or major retrofit occur without a "Digital Handover," ensuring that digital asset data is provided alongside physical keys.

This domain is fundamental to safeguarding the future of institutional investments. By integrating data exchange standards into construction contracts, the CFaO prevents the creation of expensive, disconnected "dumb" buildings. This strategy allows the university to refine its space portfolio on an ongoing basis—using spatial analytics to identify underused areas, reducing the need for new construction, and tackling the backlog of deferred maintenance. Ultimately, it links long-term capital projects and real estate portfolios to the university's academic ambitions, carbon-reduction targets, and digital-twin initiatives.

Smart Infrastructure and OT Convergence

This is the Facilities side of the IT/OT convergence bridge. While the CTO can develop the most secure network globally, it is ineffective if the physical plant cannot connect to it. Therefore, the CFaO should lead the bold modernization of Operational Technology by systematically replacing pneumatic thermostats with IP-enabled digital controllers, upgrading central plants to communicate with algorithmic demand-response systems, and deploying dense sensor networks throughout the built environment. The CFaO must view these mechanical systems not as isolated utilities but as active nodes within the enterprise network. This involves modernizing legacy physical systems—such as HVAC, lighting, physical access controls, and energy microgrids—turning them into an IoT-enabled, edge-computing platform.

This convergence is essential for institutional intelligence. When these systems are modernized, the CFaO generates large volumes of real-time telemetry that the CTO's AI systems depend on. This ongoing data stream fuels predictive maintenance, energy optimization, and the real-time resilience dashboards that campus leadership use during crises.

Lifecycle Sustainability

Sustainability should transition from a marketing tactic to a fundamental part of the Facilities organization's financial and operational decision-making. The CFaO has a crucial role in guiding the institution away from prioritizing only the "lowest initial cost' in procurement. Instead, they evaluate the total cost of ownership for systems like chiller plants over 30 years, factoring in energy prices, carbon taxes, and maintenance expenses. They also track the "embodied carbon" in construction materials such as concrete and steel, as well as the operational carbon of the finished building. This approach helps the campus move toward achieving absolute net-zero emissions.

This domain transforms the Facilities department from a mere cost center into a strategic financial driver. By integrating life-cycle cost analysis, embodied-carbon tracking, and net-zero goals into every stage—from design and procurement to daily operations—they ensure comprehensive sustainability. Through significant reductions in utility consumption via smart optimization, the CFaO converts energy savings into self-funded capital for future modernization efforts. Additionally, rigorous lifecycle sustainability measures generate auditable ESG (Environmental, Social, and Governance) metrics, enhancing the institution's brand reputation, attracting climate-conscious students, and enabling access to specialized federal and philanthropic green grants.

Data-Driven Facilities Operations

For decades, facilities operations have relied on a work-order queue: when something breaks, a campus user reports it, and a technician is sent out. The CFaO transforms this approach by using AI fault-detection diagnostics to alert the maintenance team weeks before a vibrating pump bearing actually fails. Additionally, the CFaO equips frontline workers with mobile technology, so technicians arrive at the site with the diagnostic history, digital manuals, and needed parts ready.

This operational shift significantly reduces Operating Expenses (OPEX) by preventing catastrophic equipment failures and cutting emergency overtime costs. It also greatly extends the lifespan of costly capital assets. Most importantly, it enhances occupant wellbeing—aligning with the "Building Management Hierarchy of Needs." When temperature, air quality, and lighting are consistently maintained at optimal levels, faculty and students can concentrate fully on their academic work without disruptions caused by a failing physical environment. This cultural and operational transformation moves away from reactive "break-fix" maintenance toward predictive, self-optimizing systems driven by IoT sensors, AI analytics, and mobile work management.

Stakeholder Co-Creation

Designing a smart campus requires collaboration beyond facilities engineers alone. The CFaO should engage the academic community by partnering with the College of Engineering to use the campus microgrid for research, or with the Computer Science Department to enable students to develop machine learning algorithms using anonymized HVAC telemetry. Additionally, the CTO and the Provost must be engaged to ensure that classroom designs can support emerging high-density pedagogical technologies.

By treating the campus as a "Living Lab," the CFaO breaks down the traditional divide between "Operations" and "Academics." This collaborative approach speeds up the adoption of new technologies by engaging the campus community as active stakeholders rather than passive users. It ensures that physical spaces continually adapt to support innovative research and modern teaching methods, keeping the campus environment aligned with the university's core mission. This is achieved through direct partnerships with students, faculty researchers, IT leaders, and industry vendors, using "living-lab" projects and iterative "build-measure-learn" cycles.

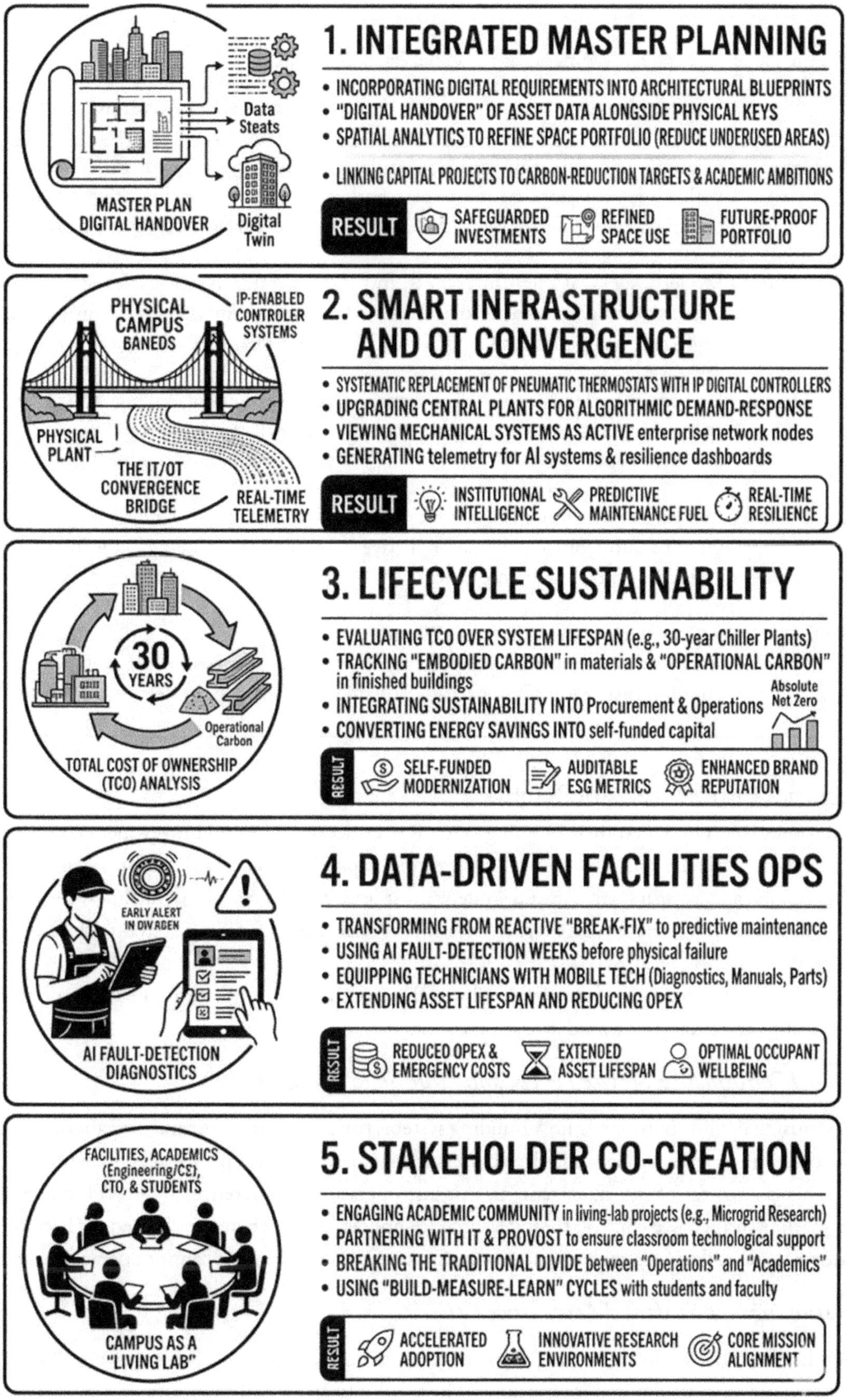

Figure 7. The Strategic Domains of the Chief Facilities Officer

Exhibit 7. Summary of the Strategic Domains of the Chief Facilities Officer

Domain	What It Means	Value to a Smart Campus
Integrated Master Planning	Align capital projects and space portfolios with academic vision, carbon targets, and digital-twin roadmaps.	Future-proofs investments, reduces deferred maintenance, and hard-wires data exchange into every new build or retrofit.
Smart Infrastructure and OT Convergence	Modernize HVAC, lighting, access, and micro-grid systems into an IoT-ready, edge-computing platform.	Generates the real-time data that powers predictive maintenance, energy optimization, and resilience dashboards.
Lifecycle Sustainability	Embed life-cycle cost, embodied-carbon, and net-zero metrics into design, procurement, and operations.	Converts energy savings and carbon credits into self-funding capital; strengthens ESG brand and grant eligibility.
Data-Driven Facilities Ops	Use sensors, AI, and mobile work management to shift from reactive to predictive and self-optimizing maintenance.	Lowers OPEX, extends asset life, and elevates occupant wellbeing—advancing the Building Management Hierarchy of Needs.
Stakeholder Co-Creation	Partner with students, faculty, IT, and vendors through living-lab projects and "build-measure-learn" loops.	Accelerates the adoption of emerging tech and ensures spaces serve evolving pedagogies and research models.

Chief Facilities Officer Decision Rights

Most campuses have developed their building systems over decades of capital investments, vendor platforms, and local "fixes that worked." This history has built essential capabilities but also led to inconsistency. The manifesto emphasizes that a successful smart campus strategy must prioritize the realities of facilities: ensuring reliability, disciplined commissioning, operator usability, and workforce enablement, all integrated into the scope rather than treated as afterthoughts.

From the CFaO's perspective, the essential decision rights safeguard safe operations, maintain performance, and enable the frontline team to operate modernization confidently.

1) Operational Performance, Service Levels, and "Safe State" Constraints

Facilities must maintain control over how building systems function in both normal and abnormal situations—such as comfort settings, ventilation standards, lab pressure regimes, life-safety protocols, and definitions of an "unsafe state." This is a matter of risk management, not just operational preference. While OT modernization can enable automation and remote control, these features are only effective when constrained by Facilities. The core message prioritizes reliability—since the institution cannot automate systems, it cannot operate predictably.

2) Commissioning Standards and Acceptance Criteria

Facilities should establish commissioning as an official institutional discipline, clarifying what "done" means for controls modernization, retrofits, and integrations. The manifesto specifically suggests creating a campus controls standard that sets naming conventions, minimum sensor packages for each space type, commissioning criteria, documentation standards, and integration patterns. This approach enables incremental modernization without disrupting operations.

In practice, facilities acceptance criteria should mandate verified sequences, validated setpoints, calibrated alarm thresholds, and a documented rollback or safe-mode procedure before deeming systems ready for production.

3) Maintenance Workflows, Operator UX, and Field Adoption

Facilities must control the workflows through which modern OT delivers value, including rounds, responses, preventive maintenance, escalation, contractor coordination, and documentation. The core message is clear: operator experience and training are essential, not optional, because confusing interfaces, unmanaged alarm volumes, and analytics that don't prioritize work hinder adoption and cause workarounds. A Facilities leader's primary responsibility is to insist on tools and integrations that lower cognitive load, suggest the next best actions, and align with real-world work practices.

4) Asset Lifecycle Decisions and Capital Renewal Sequencing

Facilities should develop a lifecycle strategy that determines when to modernize controls versus replacing systems, how to prioritize deferred maintenance based on its impact, and how to sequence investments across central plants, residence halls, research environments, and high-occupancy learning spaces. The manifesto presents deferred maintenance as a strategic risk amplifier that weakens sustainability and occupant experience, emphasizing the importance of prioritizing renewal efforts where the mission impact is greatest. This right is crucial because no matter how sophisticated the data architecture is, it cannot make up for equipment failures, uncontrolled system drift, or systems that are difficult to maintain reliably.

5) Point Accuracy, Calibration, and Operational Context for the Data Estate

Facilities must ensure "truth at the source' by maintaining point accuracy, sensor calibration, operation sequences, operational context, and maintenance history, as these factors determine whether analytics and digital twins are reliable or misleading. The manifesto's partnership model clearly divides responsibilities: the CTO manages data architecture and access controls, while Facilities provides the necessary details, such as point accuracy and maintenance records. Both parties then collaborate to uphold data quality standards.

6) Documentation as an operational backbone, not a project artifact

Facilities should ensure that modernization includes comprehensive, usable documentation such as as-built drawings, sequence descriptions, TAB/commissioning reports, point lists, system diagrams, and runbooks. These materials should facilitate knowledge transfer beyond just one technician or vendor. The manifesto emphasizes the importance of runbooks and training pathways within its scope, explicitly stating that if the frontline team cannot operate the project with confidence, the project is incomplete.

What the CTO Should Own—and Why It Matters to Facilities

The Facilities leader's capacity to ensure safe and reliable operations is increasingly tied to enterprise-grade controls that extend beyond individual buildings. These include identity management, segmentation, remote-access governance, integration standards, and data-platform patterns. The manifesto clarifies this partnership approach: as systems become more sensor-rich and connected to the cloud, facility results—such as energy optimization, IAQ assurance, predictive maintenance, and resilience—rely heavily on integration, IAM, segmentation, data governance, and disciplined service management.

CTO decision rights that protect Facilities operations

- **Identity and Access Management (IAM):**
  Ensures only the right people and vendors have the right level of access, with auditability and least privilege—reducing "mystery changes" and uncontrolled exposure.
- **Segmentation patterns and remote access architecture**
  Reduces blast radius and prevents IT incidents from becoming building incidents; enforces safe, monitored, time-bound vendor access.
- **Integration and data platform standards**
  Prevents one-off, brittle integrations that create cascading failures and operator burden; makes successful patterns reusable across the portfolio.
- **Enterprise cybersecurity monitoring and incident coordination (with the Information Security Officer)**
  Ensures OT signals are visible, triaged, and responded to without disrupting safety-critical operations—supporting continuity under disruption.

Why this matters to Facilities

Without CTO guardrails, Facilities take on unmanaged risks, including vendor exposure, inconsistent authentication, fragile integrations, and unclear accountability. Strong CTO guardrails, however, provide Facilities with greater operational confidence, leading to fewer surprises, clearer ownership, safer connectivity, and a scalable path to analytics and digital twins that enhance uptime rather than hinder it.

Translating OT Modernization into Cabinet and Board Narratives: The CFaO Perspective

For decades, the CFaO has struggled with a major communication challenge in higher education: explaining the physical plant's condition to the Board of Trustees. To a Board that is focused on finances, the Facilities department often appears as a never-ending source of deferred maintenance, sudden capital crises, and unappealing utility bills. When a CFaO requests $15 million to replace an old underground steam loop or to upgrade pneumatic HVAC controls, the audience's attention tends to fade. Securing funding is especially tough when competing with the Provost's proposal for a new academic program or the Athletic Director's desire to upgrade a stadium for revenue.

To secure the substantial funding needed for an AI-enabled, sustainable smart campus, the CFaO needs to shift its language fundamentally. They should move away from speaking like a mechanic requesting parts and adopt a tone more like that of a portfolio manager focused on optimizing institutional assets.

Translating Operational Technology (OT) modernization for the Cabinet and the Board involves shifting the focus from handling decay to emphasizing yield generation, risk mitigation, and avoiding capital debt. This is how the CFaO converts technical facility requirements into persuasive strategic priorities at the board level.

1. The Deferred Maintenance Narrative: From "Sunk Cost" to "Predictive Capital Allocation."

The Trap

Approaching the Board with a spreadsheet showing a $200 million deferred maintenance backlog and asking for a massive budget increase simply because assets have exceeded their 30-year expected lifespan.

The Translation

The CFaO frames OT modernization—specifically the deployment of IoT sensors and AI fault detection—as a precision scalpel for the capital budget.

"We are no longer replacing multi-million-dollar chillers simply because a manual says they are old. By modernizing our OT infrastructure, we are deploying an AI-driven triage system. Vibration and acoustic telemetry will tell us exactly which assets have another 5 years of life left and which are failing today. We are asking for an investment in sensors and analytics so we can flatten our deferred maintenance curve, avoid catastrophic emergency replacements, and deploy your capital with algorithmic precision."

2. The Sustainability Narrative: From "Utility Bills" to "The Net-Zero Financial Engine."

The Trap

Trying to explain the intricacies of BACnet protocols, SEER ratings, or the mechanical friction of legacy air handlers to justify an OT upgrade.

The Translation

The CFaO translates smart HVAC and microgrid modernization into the language of ESG (Environmental, Social, and Governance) leadership and hard financial returns.

"Passive energy conservation is exhausted; to hit our Board-mandated net-zero targets, our buildings must become active participants in the energy market. By upgrading our building automation systems and integrating AI, we are turning the campus into a Grid-Interactive Virtual Power Plant. This modernization allows us to autonomously shed load during peak utility pricing, generating millions in direct utility savings. Furthermore, it provides the auditable, investment-grade carbon data required to secure federal green grants and attract climate-conscious CTO students."

3. The Real Estate Narrative: From "Build-to-Solve" to "Capital Debt Avoidance."

The Trap

- Supporting requests for new, $50 million building projects simply because departments claim they are out of space, relying on anecdotal evidence and static scheduling software.

The Translation

The CFaO uses the deployment of spatial intelligence and occupancy sensors to reframe the institution's entire real estate strategy.

"Before we issue bonds and take on decades of debt to pour a single yard of new concrete, we must fully utilize the concrete we already own. Modernizing our spatial OT gives us heat maps of actual human presence. If our data proves that a $30 million administrative building is occupied at only 35% capacity during the week, we will not build a new one. We will use this spatial intelligence to densify, reallocate, and optimize, saving the institution tens of millions in avoided capital debt and future operational overhead."

4. The Risk Narrative: From "Broken Pipes" to "Cyber-Physical Continuity."

The Trap

Describing a failing legacy control system as a maintenance headache that causes hot/cold calls from faculty.

The Translation

The CFaO aligns directly with the CTO, elevating physical infrastructure failure and OT cybersecurity to the level of enterprise risk management.

"A failure in our legacy operational technology is no longer just a facilities issue; it is a critical threat to institutional continuity. A frozen pipe in our bioscience facility destroys a $5 million grant and decades of research. An unpatched, unsegmented building automation system is a backdoor for a ransomware attack that could shut down

residence halls. We are modernizing and securing our OT not just to keep the lights on, but to build an unbreakable cyber-physical perimeter that guarantees the safety of our students and the continuity of our research enterprise."

The Ultimate Shift: The CFaO as a Strategic Partner

By using these translations, the Chief Facilities Officer reshapes their relationship with the Board of Trustees. They move from being seen as the person delivering bad news about aging infrastructure to becoming a strategic partner. They now focus on using advanced technology to safeguard the endowment, cut operational costs, and realize the university's top academic and environmental goals.

Chapter Summary

Chapter 5 presents the built environment as a key strategic platform—no longer just a set of mechanical systems to maintain, but a dynamic, cyber-physical asset that enhances institutional resilience, sustainability, financial management, and the campus experience. It advocates that the CFaO should go beyond traditional capital planning and reactive upkeep to design a fully integrated, interoperable, and data-rich environment that sustains the university's long-term mission.

The chapter starts by identifying four fundamental realities influencing higher education operations today: the growing deferred-maintenance crisis, strict carbon-reduction requirements, ongoing underuse and pandemic-related changes in space utilization, and the swift rise of digital risks as OT systems integrate with the institutional network. These challenges render traditional methods impractical, necessitating a strategic shift toward predictive, data-driven, and digitally-enabled facility management.

To address this moment, the CFaO's plan centers around five key strategic areas:

1. **Integrated Master Planning** — linking physical renewal, carbon commitments, digital-twin readiness, and space optimization into a coordinated long-term blueprint that prevents the creation of "dumb" or disconnected buildings.
2. **Smart Infrastructure and OT Convergence** — modernizing HVAC, lighting, utility, and access systems into an open-protocol, IoT-enabled platform that consistently generates the telemetry required for analytics, predictive maintenance, and resilience.
3. **Lifecycle Sustainability** — shifting from lowest-cost procurement to total lifecycle valuation in capital decisions, ensuring the campus advances its net-zero goals while maximizing long-term financial return and carbon transparency.
4. **Data-Driven Facilities Operations** — using sensors, AI diagnostics, digital twins, and mobile work management to pivot from reactive repairs to predictive, proactive maintenance that stabilizes operations, extends asset life, and enhances occupant wellbeing.
5. **Stakeholder Co-Creation** — positioning the campus as a "living laboratory" by partnering with researchers, IT, students, and industry to accelerate innovation, improve adoption, and align facilities strategies with academic and institutional priorities.

This integrated approach positions the CFaO as an institutional architect, coordinating capital investments, physical systems, data infrastructure, and workflows into a cohesive operational framework. Chapter 5 emphasizes that a smart campus is not simply about acquiring technology; it develops when the built environment is intentionally designed and managed as a strategic, interoperable, human-centered platform that fosters continuous improvement and aligns with mission goals.

Chapter 6

Sustainability as an Ecological Operating System

A sustainable smart campus isn't about just achieving targets;
it's about maintaining long-term viability

Sustainability in higher education is undergoing a fundamental shift. What was once seen as an aspiration—marked by commitments, targets, and symbolic actions—is now becoming tangible infrastructure. Both public institutions and private markets are shifting from defining sustainability by promises to emphasizing how they design, operate, and manage their physical environments over time. In the context of the smart campus, this change places the built environment—encompassing buildings, systems, data, and standards—at the heart of both institutional and socio-economic sustainability.

This change indicates a broader maturation. Sustainability is no longer just an add-on to campus operations; it is evolving into a fundamental aspect of how campuses operate. Key areas such as energy systems, space use, asset lifecycle management, indoor environmental quality, and capital planning are now recognized as essential tools for achieving long-term resilience, affordability, and trust. In this framework, the smart campus transforms sustainability from a set of separate initiatives into an integrated, operational discipline integral to the institution's daily operations.

The contemporary sustainability transformation favors execution over signaling. In higher education, this mirrors a shift away from discrete, programmatic sustainability efforts toward standards-based operational systems that shape daily decisions. Frameworks such as ISO 41001 (Facility Management), ISO 50001 (Energy Management), LEED, WELL, and related benchmarks increasingly serve not as labels of intent but as performance guardrails—defining how campuses plan, operate, and measure success.

Smart campus platforms serve as the crucial link enabling this transition. They integrate building systems, energy data, asset records, and operational workflows, turning abstract sustainability objectives into concrete, measurable, auditable, and repeatable practices. Sustainability shifts from being an occasional report to an ongoing activity, ensuring long-term cost management, regulatory compliance, and fiduciary confidence as benefits of effective design.

This chapter builds on this foundation, examining how smart campus architectures transform sustainability from intent into enduring capability—ensuring that campuses remain resilient, inclusive, and economically viable in a world defined by constant change.

The voice stays manifesto-level, executive, and integrative—consistent with how Chapter 6 already treats sustainability as a systems concept rather than a program.

Scoping Environmental Sustainability

Sustainability on a modern campus begins with understanding **Greenhouse Gas (GHG) emissions**—the primary drivers of global climate change, and the most universally recognized measure of an institution's environmental impact. In the context of an AI-enabled smart campus, GHG emissions are not merely compliance indicators or reporting requirements; they are **operational signals**, **planning constraints**, and **strategic levers** that influence energy procurement, building performance, capital planning, risk management, and even research competitiveness.

To manage these emissions effectively, institutions worldwide rely on the **GHG Protocol**, the global standard for carbon accounting. The Protocol defines how emissions are categorized, measured, and reported across three scopes:

- **Scope 1** —
  Direct emissions from campus-controlled sources (boilers, vehicles, generators, refrigerants).
- **Scope 2** —
  Indirect emissions from purchased electricity, steam, heating, and cooling.
- **Scope 3** —
  All other indirect emissions across the value chain (commuting, procurement, construction, waste, food systems, business travel).

These scopes are important because they influence not only reporting but also the development of a sustainable smart campus. The scope boundaries identify which systems need instrumentation, which workflows require monitoring, which partners need governance, and which digital capabilities—such as digital twins, semantic tagging, and automated emissions modeling—must be established to ensure accuracy and credibility.

In an AI-enhanced environment, the GHG Protocol evolves from a mere accounting tool to a comprehensive technical framework. Achieving real-time emissions monitoring depends on integrated data repositories, interoperable operational technology systems, predictive algorithms, and auditable data governance—fundamental elements that ensure safety, reliability, and resilience within a unified campus setting. AI-driven models project emissions based on operational adjustments, development projects, climate variations, changes in occupancy, and energy market dynamics. Digital twins facilitate simulations of retrofit options, electrification plans, and load management strategies. Semantic layers guarantee that each meter, sensor, and building asset conforms to a unified ontology and data architecture.

This synergy—GHG Protocol integrated with Smart Campus Architecture—allows institutions to view decarbonization as a continuous journey rather than a one-time compliance project. Emissions are tracked, improved, and confirmed regularly. Sustainability goals are embedded in building automation, procurement, transportation, waste management, and investment strategies. With transparent, audit-ready, and machine-verifiable data, trust is strengthened among students, researchers, funders, regulators, insurers, and the broader community.

Ultimately, mastering GHGs and the GHG Protocol goes beyond fulfilling environmental duties—it is about creating a future-ready, reliable, and resilient smart campus. By integrating carbon accounting with the cyber-physical systems that characterize today's universities, leaders can make sustainability quantifiable, manageable, and seamlessly embedded into everyday campus activities.

Carbon Management and Decarbonization Pathways

A truly sustainable smart campus treats **carbon as data**—continuous, granular, auditable, and actionable. Decarbonization begins by unifying **GHG Scopes 1, 2, and 3** within the institutional data estate, creating a single view of emissions sources across energy, operations, procurement, mobility, and construction.

For **Scopes 1 and 2**, real-time meter telemetry (kWh, steam, gas, chilled water) is integrated with **grid carbon-intensity curves**, enabling the campus to align its load with the cleanest hours of the day and to automatically modulate HVAC, storage, and microgrid assets around the carbon profile of the regional utility. For **Scope 3**, the most complex and typically largest of the scopes, the campus ingests data streams that include:

- procurement records and material declarations,

- faculty and staff commuting profiles,
- business travel logs,
- construction and demolition emissions,
- waste and recycling operations, and
- Vendor-reported upstream emissions.

AI models then **forecast future emissions trajectories** and evaluate the impact of retrofits, electrification plans, utility tariff changes, enrollment shifts, and climate scenarios. Through simulation and modeling, the system ranks intervention pathways by **cost per metric ton of CO_2-equivalent avoided, providing leaders with a financially responsible maturity roadmap to reach net-zero** objectives. Over time, decarbonization ceases to be a compliance exercise. It becomes a **dynamic operational intelligence function**—one that strengthens institutional antifragility, reduces risk, and demonstrates accountable stewardship to regulators, funders, and the campus community.

Embodied Carbon and Circular Materials

Operational carbon is only half the story; the built environment hides a vast amount of **embodied carbon** in the materials used to construct, renovate, and maintain campus facilities. To confront this, the smart campus requires **Environmental Product Declarations (EPDs)** for every significant material purchase, ensuring transparency regarding the emissions associated with extraction, manufacturing, transport, installation, and disposal.

Each asset—steel beams, mechanical equipment, glazing, flooring, piping—is cataloged in Material Passports embedded in BIM files or in the campus Digital Twin. These passports store material type, mass, embodied CO_2e, reuse potential, disassembly instructions, and end-of-life pathways. As renovation or demolition approaches, the Digital Twin simulates deconstruction scenarios to identify which components can be reclaimed, sold, or repurposed.

This approach transitions the campus from a **linear "take–make–waste" model** to a **circular system** in which materials retain value beyond their first life. Division 25 procurement standards reinforce circularity by requiring open, interoperable, and repairable systems, reducing integration waste, and preventing vendor lock-in. This is sustainability not as symbolism, but as **infrastructure-level policy** embedded within the cyber-physical architecture of the smart campus.

Biodiversity, Land, and Ecosystem Health

Sustainability extends beyond buildings and utilities; it encompasses the health of the campus ecosystem itself. Using remote sensing, GIS layers, and IoT environmental sensors, the campus can monitor:

- canopy coverage and species health,
- pollinator and habitat corridors,
- soil moisture patterns and infiltration rates,
- heat-island hotspots,
- erosion, runoff, and microclimate variations.

These data streams feed into the Digital Twin, helping sustainability teams design landscapes that support biodiversity while reducing operational loads. For example, Smart irrigation systems adjust watering cycles based on real-time moisture telemetry and forecasted rainfall, protecting native vegetation while conserving water. And strategic shade planning—native trees, green roofs, vines, and reflective materials—reduces cooling demand in buildings, enhances outdoor comfort, and improves resilience during extreme heat events.

Campus land becomes both a **carbon asset** and a **living environmental laboratory**, reinforcing the connection between ecological health and operational performance.

Climate Resilience and Environmental Risk

The smart campus needs to address climate volatility rather than merely withstand it. Digital twins enable the simulation of floods, wildfire smoke intrusion, heat waves, polar vortex occurrences, and grid disruptions. This allows the institution to evaluate and improve its operational and emergency response plans across different stress scenarios.

Critical labs, residence halls, and data centers can be assessed for their capacity to sustain safe conditions during failure scenarios. IAQ telemetry improves wildfire smoke preparedness by enabling real-time adjustments to filtration and pressurization strategies. "Safe modes" for HVAC, water systems, and life-safety equipment help buildings degrade gracefully under stress, ensuring acceptable, minimal levels of comfort, air quality, and containment.

Similarly, microgrid-aware controls allow the campus to isolate itself electrically during a grid disturbance, shedding non-critical loads while keeping essential functions operational. By turning these responses into executable playbooks, the campus adopts a proactive antifragility approach—absorbing shocks, learning from them, and enhancing system performance after each event.

Sustainable Mobility and Transportation

Mobility stands out as one of the most visible and impactful aspects of campus sustainability. How individuals move across, to, and beyond the campus influences carbon emissions, land use, safety, accessibility, and the daily experiences of students, faculty, staff, and visitors. In a smart campus, transportation shifts from disjointed services to a cohesive mobility ecosystem that emphasizes low-carbon travel, alleviates congestion, and promotes equity and access. Achieving sustainable mobility requires more than a single mode or policy; it depends on coordinated planning, real-time data, and intelligent systems that synchronize transportation demand with the institution's climate, health, and community-related goals.

Electrified Campus Fleet

Fleet decarbonization starts with telematics, providing real-time data on charge levels, routes, vehicle duty cycles, charging habits, and maintenance. AI leverages this information to optimize charging times, coordinating with microgrid dispatch, demand-response signals, and periods of low carbon intensity.

Predictive routing reduces mileage and idle time while extending battery life. Maintenance schedules shift from **time-based** to **condition-based**, lowering costs and extending asset lifespan. Over time, the campus fleet becomes a **distributed energy asset**, capable of coordinated discharge during emergencies or peak-load events.

Human-Centered Mobility

Human mobility should be optimized not just for efficiency but also for experience and safety. Analytics-based models guide demand-responsive shuttle routes that adjust service frequency dynamically based on occupancy levels, course schedules, and special events.

Micromobility hubs, including bikes and scooters, are strategically located to align with travel patterns and campus topography. Pedestrian pathways are carefully evaluated for safety, lighting, shade, and congestion. Parking policies are used as tools for sustainability: premium spots are reserved for electric vehicles, carpools, or short-term needs, with incentives promoting shifts in transportation modes. Overall, mobility is managed as an ecosystem that lowers emissions and enhances campus access, comfort, and safety.

Commuter Emissions

Commuting typically ranks among the main sources of Scope 3 emissions, but it has historically been the least quantified. The smart campus employs anonymous travel surveys, transit pass data, license-plate analytics (when privacy policies permit), and geospatial modeling to create a precise emissions inventory for student, staff, and faculty commutes. AI identifies behavioral clusters—who drives, who carpools, who uses transit—and simulates the impact of targeted policy changes such as:

- discounted transit passes,
- parking price restructuring,
- off-peak campus scheduling,
- shuttle frequency optimization,
- incentives for carpooling or micromobility,
- employer-supported remote work arrangements.

These interventions are tracked as **Scope 3 emissions trends** within the sustainability dashboard, demonstrating progress toward institutional climate commitments. As commuting patterns evolve, the model adapts, ensuring that emissions reduction strategies remain effective and relevant.

Reframing Sustainability

Sustainability on the modern campus can no longer be understood as a collection of well-intentioned initiatives, isolated targets, or compliance exercises. Those approaches—while often necessary—tend to treat sustainability as an adjunct to operations rather than as a core design principle. In an era defined by climate volatility, deferred maintenance, financial pressure, and accelerating technological change, this framing is insufficient. Sustainability must instead be understood as an **ecological operating system**—a continuous, adaptive framework that governs how the institution senses conditions, allocates resources, makes tradeoffs, and learns over time. Viewed this way, sustainability is not something the campus *does*. It is how the campus **functions**.

An ecological operating system connects energy systems, buildings, infrastructure, data, people, and governance into a living whole. It emphasizes feedback loops rather than static plans, optimization rather than episodic progress, and long-term stewardship rather than short-term gains. Just as natural ecosystems maintain balance through constant sensing and adjustment, a sustainable campus depends on real-time insight, contextual intelligence, and intentional design to remain stable under stress—and to improve through disruption.

This reframing is essential because the forces shaping higher education are no longer linear or predictable. Climate events disrupt operations. Energy markets fluctuate. Supply chains fracture. Workforce capacity tightens. Regulatory and public expectations intensify. In this environment, sustainability cannot be achieved through one-time improvements or annual scorecards alone. It requires an operating model that can continuously adapt while preserving safety, trust, affordability, and mission performance.

Importantly, this operating system is not limited to environmental performance. While energy, water, and emissions remain critical signals, they represent only one layer of a larger system. Sustainability also governs how capital is deployed, how buildings evolve over decades, how materials are sourced and recovered, how data is trusted and shared, and how decisions are made across institutional silos. When these elements are fragmented, sustainability efforts stall. When they are integrated, sustainability becomes self-reinforcing.

The ecological operating system concept also resolves a persistent tension in campus sustainability: the false choice between environmental responsibility and operational reliability. Properly designed, sustainability enhances reliability rather than competing with it. Real-time energy optimization stabilizes systems. Predictive maintenance reduces failures. Carbon-aware planning aligns long-term financial decisions with climate realities. Transparency builds trust with stakeholders. Sustainability, in this context, is not a constraint—it is a resilience strategy.

This chapter extends that premise by first connecting sustainability to daily operational performance, where sensing, automation, and analytics deliver quick, measurable benefits. It then broadens to consider lifecycle and economic factors, illustrating how sustainability should guide long-term capital planning, material use, and supply chains. The core message is that a campus is genuinely sustainable only when sustainability is integrated into its operational principles, supported by governance, and driven by data and intelligence.

In the chapters that follow, this ecological operating system will be further extended through discussions of digital twins, institutional data estates, and AI-enabled decision-making. Sustainability is not an endpoint in this model. It is the condition that allows the campus to remain adaptable, trustworthy, and mission-aligned in an increasingly unstable world.

Sustainability Embedded in Operations

If sustainability is an ecological operating system, then **operations are its nervous system**. Operational sustainability is where sensing becomes insight, insight becomes action, and action becomes measurable impact. It is the layer at which sustainability stops being aspirational and becomes an everyday institutional capability.

Most campuses first encounter sustainability through operations—energy management, water conservation, indoor environmental quality, emissions reporting. Historically, these efforts focused on efficiency programs, retrofits, and policy mandates. While valuable, such approaches were often periodic and reactive, producing incremental gains but limited systemic change. The smart campus reframes this layer entirely. Operational sustainability becomes a **continuous, data-driven discipline**, embedded into daily decision-making rather than managed as a discrete initiative.

At the core of operational sustainability is the ability to **sense the campus in real time**. Buildings, utilities, and infrastructure now generate continuous streams of telemetry—temperature, airflow, occupancy, energy demand, water flow, equipment performance. When these signals remain isolated within individual systems, their value is constrained. When integrated into a shared operational context, they form a living picture of campus conditions—one that can be acted upon immediately and learned from over time.

This shift transforms sustainability from static targets to **dynamic optimization**. Energy systems no longer rely solely on fixed schedules or seasonal assumptions; they respond to actual occupancy, weather patterns, and demand signals. Water systems detect anomalies before losses escalate. Indoor air quality becomes observable and adjustable, supporting health, comfort, and cognitive performance rather than simply meeting minimum standards. In this model, sustainability outcomes are achieved not through sacrifice or restriction, but through intelligence.

Operational sustainability also reframes reliability. In traditional models, sustainability and uptime were often seen as competing priorities—efficiency versus safety, optimization versus resilience. Real-time operations resolve this tension. Predictive insights allow systems to operate closer to optimal conditions while reducing failure risk. Equipment that signals early degradation can be serviced before disruption occurs. Energy optimization strategies can be balanced against life-safety requirements and research criticality. Sustainability and resilience become reinforcing rather than adversarial.

Crucially, operational sustainability depends on **trustworthy data and disciplined governance**. Real-time optimization is only as reliable as the data feeding it. Sensors must be calibrated. Perhaps most importantly, reliable and trusted sustainability data becomes **strategic capital**. When carbon, energy, water, materials, and lifecycle intelligence are governed as part of the institutional data estate:

- Sustainability leaders gain decision authority, not just reporting responsibility
- Financial leaders gain confidence in long-horizon investments
- Facilities shift from reactive stewardship to portfolio optimization
- Technology leaders enable intelligence, not dashboards
- Boards see sustainability as risk mitigation and value creation

Moreover, data must be contextualized. Decision rules must be explicit and auditable. Without these guardrails, automation erodes trust, and insight becomes noise. With them, operational intelligence scales safely and transparently across campus portfolios and enables the institution to operationalize key areas like:

- Governance and decision rights
- Data estate design
- Digital twin–driven planning
- AI-assisted investment modeling
- Long-term institutional antifragility

These factors further reinforce the cultural foundation for wider sustainability transformation. When faculty have stable classroom environments, students observe clear performance improvements, and operators trust data-driven recommendations, sustainability moves beyond theory. It becomes concrete, credible, and self-sustaining.

Operational sustainability alone is not enough. While improving current performance is essential, it does not guarantee an institution's long-term survival. To truly ensure sustainability, real-time efficiency must be complemented by lifecycle insights, economic management, and material consciousness. The systems monitoring energy and environmental data also provide the information needed to assess long-term costs, carbon footprint, and asset lifespan.

Operational sustainability should be seen not as an endpoint, but as the fundamental basis of the ecological operating system. It offers the visibility, discipline, and feedback needed to extend sustainability efforts beyond daily operations, touching on areas such as capital planning, materials, supply chains, and circular value creation. This wider perspective is where sustainability shifts from just operational excellence to a strategic institutional priority.

Circular Economics and Lifecycle Intelligence

If sustainability as an ecological operating system reframes *how* the campus senses, responds, and optimizes in real time, then circular economics defines *how value is created and preserved over time.* Together, they move the institution beyond efficiency toward stewardship—where resources, assets, and capital are continuously cycled rather than consumed.

Traditional campus economics are fundamentally linear. Materials are extracted, procured, installed, depreciated, and ultimately discarded. Sustainability initiatives layered onto this model often focus on operational efficiencies—energy reduction, water savings, or waste diversion—while leaving the underlying economic logic unchanged. In contrast, a smart campus designed around circular economic principles treats buildings, infrastructure, and materials as long-lived assets within a regenerative system, governed as carefully as financial capital.

Sustainability as an Economic Operating Model

Within the ecological operating system, sustainability is no longer framed as compliance, aspiration, or reporting obligation. It becomes an economic operating model—a discipline that reshapes how decisions are evaluated, sequenced, and governed across the institution. In this model, sustainability is not adjacent to finance or operations; it is woven into them.

The transformation begins with visibility. Sensors, analytics platforms, digital twins, and governed data estates allow institutions to observe their campus not only as a collection of buildings and systems, but as a portfolio of long-lived assets whose environmental and economic performance are inseparable. Energy consumption, asset degradation, maintenance cycles, material flows, and carbon exposure are no longer hidden within siloed systems or retrospective reports. They become observable, comparable, and—critically—modelable over time.

This visibility changes the basis of decision-making, and the reframing has practical implications:

- Capital projects are evaluated on **total lifecycle value**, not lowest first cost.
- Deferred maintenance is treated as **compound risk**, not temporary savings.
- Energy, water, and materials are managed as **dynamic inputs** whose costs, carbon intensity, and resilience impacts vary over time.

Capital investments are no longer justified primarily by the lowest first cost or short-term efficiency gains. Instead, they are evaluated through a lifecycle lens that accounts for total cost of ownership, resilience under stress, carbon exposure, and adaptability over the long term. A cheaper system that degrades faster, locks the institution into proprietary dependencies, or amplifies operational volatility is no longer seen as financially prudent. Sustainability, in this framing, becomes a mechanism for protecting long-term value rather than an external constraint placed upon it.

Deferred maintenance, long treated as an unfortunate but manageable byproduct of capital scarcity, is reclassified as compound risk. Aging assets degrade nonlinearly, concentrating cost, emissions, and failure into short windows that destabilize budgets, operations, and trust. Within an economic operating model, deferral is no longer invisible. Its future consequences—emergency replacement costs, service disruption, energy inefficiency, and carbon spikes—are surfaced explicitly. Leadership can then choose whether deferral is a deliberate, time-bound strategy with known risk, or an unconscious transfer of liability to future administrations.

Energy, water, and materials are similarly reinterpreted. Rather than static utilities purchased at fixed rates, they are treated as dynamic economic inputs whose prices, carbon intensities, availabilities, and resilience implications fluctuate over time. A kilowatt-hour consumed during peak demand under a stressed grid carries a different financial and environmental cost than one shifted to a low-carbon or off-peak interval. Water use during drought conditions poses risks that extend beyond utility expenses to community trust and regulatory exposure. Material choices influence not only embodied carbon at construction but also future repairability, reuse, and end-of-life value decades later.

When these variables are integrated into shared models—supported by digital twins and governed analytics—the campus gains the ability to trade deliberately across cost, carbon, reliability, and mission impact. Sustainability ceases to be argued as a moral imperative or long-term aspiration. It becomes an evidence-based optimization problem. Leaders are no longer asked to accept sustainability on faith; they are given tools that show how environmental performance and economic performance converge when managed across the full lifecycle.

In this sense, the smart campus reframes sustainability from a values conversation into a management discipline. It allows institutions to optimize not for lowest cost or highest efficiency in isolation, but for durability, adaptability, and total value over time. The result is a campus that not merely reduces harm but also allocates capital more intelligently, absorbs volatility more gracefully, and sustains its mission with greater financial and operational confidence.

Circular Economy on Campus

A circular campus rejects the linear logic of **take–make–waste** and replaces it with a closed-loop institutional metabolism. In this model, buildings, infrastructure, and systems are no longer treated as disposable capital artifacts but as **long-lived, evolvable assets** whose value compounds over time through stewardship, adaptability, and intelligent reuse.

This is not a facilities trend, nor a sustainability initiative bolted onto existing capital practices. The circular campus is a **re-architecture of how institutions design, procure, operate, and retire the physical estate.** It requires leadership willingness to replace convenience with intentionality and to treat material flows, asset lifecycles, and embodied carbon as first-class strategic concerns—not downstream reporting metrics.

A circular campus is maintained not through virtue signaling or sporadic retrofits, but by systems that render waste economically unviable, make lock-in unlikely, and eliminate the need for premature replacement. Essentially, circularity relies on three core capabilities: visibility into the lifecycle, designing for disassembly and reuse, and capital planning guided by feedback.

Lifecycle Visibility

You cannot govern what you cannot see across time. Circularity begins with **institutional clarity about what exists, where it is, and how it ages.** Assets must be digitally represented not as static inventory records, but as living entities—rich with metadata describing material composition, operational context, maintenance history, energy performance, carbon intensity, and remaining useful life.

Digital twins and integrated asset platforms provide this visibility by collapsing the traditional divide between design data, operational telemetry, and maintenance systems. In doing so, they replace fragmented documentation and tribal knowledge with **decision-grade institutional memory.** Every air handler, lighting fixture, battery array, and structural system becomes part of an observable fabric—continuously updated, auditable, and actionable.

Lifecycle visibility shifts the campus from **forensic management**—discovering problems after failure—to **anticipatory stewardship**, where degradation, inefficiency, and obsolescence are detected early and addressed deliberately. It also lays the groundwork for accountability: sustainability claims stop being estimates, and capital justification stops being anecdotal. Without lifecycle visibility, circular economy principles collapse into aspiration. With it, they become an executable policy.

Design for Disassembly and Reuse

Buildings should evolve, not entomb decisions. Traditional campus construction prioritizes upfront cost and aesthetic ambition, often at the expense of future adaptability. Systems are custom-fit, tightly coupled, and proprietary—optimized for day-one delivery while quietly engineering future waste.

A circular campus reverses this logic. Campus standards evolve to favor **modularity, standardized interfaces, open protocols, and deliberate separation of structure from systems.** Mechanical, electrical, digital, and control components are designed to be swapped, upgraded, repaired, or redeployed without requiring wholesale demolition or vendor captivity.

Design for disassembly restores **optionality**—the most undervalued asset in capital planning. It allows institutions to adapt to pedagogical change, technology evolution, and climate realities without writing off decades of embedded value. It also weakens the grip of opaque vendor lifecycles that force premature replacement under the guise of "end of support."

In a circular campus, procurement decisions are no longer just about price and performance. They are about **future agency**: Who controls upgrades? What data remains accessible? Which components can be reclaimed, refurbished, or reused across the portfolio?

Durability, repairability, and openness become strategic virtues, not engineering inconveniences.

Feedback-Driven Capital Planning

Time-based replacement is an admission of ignorance. Linear campuses replace assets on schedules. Circular campuses replace assets based on **evidence**. Feedback-driven capital planning uses condition, performance, and risk signals—derived from lifecycle visibility—to inform when an asset should be repaired, upgraded, redeployed, or retired. This approach dismantles the false precision of calendar-based depreciation and replaces it with **situational intelligence**.

Some systems warrant early replacement due to inefficiency or carbon intensity. Others merit extended life through targeted intervention, recommissioning, or component-level upgrades. Still others become candidates for secondary use—repurposed into lower-demand contexts where value can continue to accrue.

This model fundamentally changes the economics of campus stewardship. Capital planning becomes a **continuous feedback loop** rather than an episodic budgeting exercise. Waste is reduced not through moral appeal, but because the institution has engineered waste out of its decision pathways.

Most importantly, disruption becomes rarer. Asset failure ceases to be the trigger for investment. Insight becomes the trigger.

Circularity as Economic Architecture

Sustainability is what happens when systems are aligned. In a circular campus, sustainability improvements are not special projects, pilot programs, or annual reporting exercises. They are the **emergent property of an institution that has aligned incentives, architecture, and information flows.**

When assets are visible across their full lifecycle, when systems are designed to evolve rather than expire, and when real conditions, not assumptions, guide capital decisions, sustainability ceases to be performative. It becomes structural.

This is the critical move of the circular campus: shifting sustainability from a set of goals to a **governing logic**—one that treats materials, energy, and capital as resources to be stewarded intelligently rather than consumed expediently. Circularity is not a restraint, and sustainability improvements are not episodic projects; they are the emergent result of better economic architecture.

Material Passports and the Digital Record of Value

Material passports are essential for circular campus economies and are integrated within BIM, CMMS, and the campus digital twin. They serve as long-lasting, machine-readable records detailing an asset's composition, deployment location, installation method, performance history, and potential for reuse, refurbishment, or recovery at the end of its service life.

Unlike traditional specification documents or as-built drawings, material passports persist across decades of operation. They travel with the asset as systems are renovated, spaces are repurposed, and institutional priorities evolve—preserving value that would otherwise be lost through fragmentation, staff turnover, or vendor dependency. For the institution, material passports enable four strategic capabilities:

- **Embodied carbon accounting at the asset and building level**
  Grounding decarbonization strategies in verifiable material data rather than modeled estimates or retrospective audits.
- **Residual value tracking**
  Reframing components—not as sunk costs—but as recoverable resources whose financial and material worth can be assessed, retained, and reinvested.
- **Predictive deconstruction and reuse planning**
  Allowing future renovations or retirements to be informed by real material composition, condition, and secondary-market potential rather than assumption or disposal convenience.
- **Transparent and enforceable procurement**
  Suppliers are evaluated not only on price and performance but also on durability, modularity, repairability, and end-of-life responsibility.

Over time, the campus itself becomes a **material bank**—a curated portfolio of valuable components whose worth increases through informed stewardship rather than depreciation through neglect. This is not theoretical. As supply chains grow more volatile, materials become more constrained, and carbon becomes more regulated, institutions with accurate material intelligence gain **resilience, negotiating leverage, and long-term cost predictability**.

Supply Chains, Risk, and Circular Resilience

Circular economics fundamentally reshapes how institutions understand **supply-chain risk**. Linear procurement assumes continuous access to external materials, predictable pricing, and stable logistics. In practice, these assumptions no longer hold. Campuses are increasingly exposed to **price volatility, extended lead times, vendor concentration, and geopolitical disruption**—risks amplified by climate events and accelerating global demand for scarce materials.

Circular systems reduce this exposure by design. By **extending asset lifetimes, enabling component reuse, and localizing value creation**, the institution lowers its dependency on fragile external supply chains. What was once treated as waste becomes internal capacity; what was once a procurement emergency becomes a managed transition.

Within an ecological operating system, supply-chain risk is no longer treated as a purchasing problem. It becomes an **enterprise risk domain**—modeled, monitored, and governed alongside climate, cyber, financial, and operational risk. Key shifts enable this resilience:

- **Supply-chain risk is formally modeled alongside climate, cyber, and operational risk**, allowing leaders to understand cascading impacts rather than isolated failures.
- **Alternative sourcing, reuse, and substitution scenarios are simulated within the digital twin**, enabling institutions to test responses to shortages, delays, or cost shocks before they occur.
- **Vendors are evaluated not merely on price and availability, but on lifecycle responsibility, resilience posture, interoperability, and data transparency**, aligning procurement with long-term institutional stability rather than short-term savings.

This approach moves the campus from **reactive procurement to strategic resource governance**. Decisions shift upstream—from emergency purchasing and vendor escalation to portfolio-level planning that balances resilience, sustainability, and fiscal stewardship. In an era of constrained capital, tightening material markets, and persistent volatility, circular resilience is not an optimization.

Measuring the Value of Sustainability

Sustainability cannot be managed by aspiration alone. Like safety, reliability, or financial performance, it must be made visible, comparable, and actionable through measurement. On a smart campus, measuring circular value is not about creating a new reporting burden; it is about revealing forms of value that already exist but are routinely ignored by traditional accounting and capital planning models.

Leading institutions expand their measurement frameworks beyond annual operating costs or upfront capital expense to capture how value persists—or erodes—over time. For example:

- Avoided carbon from extended asset life
- Deferred capital through condition-based maintenance
- Materials recovered, reused, or redeployed
- Reduced exposure to volatile energy and materials markets

One of the most consequential indicators is avoided carbon resulting from extended asset life. Every year that a building system, structural component, or piece of equipment operates effectively beyond its assumed replacement horizon represents carbon that was never emitted. This avoided carbon is not theoretical; it is the measurable consequence of retaining embedded materials, embodied energy, and construction effort that would otherwise need to be replaced.

Closely related is the concept of deferred capital through condition-based maintenance. When assets are monitored, tuned, and intervened upon based on real conditions rather than calendar schedules, institutions delay or eliminate the need for large, disruptive capital projects. This deferral is not neglect; it is precision. Capital that would have been spent reactively can be redirected strategically, smoothing expenditure curves and reducing exposure to fiscal shocks that often accompany emergency replacements or accelerated modernization programs.

Circular value also becomes tangible through tracking materials recovered, reused, or redeployed across the campus portfolio. Doors, fixtures, mechanical components, structural materials, and even entire modular units can often be harvested and reintegrated when lifecycle data and material specifications are known in advance. What was once treated as demolition waste becomes an internal material supply chain. Over time, the campus itself becomes a managed inventory of recoverable value rather than a collection of disposable projects.

Another critical dimension of circular measurement is reduced exposure to volatility in energy and materials markets. Institutions that extend asset life, optimize energy demand, and reuse materials are less dependent on external price swings, supply chain disruptions, and geopolitical shocks. This reduced exposure is a form of risk mitigation that traditional sustainability metrics rarely capture, yet it directly influences long-term financial stability. Circular campuses are not just greener; they are more economically resilient.

Taken together, these measures close the loop between sustainability, finance, and operations. They operationalize the ecological operating system introduced earlier in the Manifesto by aligning day-to-day decisions with long-term consequences. In this model, improvement is not driven by urging individuals to make better choices under pressure. Instead, performance improves because the system itself learns—detecting patterns, validating outcomes, and reinforcing behaviors that preserve and drive value over time.

Sustainability as Fiscal Discipline

Sustainability is frequently viewed as an aspect of environmental responsibility or institutional values. On the smart campus, however, it often reflects fiscal discipline even more. Institutions embracing lifecycle thinking are not just opting for greener solutions; they are making more financially sound decisions. They recognize that buildings, infrastructure, and energy systems operate based on long-term financial and physical principles that cannot be dismissed with short-term budgeting.

An institution that recognizes the full lifecycle of sustainability asks fundamentally different questions than one operating under annual budget cycles or five-year capital plans. For example:

- What is the **total cost of ownership** over 30–50 years, not five?
- How does deferred maintenance amplify risk and carbon simultaneously?
- Where does replacing a component destroy embedded value?
- What assets should be extended versus retired?
- How does energy optimization interact with asset degradation?

Instead of focusing narrowly on upfront cost, it considers the total cost of ownership over thirty to fifty years—the actual lifespan of campus assets. It recognizes that decisions made at procurement reverberate across decades of operations, maintenance, energy consumption, and eventual decommissioning. What appears inexpensive today often proves costly over time when energy inefficiency, accelerated wear, or compatibility issues surface years later.

Deferred Maintenance

Deferred maintenance is central to this issue. While postponing repairs might seem financially wise initially, helping institutions handle short-term costs, it gradually turns into a growing liability. Older systems become less efficient, use more energy, and fall out of calibration, raising operational costs and increasing carbon emissions. Both financial and sustainability risks escalate concurrently. For example, a chiller that isn't properly maintained not only wastes energy but also risks imminent failure, especially as conditions become more unpredictable, costly, and volatile.

Sustainability lifecycle thinking shifts how institutions evaluate replacement choices. Not every aging part needs removal just because it hits an arbitrary end-of-life point. Often, replacing components wastes embedded value—materials, energy, and capital—that could be preserved through refurbishment or targeted upgrades, reducing costs and carbon footprint. A campus that adopts a lifecycle discipline can differentiate between assets suitable for extension, stabilization, or retrofitting and those that truly need retirement. This subtlety is often overlooked in blanket modernization efforts that prioritize novelty over necessity.

Energy optimization further complicates this picture. Aggressive efficiency measures that fail to account for asset condition can unintentionally accelerate degradation. Systems pushed beyond their design tolerances may appear efficient in the short term while imposing hidden maintenance costs or shortening lifecycle expectancy. A sustainable campus balances efficiency gains with equipment health, recognizing that true optimization operates across energy performance, reliability, and longevity simultaneously.

AI-Enabled Lifecyle Modeling

AI-enabled lifecycle modeling dramatically improves the institution's ability to navigate these trade-offs. By integrating asset data, maintenance history, energy performance, and operational context, smart campuses can simulate future outcomes before committing capital. Leaders can model how decisions impact cost, carbon, risk, and resilience over decades rather than fiscal quarters. The result is not perfection, but foresight.

When lifecycle sustainability is treated as fiscal discipline, institutions experience fewer emergency replacements, reduced volatility in capital expenditure, and more predictable long-term financial planning. Sustainability ceases to be an aspirational overlay and becomes a stabilizing force—one that aligns environmental stewardship with financial stewardship, and replaces reactive spending with deliberate, resilient investment.

AI-enabled lifecycle modeling allows institutions to simulate these questions before committing capital. The result is fewer emergency replacements, lower volatility in capital spending, and more predictable long-term financial planning.

From Sustainability to Stewardship

This approach to sustainability marks a significant change in how institutions view the issue. Instead of being confined to facilities or environmental teams, sustainability now acts as a universal economic language spanning finance, IT, planning, and operations. Circular economics provides a framework to preserve long-term value for students, faculty, and infrastructure across generations.

In this context, a smart campus isn't just about efficiency or low carbon; it's about durability. It withstands waste, absorbs shocks, and increases value—material, financial, and environmental—over the long term. This durability is the real benefit of approaching sustainability as an ecological operating system rather than a series of isolated projects.

The Institutional Mandate for Sustainability

Sustainability as an institutional operating mandate
reshapes strategy, investment, leadership accountability, and builds antifragility.

Sustainability is now a concrete part of daily operations, seen as an ecological system integrated with circular economics, governed and driven by continuous learning. The hardest part is institutionalizing it. Success depends less on technology, ambition, or funding, and more on leadership's commitment—whether they view it as optional or essential. For many organizations, this stage marks a shift from treating sustainability as a strategy or initiative to recognizing it as a core responsibility or mandate.

Sustainability as an Executive Obligation

The AI-enabled sustainable smart campus exposes one of higher education's most persistent structural failures: fragmented accountability. Energy management resides in Facilities. Data governance lives in IT. Capital planning sits in Finance. Risk is assigned to Audit and Compliance. Sustainability, however, has traditionally belonged everywhere and nowhere at once—invoked rhetorically, delegated tactually, and evaluated inconsistently.

This fragmentation is not simply an organizational inconvenience; it is a systemic vulnerability. Sustainability outcomes emerge from the interaction of capital decisions, operational practices, data visibility, and risk tolerance. When these domains are managed independently, institutions optimize locally while underperforming globally. Energy efficiency initiatives are undermined by capital deferral. Digital innovation is pursued without regard for physical lifecycle constraints. Financial prudence today produces environmental and fiscal instability tomorrow.

The ecological operating system demands a fundamentally different posture. Sustainability becomes an executive obligation—not as a single line item or portfolio, but as a shared condition of leadership. No single executive owns sustainability outcomes outright. Instead, every executive leader is accountable for how decisions within their domain either strengthen or weaken the system as a whole. Facilities leaders influence carbon and reliability. Technology leaders govern the data and platforms that make sustainability observable. Finance leaders shape lifecycle outcomes through capital timing, depreciation models, and investment horizons. Risk leaders define which vulnerabilities are tolerated and which are mitigated by design.

This reframing shifts the institution's central question. Instead of asking, "Who is responsible for sustainability?" leaders must ask, "How does this decision affect the long-term health, resilience, and adaptability of the campus system?" The answer cannot be aspirational or anecdotal. It must be legible, modeled, and reviewable—expressed through shared data, common frameworks, and governance mechanisms that enable evaluation of sustainability impacts alongside cost, risk, and mission performance.

In this model, sustainability is no longer dependent on individual champions or moral suasion. It is embedded in executive behavior through architecture, incentives, and decision rights. Sustainability persists not because leaders try harder, but because the system itself makes tradeoffs visible and consequences unavoidable.

Sustainability as Institutional Memory

At maturity, sustainability evolves beyond governance and becomes institutional memory. It is no longer a set of goals renewed with each strategic plan cycle, nor a collection of initiatives vulnerable to leadership turnover. Instead, it becomes a durable record of how the institution has built, operated, and learned over time. The campus remembers:

- What was built, with what materials, and why
- How systems performed under stress
- Which choices created resilience—and which created fragility
- How value was preserved or destroyed over time

A campus with institutional memory remembers what was built, with what materials, and why. It retains knowledge of design assumptions, tradeoffs, and constraints—rather than losing that context when project teams disband or vendors change. It remembers how systems performed under stress: during heat waves, storms, pandemics, and supply chain disruptions. These experiences are not archived as anecdotes but captured as data, models, and operational insight.

Crucially, the institution also remembers which choices produced resilience and which quietly introduced fragility. It can trace how deferred maintenance compounded risk, how short-term savings destroyed long-term value, or how disciplined lifecycle investment stabilized both costs and carbon. This memory allows leaders to see beyond individual projects to patterns of consequence—recognizing recurring failure modes and reinforcing practices that consistently preserve value.

This institutional memory is not human alone. It is encoded in the campus digital twin, asset models, maintenance histories, and sustainability metrics that outlive any single leader or team. Governance ensures that this knowledge is consulted rather than bypassed. Architecture ensures it remains accessible and interoperable. Together, they allow the institution to learn across decades rather than relearning the same lessons each capital cycle.

When sustainability functions as institutional memory, executive obligation deepens. Leaders are no longer free to make decisions in isolation from history or consequence. The campus remembers—and the system responds accordingly. In this way, sustainability becomes not just an environmental commitment, but an enduring expression of organizational intelligence.

Institutionalizing Sustainability Through Decision Interfaces

Sustainability does not become real through vision statements, commitments, or strategic plans alone. It becomes real at decision interfaces—the specific moments where tradeoffs are made, priorities are set, and resources are committed. These are the points at which intentions encounter constraints and where institutions silently declare what they truly value.

In most organizations, sustainability efforts fail not because leaders disagree with the goal, but because sustainability is external to the decisions that actually shape the campus. It is discussed after capital commitments have been made, after procurement specifications have been written, or after systems have been deployed. By then, outcomes are already locked in.

Mature institutions reverse this pattern by embedding sustainability directly into the mechanisms through which decisions are evaluated and approved, and institutionalized through:

- **Capital planning gates** that evaluate lifecycle cost, carbon, and resilience together
- **Procurement standards** that require interoperability, repairability, and material transparency
- **Architecture review processes** that reject short-term efficiency gains that introduce long-term fragility
- **Operational dashboards** that surface environmental stress as early risk signals

Capital planning becomes an integrative exercise rather than a financial one-dimensional gate. Projects are assessed not only on upfront cost or funding availability, but on lifecycle cost, carbon exposure, operational resilience, and long-term adaptability. Sustainability is no longer an add-on justification; it becomes part of the definition of a viable investment.

Procurement is similarly transformed. Instead of optimizing for the lowest bid or fastest delivery, institutions require interoperability, repairability, and material transparency as conditions of participation. Vendors are evaluated on how well their products support long asset lifecycles, data openness, and eventual reuse or recovery. This subtly but decisively reshapes markets—rewarding suppliers who align with institutional durability and penalizing those who externalize future cost and risk.

Architecture review processes play an equally critical role. Many sustainability failures originate in well-intentioned efficiency gains that introduce hidden fragility—systems that perform optimally under ideal conditions but degrade rapidly, resist integration, or lock institutions into proprietary dependencies. A disciplined architecture function rejects short-term optimization that compromises long-term control, insisting instead on designs that remain observable, governable, and evolvable over decades.

Operational visibility completes the loop. Dashboards that surface environmental stress—rising energy intensity, abnormal system cycling, declining indoor air quality, or climate-driven performance drift—allow sustainability to function as an early risk signal rather than a retrospective report. These signals inform maintenance prioritization, capital forecasting, and executive situational awareness, turning sustainability into a management input rather than a reporting obligation.

When these decision interfaces are aligned—capital, procurement, architecture, and operations—sustainability no longer depends on persuasion, moral appeal, or individual heroics. It becomes the default outcome of disciplined decision-making. Leaders are not asked to care more; they are asked to decide more clearly, using systems that make long-term consequences visible at the moment choices are made.

Avoiding the Sustainability Collapse Pattern

The Manifesto repeatedly warns against a failure mode that feels familiar precisely because it is so common. Institutions make bold and sincere sustainability commitments, often under public pressure or moral conviction. These commitments are followed by a flurry of visible pilot projects—solar arrays, dashboards, smart buildings, living labs. For a time, momentum feels real.

Then the pattern asserts itself. Funding becomes episodic rather than structural. Governance becomes advisory rather than decisive. Metrics proliferate, but trust in the data erodes. Leaders move on, staff rotate, and the pilots quietly age into exceptions that no longer scale. Sustainability does not fail loudly; it decays administratively. The institution does not abandon its goals—it simply loses the capacity to sustain them. This is the sustainability collapse pattern.

What makes this pattern especially damaging is not the absence of effort, but the absence of integration. Sustainability collapses not because institutions lack ambition, talent, or technology, but because sustainability remains orthogonal to core operations rather than embedded within them.

An ecological operating system fails only under specific conditions when:

- Sustainability is decoupled from core operations
- Data is untrusted or inaccessible
- Governance is advisory rather than authoritative
- Incentives reward short-term performance over long-term durability

It fails when sustainability is treated as a thematic overlay rather than an operational principle—something discussed in reports but not enforced in capital planning, procurement, or architecture decisions. It fails when data is fragmented, untrusted, or inaccessible across domains, preventing leaders from seeing how environmental performance, financial exposure, and operational risk interact. It fails when governance bodies can advise but not constrain, recommend but not decide. And it fails when incentives reward short-term optimization—lowest first cost, fastest delivery, visible wins—over long-term durability, adaptability, and system health.

The Antidote to Collapse is Integration.

Institutions that avoid this cycle do not rely on heroic sustainability leadership or perpetual persuasion. Instead, they redesign the system itself so that sustainability outcomes compound naturally over time. Decisions made in Facilities reinforce Finance. Architecture choices reduce future operational volatility. Data collected for maintenance improves carbon modeling. What begins as environmental stewardship becomes operational stability, fiscal predictability, and cultural coherence.

In these institutions, sustainability stops behaving like a program with its own lifecycle. It becomes a property of the institution's operating model.

Compounding returns emerge not because individuals work harder, but because the system learns. Capital investments generate fewer surprises. Deferred maintenance becomes visible before it becomes a crisis. Energy volatility is dampened rather than absorbed reactively. Governance conversations shift from exception management to pattern recognition.

Most importantly, sustainability remains durable across leadership change. When the operating system is ecological—when sustainability is encoded in data, architecture, governance, and incentives—the institution need not rediscover its values every decade. It remembers them operationally. Moreover, institutions that institutionalize sustainability experience compounding returns—not just environmental gains, but operational stability, financial predictability, and cultural coherence.

Sustainability and Institutional Survival

The Manifesto has been building towards but has not yet directly stated that: **sustainability is no longer a statement of values—it is a strategy for institutional survival.** For much of higher education's recent history, sustainability has been framed as a moral commitment, a reputational signal, or an environmental obligation distinct from the core mechanics of running a university. That framing is no longer sufficient. The external conditions under which institutions operate have changed too profoundly—and too permanently—for sustainability to remain aspirational or optional.

Higher education now operates in an environment defined by simultaneous, compounding pressures. Climate volatility introduces physical risk, operational disruption, and regulatory scrutiny. Capital scarcity limits an institution's ability to absorb failure or inefficiency. Demographic shifts and enrollment pressure destabilize long-standing financial models. Energy markets fluctuate unpredictably, turning utilities from a manageable expense into a strategic exposure. Workforce shortages—particularly in skilled trades, IT, and facilities—erode institutional capacity at precisely the moment complexity increases. At the same time, cyber-physical risk blurs the boundary between digital disruption and physical consequence, raising the stakes of failure across every connected system.

Institutions that approach sustainability as an overlay—a layer added after decisions are made, or a program parallel to core operations—struggle under these combined stresses. Their systems are brittle, their responses reactive, and their planning horizons short. By contrast, institutions that treat sustainability as an operating system—embedded in governance, architecture, data, and incentives—demonstrate an ability to adapt under pressure. They do not eliminate volatility, but they dampen its impacts. They do not predict every disruption, but they learn from each one.

In this context, sustainability becomes inseparable from the institution's most fundamental obligations of:

- Mission continuity
- Research reliability
- Student and staff wellbeing
- Financial resilience
- Public trust

Mission continuity depends on the reliability of physical infrastructure, energy systems, and digital platforms amid increasing disruption. Research reliability depends on stable environments, predictable utilities, secure data flows, and resilient operational support. Student and staff well-being is shaped not only by pedagogy and policy, but by indoor environmental quality, safety systems, thermal comfort, and institutional responsiveness during stress. Financial resilience depends on reducing long-term exposure to volatility, avoiding catastrophic failure, and making capital decisions that preserve value rather than destroy it. Public trust depends on transparency, accountability, and the visible alignment between stated commitments and lived outcomes. Sustainability, in other words, is no longer about achieving targets in isolation. It is about **maintaining institutional viability over time**.

The AI-driven sustainable smart campus exemplifies this change. Its goal is not sustainability for the sake of rankings, reports, or symbols—although these might be incidental benefits. Instead, it seeks sustainability because unmanaged complexity now poses a real existential threat. By evolving into a governed, observable, and adaptive system, the campus can operate effectively in conditions that would otherwise be overwhelming. Additionally, the smart campus aims for sustainability not merely to achieve decarbonization goals but to ensure its continued viability.

The Stewardship Compact

The culmination of smart campus sustainability is not just a technical framework, an architectural diagram, or a maturity model. It is a **stewardship compact**—an implicit but binding contract between current leadership and the future generations of students, staff, and faculty who will inherit the consequences of today's decisions.

The Campus is a Living System, not a Static Asset

This compact begins with a fundamental reframing of what a campus is. A university campus is not a static asset to be optimized, depreciated, and eventually replaced. It is a living system—dynamic, interdependent, and continuously shaped by use, stress, and adaptation. Buildings age. Infrastructure drifts. Energy systems respond to climate. People change how space is occupied and valued. Treating the campus as if it were inert invites fragility; stewarding it as a living system enables resilience and leads to antifragility.

Decisions Shape Environmental, Financial, and Cultural Conditions

Within that frame, time takes on a different significance. Decisions made today—often justified by short-term cost savings or delivery speed—shape environmental performance, financial exposure, and institutional culture for decades. A choice of materials, an architectural shortcut, a deferred maintenance decision, or an opaque procurement contract does not expire at the end of a fiscal year. It accumulates. It constrains or enables future action. Stewardship, therefore, is not an abstract moral stance; it is the discipline of acknowledging long horizons and acting responsibly within them.

Stewardship Requires Visibility, Humility, and Continuous Learning

True stewardship also requires humility. No generation of leaders possesses complete foresight. Climate conditions change. Technologies evolve. Assumptions prove wrong. What distinguishes stewardship from control is the willingness to learn continuously—by observing system behavior, commissioning outcomes, capturing data, and revisiting prior decisions without defensiveness. Institutions that treat sustainability as something to be "completed" eventually stop paying attention. Institutions that treat it as a practice remain alert.

This is why the Manifesto insists that sustainability is never finished. It is exercised. It is refined. It is tested under stress. Each capital cycle, operational disruption, or unexpected failure becomes an opportunity to either repeat the past or encode learning for the future.

The ecological operating system exists precisely to honor this stewardship compact. Its purpose is not to prescribe answers, but to make consequences visible. It surfaces the downstream effects of decisions that would otherwise remain hidden—linking carbon to capital, maintenance to risk, and performance comfort. It makes tradeoffs explicit rather than implicit, forcing institutions to confront what is being gained, what is being lost, and what is being deferred. And it makes learning unavoidable by capturing operational truth in data, models, and governance processes that persist beyond individual roles and administrations.

Sustainability as an Economic Flywheel

Sustainability on the smart campus should not be viewed merely as an accounting task or a moral duty justified solely by external pressure. When seen narrowly—as compliance reports, energy dashboards, or a collection of "green projects"—it is likely to struggle against more immediate operational and academic priorities. The Manifesto promotes a different view: by designing sustainability as an ecological operating system, it can serve as an economic engine rather than a burden for the institution. As an ecological system, sustainability is integrated into how campus assets are conceived, procured, operated, financed, adapted, and eventually retired. This approach redefines how value is preserved and regenerated over decades—the critical time span for higher education infrastructure.

From Linear Spend to Circular Value

Most institutions still follow a linear economic approach. In traditional campus economics, sustainability efforts are usually seen as exceptions—such as pilot projects, grant-funded demonstrations, or isolated upgrades justified only by energy savings. This perspective underestimates the true potential of sustainability. Capital is directed toward the construction or purchase of assets, but performance declines often go unnoticed until they become critical. Deferred maintenance quietly accumulates, leading to eventual replacement under urgent circumstances—sometimes at the worst possible time.

The financial results are predictable: increasing lifecycle costs, decreasing reliability, and a growing backlog that limits institutional flexibility. The smart campus introduces a materially different economic logic- **circular economic logic**-- enabled by digital systems:

- Assets are designed for longevity, adaptability, and recoverability
- Performance is continuously measured, not assumed
- Maintenance is predictive, condition-based, instead of reactive
- Materials, components, and embodied carbon are treated as tracked assets
- End-of-life becomes a transition, not a disposal event

By instrumenting the physical environment, governing data across lifecycle stages, and using AI to anticipate change rather than react to failure, the institution begins to behave as a **steward of value over time rather than a consumer of capital**. Assets are no longer treated as disposable endpoints; they become evolving participants in a managed system whose purpose is continuity, resilience, and renewal.

In this model, sustainability generates returns not just through efficiency, but through **capital deferral, risk reduction, reuse, and reinvestment**. This is where circular economy principles cease to be abstract theory and become operational economics.

Material Passports as Institutional Memory

Material passports are often discussed in sustainability circles as documentation artifacts. On a smart campus, they function as something far more consequential: a form of **institutional memory embedded in physical assets**. It is an **economic record** that captures:

- Material composition and provenance
- Embodied carbon and environmental impact
- Toxicity and circularity potential
- Disassembly pathways and recovery value
- Service-life expectations and reuse scenarios

Therefore, a material passport details the composition of an asset, its material origins, the environmental and carbon impacts associated with it, its disassembly process, and its potential future recoverable value.

Integrating material intelligence into building information models, maintenance systems, capital planning processes, and the campus digital twin fundamentally transforms the character of the built environment. Buildings cease to be opaque containers of sunk costs and instead become material banks—repositories of known resources that can be evaluated, modeled, and intentionally selected for future use. When incorporated into BIM, CMMS, capital planning, and the campus digital twin, material passports turn materials from sunk costs into valuable future options.

This change affects both the economy and impact. Renovation becomes a strategic extraction of value rather than just costly demolition. Replacement choices are guided not only by current performance but also by the remaining usefulness of existing assets. As a result, capital planning shifts from reactive crises to ongoing strategies focused on maintaining value.

Lifecycle Sustainability as Financial Discipline

Lifecycle sustainability is often framed as an environmental ethic—a statement of values about stewardship and responsibility. Within *The Smart Campus Manifesto*, it is also positioned as something more pragmatic and, in many ways, more demanding: a discipline of institutional finance. On a modern campus, the largest drivers of cost, risk, and emissions are long-lived assets—central utility plants, research facilities, residence halls, data infrastructure, and distribution systems whose lifespans extend across generations of leadership and budget cycles.

These assets do not fail in neat, linear ways. Degradation is slow, invisible, and frequently tolerated—until it is not. When failure finally occurs, it concentrates cost, carbon, and operational risk into narrow windows that destabilize budgets, disrupt mission-critical activities, and force reactive decision-making. Emergency replacements, deferred capital piled on top of deferred maintenance, and crisis-driven procurement are not anomalies; they are symptoms of institutions that manage sustainability as an aspiration rather than as a financial system.

AI-enabled lifecycle modeling changes the nature of this problem. Instead of asking how to minimize annual operating expenses or meet next year's efficiency target, leaders can ask fundamentally different questions: Where does risk accumulate over time? Which components are approaching nonlinear failure thresholds? What intervention today reduces volatility ten, twenty, or thirty years out? These models surface tradeoffs that traditional budgeting obscures, allowing institutions to shift from short-term optimization to long-horizon stability.

This reframing has profound financial consequences. Efficiency gains are no longer pursued in isolation, stripped of their downstream impacts. An energy retrofit that increases system complexity but shortens asset life becomes visible as a net loss. A maintenance investment that modestly raises near-term operating cost but prevents stranded assets and emergency replacements becomes defensible—not just environmentally, but fiscally. Sustainability decisions move from being discretionary enhancements to being tools of risk management.

In this context, sustainability becomes a stabilizing force within the institution's financial architecture. Capital cycles are smooth rather than spiky. Emergency replacements decline as predictive insight replaces surprise. Deferred maintenance is no longer a hidden liability quietly compounding; it becomes a managed variable, explicitly modeled, prioritized, and governed. The institution gains the ability to pace renewal instead of being forced into it.

The Sustainability Flywheel

As data maturity increases, sustainability begins to behave less like a discrete program and more like a flywheel—self-reinforcing, compounding, and increasingly difficult to reverse. The shift is subtle at first. Improved sensing reveals early signs of inefficiency, drift, or degradation that once remained hidden. Predictive models translate those weak signals into foresight. Timely intervention preserves asset health, reduces waste, and avoids the downstream cost and disruption associated with late-stage failure. What changes most profoundly is not the technology, but the questions leaders can ask:

- What is the **total cost of ownership** over 30–50 years, not five?
- How does deferred maintenance amplify risk and carbon simultaneously?
- Where does replacing a component destroy embedded value?
- What assets should be extended versus retired?
- How does energy optimization interact with asset degradation?

Lifecycle-aware institutions stop optimizing for the next budget cycle and begin optimizing for decades. They examine the total cost of ownership over thirty or fifty years rather than five. They see deferred maintenance not as a neutral accounting artifact but as a force that amplifies both carbon exposure and operational risk. They recognize that replacing components too early can destroy embedded economic and environmental value just as surely as waiting too long. Decisions about extending versus retiring assets, or about aggressive energy optimization that accelerates system wear, become explicit tradeoffs rather than accidental outcomes.

This shift transforms sustainability from a primarily defensive posture—focused on minimizing harm—into a regenerative operating model. This flywheel transforms sustainability from a defensive posture ("reduce harm") into a **regenerative operating model**. Over time, institutions experience:

- Reduced deferred maintenance growth

- Measurable carbon reduction with verified data
- Improved resilience against price volatility and supply-chain shocks
- Lower exposure to stranded assets
- Stronger audit, ESG, and accreditation postures
- Increased trust from boards, regulators, and students

Each cycle of sensing, modeling, and intervention improves the institution's ability to act earlier and with greater precision. Over time, the effects compound. Deferred maintenance growth slows and then stabilizes. Carbon reduction becomes measurable with trusted, auditable data. Exposure to price volatility and supply-chain shocks diminishes as asset strategies become more deliberate. The risk of stranded assets declines as lifecycle intelligence informs procurement and capital planning. External confidence strengthens as audit readiness, ESG reporting, and accreditation narratives move from aspirational claims to demonstrable performance. Boards, regulators, faculty, and students observe not just progress, but institutional competence.

Crucially, the gains generated by this flywheel are not extracted; they are reinvested. Financial savings, avoided costs, and operational efficiencies are reinvested in deeper modernization, richer data environments, and more capable predictive systems. Each flywheel increases institutional capacity, making the next turn faster and more effective than the last.

At this stage, sustainability no longer needs to be justified on a project-by-project basis. It becomes self-reinforcing because it consistently produces value across domains that leadership already prioritizes: financial stability, operational reliability, risk reduction, and trust. What begins as energy optimization evolves into portfolio optimization. What begins as carbon accounting matures into capital intelligence.

From Metrics to Strategic Capital

Perhaps the most consequential shift enabled by the AI-enabled sustainable smart campus is the transformation of sustainability data from a retrospective reporting output into a strategic form of institutional capital. When carbon, energy, water, materials, and lifecycle performance are treated merely as metrics to be disclosed—annually summarized, externally benchmarked, or compliance-driven—they remain peripheral to decision-making. They inform narratives, not outcomes.

That posture changes fundamentally when sustainability data is governed as part of the institutional data estate. In this model, environmental and lifecycle signals are no longer isolated in sustainability reports or dashboards owned by a single office. They are normalized, trusted, and integrated alongside financial, operational, and risk data—available at the same moment decisions are made, not after consequences are absorbed.

This integration repositions sustainability from a communications function to a decision function.

Finance leaders gain confidence not simply because more data exists, but because uncertainty is reduced over long horizons. Capital planning becomes lifecycle-aware rather than project-bounded. Volatility from deferred maintenance, energy price swings, climate exposure, and asset failure can be modeled, stress-tested, and managed explicitly. Sustainability data becomes a tool for smoothing capital cycles and protecting institutional balance sheets, not just demonstrating intent.

Facilities organizations experience an equally profound shift. When environmental performance, asset health, and system efficiency are observable in near real time, the culture moves away from heroic firefighting toward predictive stewardship. Maintenance decisions are informed by impact and risk, not just urgency. Asset extension, replacement, and retirement become strategic acts tied to institutional resilience rather than reactive responses to failure. The facility's mission evolves from keeping systems running to preserving long-term value.

Technology leaders, in turn, transition from delivering dashboards to enabling intelligence. Their role is no longer to visualize sustainability metrics, but to ensure those metrics are interoperable, governed, secure, and actionable across systems. Digital twins, analytics platforms, and AI models become instruments of institutional learning rather than isolated innovation projects. Sustainability insight becomes embedded in workflows rather than presented as an overlay.

Most importantly, governing boards see a different institution emerge. Sustainability is no longer framed solely as an aspiration, a brand, or a moral obligation. It is understood as a disciplined strategy for risk reduction, value preservation, and long-term adaptability. When environmental and lifecycle data inform capital allocation, infrastructure strategy, and operational priorities, sustainability becomes legible in the language of governance: risk, return, resilience, and trust.

Sustainability as Institutional Learning

As an institution matures, sustainability transitions from a goal to a developing capacity—a skill to be learned across generations. The campus transforms into a remembering system that preserves knowledge of its history: what was constructed, using which materials and assumptions, how those systems functioned over time, where value was maintained through careful stewardship, and where vulnerability built up due to short-term choices or delayed maintenance.

This memory is not anecdotal or tribal. It is encoded deliberately—in asset records, lifecycle models, digital twins, performance data, governance decisions, and operating practices. Together, these elements form a durable institutional narrative that persists beyond individual projects, staff turnover, or leadership transitions. Each capital cycle does not start from zero. Each crisis does not require reinvention. The institution evolves because it remembers.

This is the deeper economic promise of sustainability when treated as an ecological operating system rather than a portfolio of initiatives. The smart campus does not simply consume fewer resources or emit less carbon in a given year. It learns how resources behave over time. It learns which interventions compound value and which quietly destroy it. It learns that early signals of stress—energy inefficiency, asset drift, indoor environmental degradation, and supply chain fragility—foreshadow more severe financial and operational consequences if left unaddressed.

Learning, in this sense, is cumulative. It compounds. Decisions made today are informed by the observed consequences of decisions made decades earlier. Failures become data rather than reputational events. Successes become patterns that can be replicated, scaled, and governed. Sustainability maturity is expressed not in perfection, but in reduced surprise and increased foresight.

This reframes sustainability as an institutional competence rather than a moral stance. The value is not only environmental, though that value is real and measurable. It is also financial, operational, and cultural. Institutions that learn systematically from their physical, digital, and environmental systems spend less time reacting and more time anticipating. They experience fewer shocks, smoother capital cycles, and greater confidence in long-horizon planning. They preserve optionality in a world defined by volatility.

Sustainability as a Smart Campus Design Principle

In higher education, sustainability's material impact is most visible in the campus building portfolio. Buildings are no longer passive containers for academic activity; they are active socio-economic infrastructure. Intelligent buildings—modeled through BIM, governed through digital twins, and operated via integrated building management systems—allow institutions to optimize energy and maintenance spending, extend asset lifecycles, and improve the reliability of learning and research environments.

As sustainability moves from aspiration to execution, its center of gravity shifts from policy statements to the systems that shape daily institutional decisions. On the smart campus, this shift is most evident in how universities procure, govern, and operate their physical assets—and in how technology reshapes who can participate effectively in that transformation. Supply chains, data standards, analytics platforms, and governance models become the mechanisms through which sustainability is scaled, enforced, and normalized. Together, these forces redefine sustainability not as a collection of initiatives, but as a durable capability embedded in how the institution designs value, manages risk, and allocates resources over time.

Socioeconomic Sustainability

Socioeconomic sustainability on the smart campus is realized through disciplined stewardship rather than reactive intervention. Deferred maintenance is reduced through predictive insight, not emergency response. Capital planning becomes evidence-based rather than episodic. Infrastructure investment is guided by measurable performance and risk reduction, strengthening academic continuity, research integrity, and institutional credibility amid sustained fiscal, environmental, and regulatory pressure.

Beyond assets and balance sheets, socioeconomic sustainability fundamentally encompasses the well-being, productivity, and inclusion of the campus community. As standalone DEI and wellness programs encounter scale and durability limits, responsibility increasingly shifts to the built environment itself. Standards such as WELL, Fitwel, and accessibility codes embed equity structurally into campus design—through indoor air quality, lighting, thermal comfort, safety, and universal access—making inclusion a function of space, systems, and operations rather than an optional programming add-on.

Smart campus architectures reinforce these outcomes through occupancy insight, utilization analytics, and transparent governance spanning facilities and IT domains. In this model, inclusion is no longer a voluntary add-on or aspirational policy statement. It becomes a designed condition of the campus experience—sustained by data, standards, and accountable operations, and reinforced through continuous measurement and adaptation.

Supply Chains, Procurement, and Asset Governance

Just as private markets now enforce sustainability through supply chains, universities exercise profound socio-economic influence through procurement, capital delivery, and asset governance. Every capital project, service contract, and vendor relationship encodes long-term environmental, financial, and social consequences. In this context, procurement is no longer a back-office function—it is a strategic control surface for sustainability, risk, and institutional credibility.

Smart campuses operationalize this influence by treating assets not as one-time purchases, but as lifecycle commitments. Digital asset management systems aligned with BIM, COBie, and ISO data standards enable institutions to move sustainability "left" into design, sourcing, and construction—where the majority of environmental and financial impact is determined. Embodied carbon, material provenance, maintainability, and end-of-life recovery can be evaluated before contracts are executed, rather than measured after consequences are locked in.

Vendor relationships also evolve under this model. Standards-based asset data and interoperable platforms allow institutions to enforce accountability across decades-long lifecycles—ensuring that performance claims, maintenance obligations, cybersecurity requirements, and sustainability commitments persist well beyond commissioning. Procurement shifts from lowest-bid optimization to value-based stewardship, where vendors are selected and governed based on transparency, long-term performance, and alignment with institutional standards.

This governance transformation reduces structural risk. Deferred maintenance is no longer an unknowable future liability but a measurable outcome of procurement and design decisions. Lifecycle cost models replace reactive budgeting. Asset data continuity—from design through operations—enables predictable capital planning, stronger auditability, and resilience against supply chain volatility, regulatory change, and workforce turnover.

At scale, smart campus infrastructure converts procurement into a sustainability force multiplier. Local purchasing decisions aggregate into measurable reductions in carbon exposure, energy intensity, and operational waste. Regional suppliers are incentivized to modernize practices to meet data, reporting, and performance expectations. Universities become anchors of sustainable economic behavior—not by policy mandate alone, but through disciplined, data-driven execution embedded in how assets are specified, delivered, and governed.

In this architecture, asset governance is a strategy. Procurement is instrumentation. And the campus becomes a durable socio-economic platform—one capable of aligning environmental responsibility, financial stewardship, and mission continuity through every decision that shapes the built environment.

Technology as an Equalizer

Advanced analytics, digital twins, and AI-enabled reporting have fundamentally lowered the barrier to sustainability maturity across higher education. Capabilities that once required large capital programs, specialized teams, and bespoke integrations are now accessible through shared platforms, interoperable standards, and scalable cloud-edge architectures. As a result, institutions with constrained financial, technical, or staffing resources can achieve levels of visibility, compliance, and operational insight previously available only to the most well-funded campuses.

This shift rebalances the sustainability landscape. Measurement no longer depends on manual audits or episodic studies; energy, carbon, water, space utilization, and asset performance can be observed continuously through governed data pipelines. Digital twins translate this data into decision-grade context, while AI-enabled reporting automates normalization, validation, and disclosure. What once required specialized sustainability offices and external consultants can now be embedded directly into daily operations.

Critically, this democratization is not about doing more with less—it is about doing *equally well* with fewer structural advantages. Standards-based platforms ensure that a regional public university can report against the same benchmarks, comply with the same frameworks, and optimize against the same objectives as a flagship research institution. Sustainability performance becomes comparable, auditable, and defensible regardless of institutional size or prestige.

Technology also compensates for workforce constraints. As sustainability expectations expand—often faster than staffing levels—AI-driven analytics and automation absorb the administrative burden of data collection, reconciliation, and reporting. Staff are freed to focus on judgment, prioritization, and coordination rather than spreadsheets and compliance mechanics. In this way, technology acts as a workforce amplifier, not a replacement—raising institutional capacity without escalating overhead.

At a sector level, this equalization accelerates collective progress. Shared data standards, interoperable architectures, and repeatable patterns reduce fragmentation and prevent sustainability from becoming a zero-sum competition among institutions. As more campuses reach baseline maturity, peer learning increases, benchmarks stabilize, and innovation shifts from rhetoric to execution.

Ultimately, technology transforms sustainability from a resource-dependent aspiration into an achievable institutional capability. By lowering barriers to insight, governance, and accountability, smart campus platforms enable sustainability to scale *horizontally* across higher education—not just vertically among elite institutions. The result is a more equitable, resilient, and collectively responsible academic landscape, aligned around shared standards and shared outcomes.

Sustainability is a Design Principle

Taken together, these shifts point to a clear and unavoidable conclusion: on the smart campus, sustainability is no longer a policy objective or a specialized program—it is a design principle. Like safety, reliability, or accessibility, sustainability must be engineered into the campus from the outset, expressed through architecture rather than aspiration. When sustainability is designed, it becomes systemic rather than episodic, durable rather than performative.

As a design principle, sustainability is manifested first through **interoperable systems**. Energy, facilities, space, finance, and digital platforms are no longer optimized independently; they are structured to work as a coordinated whole. Interoperability ensures that decisions made in one domain—such as procurement or scheduling—are visible and accountable across others. This integration replaces siloed efficiency gains with system-level optimization, where trade-offs are explicit, and outcomes are measurable.

Second, sustainability is embedded in standards-based asset models that persist throughout the full lifecycle of the built environment. Buildings are specified, delivered, operated, and renewed using consistent data structures that carry intent forward over decades. Embodied carbon, energy performance, maintainability, accessibility, cybersecurity posture, and end-of-life considerations are not retrofitted concerns; they are encoded at design and preserved through governance. Sustainability thus becomes repeatable and auditable, rather than dependent on individual champions or transient funding.

Third, sustainability as design depends on **governance frameworks that align people, place, and capital**. Decision rights, accountability, and prioritization are made explicit—ensuring that sustainability outcomes survive leadership transitions, budget cycles, and vendor changes. Governance transforms sustainability from an advocacy function into an operational discipline, one that balances near-term constraints with long-term institutional risk and value creation.

When sustainability is designed intentionally, it scales without friction. New buildings inherit the same principles as legacy assets. Smaller institutions leverage the same frameworks as larger ones. Compliance, reporting, and optimization become byproducts of how the campus operates; not additional work layered on top. This design-centric approach eliminates the constant reinvention that so often undermines sustainability programs.

Most importantly, designing for sustainability produces campuses that perform better under stress. Energy volatility, climate events, workforce shortages, and financial pressure do not derail progress—they reinforce it. The campus becomes more predictable, more resilient, and more trustworthy precisely because sustainability is intrinsic to its operating model.

In this sense, sustainability on the smart campus is not an endpoint or a metric. It is an organizing logic—a way of structuring systems, data, and decisions to improve institutional performance over time. When treated as a design principle, sustainability ceases to be fragile or optional. It becomes foundational, enduring, and inseparable from the mission of teaching, learning, research, and stewardship.

Governance, Measurement, and Trust

If circular economics defines how value is preserved and compounded, governance defines how that value is protected over time. Without clear decision rights, measurement discipline, and transparent accountability, even the most technologically advanced sustainability strategies degrade into disconnected projects and short-lived gains. The ecological operating system, therefore, requires governance not as a bureaucracy but as an enabling architecture.

In the smart campus, sustainability succeeds only when it is *governable, auditable, and repeatable.* This section describes how institutions operationalize that discipline—turning sustainability from intent into durable institutional capability.

Governance as the Enabling Architecture

The Manifesto emphasizes that governance is not just an overlay after transformation; it is the fundamental structure that enables it. This is particularly clear in sustainability, where energy systems, building controls, material sourcing, procurement, data sharing, cybersecurity, and capital planning all intersect, often across different organizational boundaries. Without shared authority and common constraints, optimizing one area can cause vulnerabilities in another. Effective sustainability governance ensures:

- **Clear decision rights** across IT, Facilities, Sustainability, Finance, Procurement, and Risk
- **Non-negotiable standards** for interoperability, data quality, cybersecurity, and lifecycle accountability
- **Explicit exceptions and waivers**, time-bound and auditable, rather than silent deviations

Moreover, without governance, sustainability becomes accidental. Facilities optimize energy efficiency without visibility into cybersecurity risk. Procurement drives down upfront cost without accounting for lifecycle performance or material recovery. IT enables analytics without shared constraints on data quality, ownership, or ethical use. Finance evaluates projects in isolation, disconnected from carbon exposure or deferred maintenance risk. Each decision may be locally rational—and systemically damaging.

Effective sustainability governance exists to prevent this pattern. It establishes shared authority and enforced constraints across Technology, Facilities, Sustainability, Finance, Procurement, and Risk—not to slow progress, but to make it durable. Clear decision rights ensure that sustainability choices are evaluated holistically rather than sequentially. Technology does not enable connectivity without the assurance of safety and the facilities ' uptime requirements. Procurement does not select systems that cannot be integrated, maintained, or audited over the long term. Sustainability goals are not aspirational overlays but embedded criteria within capital, operational, and architectural reviews.

Optimization without governance is how efficiency turns into fragility.

At its core, governance defines what is non-negotiable. Interoperability standards prevent vendor lock-in that would trap future leaders. Data quality and lineage requirements ensure that sustainability metrics are trusted enough to drive financial and operational decisions. Cybersecurity baselines protect connected energy and building systems from becoming new attack surfaces. Lifecycle accountability forces visibility into embodied carbon, maintainability, repairability, and end-of-life pathways—before contracts are signed, not after assets fail.

Equally important is how governance handles exceptions. In mature institutions, deviations are explicit, time-bound, justified, and reviewable. They are not silent accommodations made under pressure. This discipline preserves institutional memory, prevents normalization of fragility, and ensures that short-term expedience does not quietly erase long-term resilience.

Many institutions operationalize this model through a Smart Campus or Sustainability Steering Committee, co-led by Technology and Facilities leadership and anchored by formal authority rather than advisory influence. When properly structured, this body does not exist to approve projects—it exists to protect the system. Its role is to ensure that sustainability investments are evaluated simultaneously for environmental impact, operational resilience, safety, financial performance, cybersecurity posture, and long-term adaptability. This integrated review does not dilute sustainability—it strengthens it.

When governance functions as an enabling architecture, sustainability scales safely. The institution avoids innovation theater—isolated pilots that cannot be sustained. It avoids brittle integrations that collapse under stress. It avoids the slow erosion of trust that occurs when systems behave unpredictably or when metrics are contested. Governance becomes the mechanism by which sustainability moves from intention to institution—preserved across leadership changes, funding cycles, and technological evolution.

From Reporting to Continuous Measurement

Traditional sustainability programs rely heavily on retrospective reporting. Annual energy summaries, carbon inventories, benchmarking reports, and certification cycles have become the lingua franca of environmental accountability in higher education. These instruments remain necessary—they establish baselines, support regulatory compliance, and communicate intent—but they are fundamentally backward-looking. They describe what happened long after decisions were made and consequences were incurred.

An ecological operating system requires a different temporal posture. On a smart campus, sustainability measurement shifts from reporting to real-time observability. Digital twins, integrated data estates, and analytics platforms allow institutions to measure sustainability as it happens:

- Energy, water, and carbon performance adjusted for occupancy and conditions
- Asset health and lifecycle extension, expressed in avoided cost and avoided carbon
- Indoor environmental quality is tied directly to wellness, productivity, and learning outcomes

Instead of asking whether targets were met last year, leaders can see how the campus is performing right now—and why.

In this environment, energy, water, and carbon performance are measured in context rather than in isolation. Consumption is interpreted in light of occupancy patterns, weather conditions, operational schedules, and system constraints. Asset health and lifecycle extension are no longer inferred through age-based assumptions but modeled through live telemetry—translated into avoided emergency replacements, deferred capital expenditure, and avoided embodied carbon. Indoor environmental quality becomes legible as an operational variable, directly connected to wellness, cognitive performance, teaching effectiveness, and research reliability rather than treated as an abstract comfort metric.

This measurement is operational, not symbolic. Variance is no longer an embarrassment hidden in aggregate reports; it becomes an early signal. Stress is no longer treated solely as a disruption to be absorbed; it becomes data. When systems drift, degrade, or operate outside expected bounds, the campus does not simply recover—it learns. That learning feeds preparation, redesign, and continuous improvement rather than postmortem explanation.

Equally important is where this insight lives. Sustainability data in a smart campus is not sequestered in a single office or dashboard. It is shared across domains and translated into the languages leaders already use. Facilities teams see predictive maintenance opportunities and reduced failure risk. Finance sees capital deferral, reduced volatility, and improved confidence in forecasting. Academic leadership sees learning environments that are healthier, more stable, and more reliable. Sustainability becomes legible because it is expressed in outcomes that those leaders are already accountable for managing.

Antifragility and Continuous Improvement

The ecological operating system introduced earlier in this chapter is designed to do more than absorb disruption. Resilience, while necessary, is insufficient. A resilient system returns to baseline after stress. A sustainable system at maturity must go further—it must become antifragile, improving because of stress rather than merely surviving it.

In an AI-enabled smart campus, antifragility is not a metaphor. It is observable and measurable over time:

- Systems predict failure earlier over time
- Energy optimization improves during volatility rather than degrades
- Maintenance effort shifts steadily from reactive to predictive
- Capital is redirected from emergency response to strategic renewal

Systems begin to anticipate failure earlier and with greater precision, shrinking the gap between signal and action. Energy optimization does not degrade under volatile conditions such as heat waves, grid instability, or fuel price shocks; instead, it improves as models learn from real-world stress. Maintenance organizations gradually escape the cycle of reactive response, shifting effort away from emergency repair and toward predictive and condition-based intervention. Capital dollars follow this shift, moving steadily from crisis replacement toward deliberate renewal and strategic reinvestment.

> ***"Sustainable campuses do not recover from stress—they improve because of it."***

What distinguishes antifragility from incremental improvement is not any single metric, but the direction of change. Governance bodies, therefore, track these signals as directional indicators rather than static targets. The question is not whether performance fluctuates—volatility is expected in complex systems—but whether volatility produces learning. A sustainable campus is one whose post-disruption state is measurably stronger than its pre-disruption state.

This perspective alters the practice of sustainability leadership. Success is no longer framed as compliance with a predefined outcome or the achievement of a fixed target. Instead, leaders ask what was learned, and whether the institution emerged from stress with clearer insight, better models, and stronger operating discipline. Did the system adapt? Did it become more predictable, more efficient, and more trustworthy under pressure?

Transparency and Institutional Trust

As campuses deploy increasingly granular sensing—tracking occupancy, energy use, air quality, mobility patterns, and material flows—**trust becomes inseparable from the success of sustainability efforts**. This includes:

- Clear communication about what data is collected and why
- Purpose-limited use of data tied explicitly to safety, sustainability, and performance
- Strong privacy safeguards and ethical review for AI-assisted optimization
- Public-facing dashboards that build confidence through visibility, not surveillance

The same technologies that enable unprecedented operational insight can, if poorly governed, erode confidence and provoke resistance. In this context, transparency is not a public-relations strategy; it is a system design requirement.

An ecological operating system must be transparent by design. This begins with clear, plain-language communication about what data is collected, how it is used, and—critically—what it is not used for. When sensing expands quietly or explanations lag deployment, institutions create ambiguity. Ambiguity invites suspicion. Even well-intentioned sustainability programs can stall when community members are left to infer purpose rather than understand it explicitly.

Equally important is **purpose limitation**. Data collected for sustainability must be rigorously tied to safety, environmental performance, operational reliability, and institutional stewardship. When datasets begin to drift into secondary or undefined uses, the institution weakens its own legitimacy. Purpose limitation is not a technical constraint; it is an ethical boundary that must be reinforced through governance, policy, and auditability. The question is not simply whether the institution can use the data, but whether it should—and whether that use aligns with clearly articulated institutional values.

As AI-assisted optimization becomes more prevalent, ethical review moves from an abstract concern to a daily operational necessity. Models that recommend energy curtailment, space reallocation, or automated control actions can subtly yet consequential influence comfort, access, and experience. Ethical governance ensures these systems remain explainable, contestable, and aligned with human judgment. Human-in-the-loop design is not a conservative compromise; it is a trust-preserving discipline that acknowledges both the power and the limits of automation.

Transparency also extends outward. Public-facing dashboards and shared performance indicators shift sustainability from a closed technical exercise to a visible institutional practice. When designed thoughtfully, these dashboards do not surveil individuals; they illuminate systems. They show how buildings respond to weather, how energy demand shifts with occupancy, and how investments reduce volatility over time. Visibility builds confidence not because it reveals everything, but because it reveals the right things—clearly, consistently, and without defensiveness.

The relationship between trust and usability is direct. When sustainability data is trusted, it becomes a shared asset. It informs decision-making, enables collaboration, and sustains momentum across organizational boundaries. When it is opaque, poorly governed, or ethically ambiguous, resistance grows. Progress stalls not through protest, but through disengagement.

This is why governance sits at the center of trust. Transparency, ethics, and accountability cannot be retrofitted after deployment; they must be encoded into architecture, policy, and decision rights from the beginning. In mature institutions, trust is not managed—it is produced continuously by systems that are legible, bounded, and worthy of confidence.

Sustainability as a Managed Institutional Capability

Sustainability is no longer understood as a program, an office, or a portfolio of aspirational goals. In a mature smart campus, it becomes a **managed institutional capability**—embedded into governance, architecture, economics, and culture in the same way that safety, finance, and academic quality are embedded. In an AI-enabled sustainable smart campus:

- Sustainability decisions are evidence-based, not anecdotal
- Tradeoffs are explicit and modeled, not hidden or deferred
- Value compounds over time—financially, environmentally, and socially
- Trust is maintained even as systems become more autonomous

Sustainability decisions are grounded in evidence rather than anecdote. Leaders no longer rely on intuition, isolated studies, or optimistic projections when making tradeoffs. Instead, decisions are informed by operational data, lifecycle models, and observable system behavior. Sustainability becomes something the institution can reason about, test, and revisit—not something it must periodically re-argue from first principles.

This shift makes tradeoffs unavoidable and, therefore, transparent. When sustainability functions as a capability rather than an add-on, it exposes the real costs and consequences of choices that were previously deferred or obscured. Short-term savings are weighed against long-term volatility. Capital efficiency is evaluated alongside carbon impact and resilience. What changes is not the presence of tradeoffs—those always existed—but whether the institution is willing and able to name them explicitly.

Over time, this clarity allows value to compound rather than dissipate. Financial stability improves as emergency replacements decline and capital planning becomes more predictable. Environmental performance improves as systems are tuned continuously rather than corrected episodically. Social value grows as students, staff, and faculty experience a healthier, more reliable, and more trustworthy campus. These gains are not linear; they reinforce one another across operational cycles.

Crucially, trust is maintained even as systems become more autonomous. As AI-assisted optimization, sensing, and closed-loop controls become commonplace, the legitimacy of sustainability efforts depends less on persuasion and more on governance. A managed capability operates within defined decision rights, ethical boundaries, and transparent accountability. The community does not need to trust individuals or vendors; it trusts the system because the system is legible, constrained, and reviewable.

This is what allows sustainability to endure leadership transitions, budget cycles, and technological change. Programs depend on champions. Mandates depend on authority. Capabilities persist because they are woven into how decisions are made, how assets are managed, and how performance is evaluated. The ecological operating system does not fade when priorities shift; it continues to produce a signal because it is part of the institution's operating fabric.

At this stage of maturity, the smart campus is no longer defined primarily by its greenness or efficiency. It is defined by durability. It is governed with intention rather than exception. It is measured with clarity rather than rhetoric. It is trusted by the community it serves because its behavior is consistent over time.

Chapter Summary

Sustainability is no longer a values conversation; it is a survival strategy.

Chapter 6 reframes sustainability from an aspirational agenda to a governing discipline for institutional survival. It argues that the AI-enabled sustainable smart campus is not sustained by targets, programs, or isolated initiatives, but by the intentional design of the campus as an ecological operating system—one that integrates data, governance, economics, and learning over the long term.

The chapter presents six closely linked claims. Firstly, sustainability is not just a program; it functions as an operating system. Viewing it as an overlay causes it to compete with main operations, budgets, and priorities. In contrast, considering it an operating system influences decision-making processes by default. Energy, carbon, capital, risk, and experience are no longer optimized in isolation but managed as a unified, interconnected system. This system's performance can be continuously monitored, modeled, and enhanced.

Second, sustainability is emerging as a discipline of lifecycle finance, not just an environmental concern. Campus assets with long lifespans pose financial, operational, and carbon risks when they unexpectedly fail. Using AI-enabled lifecycle modeling, institutions can shift their focus from reducing short-term costs to minimizing long-term volatility. Deferred maintenance becomes a controllable element instead of an ongoing liability. Sustainability helps stabilize capital flows, lower emergency expenses, and protect embedded value—making the institution more predictable, manageable, and resilient.

Third, the chapter defines sustainability as an executive responsibility, upheld through governance rather than assigned to a single department. Fragmented accountability—where energy, data, capital, and risk are managed separately—leads to systemic vulnerability. An ecological operating system requires shared authority, clear decision-making rights, and strict standards across departments such as Technology, Facilities, Finance, Procurement,

Sustainability, and Risk. Sustainability is effective not through unanimous agreement but because decision-making interfaces reveal and make unavoidable the necessary trade-offs.

Fourth, Chapter 6 redefines how we measure sustainability, shifting from simple reporting to ongoing learning and antifragility. Retrospective metrics fall short in a volatile environment. With real-time observability—using digital twins, integrated data estates, and analytics—variance becomes a meaningful signal, and stress turns into valuable data. Organizations now focus not just on whether targets are achieved but on whether their systems learn, adapt, and improve. Sustainability maturity is demonstrated by continuous, directional progress following disruptions rather than static performance.

Fifth, sustainability evolves into institutional memory and learning. When mature, the campus retains knowledge of what was constructed, its performance, decisions that maintained value, and those that introduced fragility. This memory—stored in data, models, governance, and practices—continues through leadership changes and funding cycles. The institution avoids repeating the same errors because the outcomes are remembered, not rediscovered. In this way, sustainability grows as a body of knowledge, not merely as a savings account.

Finally, the chapter explicitly states a key point the Manifesto has emphasized: sustainability is an essential strategy for institutional survival. Universities today face challenges like climate volatility, limited capital, energy issues, enrollment pressures, workforce constraints, and rising cyber-physical risks. Those who see sustainability as just an added feature struggle to cope with these pressures. Conversely, institutions that integrate sustainability into their core operations adapt more effectively. In this setting, sustainability becomes crucial to maintaining mission continuity, ensuring research reliability, supporting human well-being, strengthening financial resilience, and preserving public trust..

A sustainable smart campus, ultimately, is not defined by whether it meets this year's carbon target, energy benchmark, or reporting requirement. It is defined by whether it remains **viable, trustworthy, and governable over decades of change.** Targets are milestones; viability is the journey. An institution that treats sustainability as a checklist will repeatedly fall behind as standards shift, markets become volatile, and risks compound. An institution that views sustainability as an integral operating principle—woven into governance, architecture, data, and decision-making—develops the capacity to adapt seamlessly and achieves antifragility. Within this context, the sustainable smart campus is not designed for a single goal but for resilience: the ability to learn, adjust, and stay mission-focused amidst ongoing volatility. Therefore, sustainability is not the final goal of smart campus transformation; instead, it is the foundation that fosters momentum and enables the institution to flourish.

Chapter 7
The OT Architecture

Facilities leaders design and leverage interoperable systems, lifecycle sustainability, and workforce modernization to realize the smart campus vision as strategic, operational, and durable.

To implement the AI-enabled sustainable smart campus strategy, the CFaO needs to turn the overarching vision into practical infrastructure. While the CTO develops the digital enablement layer, the CFaO is responsible for designing the physical environment that can receive, produce, and respond to digital intelligence. The CFaO's role has shifted from merely pouring concrete and laying pipe to designing a cyber-physical platform. To succeed, they must concentrate on four key Architectural Aims and uphold five essential Execution Points.

The CFaO's Architectural Aims (The "What")

The architectural aims specify the desired final configuration of the physical systems. They signify a transition from closed, proprietary hardware to open, integrated ecosystems.

The Interoperable Physical Plant (Silo Eradication)

Historically, facilities infrastructure consisted of isolated systems, each with its own proprietary language. HVAC, lighting controls, and fire alarms often need separate vendor contracts to generate reports. The CFaO's main goal is seamless interoperability, which requires the physical plant to rely on open protocols like BACnet/IP or MQTT. This ensures that devices such as smart thermostats, occupancy sensors, and central chillers can communicate directly without relying on costly, fragile middleware.

The Cyber-Physical Data Estate (The OT Semantic Layer)

Creating data is simple; making it useful is more challenging. The CFaO needs to design a consistent physical data infrastructure. This entails implementing a strict "semantic layer" or ontology, like Project Haystack or Brick Schema, across all facilities. When the architecture guarantees that "Building 4, Room 101" is labeled the same way in the Building Management System (BMS), space scheduling tools, and energy dashboards, the campus can then leverage AI to identify cross-domain patterns and areas for optimization.

The Grid-Interactive Energy Architecture

The architecture should move away from viewing the campus as just a passive energy user. Instead, the CFaO intends to develop a proactive, bidirectional energy system. This involves integrating on-site energy sources, such as solar panels, thermal, and battery storage, along with advanced load-shedding algorithms. The result is a physical plant that constantly interacts with the external utility grid to reduce both carbon emissions and costs.

The Context-Aware Human Interface

The physical design should respond to human presence rather than adhere to strict schedules. This involves integrating spatial intelligence—such as IoT occupancy sensors, environmental quality monitors, and Bluetooth wayfinding beacons—into the built environment. The architecture needs to support spaces that automatically adjust lighting, airflow, and temperature based on current usage, ensuring a seamless, health-enhancing experience for students and faculty.

The Required Execution Points

To meet these architectural goals, the CFaO needs to implement strict execution steps that fundamentally alter how the facilities organization handles procurement, security, and operations.

1. Mandating API-First, Digital-Twin-Ready Procurement

The CFaO needs to revise the capital procurement guidelines. No physical infrastructure—like large air handlers or basic smart locks—must be purchased unless it meets strict API connectivity standards. Vendors must also provide a digital, semantic model of their equipment during installation. The CFaO ensures this by making the "Digital Handover" a legal component of the physical construction contract.

2. Enforcing IT/OT Micro-Segmentation (The Security Alliance)

The CFaO cannot execute this architecture in a vacuum; they must forge an ironclad alliance with the CTO/CTO. The required execution point here is joint governance over the cyber-physical perimeter. The CFaO must actively participate in architecting network micro-segmentation, ensuring that vulnerable operational technology is placed in isolated "zones and conduits" far removed from public student Wi-Fi and enterprise IT networks, and using Zero-Trust access controls for all third-party maintenance vendors.

3. Deploying the Facilities "Control Tower."

The CFaO should relocate the operational team from the mechanical rooms to the analytical command center. This step involves merging all physical data streams into a single Digital Twin or smart campus dashboard. This "Control Tower" serves as the central hub where AI fault-detection algorithms generate predictive maintenance alerts, enabling the CFaO to assign labor effectively before a major failure occurs.

4. Rewriting the Capital Planning Calculus

Execution involves shifting the approach to campus growth calculations. The CFaO must incorporate Life-Cycle Cost Analysis (LCCA) and embodied carbon tracking into every capital project. Rather than simply opting for the HVAC system with the lowest initial cost, the decision should prioritize the system that offers the lowest Total Cost of Ownership over 30 years, considering predictive maintenance, energy efficiency, and carbon pricing.

5. Upskilling the "Wrench to Keyboard" Workforce

The key to successful execution lies in human infrastructure. The physical systems will fail if the staff cannot operate them. The CFaO needs to implement a thorough change management and training plan. Master electricians and HVAC technicians must be trained to understand IP addresses on networks, basic data analytics, and mobile work-order systems. The CFaO should foster a culture that views both the wrench and the keyboard as equally essential tools for maintaining the cognitive campus.

How Facilities Success is Measured on a Smart Campus

The core principle of facilities management stays the same: success is experienced before it is officially recognized. People recognize when spaces are consistently comfortable, the air is fresh, classrooms are prepared immediately, and there are no disruptions. Yet, in the age of AI-powered sustainable smart campuses, the approach to achieving these human-centered results—and how the CFaO measures them—has undergone a significant transformation.

Historically, the facilities scorecard was a lagging indicator. It measured what had already happened: how much energy was consumed last month, how many work orders were closed, and how many times a critical asset broke down.

Today, the integration of Operational Technology (OT), IoT telemetry, and Artificial Intelligence requires CFaOs to move from a reactive scorecard to a predictive and dynamic dashboard. These metrics no longer measure past performance; they also quantify risk avoidance through algorithms and actively contribute to the creation of institutional value.

Here is how the traditional measures of facilities success should be reevaluated and enhanced for the cognitive campus:

From "Comfort Complaints" to the "Frictionless Human Experience."

- **The Traditional Measure**
  Tracking the volume of hot/cold comfort complaints, time-to-resolution, and basic, static CO_2 monitoring.
- **The Smart Campus Evolution**
  The new metric of success is the *absence* of the complaint. Success is measured by the system's ability to adjust the environment before a human ever feels uncomfortable autonomously.
- **Key Metrics**
  Predictive Comfort Yield (the percentage of spaces dynamically conditioned based on predictive weather and real-time Wi-Fi occupancy data rather than static schedules).
  Multi-Factor IAQ Health Index (moving beyond basic CO_2 to track volatile organic compounds [VOCs], particulate matter, and humidity in real-time, correlating these factors to optimal cognitive performance environments).

From "Operational Responsiveness" to "Predictive Avoidance."

- **The Traditional Measure**
  Work order cycle times, the balance of preventive versus corrective maintenance, and the trajectory of the deferred maintenance backlog.
- **The Smart Campus Evolution**
  Fixing something quickly after it breaks is no longer the gold standard; predicting its failure is. The CFaO measures the system's intelligence and the optimization of human labor.
- **Key Metrics**
  Algorithmic Failure Avoidance (quantifying the number of catastrophic failures prevented by AI-driven acoustic and vibration fault detection).
  Mean Time to Predict (MTTP) (replacing Mean Time to Repair, tracking how far in advance the digital twin flagged an anomaly).
  Dispatch Optimization Rate (the percentage of work orders where technicians arrived with the exact diagnostic data and parts required, eliminating exploratory "truck rolls").

From "Energy Usage" to "Algorithmic Carbon Management."

- **The Traditional Measure**
  Energy Use Intensity (EUI), monthly utility bills, and static, annualized emissions reporting.
- **The Smart Campus Evolution**
  Sustainability is no longer a static monthly report; it is a real-time financial trading floor. The CFaO measures the campus's ability to act as a Grid-Interactive Virtual Power Plant.
- **Key Metrics**
  Automated Demand Response Yield (the financial value generated and megawatts shed algorithmically during peak grid pricing events without human intervention).
  Dynamic Carbon Intensity (tracking energy consumption against the real-time carbon density of the local utility grid, shifting loads to hours when power is greenest).
  Net-Zero Velocity (real-time tracking of Scope 1 and Scope 2 emissions reductions driven directly by AI optimization).

From "Asset Reliability" to "Cyber-Physical Resilience and Yield."

- **The Traditional Measure**
  Frequency of unplanned downtime and Mean Time Between Failures (MTBF) for critical assets.
- **The Smart Campus Evolution**
  Reliability must now encompass the digital security of the physical plant and the financial yield of asset longevity.
- **Key Metrics:**
  Digital Twin Accuracy (the continuous semantic alignment between the virtual models and the physical reality, ensuring AI decisions are based on perfect data).
  OT Zero-Trust Compliance (the percentage of physical building controllers and vendor access points governed by strict network micro-segmentation and identity management).
  Capital Debt Avoided (quantifying the millions of dollars saved by extending the life of capital assets through continuous commissioning, rather than age-based replacement).

By redefining success through these advanced metrics, the CFaO completely changes the narrative delivered to the Cabinet and the Board of Trustees. These measures do more than evaluate operational performance; they create a powerful, forward-looking language. The CFaO is no longer reporting on the cost of keeping the lights on; they are reporting on the algorithmic defense of the endowment, the absolute security of the research enterprise, and the strategic realization of the university's highest sustainability ambitions.

Chapter Summary

This chapter transforms the Chief Facilities Officer's strategic vision into a tangible architectural and operational plan for large-scale OT modernization. It highlights the CFaO's role as the architect of the physical-digital platform necessary for predictive maintenance, resilience, decarbonization, and a seamless human experience. The chapter emphasizes that the built environment can no longer be a set of standalone proprietary systems and isolated control layers; instead, it must function as an interoperable, data-rich cyber-physical ecosystem.

The chapter describes four architectural goals shaping the future campus: a connected physical infrastructure based on open standards; an integrated cyber-physical data platform with a common semantic layer; an energy system that interacts with the grid to support load management and sustainability; and a human interface that adapts to real-time occupancy, health metrics, and environmental factors. These goals transform the campus from a static setup into a dynamic, measurable, and intelligent environment.

To achieve these objectives, the chapter outlines five key execution imperatives: (1) integrating API-first and digital-twin readiness requirements into every procurement process; (2) collaborating with the CTO to enforce IT/OT micro-segmentation, securing the cyber-physical boundary; (3) creating a unified operational "control tower" that combines OT telemetry with AI insights; (4) redefining capital planning by emphasizing lifecycle cost analysis and embodied carbon as primary decision factors; and (5) enhancing workforce skills from solely mechanical tasks to a hybrid of digital and mechanical expertise. These points underscore that modernization is not merely about adopting new technology but about fundamentally re-engineering how facilities operate, are governed, and are managed.

Finally, the chapter redefines success for Facilities within a smart campus setting. The scorecard moves away from lagging indicators like energy consumption, work order counts, and reactive repairs, towards predictive, value-adding metrics such as failure prevention, real-time carbon optimization, demand-response performance, digital-twin precision, and avoided capital debt. This shift positions Facilities not merely as a cost center but as a strategic driver of measurable value for the institution.

Part IV

Planning, Organizing, and Leading

Chapter 8
Smart Campus Leadership

Effective smart campus leadership starts when silos are eliminated, establishing integrated CTO–CFaO perspectives, collaboration, and governance that foster operational clarity and trust, creating aligned priorities, language, culture, and decision-making across the institution.

An AI-enabled Sustainable Smart Campus is often seen as a technological project, but institutions usually do not fail due to technical problems. Instead, their setbacks tend to stem from difficulties in coordinating among various parts of higher education, such as managing separate budgets, differing professional cultures, decentralized decision-making processes, and different criteria for "success."

Smart campus initiatives broadly impact institutions across physical and digital infrastructure, as well as student affairs, academics, research, administrative operations, auxiliary enterprises, and long-term capital planning. Because no single leader oversees all these areas, effective leadership depends on a systemic approach rather than the personal qualities of individual leaders.

This chapter emphasizes that cross-silo alignment is crucial, not just a minor or 'soft" issue; it serves as the primary mechanism for responsible institutional modernization. Poor alignment leads to problems such as uncontrolled pilot growth, vendors acting as de facto architects, inconsistent security measures, and disputed data rather than trusted information. Conversely, strong alignment helps the institution promote innovation while ensuring safety, privacy, reliability, and adaptability over the long run.

This manifesto clearly defines the key elements to develop early: a mission-aligned vision backed by a top sponsor, an inclusive governance structure, and a stakeholder engagement and communication plan focused on ongoing refinement. This chapter translates these requirements into a practical leadership model.

When Worlds Collide

The longstanding separation between Information Technology and Facilities Operations essentially amounts to a language barrier. For many years, the CTO and the CFaO have worked in separate worlds, using different terminologies related to risk, architecture, timelines, and success. To effectively create an AI-driven, sustainable smart campus, this communication gap needs to be bridged.

The merging of IT and OT calls for thorough cross-domain learning. The CTO needs to grasp the physics behind the built environment, while the CFaO must understand digital governance within the network. When these leaders do not comprehend each other's core realities, modernization projects can get delayed, costs may escalate, and weaknesses could arise in the cyber-physical system. Closing this gap involves addressing several key areas of past misalignment.

Figure 8. Strategic Alignment Between the CTO and CFAO

Agile Obsolescence vs. Decadal Permanence

Before they can synchronize on technology, the CTO/CFaO needs to agree on frequency and timing. The biggest gap between IT and Facilities is the vastly different speeds at which they function.

The CTO operates in the realm of continuous integration and swift obsolescence. In IT, technology becomes outdated within months, and software updates are rolled out in days or hours. A server architecture is deemed legacy after three to five years, while AI models develop weekly. The IT planning process is naturally agile, quickly adjusting to new digital capabilities.

In contrast, the CFaO's timeline is almost geological, whereas IT operates on a much shorter scale. Facilities leaders plan based on 10-year campus master plans and utilize multi-year capital funding windows. When they pour concrete, install a central chiller plant, or lay underground steam loops, those systems are typically expected to function reliably for 30 to 50 years.

This temporal conflict causes significant tension. The CTO aims to patch or replace a vulnerable system quickly. Still, the CFaO cannot simply reboot a building or replace a 20-year-old HVAC controller without a lengthy capital approval process. To address this gap, the CTO needs to design digital overlays and API gateways to securely connect outdated mechanical hardware to modern AI models. At the same time, the CFaO must revise its long-term plans, recognizing that digital components that control the buildings will become obsolete much faster than mechanical parts. This requires ongoing operational funding, rather than occasional large capital investments.

Kinetic Uptime vs. Information Security

Beyond just clock speed, the core mission and risk models of the two organizations often conflict. The CTO assesses risk using the CIA triad: Confidentiality, Integrity, and Availability. In IT, when a system is compromised, the usual response is to quickly isolate it, take it offline, and apply patches.

The CFaO operates under constant obligations related to life safety, code compliance, and research continuity, making system downtime almost impossible. If the CTO attempts to disconnect a compromised Building Management System (BMS) without considering the physical environment, they could unintentionally cut off negative air pressure in a hazardous materials lab or turn off ventilation in a crowded lecture hall. Conversely, the CFaO needs to understand that cybersecurity, privacy policies, and audit standards are not bureaucratic obstacles created by IT to hinder progress; they are critical requirements. The facilities leader must learn how to implement zero-trust segmentation and endpoint hardening in OT environments without disrupting physical system uptime.

Bridging the Infrastructure Gap

A smart campus cannot be achieved without architects understanding each other's building materials. The CTO needs to be proficient in key OT systems (BMS, EMS, SCADA, lighting controls) and their often-outdated protocols, such as BACnet, Modbus, and LonWorks. Recognizing that deferred maintenance in the physical plant isn't just a budget concern, but a significant technology risk is essential. Aging equipment and piecemeal controller upgrades greatly increase the cyber-attack surface and complicate digital integration.

At the same time, the CFaO must acknowledge that the enterprise network is a tightly governed, shared resource. Facilities leaders need a good understanding of VLANs, firewalls, and quality of service (Quality of Service) to understand why connecting a smart meter to an arbitrary wall port or making minor network adjustments might disrupt essential building controls or leave the university vulnerable to ransomware attacks.

Data Estates and Digital Twins

The concept of the smart campus—featuring predictive maintenance, energy optimization, and digital twins—depends entirely on converting physical assets into structured digital data. The CTO needs to understand how controls modernization works, including point mapping, tagging strategies, and the physical commissioning process required to gather accurate data from equipment such as chillers or air handlers.

Meanwhile, the CFaO must grasp the structure of the enterprise data ecosystem, understanding data normalization, time-series storage, and API event streams. The CFaO should also adopt an operational analytics perspective, recognizing that raw fault-detection data becomes useful only when properly integrated into the CMMS (work order system) to enable effective dispatch, prioritization, and ongoing physical improvements.

Safety, Privacy, and Service Management

Ultimately, managing a converged campus necessitates integrating IT service management with the execution of physical work orders. The CTO must have a thorough understanding of the safety and regulatory constraints governing the physical infrastructure. While IT teams often embrace a "fail fast" mentality and beta testing, the CTO needs to recognize the strict life-safety boundaries that prohibit certain mechanical systems from being experimented on in a live environment.

In return, the CFaO needs to respect the privacy and compliance considerations related to the digital technologies they implement. When deploying large-scale IoT sensor networks or sophisticated video analytics to monitor space usage, the facilities leader must understand principles such as data minimization, data retention policies, and the governance needed to maintain trust within the campus community. They should adopt IT's disciplined approach to service management—using formal change control, release procedures, and Service Level Agreements (SLAs) to ensure that digital updates to building controllers do not cause unintentional physical outages.

Mutual understanding—the recognition of the natural tension between a three-month agile software sprint and a thirty-year capital infrastructure plan—alone does not provide the resources, structures, or agreement needed to create a cohesive cognitive campus. To turn this cross-sector awareness into institutional action, these leaders must fundamentally change their approach at the cabinet level. They cannot continue acting as independent authorities negotiating across boundaries. Instead, they must combine their power, budgets, and influence to manage the cyber-physical ecosystem as a unified asset. This shift requires moving from tactical cooperation to a shared executive mandate—an important transformation that involves adopting Leadership as Strategic Integration.

The Bilingual Service Catalog

The **Bilingual Service Catalog** is a foundational operating artifact for an AI-enabled, sustainable, smart campus. Its purpose is to bridge the long-standing language, priority, and accountability gaps between **Technology (IT)** and **Facilities (OT)** by deliberately describing services in *both* operational dialects—digital and physical—at the same time. Rather than forcing one domain to adopt the vocabulary, metrics, or mental models of the other, the catalog creates a **shared, bidirectional translation layer** that allows each community to see how its work contributes to institutional outcomes.

At its core, the Bilingual Service Catalog reframes campus capabilities not as systems or assets, but as **services that produce outcomes.** Each service is expressed simultaneously in two languages:

- **The IT language**, emphasizing availability, cybersecurity posture, data integrity, interoperability, identity, observability, and lifecycle management.
- **The Facilities language**, emphasizing uptime, safety, comfort, regulatory compliance, maintainability, energy performance, and continuity of operations.

By presenting a single service through both lenses, the catalog makes explicit the dependencies, tradeoffs, and shared responsibilities that typically remain implicit—and therefore contested.

A traditional IT service catalog answers questions such as Who owns this service? What are its SLAs? What systems support it? A traditional Facilities asset register answers questions such as What equipment is installed? Who maintains it? What are the failure modes? The Bilingual Service Catalog deliberately goes further. For every service, it documents:

- **The user-visible outcome** (what the campus community experiences)
- **The operational promise** (what "good" looks like in physical terms)
- **The digital promise** (what "good" looks like in cyber and data terms)
- **The shared risk model** (what happens when things fail, and why)
- **The joint accountability model** (who decides, who operates, who escalates)

This dual framing eliminates the false dichotomy of "IT services" versus "Facilities systems" and replaces it with **institutional services that span both domains.**

Why the Bilingual Service Catalog Matters

The Bilingual Service Catalog plays several critical roles in an enduring smart campus operating model:

- **Governance and Decision Rights**
  It makes decision boundaries explicit. Facilities retains authority over how buildings operate safely; IT retains authority over how connected systems are secured, integrated, and governed. The catalog documents these boundaries service by service, preventing ambiguity during incidents, audits, or funding debates.
- **Portfolio and Funding Alignment**
  By describing services in outcome-based terms, the catalog allows capital and operating investments to be evaluated against shared institutional value—reliability, sustainability, resilience, and experience—rather than against siloed departmental budgets.
- **Risk Translation**
  Cyber risk and physical risk are expressed together. A network outage is no longer abstract, and a chiller failure is no longer "just facilities." Each risk is documented in terms of its downstream institutional impact.
- **Operational Clarity**
  Operators, engineers, and service managers can see how their daily work contributes to a broader service promise. This reduces friction, handoff failures, and the creation of "shadow OT" or ad-hoc workarounds.
- **Data and Digital Twin Enablement**
  The catalog serves as the foundation of the institutional data estate. Services specify the required data, its quality standards, and its applications—ensuring that analytics and digital twins are practical and operational rather than merely theoretical.

Exhibit 8. The Bilingual Service Catalog: IT-to-Facilities Mapping

IT Technical Service/Requirement	Facilities Operational Outcome	Value Proposition (The "Why")
Microsegmentation (Software-Defined Perimeter)	Zonal Life-Safety Isolation	Prevents a breach in a public Wi-Fi kiosk from migrating to the elevator control system or fire suppression network.
Low-Latency Edge Computing	Real-Time Structural Health Monitoring	Enables vibration sensors on bridges or high-rise labs to trigger immediate shut-offs during seismic events or structural anomalies.
Semantic Data Tagging (e.g., Brick/Haystack)	Universal Asset Interoperability	Allows a new HVAC controller from Vendor A to "talk" to an existing lighting system from Vendor B without custom manual programming.
Predictive Analytics and AI Inference	Condition-Based Maintenance (CBM)	Shifts from "Fix it because it's Tuesday" to "Fix it because the motor's acoustic signature indicates a bearing failure in 48 hours."
Digital Twin Synchronization	Virtual Commissioning and Shadow Testing	Allows Facilities to "test" a new cooling sequence in a virtual model to see if it causes a condensation spike before applying it to the physical building.
High-Availability (HA) Clusters	Research Continuity and Zero-Downtime Labs	Ensures that ultra-low-temperature freezers and sensitive lab equipment remain monitored and powered during primary network or server outages or failures.
Identity and Access Management (IAM)	Dynamic Space Personalization	Automatically adjusts lighting, temperature, and ergonomic settings the moment a specific faculty member's badge or phone enters a shared office.

Example: "Healthy Learning Environments" as a Bilingual Service

Rather than listing HVAC, BAS, sensors, networks, and analytics as separate line items, the catalog defines a service such as **Healthy Learning Environments**, described in two coordinated views:

- **Facilities view:**
    - Indoor air quality within defined thresholds
    - Reliable ventilation and thermal comfort during occupied hours
    - Compliance with safety and health standards
    - Maintainable systems with predictable failure modes
- **IT view:**
    - Secure, segmented connectivity for building systems
    - Trusted telemetry feeding analytics and reporting
    - Identity-controlled access for operators and vendors
    - Data quality is sufficient for AI, dashboards, and verification

Both descriptions refer to the *same service*, governed by a single service-owner model, with clearly defined rights over operational and architectural decisions.

From Catalog to Capability

The Bilingual Service Catalog is not a fixed document; it is a dynamic, evolving artifact that grows as services develop, technologies advance, and organizational priorities shift. When integrated into standard operating procedures—such as procurement, change management, incident response, sustainability reporting, and performance measurement—it helps ensure that IT/OT convergence remains consistent, manageable, and sustainable.

Thus, the Bilingual Service Catalog plays a crucial role in establishing the smart campus by converting strategy into practical steps, synchronizing digital and physical environments, and making convergence a normal part of daily life rather than just an ideal.

Leadership as Strategic Integration

Mastering cross-domain knowledge enables the CTO/CFaO to go beyond their usual silos, creating a common architectural language and mutual understanding essential for managing IT/OT convergence. However, building this shared vocabulary is only the first step toward transformation.

Mutual understanding—including recognizing the challenge of aligning a short-term three-month agile software sprint with a long-term thirty-year capital infrastructure plan—does not automatically translate into the ability to fund, develop, or ensure a cognitive campus. To turn this cross-disciplinary insight into a tangible institutional change, these leaders need to fundamentally reshape their approach at the cabinet level. They must move away from acting as isolated sovereigns negotiating across boundaries. Instead, they should combine their authority, budgets, and influence to collaboratively manage the cyber-physical ecosystem as a unified, indivisible asset.

This involves transforming their partnership from simple tactical cooperation into a unified executive mandate, marking a significant shift towards viewing Leadership as Strategic Integration. In this new approach, the CTO/CFaO moves beyond their operational roles to act as the Integrative Architects, shaping the AI-enabled sustainable smart campus. Together, they co-create the university's cyber-physical future.

To execute this level of strategic integration, these leaders must actively govern, fund, and culturally unite the campus ecosystem across six critical domains.

Domain 1: Co-Governance and the Unified Capital Strategy

The conventional higher education funding approach—where IT requests a technology budget for servers and software, while Facilities submits a separate capital budget for chillers and deferred maintenance—undermines the development of a smart campus.

As Integrative Architects, the CTO/CFaO should promote a combined cyber-physical capital strategy. They need to present a unified ledger to the Chief Financial Officer. When Facilities requests a multi-million-dollar HVAC upgrade, IT must also co-sponsor the initiative by specifying the edge-compute, IoT sensors, and network capacity needed to make the HVAC system "smart."

Moreover, co-governance demands the courage to allocate funds to the unseen digital infrastructure. Convincing a Cabinet or Board to invest in API gateways, data lakes, and middleware is challenging because they tend to favor projects with clear, visible milestones, such as ribbon-cutting ceremonies. The Integrative Architects must persuade the institution to reconsider its financial priorities, shifting from focusing solely on initial construction costs to a comprehensive 30-year Total Cost of Ownership (TCO). This revised TCO should include the financial benefits of predictive maintenance savings, carbon pricing, and the energy efficiency enabled by digital infrastructure, which is often less visible but equally valuable.

Domain 2: Enterprise Cyber-Physical Risk Management

As the campus develops into a unified, interconnected cognitive entity, digital threats can lead to serious physical consequences. A compromised system isn't just about data theft anymore; it can cause frozen pipes, interrupted research, or gaps in physical security.

The CTO/CFaO should collaborate to lead a new approach to enterprise risk management. This involves elevating Operational Technology (OT) security concerns to the Board of Trustees, helping them understand that an unpatched building automation system is just as risky as a compromised financial database. Practically, this calls for establishing a Unified Incident Command. The IT Security Operations Center (SOC) must no longer operate in isolation; instead, it should coordinate with Facilities dispatch and Campus Police to develop a unified cyber-physical incident response plan. These leaders must consistently balance risk—protecting legacy operational technology with Zero Trust micro-segmentation—while ensuring they do not compromise life-safety or research continuity in the physical plant.

Domain 3: Governing the Technology Stack of the "Sensory Campus."

An AI-powered smart campus functions as a sensory system, constantly monitoring occupancy, movement, energy consumption, and human activity to enhance the environment. Although this promotes significant efficiency, it also introduces a deep paradox of surveillance.

For the CTO/CFaO acting as Integrative Architects, leadership involves turning the complex ethical issues of ongoing monitoring into a clear, practical framework. They need to move beyond abstract ethical discussions to focus on detailed planning, execution, and ongoing oversight of the cyber-physical system. Trust must be built into the architecture through disciplined practices.

Planning: Cataloging the Cyber-Physical Baseline

You can't govern what you're unaware of. Before adding new AI layers, leadership must first thoroughly identify and document all existing Operational Technology (OT) and legacy IoT devices. Over the years, academic departments and facilities teams have often implemented "Shadow OT"—such as untracked smart cameras, standalone environmental sensors, or proprietary building access systems. The CTO/CFaO needs to collaboratively create a comprehensive map of this environment, establishing a clear baseline of current data collection, protocols used, storage locations, and access permissions.

Implementing: Standardization and Data Minimization

In the future, the Integrative Architects should implement strict standardization across all OT procurement and deployment. Vendors should no longer have the authority to determine the campus technology stack through closed, proprietary systems. The CTO/CFaO needs to develop enforceable policies to guarantee that every new sensor, controller, or AI application adheres to enterprise interoperability and security standards.

Importantly, this implementation stage involves embedding ethical principles into the system design via data minimization. Leadership needs to ensure the technology setup collects only the essential telemetry required for environment optimization, nothing extra. They should establish and enforce strict data retention policies, making sure spatial and behavioral data are aggregated, anonymized at the source, and automatically deleted to safeguard individual privacy.

Monitoring: Algorithmic Transparency and Continuous Auditing

Governance extends beyond deployment; ongoing lifecycle monitoring of the technology stack is essential. The CTO/CFaO needs to collaboratively oversee not only the sensors' uptime but also the ethical integrity of the AI models that handle their data.

If an algorithmic system autonomously opts to dim lights, lock doors, or modify ambient temperatures through spatial analytics, its decision-making process must be consistently reviewed. The underlying models should be fully transparent, explainable, and regularly evaluated for algorithmic drift or bias that could unfairly affect certain student groups or hinder specific academic workflows.

Domain 4: The Cultural Merger, Workforce Evolution, and Change Management

The most challenging architecture to integrate isn't digital or physical; it's human. The CTO/CFaO must blend two fundamentally different operational cultures: IT's approach of "fail fast and iterate" and Facilities' method of "measure twice, cut once, and ensure it lasts fifty years."

Strategic integration demands joint leadership to enhance workforce skills by systematically connecting technical and operational expertise. Network engineers should learn about the safety-critical aspects and physical principles of the central plant. Meanwhile, master electricians and HVAC technicians need training in IP networking, cybersecurity basics, and data analytics.

Importantly, these leaders need to address fears about automation. As AI handles routine building-optimization tasks, employees will worry less about job security. The Integrative Architects should guide a change-management message that frames these tools as empowering innovations that reduce drudgery and physical hazards, enabling the human workforce to focus on high-value, complex, and strategic problem-solving.

Domain 5: Academic Alignment and the Campus as a Living Laboratory

The smart campus should not serve as a quiet, background utility; instead, it should actively contribute to the university's primary goals of research and teaching. The CTO/CFaO should work closely with the Provost and academic deans to dismantle remaining silos. Leading in this area involves exposing students and faculty to the cyber-physical infrastructure, turning the physical environment into a dynamic "Living Laboratory."

This entails creating secure policies and data sandboxes that enable engineering students to analyze real-time microgrid telemetry or allow computer science classes to train machine learning models on anonymized HVAC data. In doing so, the smart campus serves as a tangible example of the university's dedication to climate action, technological advancement, and hands-on learning, transforming campus operations into a compelling recruitment and educational resource.

Domain 6: The Unified Boardroom Narrative

Ultimately, strategic integration is about mastering executive communication. The CTO/CFaO needs to present a unified voice to the President, the Cabinet, and the Board of Trustees. When communicating with the Board, they should avoid technical jargon—such as BACnet protocols or server virtualization—and instead, focus on explaining how their joint technology efforts affect the institution's bottom line. They must describe how IoT sensors, digital twins, and network micro-segmentation help reduce enterprise risk, improve operational efficiency, and prevent large capital debt.

By delivering this unified narrative, they reinforce their role as Integrative Architects. They are now seen not just as overseers of costly departments but as key strategic partners in shaping the university's lasting resilience, sustainability, and technological leadership.

In complex organizations, leadership goes beyond just setting a direction. It entails integration—linking strategy with daily operations, policy with execution, and fostering innovation alongside managing risks. While a college President might promote a vision of a "connected campus" and a VP of Facilities oversees sensor installations, the CTO/CTO bears a distinct leadership responsibility that sets them apart from other executives.

Relationship Building With and Through Campus Leaders

The CTO/CFaO can greatly benefit from exchanging insights. They often have extensive relationships with other executives in the organization. Key executive leaders should be involved regularly from the beginning of a smart campus project. For instance, some institutions have Chief Sustainability Officers who frequently work with CFaOs on energy management and reaching carbon neutrality. Chief Cybersecurity Officers also collaborate with the CTO to develop and implement cybersecurity strategies. These are clear examples, but the main point is that as CFaOs and CTOs increasingly lead smart campus initiatives together, they need to broaden their network of influencers and collaborators to succeed.

The CTO/CFaO should introduce their CXO colleagues to their key relationships, ongoing discussions, challenges, and goals for improving efficiency and capabilities on modern college campuses. Additionally, they should work to extend their network and influence on other executives outside their usual circles.

Intentionally expanding communication among executive leaders to share a common vision, concerns, and aspirations for outcomes that drive transformation. This promotes alignment with institutional strategies and priorities that support fiscal management, curriculum development, and student success, linked to a sustainable, smart campus. While executive leaders have diverse perspectives, their collaboration fosters transformational opportunities for the institution. By working together, they ensure that technology, infrastructure, and financial planning are aligned, culminating in a cohesive and comprehensive master plan.

If your institution is about to create a comprehensive master facilities or strategic plan, the guiding firm will likely start by building relationships and creating opportunities similar to those previously mentioned. This approach provides valuable insights for establishing connections focused on sustainable, smart campus initiatives. Beyond formal planning processes such as master planning, institutional leaders can implement the following tactics to enhance relationships with their executive colleagues and stakeholders.

- **Regular Interactions and Communication**
  Regular meetings and collaborative planning sessions enable all CXO participants to understand each other's perspectives, strengths, and expertise. Using clear communication and collaboration channels reduces misunderstandings and fosters a shared vision with aligned goals.
- **Create Cross-Functional Teams**
  Creating cross-functional teams with members from IT, facilities, sustainability offices, and other departments can break down silos and encourage collaboration. These teams can concentrate on specific projects to ensure both technical and operational needs are met.
- **Define Shared Objectives**
  Aligning goals and objectives among CXOs ensures that investments in time, talent, and resources generate genuine value. This requires close collaboration to identify and prioritize initiatives that support the facility's mission.
- **Leverage Modern Collaboration Technology**
  Utilizing collaboration and video conferencing tools improves idea sharing and communication across departments, especially in large or geographically spread organizations. Tools like Microsoft© Teams are efficient for managing projects, facilitating communication, and sharing data between departments.
- **Shaping Culture and Promoting Diversity**
  Fostering a culture that values learning, sharing, opportunity, and adaptability helps CXOs facilitate teamwork in an environment seen as "glass half full." Leaders should encourage open dialogue and knowledge sharing within their teams to boost creativity, innovation, and collective support. Embracing

diverse perspectives, skills, and personal experiences promotes innovative thinking and effective problem-solving, which are essential for addressing the complex challenges of a smart campus. It is crucial to develop a culture rooted in mutual respect and trust, recognizing the expertise and contributions of all departments. Our institutional traditions significantly influence the culture, so to implement cultural change, consider which activities, recognitions, or "rituals" should be adjusted to reflect a new mindset, approach, and identity.

- **Participating in Governance Frameworks**
  Engaging in governance structures helps develop a shared understanding and guarantees that smart campus initiatives are effectively managed and aligned with organizational objectives. This involves defining clear roles and responsibilities and creating processes for decision-making and accountability. Strong governance facilitates faster, more informed decisions, which is crucial for the evolving nature of smart campus projects. Furthermore, governance fosters ongoing feedback and improvements, enabling ongoing adjustments and enhancements to the plan.
- **Participating in Workshops and Conferences**
  When CXOs participate in seminars, workshops, and conferences together, they engage in social learning and relationship building. These experiences can foster meaningful dialogue.
- **Celebrating Success**
  Nothing bonds people together like success! Recognize and celebrate collaborative successes to build morale and encourage ongoing cooperation.

Ultimately, strong relationships across the executive landscape transform smart-campus modernization from just a technical goal into an institution-wide capability. By building trust, aligning objectives, and fostering shared understanding among cabinet leaders, the CTO/CFaO broadens the initiative's strategic scope and operational resilience. This relational infrastructure ensures decisions mirror mission priorities, risks are collectively understood, and opportunities for academic, operational, and financial innovation are identified early. When leaders work together deliberately, the smart campus evolves from a mere technology project into a coordinated effort rooted in shared stewardship and institutional purpose.

Executive Sponsorship of the Smart Campus Initiative

A transformation of this scale cannot be led from the middle of the organizational chart. It demands a strong Executive Sponsor. Even if the CTO/CFaO are aligned and effective as Integrative Architects of the AI-enabled sustainable smart campus, they still need backing from the institution's top executive—the ultimate authority over the enterprise strategy and capital reserves.

Within a smart campus setting, executive sponsorship isn't just a ceremonial post or a formal signature. It represents a highly engaged, decisive, and cross-departmental mandate. The sponsor needs enough institutional authority to shield the vision from immediate budget concerns, mediate disputes between IT and Facilities, and keep the cyber-physical transformation firmly aligned with the university's academic goals.

The Trap of Departmental Sponsorship

A common and critical error in smart campus projects is having the CTO or CFaO sponsor the initiative themselves. When the CTO sponsors it, the AI-enabled Sustainable Smart Campus is quickly seen as merely an "IT Project." This leads facilities and academic departments to be suspicious, thinking it is a digital mandate that ignores the campus's physical and educational realities. The project risks becoming just an exercise in deploying technology without a clear purpose.

If the CFaO sponsors the initiative, it is likely to be labeled as a "Facilities Project" or merely a high-profile energy-saving effort. Such branding may hinder its ability to obtain the substantial enterprise data governance, cybersecurity resources, and academic collaboration needed to transform it from a simple building automation upgrade into a genuine cognitive campus.

The CTO/CFaO must be the *drivers* and the *architects*, but they cannot be the ultimate sponsors. The sponsor must sit above the divide.

Who Should Be the Executive Sponsor?

It is common to assume that the President or Chancellor is the primary executive sponsor. On some smaller campuses, this might be a good fit. However, at other institutions, these leaders may be less involved in campus operations, focusing more on legislative work and fundraising. Additionally, many top institutional leaders prefer to delegate operational responsibilities to others, often due to political considerations and governance structures.

Therefore, the best executive sponsor for an AI-enabled sustainable smart campus is usually the Executive Vice President of Administration / COO or, depending on the organization, the CFAO. These leaders are well-placed to support the transformation for three key reasons:

- **1. Unified Capital Authority**
  The EVP/COO controls the purse strings for both the digital and the physical realms. They have the authority to merge the IT capital budget with the Facilities deferred maintenance budget. When the CTO/CFaO proposes an integrated 30-year Total Cost of Ownership (TCO) model that funds "invisible" digital infrastructure to generate physical energy savings, the EVP/COO is the only leader with the purview to approve and defend that blended financial strategy to the Board of Trustees.
- **2. Enterprise Risk Ownership**
  The EVP/COO is ultimately responsible for institutional risk and business continuity. They understand that a failure in Operational Technology (OT) cybersecurity or a catastrophic breakdown of the physical plant is not an isolated departmental failure; it is an existential threat to the university's bond rating, research grants, and public reputation. They provide the executive "air cover" required to enforce mandatory IT/OT convergence and strict security standardization across the enterprise.
- **3. Tie-Breaking Authority**
  The merger of IT and Facilities cultures will inevitably produce friction. When the CTO demands taking a legacy system offline for emergency patching, and the CFaO demands keeping it online for research continuity, the EVP/COO serves as the ultimate arbiter, making decisions based on overarching institutional priorities rather than departmental preferences.

The Indispensable Co-Sponsor: The Provost

While the EVP/COO provides operational and financial support, they cannot independently endorse the initiative. To ensure smart campus development does not become a mere efficiency exercise, it must also be co-sponsored by the Provost (Chief Academic Officer).

The Provost protects the core principles of the academy. As a co-sponsor, they guarantee that the physical and digital systems facilitate teaching, support faculty, and enhance student experiences. The Provost serves as a vital link, transforming the smart campus into a "Living Laboratory" by ensuring data from microgrids and AI platforms is securely available to academic research and practical learning.

The Role of the Sponsor in Maintaining Momentum

The Executive Sponsor holds a crucial ongoing role. While architects—the CTO and CFAO—implement the strategy in stages, achieving an AI-enabled, sustainable, smart campus generally takes years, outlasting typical technology cycles and often extending beyond individual presidencies.

The sponsor's primary duty is to sustain momentum over the decades. They should embed the vision firmly into the university's master plan, capital allocation strategies, and board reports to make the transformation permanent. By championing the Integrative Architects, the sponsor helps the institution remain committed, fostering a resilient, cognitive environment that benefits future generations of students.

Crafting the Shared Vision

The key starting point for an AI-enabled sustainable smart campus is not simply buying sensors or setting up servers, but creating a shared story. For too long, the CFaO and the CTO have followed separate, often conflicting strategies. Facilities focus on long-term building durability over thirty years, while IT aims for short-term technology updates every three years. Facilities emphasize physical robustness, IT values digital flexibility.

To build a unified cognitive campus, these two leaders, ideally working together with the executive sponsor and colleagues, must go beyond their individual roles to jointly craft a shared vision. This requires a deliberate summit, akin to a "Constitutional Convention," in which the CTO/CFaO explores the interrelation between digital and physical systems. They need to understand that the campus can't be genuinely "smart" if the network security is weak, and the network can't enhance the campus if mechanical systems are unresponsive and unaware.

Executive Visioning Retreat

To establish an institutional vision, it is helpful to get leaders into the same room. Organize a high-level retreat or workshop with the university's executive leaders to educate them on the AI-Enabled Sustainable Smart Campus, then work collaboratively to **craft the smart campus** vision **and guiding principles**.

Consider taking this vision team off-site to minimize campus distractions and emphasize the project's importance. Encourage open discussions about long-term goals, institutional challenges, and how a sustainable smart campus can provide solutions. The CFaO and CTO should come prepared to share their own visions. Even if it requires courage and is likely not to be adopted as initially presented, it is usually more effective to have people respond to a draft and refine it than to create one from scratch. However, the process should result in a draft vision statement that explicitly ties technology and sustainability initiatives to the institution's mission and strategic priorities.

By involving the entire leadership team in developing the vision, you secure early support and foster a common purpose. Investing time upfront in co-creating the vision results in greater consensus and reduces obstacles later. It also highlights diverse perspectives, including academic needs, student experience, and operational constraints, that the vision must account for.

The Crucible of Vision Writing

Creating the vision involves translating various concepts. Leaders need to unify their terminology, like combining "uptime" with "occupant comfort" and "cybersecurity" with "physical safety." Below are five collaborative vision statements representing this unified leadership, each highlighting a particular strategic goal of the integrated organization. One or more of these statements could serve as effective starting points for the processes outlined in the manifesto.

The Vision "Platform" (Focus on Interoperability and Outcomes)

This approach redefines the campus not as a collection of buildings, but as a "platform." It explicitly links technical inputs (connecting utilities) to institutional outputs (learning experience).

"Together, we will build a secure, interoperable smart campus platform—connecting buildings, utilities, and services—so the university can measurably improve safety, sustainability, reliability, and the learning experience."

The "Operational Excellence" Vision (Focus on Modernization and Trust)

This version directly addresses the "run-to-failure" culture by promising "continuous" rather than "episodic" improvement. It frames the convergence as a mechanism for building trust in the institution's operations.

"We will modernize the campus as one system—combining operational excellence with digital governance—so performance improves continuously rather than episodically, and trust grows with every phase."

The "Foundation for Innovation" Vision (Focus on Data and Security)

This is the "Architect's Vision." It prioritizes the prerequisites (data estate, security) required to deploy advanced technologies (AI, Digital Twins), ensuring the Board understands that basic reliability must precede advanced innovation.

"We will align IT and Facilities around a common data estate, cybersecurity baseline, and operating model—so analytics, AI, and digital twins scale safely on a foundation of reliability and stewardship."

- **Why this works:**

The "Human-Centric" Vision (Focus on Experience and Ethics)

This version places the "user" at the center of the technology. It proactively addresses the "surveillance paradox" by enshrining privacy and transparency as non-negotiable architectural requirements.

"We will design for people first—operators, students, faculty, and staff—by making privacy, transparency, usability, and accessibility architectural requirements of every smart campus investment."

The "Resilience" Vision (Focus on Sustainability and Continuity)

This approach frames sustainability not as a passive goal, but as a "managed capability." It speaks the language of risk management, promising a campus that remains functional and efficient even during climate or operational disruptions.

"We will deliver decarbonization and resilience as managed capabilities—verified through measurement, strengthened through continuous commissioning, and enabled through secure digital infrastructure—so the campus performs under both normal conditions and disruption."

Ultimately, the shard vision will be shaped by the institution's mission, culture, and strategic goals. The resulting vision statement should go beyond a mere slogan; it must serve as a Declaration of Interdependence—a binding commitment to the Board, the Cabinet, and the community that the digital and physical assets now form a unified whole.

It is written in the tone and structure of the manifesto, grounded in higher-education best practices, and addresses alignment, gaps, and planning discipline.

Aligning Vision with Strategy

A unified vision only becomes actionable when it is coherently mapped to the institution's existing strategic plans, mission statements, and long-range capital and academic priorities. Aligning a smart campus vision with institutional strategy requires more than an inspiring statement of intent; it demands a disciplined, structured planning practice. Visioning is inherently creative and aspirational, but formal planning is operational, sequenced, and rigorous. Institutions that succeed treat alignment not as a single milestone, but as an enduring practice of translation, integration, and verification.

In higher education, the most successful smart campus initiatives are those that explicitly anchor technological modernization to institutional intent—linking AI, data governance, and cyber-physical transformation directly to goals such as student success, research excellence, sustainability commitments, financial resilience, and community engagement. The CTO and CFAO, as Integrative Architects, must ensure that the vision for an AI-enabled, sustainable, smart campus is not positioned as a parallel or competing strategy, but as the operational engine that enables the institution's existing strategy to succeed.

Best Practices in Higher Education Strategic Alignment

Link to the Mission and Institutional Strategic Plan

Institutions that excel in modernization start by identifying the strategic pillars—such as access, enrollment growth, faculty innovation, student experience, research competitiveness, or sustainability—and explicitly articulate how the smart campus enables each one. For example, AI-driven operations can be aligned with institutional commitments to climate action; digital twins can be mapped to research and STEM priorities; and cyber-physical governance can support risk-management goals already endorsed by the Board.

Identify Strategic Themes

The first step in this process is to distill the smart campus vision into a coherent set of strategic themes that map directly onto the institution's existing priorities. Rather than leaving the vision abstract, leaders identify four to six core themes—such as Operational Excellence, Sustainability as Capability, Cyber-Physical Resilience, Human-Centered Experience, or AI-Enabled Decision Architecture—that clearly connect modernization efforts to the university's mission, academic goals, and long-range plans.

Identify and link Outcomes.

Once these themes are pinpointed, the institution must define measurable outcomes to transform broad goals into specific, actionable commitments. Each theme needs clear, quantifiable indicators aligned with KPIs and scorecards, such as carbon-reduction paths, avoided downtime, predictive-maintenance accuracy, indoor-environmental-quality metrics focused on equity, energy savings, and research continuity.

The most effective strategies turn the smart campus into tangible results that are recognizable to senior leadership and the Board: lower operating costs, ongoing research, better retention through healthier environments, increased resilience, and real progress in decarbonization. This approach refocuses effort from mere products to a clear purpose—and from individual projects to overall institutional capabilities. Without measurable outcomes, alignment risks remaining superficial rather than operational.

Link Governance

A unified smart campus strategy also requires cross-functional ownership. No single division can deliver or maintain cyber-physical transformation on its own. Academic affairs, sustainability, IT, facilities, finance, research, and student affairs must share stewardship of goals, resources, and risks. Plans that centralize ownership in one unit ultimately fragment; plans that institutionalize shared accountability endure.

To ensure this shared ownership translates into durable progress, the formal plan must articulate the enterprise structures that hold the work together. Governance models, clear decision rights, architectural standards, cybersecurity guardrails, risk frameworks, and change-management pathways form the operating system of the smart campus. These structures move the institution from episodic progress to repeatable, scalable capability.

Institutions that avoid drift or fragmentation use formal governance bodies (e.g., Smart Campus Steering Committees, IT/Facilities Portfolio Councils) to ensure alignment is continuously renewed as priorities evolve. The smart-campus vision must survive leadership turnover, new strategic plans, and emergent opportunities; governance becomes the engine that preserves strategic coherence.

Link Resource Allocation

Leading universities ensure that the smart campus strategy is woven directly into capital renewal, space planning, enrollment modeling, academic master planning, and IT strategic planning. This prevents the initiative from becoming a siloed "technology project list" and instead makes it a cross-functional enterprise strategy. The CFaO and CTO must also coordinate with strategic enrollment management, research leadership, and campus planning to ensure alignment with long-term forecasts and capital-renewal frameworks. When funding sequences, project timing, and modernization priorities are harmonized, the institution avoids fragmentation and resists the drift of vendor-driven agendas. In this model, strategic priorities—rather than opportunistic projects—determine the pace and nature of modernization.

Where Strategy May Need Realignment

Aligning the smart-campus vision to institutional strategy often reveals gaps or even contradictions that must be addressed for the initiative to be credible:

- **Misalignment between sustainability commitments and capital/funding reality.**
  Many universities publicly commit to net-zero goals but often lack the necessary operational structure, data systems, or investment strategies to reach them. The smart-campus vision highlights these shortcomings and offers a clear roadmap to operational maturity and to the real-time implementation of climate objectives.

- **Outdated space-planning assumptions.**
  The post-pandemic transition to online, hybrid learning, and research-driven flexibility, along with enrollment decreases, organizational mergers, and new partnerships, may conflict with legacy facilities plans that presume steady building occupancy. Adopting smart-campus spatial intelligence requires rethinking long-term real estate strategies.

- **Disconnection between IT strategy and physical-plant realities.**
Institutions might have strong digital strategies but still lack updated controls, sensor infrastructure, or a solid OT security stance. Aligning vision should refocus digital priorities on cyber-physical risks, data integrity, and operational resilience.
- **Inadequate governance for AI, data ethics, and cross-domain risk.**
If institutional strategy does not currently include AI governance or cyber-physical security as key enterprise risks, the smart-campus plan should catalyze efforts to bring these concerns into the strategic spotlight.

Aligning the vision for an AI-enabled, sustainable smart campus with the institution's strategy involves turning aspirations into disciplined, cross-functional practices. Successful institutions view this alignment as an ongoing operational cycle—linking the smart-campus vision to key strategic themes, setting measurable outcomes aligned with institutional KPIs, and sharing ownership among the academic, administrative, financial, sustainability, and operational sectors. Achieving this requires embedding governance, architectural standards, risk-management frameworks, and change-management structures into formal plans, ensuring that modernization becomes a lasting enterprise capability rather than just a series of initiatives. When incorporated into budget and capital-planning cycles, the smart-campus strategy enhances institutional coherence, guarantees modernization aligns with the institution's mission rather than vendor timelines, and positions the campus for long-term resilience, sustainability, and operational excellence.

The Communication Plan

With a clear vision, strategic alignment, a strong sense of purpose, and initial ideas on action themes and priorities, develop a communication plan for the executive team to consistently share the vision, progress, and importance of the smart campus initiative with all stakeholders. Leaders should agree on consistent messaging and key phrases that clearly convey goals and expectations.

In this activity, identify information outlets and plan a schedule for updates, such as quarterly town halls, monthly dashboard reports, campus news releases, or speeches on the institution's status. This should include regular email updates from the Executive Sponsor and also encompass the President/Chancellor's periodic communications with the institution, alums, and the Board. Additionally, the strategy should feature a dedicated section on the university website or intranet for smart campus news.

A transparent communication strategy maintains continuous alignment among the leadership team and builds stakeholder trust by regularly updating the community on successes and challenges. It also facilitates change management: consistent, clear communication helps reduce rumors and resistance by ensuring people feel involved and informed about the project's progress.

Planning Stakeholder Engagement

Smart campus transformation impacts daily life by altering building behaviors, service access, maintenance prioritization, and data utilization. Stakeholder engagement becomes essential as it fosters trust and enhances design. This engagement is an ongoing, iterative process supported by specific audiences, channels, timelines, feedback tools, responsible stakeholders, and measurable outcomes. The objectives of engagement include...

Smart campus initiatives change how people participate in work and learning. If institutions treat engagement merely as a communication campaign rather than a deliberate design process, resistance is predictable. People often believe that decisions are already made, that risks are not fully addressed, and that they are burdened with responsibility.

Effective engagement has two primary goals. First, it improves decision-making by gathering frontline insights, such as maintenance processes, scheduling challenges, research compliance, and student accessibility requirements. Second, it builds trust by demonstrating that feedback is genuinely valued. The 'you said / we did' approach is more than just PR; it's essential for maintaining legitimacy through good governance.

The CTO, CFAO, and executive sponsor work to foster a comprehensive understanding of the smart campus vision and its underlying principles, including institutional resilience, sustainability, experience, and stewardship. This approach creates structured opportunities for stakeholders to learn about and influence priorities and design decisions. Continuous engagement helps sustain the initiative's credibility by regularly updating stakeholders on progress and acting on feedback. Over time, this will develop ambassadors—faculty, staff, and students—who can adapt the work to their local context and promote the adoption of new practices.

Engagement strategies should be tailored to different decision-making groups in higher education. Executive leaders and the board require summaries of risks and benefits, as well as scorecard reports. Faculty governance bodies focus on clear information regarding academic impacts, data ethics, and autonomy. Facilities and operational teams benefit from engagement that emphasizes workflows, including training and hands-on problem-solving. Students want transparency, simplicity, accessibility, and insight into how data affects their experience. Research leaders need assurance of compliance, secure data access management, and continuous, governed access for approved uses.

Campus Engagement Strategy

A sustainable smart campus is not something you "roll out" to people; it is something you build with them—because the campus itself is a living social system. The CTO's engagement strategy, therefore, becomes as consequential as the architecture. When engagement is treated as an afterthought, smart campus initiatives are experienced as technology projects imposed on teaching, research, and operations. When engagement is designed as a core workstream—education, visibility, and co-design—smart campus efforts begin to feel like a shared institutional evolution: practical, mission-aligned, and locally owned.

Define the Smart Campus

Start by clearly defining the concept. Many campuses use the same terms—smart classrooms, smart buildings, smart apps, digital twin—but often mean very different things. A CTO can reduce confusion and speed up alignment by helping create a campus-specific definition of a sustainable smart campus through consensus.

This isn't about branding; it's about setting boundaries. It clarifies the scope (what's included and excluded), the purpose (why we're doing this), and the limits (what we won't do). When this definition is developed collaboratively with Facilities, Sustainability, Risk, Academic Affairs, Student Affairs, and shared governance, it has a significant impact: it transforms a technical goal into a shared story stakeholders can relate to. It also simplifies future decision-making—since proposals can be evaluated against a common purpose rather than debated from scratch each time.

Speak with One Voice: Mission, Trust, Value, Participation

Messages should consistently reinforce four themes. First, the mission: why this matters for student success, research enablement, and stewardship. Second, trust: privacy, cybersecurity, and governance guardrails. Third, value: what outcomes will be measured and reported (reliability, sustainability, experience, stewardship). Fourth, participation: how stakeholders can shape priorities and how the institution will respond. This consistency reduces rumor cycles and prevents well-intended leaders from sending conflicting signals.

Make Progress Visible

Next, make progress visible in ways that people can interpret. Often, smart campus initiatives operate "below the waterline"—networks, sensors, integrations, data pipelines—and this invisibility can pose risks. A practical approach involves using technology to demonstrate what is changing and why. This can include dashboards that translate energy and carbon data into clear results, digital signage that shows real-time improvements in comfort or availability, smart classroom demos that make the benefits of teaching tangible, and carefully selected IoT pilots that showcase value without overwhelming the institution. The goal is not to impress but to foster shared understanding. Visibility helps turn skepticism into curiosity by allowing people to see the campus's improvements rather than just believing in them.

Co-Design Collaboration

Visibility alone isn't enough. The fastest way to create resistance is to treat implementation as a sequence of technical decisions that people must accept after the fact. The CTO and CFaO's most effective engagement tool is co-design, structured collaboration in which faculty, students, and operational teams shape priorities, workflows, and service expectations. Co-design can take many forms—design charrettes for a "day in the life" student journey, cross-functional "use case clinics" where Facilities and IT jointly map a building outcome end-to-end, or short sprints where frontline trades staff validate what data is actually useful in the field. This kind of involvement builds ownership, improves requirements, and—critically—reduces the cultural tension at the heart of IT/OT convergence by honoring both realities: IT's duty to protect the institution and Facilities' duty to keep the campus safe, functional, and comfortable.

Formal Engagement

Leaders must acknowledge the formal engagement pathways that match how universities actually function. Shared governance bodies cannot be treated as a communications endpoint; they must be engaged early as design partners to avoid surprises and build trust. Likewise, unions, frontline operations leaders, and student representatives should be brought into structured listening sessions before tools are selected or workflows are redesigned. This approach allows leaders to identify advocates and concerns early, adjust plans before they harden, and make risk and privacy considerations explicit rather than implicit. It also shifts the story from "a technology project that will change your work" to "an operating model we are shaping together."

Demonstrations

Demonstration spaces—living labs—become the campus's engagement engine. A living lab is not a showroom. It is a place where people can see how OT data becomes insight, how a digital twin supports decisions, how analytics changes maintenance, and how governance and privacy guardrails are enforced in practice. When paired with tours, short "explainers," and role-based training, these spaces become contextual learning environments: Facilities teams see operational relevance, faculty see research possibilities, students see innovation in action, and executives see evidence of measurable value. In many institutions, the living lab becomes the first moment where convergence stops being abstract and becomes intuitive.

Channels and Cadence

Transparency must be predictable. Quarterly town halls (with open Q&A) provide broad visibility and legitimacy. Monthly scorecard updates—delivered via an intranet hub—translate progress into measurable outcomes and clarify what is changing next. Targeted communications (faculty-focused, operations-focused, student-focused) address local realities and reduce the "one message fits all" failure mode. Executive milestone messages, delivered jointly by the Executive Sponsor, CTO, and Facilities leader, are reserved for major decisions and visible wins to reinforce shared ownership and momentum.

Two-way feedback loops and follow-through

Feedback mechanisms must be simple and visible: listening sessions, short pulse surveys after pilots, and advisory groups for key use cases (e.g., digital classroom readiness, residence hall comfort, research environment uptime). Most importantly, feedback must have closure. The program should publish "you said / we did" summaries regularly, explaining what changed, what did not, and why. This is the difference between engagement that builds trust and engagement that breeds cynicism.

Ownership and Measurement

Engagement is accountable work. A named communications/change lead owns the calendar, channel hygiene, and editorial consistency. The Governance Steering Committee ensures transparency in decision-making and that updates reflect real progress and real constraints. Product/service owners own pilot engagement and feedback response. Engagement metrics are tracked and reviewed quarterly: participation rates, sentiment trends, response times to feedback, adoption indicators, and closure rates for addressed feedback items.

Deepen and Broaden Engagement

To deepen engagement further, the CTO/CFaO can add methods that are often underused—but highly effective on campus:

- **Create a Smart Campus Ambassadors network**
  Recruit faculty, students, and staff from various schools and units who get early briefings, take part in pilot programs, and serve as translators for their communities. This approach fosters distributed trust and diminishes the view that innovation is centralized and unclear.
- **Run "office hours" and open demonstrations**
  Create an ongoing, informal forum where everyone can ask questions, view dashboards, and learn about the collected data, its protections, and the decisions it informs. Transparency isn't just about compliance—it's a strategy for fostering adoption.
- **Launch a small grants program for "campus as a living lab" projects**
  Provide modest funding for faculty and student teams to develop use cases with campus data, pending governance approval. This approach transforms engagement into scholarly and experiential value, while also broadening the institution's innovation efforts.
- **Publish a quarterly Smart Campus Scorecard**
  It's not a marketing piece but a straightforward, consistent update that reports progress, lessons learned, costs avoided, risks identified, and next steps. Over time, this builds credibility by showcasing both achievements and realities.
- **Use persona-based engagement**
  Customize communications for various departments, such as trades staff, faculty, students, administrators, researchers, and safety teams, so they receive information about benefits and concerns in their own language rather than through generic messages.
- **Create structured feedback loops**
  After each pilot or phase, hold a brief retrospective with users and operators to record what was modified and outline what will be done differently in the next cycle. When participants see their feedback influencing decisions, their engagement naturally increases.

Taken together, these practices directly tackle the convergence challenge by shifting the focus from individual projects to the institution's broader operational evolution. Engagement serves as the crucial link in translating enabling technology into a shared institutional capability, ensuring the smart campus develops as a coherent strategy rather than a collection of disconnected systems. These activities are rich in context and geared toward executives, reflecting the strategic planning and consensus-building essential for successful leadership. By undertaking them, the leadership team not only refines their smart campus plan but also signals a new mode of operation—one rooted in strategic thinking, inclusivity, and proactivity—that can positively influence other initiatives as well.

Chapter Summary

Chapter 8 establishes governance as the decisive enabler of a successful AI-enabled, sustainable smart campus. It argues that the greatest risks to smart campus initiatives are not technical but structural—unclear decision rights, fragmented accountability, vendor-driven sprawl, and the absence of a shared operating model across IT, Facilities, Sustainability, Safety, and Academic stakeholders. Without disciplined governance, modernization efforts devolve into disconnected pilots that increase risk, cost, and complexity rather than delivering durable institutional value.

Shared decision rights matter more than organizational charts or technology choices.

The chapter reframes governance as a **strategic integration function** rather than a bureaucratic overlay. Effective smart campus governance defines *who decides, who owns, who operates, and who is accountable* across cyber-physical systems. It formalizes enterprise standards for interoperability, cybersecurity, data stewardship, AI use, and lifecycle management while preserving Facilities' authority over safe operation and uptime. Governance becomes the mechanism that reconciles competing risk models—IT's focus on confidentiality, integrity, and cyber resilience with Facilities' focus on availability, life safety, and continuity.

A central theme is the creation of **joint governance structures**, such as a Smart Campus Steering Committee, supported by clear charters, RACI models, and escalation paths. These structures ensure that technology, facilities, sustainability, and academic priorities are aligned to shared outcomes rather than optimized in isolation. Portfolio discipline replaces ad hoc projects, enabling the institution to sequence investments, reuse successful patterns, and scale only what is proven and governable.

The chapter also emphasizes governance over **data and intelligence**. As campuses deploy pervasive sensing, digital twins, and AI, governance must clarify what data is collected, for what purpose, who can access it, how it is secured, and how trust is preserved. Ethical AI, privacy protections, auditability, and transparency are treated as design requirements—not afterthoughts—because institutional legitimacy depends on them.

Chapter 9

Smart Campus Governance

Governance drives clarity around decision rights, standards, and guardrails that prevent fragmentation and ensure modernization advances safely and strategically towards antifragility.

A sustainable smart campus goes beyond leadership or advanced technology initiatives; it represents a fundamental organizational shift. This involves embedding digital intelligence into traditional physical systems, enabling them to make decisions that affect real-world aspects such as air quality, power supply, access control, lighting, and safety. Additionally, it impacts student services and offers new insights into campus life. As a result, governance is crucial—not just a routine administrative task, but the essential foundation that fosters trust in the entire system.

Organizations often associate 'collaboration' with cooperation. However, governance is a different concept; it is the structure that guides how an institution makes decisions, enforces standards, manages risks, and maintains accountability across various departments. While this chapter examines governance models, the Manifesto has discussed governance in the context of other smart campus topics. Governance is pervasive. It is a foundational thread present and active in each stage of smart campus development. It helps identify the scope and relationships between policy and technology, and between the organization and the policy, who is involved, which decisions are shared, and which are non-negotiable.

A well-designed governance framework helps resolve three recurring issues. Initially, governance tackles "pilot **technology sprawl and shadow technology**. Without effective oversight, different units might buy similar solutions, leading vendors to serve as the integration layer. This causes the organization to face double costs: one from redundant spending and another from rising technical debt.

Secondly, it protects **trust**. Connected campus systems bring new responsibilities related to data use, privacy, and cybersecurity. Strong governance helps the institution clearly communicate what data is collected, why it's collected, who can access it, and how it is securely managed.

Third, governance supports **scaling** by ensuring that standards like interoperability commitments, security baselines, data definitions, and consistent implementation patterns are enforced. It functions as the process that makes these standards mandatory, with any exceptions being transparent, justified, and managed within set timeframes.

Governance also aligns with Board expectations for overseeing enterprise risk. Regarding Cybersecurity, the NIST Framework 2.0 emphasizes "Govern" as a core function, highlighting roles, responsibilities, and oversight—terms that align closely with institutional accountability and risk management.

> ***You cannot automate what you cannot govern.***
> ***And you cannot scale what you do not trust.***

When governance is weak, smart-campus initiatives often turn into an appealing but expensive collection of disconnected pilots, competing dashboards, vendor lock-in, unclear data rights, and security exceptions justified by urgency.

On the other hand, strong governance provides the institution with more than just tools; it ensures a stable decision-making process, disciplined operations, and a reliable learning cycle—sustaining the trust of students, faculty, staff, and the board.

Governance serves as the link that integrates vision, policy, ethics, and culture, clearly connecting the success of smart campuses with robust cybersecurity and data management—particularly as IoT expands the attack surface and institutions become increasingly digital. To effectively implement governance, leaders must move from abstract principles to practical mechanisms, including decision rights, operational frameworks, and a purposeful cadence.

Governance for the Smart Campus

The most advanced sensors and AI algorithms will be ineffective without proper governance. Currently, at many universities, oversight of digital and physical realms is dangerously divided.

- **IT Governance** focuses on data security, software licensing, and network uptime.
- **Facilities Governance** focuses on capital construction, master planning, and mechanical reliability.

The Smart Campus bridges the gap between Facilities, IT, and Digital Experience. A "smart" HVAC controller is a mechanical device (Facilities) connected to the network (IT), producing data (IT/Business), and controlled via a student's mobile phone (Digital Experience). Without a unified governance system, this device poses a risk: Facilities may purchase it without IT's awareness, install it without proper security measures, and generate data without ownership.

To implement the role of the Integrative Architects, the CTO/CFaO should jointly establish a new governing body: The Smart Campus Steering Committee (SCSC). This committee does not replace existing governance structures; instead, it serves as the connecting link between them.

Integrating with Existing Governance Bodies

The Smart Campus Steering Committee (SCSC) must operate as part of the university, not independently. It embodies the institution's overall digital strategy, with its decisions impacting all areas of campus life — from student experiences to financial records.

Therefore, the Chief Technology Officer and Chief Facilities Officer should design the SCSC as an interconnected element within the university's governance system, rather than an isolated silo. To align the cyber-physical strategy with the institutional mission, the SCSC must formally collaborate with five key bodies.

The AI Governance Council (Stewards of Ethics)

The Smart Campus is poised to become the university's largest user of Artificial Intelligence, deploying algorithms for predictive maintenance, energy load management, and occupancy analysis. The SCSC must submit its high-impact algorithmic applications to the AI Governance Council for ethical evaluation. This process aims to prevent the AI systems—such as those managing building access or thermal comfort, often referred to as the "Ghost in the Machine"—from developing biases. For instance, the Council checks that energy-saving algorithms do not favor the comfort of administrative areas at the expense of reducing air quality in student residence halls.

The Digital Experience Committee (The User Interface)

The main interface for the Smart Campus has shifted from a wall thermostat to the smartphone's screen. When the SCSC approves new physical systems, such as a smart dorm lock or a mobile food-ordering kiosk, they must collaborate with the Digital Experience Committee. This coordination guarantees that physical access methods are smoothly incorporated into the university's official mobile app. It helps avoid the frustrating "app fatigue" where students have to download multiple unrelated third-party apps to manage their daily activities.

The Enterprise (ERP) Systems Steering Committee (The Data Backbone)

This long-standing body oversees major record-keeping systems, including the Student Information System (SIS), the Fiscal System, and the Learning Management System (LMS). The SCSC must integrate with these systems to facilitate Data Interoperability. For example, a smart classroom can't automatically control lights and HVAC without reading the class schedule from the SIS. Similarly, energy savings calculations rely on the Building Management System being able to send consumption data to the General Ledger. This integration keeps the "Business of the University" aligned with the "Operation of the Campus."

The Capital Projects Committee (The Funding Gate)

Traditionally, this influential committee approves new buildings based on their functional needs, visual appeal, and size. The SCSC is required to add a mandatory "Digital Readiness" review step into every capital project approval process. This introduces a strict governance checkpoint: construction cannot start, and no concrete can be poured until the cyber-physical architecture receives approval. This measure prevents the common issue of opening a stunning new building that is digitally outdated from Day One because the technology was "value-engineered" out of the design.

The Enterprise Risk Advisory Board (The Safety Net)

This board provides advice to the President and Cabinet on existential threats that span multiple departments. As the campus evolves into a programmable environment, the SCSC serves as a vital source of information about Cyber-Physical Risk to this board. A vulnerability in the stadium lighting control system is no longer just a maintenance issue; it represents a public safety crisis. The SCSC makes sure these hybrid risks are formally added to the institutional risk register, ensuring that strategies for physical safety and digital security are adequately funded and prioritized at the highest levels.

Composition of the SCSC

The Smart Campus Steering Committee (SCSC) should move away from a siloed approach. It shouldn't consist of just "IT people talking to IT people" or "Facilities people talking to Facilities people." Instead, it needs to be an orchestrated integration of diverse perspectives. To guarantee this, the SCSC should be jointly led by the CTO and the CFAO. This dual leadership is more than symbolic; it demonstrates complete equality between the digital and physical sectors, preventing either organization from overshadowing the other. However, although the Chairs set the tone, the table must be filled by the operational lieutenants who carry out the vision. These individuals bring the necessary technical, ethical, and operational skills to develop the cognitive campus.

Potential SCSC Members:

- **The Chief Information Security Officer (CISO)**
  The CISO sits as the ultimate authority on cyber-physical risk. In this forum, their mandate expands beyond data privacy to include life safety. They hold **absolute veto power** over any device, sensor, or controller that cannot be secured, ensuring that convenience never comes at the expense of enterprise safety.
- **The Director of Automation and Controls (Facilities)**
  This role serves as the voice of "physics" at the digital table. They are the expert on the mechanical realities

of the central plant and the specific industrial protocols (BACnet, Modbus) required to keep the buildings breathing. They ensure that IT security policies do not inadvertently cripple critical mechanical operations.

- **The Network Architect (IT)**
  Acting as the "digital civil engineer," this member designs the network's segmented topology. Their role is to architect the **OT VLANs** (Virtual Local Area Networks) that rigorously isolate dangerous, legacy building equipment from the student and administrative networks, preventing a hacked thermostat from becoming a gateway to the financial system.
- **The Campus Architect / Planner (Facilities):** The guardian of the physical legacy. This member ensures that the integration of technology aligns with the university's aesthetic and functional master plan. They prevent the "smart campus" from becoming a visual blight of antennas and conduit, ensuring that sensors are invisible and technology serves the architecture, not the other way around.
- **The Chief Data and Privacy Officer**
  The "conscience" of the committee. As the campus becomes a sensory apparatus collecting millions of data points on human movement, this leader is responsible for the ethical management of that telemetry. They define the rigid retention policies and privacy protections that prevent the smart campus from drifting into a surveillance state.
- **The Mobile Product Manager**
  The bridge to the user. Since the primary interface for the smart building is often the student's smartphone, this member ensures that the "building" reality aligns with the "app" experience. They advocate for a frictionless, single-sign-on experience, preventing a fragmented landscape where students are forced to download a dozen different vendor apps to navigate their day.
- **The Director/VP of Sustainability**
  The accountability officer. This member ensures that every smart investment is driving toward the university's decarbonization and ESG goals. They validate that the "smart" technology is actually delivering measurable energy reductions and not just generating digital noise.
- **The Procurement Lead**
  The enforcer. Vision without contractual teeth is hallucination. This member is responsible for translating the committee's technical standards into binding language in vendor contracts, ensuring that the "No Shadow OT" policy is enforced at the point of purchase.

Membership varies, as each institution has its own governance style and may evolve as the AI-Enabled Sustainable Smart Campus matures. Ultimately, membership should be based on the trust it fosters among the campus community and stakeholders. Hence, an inclusive membership policy is preferable over an exclusive one. This approach demands strong leadership from the SCSC to keep governance focused on action and to overcome the inherent inertia within institutions.

Responsibilities of the Smart Campus Steering Committee

The Smart Campus Steering Committee (SCSC) is more than just a passive advisory group or a monthly "status update" meeting where department heads politely exchange information. It functions as a working legislative body with binding authority, tasked with turning high-level vision into enforceable operational law. To achieve this, the SCSC has four main responsibilities that significantly impact how the university constructs, purchases, and manages its physical environment.

Defining the "Cyber-Physical Constitution" (The Smart Building Standard)

The primary duty of the SCSC is to develop, approve, and implement a standard technical specification for the built environment. In the construction sector, this is officially referred to as Division 25 (Integrated Automation). The SCSC elevates Division 25 from a generic technical document to the foundational "Constitution" of the smart campus.

This living document dictates the laws of the land for every contractor and vendor:

- **Mandatory Open Interoperability:** It explicitly mandates open-standard protocols (such as **BACnet/SC** or **MQTT**) as the only acceptable languages for building systems.
- **Prohibition of Walled Gardens:** It aggressively bans proprietary vendor ecosystems that lock the university into single-source contracts or obscure data ownership.
- **Security by Design:** It codifies the required cybersecurity hardening for every device before it is unboxed—mandating specific encryption standards, password complexity policies, and port security configurations.
- **Semantic Data Governance:** It imposes a rigorous data tagging ontology (such as **Project Haystack** or **Brick Schema**) required for Digital Twin integration, ensuring that a "temperature sensor" in one building is digitally identical to one in another.

The Procurement Gatekeeper (Stopping Shadow OT)

The SCSC functions as the main control point for the university's supply chain, enforcing a strict "No Shadow OT" policy. For many years, "Shadow OT"—where departments acquire smart devices without IT oversight—has been a major source of cyber risks.

The governing body sets a strict operational rule: devices with an IP address, a radio transmitter, or data collection capabilities cannot be purchased with a Facilities credit card, a departmental P-Card, or capital funds without SCSC approval. This safeguard stops well-intentioned academic departments or facilities managers from unintentionally purchasing 500 insecure, consumer-grade smart lightbulbs on Amazon and connecting them to the campus network, thereby safeguarding the enterprise network from potential vulnerabilities.

Synchronizing the Cyber-Physical Clock (Lifecycle Management)

The SCSC must manage the significant temporal tension between the digital and physical realms. IT hardware is like perishable produce, deteriorating every 3 to 5 years, while physical buildings are akin to geology, remaining sturdy for 50 to 100 years.

The SCSC oversees budgeting and enforcement for the "Mid-Life Digital Refresh." They need to establish a funding strategy that accounts for the obsolescence of the "smart" layer well before the physical infrastructure shows signs of wear. This guarantees that capital plans include dedicated funds to retire and replace sensors, controllers, and edge gateways 10 to 20 years after installation—avoiding the campus becoming a display of outdated, insecure technology inside modern buildings.

Unified Incident Command (The Joint Playbook)

Finally, the SCSC outlines the chain of command for situations in which the cyber-physical environment malfunctions. In a smart campus, a digital breach can lead to physical outcomes. For example, if a chemistry lab's ventilation controller is compromised, should it be treated as a Facilities emergency or an IT security incident?

The governance body addresses this by assisting the CTO/CFaO in developing and updating the Joint Incident Response Playbook. This playbook outlines the precise "rules of engagement" for cyber-physical incidents: who can disconnect a building's network, who sends physical responders to override locked doors, and how the university communicates with the public during digital threats that affect physical safety.

Critical Success Factor: The "Veto" and the "Waiver"

Effective governance structures must be equipped to handle exceptions in real-world settings, where deadlines are tight, budgets are limited, and infrastructure varies in age. Relying solely on a unanimous consensus risks causing gridlock. Therefore, the Smart Campus Steering Committee (SCSC) needs two formal mechanisms to navigate conflicts between digital security and physical needs: The Veto and The Waiver.

The Veto is the Absolute Stop

As IT and OT converge, the CISO's role expands beyond data protection to include safeguarding physical safety and operational continuity. Consequently, the governance charter should empower the CISO with sole authority to veto any system installation, connection, or procurement that poses unacceptable risks to the organization.

This power is absolute and operates regardless of the Facilities project timeline or sunk costs.

- **The Scenario:** A mechanical contractor attempts to commission a new chiller plant control system three weeks before a building opening. The CISO discovers that the system utilizes hardcoded, unchangeable administrative passwords and requires an open port to the public internet for vendor maintenance.
- **The Action:** The CISO exercises the Veto. The system is denied network access.
- **The Implication:** This may delay the building opening or force a costly change order. However, this friction is intentional. It signals to the entire vendor ecosystem and internal teams that **security is a non-negotiable life-safety requirement**, not a "nice-to-have" feature that can be traded away to meet a schedule.

The Waiver is the Pragmatic Bridge

On the other hand, the governance body needs to recognize the enduring nature of physical infrastructure. Unlike software that can be rewritten, physical assets typically last 30 to 50 years. We can't just "deprecate" a heating system because its controller becomes outdated. Hence, the CFaO should have the authority to approve a Waiver for Critical Legacy Systems.

This is not a "hall pass" to ignore security; it is a formal acceptance of risk coupled with mandatory mitigation.

- **The Scenario:** A 30-year-old boiler system is critical for heating a residence hall. Its controller runs on an obsolete operating system that cannot be patched or secured to modern standards. Replacing it would cost millions and require a six-month shutdown that the university cannot afford.
- **The Action:** The CFaO grants a Waiver, acknowledging that the asset *must* run for life safety and business continuity despite its technical obsolescence.
- **The Condition (Compensating Controls):** The Waiver triggers a mandatory set of **compensating controls** dictated by the governance body. Since the device cannot be secured, the surrounding environment is hardened. The governance body may mandate that the device be **air-gapped** (physically disconnected from the network), placed behind a rigorous **industrial firewall**, or restricted to a **locked VLAN** with zero inbound access. The Waiver allows operations to continue, but it documents the risk and traps the vulnerability so it cannot metastasize to the rest of the campus.

The CTO/CFaO establishes the SCSC, creating the institutional mechanism that turns their shared vision from a PowerPoint into an enforceable operational reality.

Reporting, Cadence, and Strategic Integration

Forming the Smart Campus Steering Committee (SCSC) and selecting the appropriate cross-functional leaders are only part of the challenge. If this committee functions in an administrative vacuum—issuing directives that the Cabinet disregards or the Board remains unaware of—it will soon become an irrelevant debating society.

For the SCSC to effectively serve as the core driver of the AI-powered sustainable smart campus, it must be integrated into top institutional leadership. The CTO/CFaO should clearly specify the reporting lines, the metrics it prioritizes, and how it processes data from the entire university ecosystem.

The Reporting Structure: Where the SCSC Elevates

Since the Smart Campus encompasses physical infrastructure, digital security, and academic continuity, the SCSC cannot be affiliated with just one department Vice President. If it reports solely to the CTO, it is seen as an IT committee; if it reports only to Facilities, it is viewed as a building committee.

The SCSC should report directly to the University Executive Cabinet through its Executive Sponsor—the Executive Vice President of Administration/Chief Operating Officer (COO)—with a dotted-line reporting relationship to the Provost.

Reporting to the COO, the SCSC directly links its cyber-physical responsibilities to the institution's financial stability and enterprise risk management. Maintaining a dotted line to the Provost ensures that operational efficiencies do not subtly compromise the academic mission or the "Living Laboratory" experience.

The Cadence of Accountability: When and What They Report

The rhythm of SCSC reporting needs to align with the institution's two clocks— the quick beat of digital risk and the slower, decade-long process of capital planning.

Quarterly Executive Briefings for the Cabinet Level Leaders

Every quarter, the CTO/CFaO delivers the Cyber-Physical Scorecard to the Cabinet. This is not just a project update; it serves as a strategic briefing emphasizing yield and risk. The SCSC reports on:

- **Algorithmic Yield and Decarbonization:** Hard data on energy savings generated by AI load-shedding, measured against the university's ESG goals.
- **Capital Debt Avoidance:** Quantification of deferred maintenance costs avoided through predictive sensor deployments.
- **The Risk Posture:** A summary of major "Vetos" exercised by the CISO (preventing dangerous shadow OT) and "Waivers" granted by the CFaO (documenting trapped risk in legacy systems).

Annual Strategic Presentations

Annually, the SCSC is required to present its results to the Cabinet and the Board of Trustees in accordance with usual practices. The report goes beyond operational metrics to emphasize fiduciary stewardship. The co-chairs detail how the integrated cyber-physical system protects the endowment, defends the research enterprise from digital threats, and preserves the physical campus as an essential asset for attracting students.

3. The Intake Mechanism: Gathering the Intelligence

To govern effectively, the SCSC must go beyond relying on its members' intuition. Instead, it functions as a clearinghouse, continuously gathering intelligence from the wider university ecosystem. This is accomplished through a structured, two-way flow of information with other governance bodies.

Taking in "Demand" from the Capital Projects Committee

The SCSC does not wait until a building is designed to review it. Instead, they analyze the 5- to 10-year Campus Master Plan. By examining the pipeline of upcoming capital projects, the SCSC can proactively plan the network density, sensor arrays, and API integrations needed. This approach allows them to incorporate the costs of "Digital Readiness" into the initial bond requests, avoiding the need to seek additional funds later.

Taking in "Ethics and Experience" from the AI and Mobile Councils

The SCSC regularly gathers policy guidance from the AI Governance Council and collects user feedback through the Digital Experience Committee. When the Mobile team reports that students find a new biometric building-access system intrusive or inconvenient, the SCSC uses that information to adjust the hardware strategy or update data privacy thresholds.

Taking in "Friction" from the Frontlines

Governance must be grounded in reality. The SCSC establishes a formal feedback loop with the frontline technicians—the network engineers and the HVAC mechanics. By reviewing aggregate work-order data and incident reports from the Joint Incident Command, the SCSC learns where its policies are creating impossible bottlenecks. If a new cybersecurity protocol prevents mechanics from resetting boilers during a freeze, the SCSC takes over this critical operation.

Ultimately, the SCSC succeeds because it functions as the central nervous system of the institution's physical framework. By providing the Cabinet with accurate and truthful information, showcasing fiduciary value to the Board, and consistently engaging with related governance bodies, the CTO/CFaO work together to make their shared vision a reliable, lasting, and well-funded institutional capability.

Data Ontology

In Operational Technology and Smart Buildings, a data ontology is a formal method for naming and defining the types, characteristics, and relationships of entities within a building's ecosystem. It functions like a "digital language" that enables a chiller from one manufacturer, a light sensor from another, and an occupancy counter from a third to communicate effectively and understand their shared environment.

Once data domains are effectively mapped and integrated using a Data Ontology (such as Brick or Haystack), they create a Unified Namespace. This stage enables data to be used in intelligent systems such as a Digital Twin.

- **Example of Convergence**
  A live sensor detects an energy spike, while static data identifies the specific motor model. Operational data confirms if a replacement is in stock, and spatial data maps the fastest route for the technician. Meanwhile, academic data ensures the repair is scheduled after the high-stakes exam happening in that zone.

Historically, OT systems such as HVAC, Lighting, and Access Control operated in isolated units using proprietary protocols. Data was often labeled inconsistently—one system might refer to a floor as "L1" while another calls it "Floor_01." An ontology advances this by going beyond basic data schemas, which only define structure, to include semantics that define meaning. It addresses:

- **What is this?**
  (e.g., A discharge air temperature sensor).
- **Where is it?**
  (e.g., Inside VAV Box 4, serving Room 202).
- **How does it relate to others?**
  (e.g., This VAV box is fed by Air Handling Unit 1).

Leveraging the data ontology enables Proactive or Condition-Based **Maintenance** rather than reactive alarms.

- **Example of applied Ontology in Action**
  Instead of waiting for a "High Temp" alarm in a room, the building's digital twin identifies the root cause upstream through its mapped relationships, creates a work ticket, and dispatches technicians proactively to rectify the problem before it impacts operations.

Core Components of a Building Ontology

A building ontology is the foundational semantic framework of a smart campus, providing a unified meaning for all physical assets, spaces, systems, and data points across various platforms, departments, and over time. Without this framework, sensors, analytics, and digital twins may rely on conflicting or scattered definitions of the campus environment. With a building ontology, the university establishes a shared language that enables HVAC systems, spatial models, work-order platforms, energy dashboards, and AI agents to interpret and interact with the built environment consistently.

A well-designed ontology defines relationships such as how spaces are arranged, how systems communicate, how assets operate, and how data is understood. It transforms raw telemetry and static records into meaningful insights by providing the conceptual foundation necessary for interoperability—across software, teams, and time. Whether based on open standards such as Brick, Haystack, or DTDL, or integrated into a hybrid institutional structure, a building ontology ensures that each data point is identifiable, traceable, and semantically consistent with the physical entity it represents.

A robust smart building ontology typically consists of three pillars:

1. **Entities (Classes):** The "nouns." These include physical assets (Boilers, Pumps), locations (Floors, Zones), and points (Temperature sensors, Fan speed commands).
2. **Relationships:** The "verbs." These define how entities interact.
   Examples: isLocatedIn, feeds, controls, isPointOf.
3. **Properties:** The "adjectives." These are static attributes of an entity, such

Building on that foundation, let's explore how these ontologies truly serve as the "brain" of a modern facility. Transitioning from basic data collection to an ontological model essentially involves creating a Knowledge Graph of the building.

From "Data Points" to "Digital Context"

In traditional OT, a data point is simply a series of numbers, like Point_42: 72.5. Without an ontology, a person must examine a spreadsheet to understand that Point_42 represents a temperature sensor in the West Wing.

With a comprehensive data ontology, that same point is enriched with semantic metadata. It doesn't merely display "72.5"; instead, it states: "I am a Room Temperature Sensor (Entity), I am currently reading 72.5°F (Value), I am located in Office 301 (Location), and I am part of VAV Terminal 12 (Relationship)."

One of the most "advanced" features of a formal ontology (like **Brick Schema**) is the ability to *infer* information.

- **Explicit Data:** "Pump A feeds Chiller B."
- **Ontological Inference:** If Chiller B is currently "Faulted," the system can automatically infer that any downstream Air Handling Units (AHUs) will soon see a rise in discharge air temperature.

Comparing Popular Ontologies: Haystack vs. Brick

As institutions modernize their cyber-physical environments, two open ontologies—Project Haystack and Brick—have emerged as the most widely adopted frameworks for giving structure and meaning to building data. Both seek to solve the same fundamental problem: the inconsistency, ambiguity, and proprietary labeling that make it difficult to normalize sensor data, interpret building systems, and scale digital twins across diverse platforms. Yet they approach this problem through different philosophical and technical lenses.

Haystack focuses on adaptable tagging and simple semantic models that enable quick annotation of real-world systems. Meanwhile, Brick provides a well-structured, graph-based schema aimed at precise modeling and sophisticated analytics. Recognizing the differences between these ontologies helps in choosing the best OT that matches an institution's maturity level, data strategy, and long-term interoperability goals.

To choose the right path for a smart building project, it helps to see how these two standards actually "talk" about a piece of equipment:

Feature	Project Haystack	Brick Schema
Core Philosophy	Tagging-based. It uses "markers" to describe things.	Relationship-based. It uses "triples" (Subject-Predicate-Object).
Logic Style	*Informal:* vav, cooling, temperature, sensor.	*Formal:* VAV_1 $\rightarrow$ hasPoint $\rightarrow$ Discharge_Air_Temp.
Best For	Quick implementation, BAS integration, and human readability.	AI/ML applications, complex Digital Twins, and cross-system automation.
Analogy	Like using Hashtags on social media to find content.	Like a Family Tree that shows exactly how everyone is related.

For CTOs, the data ontology acts as a key to prevent vendor "Lock-in." In the past, installing a proprietary lighting system meant being tied to their dashboard. By implementing an ontological layer, such as a Unified Namespace, you make your data 'portable.' This allows you to replace your dashboard provider on Friday and connect to a new AI-driven analytics platform by Monday. Since the data adheres to a standard ontology, the new software already understands what the chillers are and how the floors are configured.

Ultimately, data ontology serves as the bridge to the Autonomous Building. When a building understands its own structure, it can carry out "self-healing" tasks—such as rerouting airflow from an adjacent zone when a specific AHU fails—without any intervention from a technician.

Safety Technologies Are Force Multipliers—

Multipliers, yes - but only when integrated with human response. Safety remains a key focus in facilities modernization due to its immediate, human-centered value: safeguarding individuals, minimizing risks, and boosting trust within the campus. Technologies such as environmental hazard sensing, video analytics for situational awareness, access control for sensitive areas, and integrated alert systems can enhance the university's duty of care. This is especially important on large, complex campuses where incidents can happen rapidly and staffing resources are limited. Therefore, it is an area of governance that requires special attention and consideration.

Safety technology's promise is genuine. Environmental sensors can issue early warnings about conditions that could harm occupant health or cause operational disruptions—such as smoke, poor air quality, carbon monoxide, temperature spikes, or unusual sounds that require attention. Video analytics can identify anomalies more quickly than human monitoring alone, especially in busy or high-risk areas. When systems are integrated—combining analytics, access control, dispatch workflows, and communications—they form a seamless network that reduces the time from detection to response. This integration enhances outcomes by increasing speed, providing clearer information, and improving coordination.

Adding more sensors alone does not make a campus safer. Safety is enhanced when sensing is part of a broader operational system that includes clear decision-making authority, trained responders, rehearsed protocols, and measures to prevent overreaction. Essentially, safety technologies serve as effective "force multipliers" only when paired with a disciplined human response system. Governance is vital for guiding the effective use of technology and coordinating human efforts. Keep these questions in mind in this context.

Who Receives an Alert at 2:00 a.m.?

The initial issue concerns operational reality. Many safety incidents happen outside regular hours, when staffing is limited, but the campus remains active. The institution needs to clearly specify who should receive different types of alerts after hours—such as campus safety, on-call facilities staff, residence life duty personnel, IT security on-call, or a contracted monitoring service—and outline how escalation occurs if the first contact is unresponsive. This issue is not just about routing; it involves resilience planning. Alerts must reach someone with both the authority and capacity to respond. Without this, detection may give a false sense of security, while response efforts remain slow or inconsistent.

What Triage Logic Prevents False Positives From Becoming Panic Events?

Sensors and analytics produce signals, not certainty. A robust safety approach employs triage logic to distinguish urgent threats from uncertain anomalies, preventing the community from panicking or ignoring alerts. Triage should be layered: low-confidence signals might require verification, such as camera checks, sensor correlation, or quick patrols, while high-confidence signals demand immediate action. Institutions must set thresholds, corroboration rules, and confidence scores to ensure not every anomaly is treated as an emergency. Without proper discipline, false positives can cause reputational damage, unnecessary disruption, and alert fatigue, ultimately weakening the safety culture that the technology aims to enhance.

What Actions Are Authorized, and by Whom?

Integrated safety systems frequently enable critical operations such as locking or unlocking doors, triggering mass alerts, sounding alarms, deploying responders, or switching building systems to protective states. These significant actions require explicit governance. The institution must specify which actions are permitted at each alert level, who has authority to initiate them, and what checks are necessary before escalation. In higher education, where civil liberties, academic environment, and community trust are crucial, decision rights should be clear and proportionate. The aim is not to hinder response but to ensure it is legitimate, defensible, and consistent with institutional values.

How Are Incidents Documented, Reviewed, and Improved?

Safety modernization requires a feedback loop. Every incident—real or false—should be documented systematically: what triggered the alert, decisions made, the duration of each step, and the results. This provides a foundation for ongoing improvement: adjusting thresholds, refining workflows, retraining staff, enhancing signage or policies, and strengthening integration points. Regular review practices also build accountability and trust. When the community observes that the institution evaluates outcomes and learns from mistakes, confidence increases. Conversely, if incidents are discussed only informally, suspicion arises, and the technology risks being seen as unchecked power.

How Do We Avoid "Automation Escalation" Where a Minor Signal Triggers a Disproportionate Response?

A common overlooked risk in smart safety systems is automation escalation: when minor signals—such as noise anomalies, a sensor spike, or brief access irregularities—cause disproportionately large responses, leading to unnecessary fear or disruption. This risk arises when systems are closely linked without sufficient human judgment gates, especially during cascading automated actions. To mitigate this, institutions should implement "proportionality controls" like verification steps, staged escalation, time limits, and clear de-escalation procedures. The goal is to design systems that promote calm, competent responses while avoiding automated overreach.

The Guiding Principle: Accelerate Human Judgment, Don't Replace Decision Rights

The fundamental idea of strategic design is simple: automation can enhance the speed of human judgment, but responsible decision-making remains irreplaceable. On campus, safety transcends technical considerations and is fundamentally a social achievement grounded in legitimacy, proportionality, and trust. Consequently, security and facilities leaders must approach the human-in-the-loop model with the same thoroughness they apply to network architecture—defining roles, escalation protocols, authorization processes, training, and rehearsals.

When this human response system is clearly established and consistently practiced, safety technologies greatly improve safety by boosting situational awareness, decreasing response times, and producing better outcomes. Conversely, without it, an institution might have safety tools on campus that don't actually make it safer, interconnected systems that lack coordination, and automation that operates without accountability.

Governing Digital Identity

Another area that might need new governance is digital identity and access management. Modern access control goes beyond physical locks, serving as a campus-wide digital identity system that subtly guides movement, permissions, and user experience—making users feel welcome or restricted. It encompasses smart cards, mobile credentials, touchless entry, cloud-based permissions, visitor management, and biometric solutions. These offer clear benefits: increased user convenience, stronger security for sensitive assets, better audit trails, and faster permission updates.

Access systems based on a digital identity greatly impact the user experience because they operate at the "micro-friction" level — the small, repeated moments people face daily. When credentials fail, doors behave unpredictably, guest procedures feel strict, or accommodations are poorly handled, the campus seems unsafe. Instead, it feels hostile, overly surveilled, or uncaring. In higher education, where community trust, inclusion, and mission are crucial, these small moments serve as a test of whether modernization genuinely benefits individuals or merely exercises control over them.

Facilities transformation should approach access control as a dignity system rather than just a security measure. While security is essential, dignity influences whether it is accepted and deemed legitimate. To design for dignity, access must be dependable in real-world conditions, inherently inclusive, protect privacy, and be welcoming without seeming naïve.

Security That Works When Conditions Don't

Access control issues are most acute at the edges: a student returning late, staff carrying equipment, a researcher with gloves, a visitor during a network outage, or residence hall peak traffic after an event. If the system only functions under perfect conditions, the community perceives it as unreliable, leading people to create workarounds that compromise both security and user experience.

A dignity-first design requires operational resilience, including graceful degradation during network outages, predictable cloud-disruption responses, battery-failure contingency plans, and surge management during events and class changes. It also involves disciplined monitoring and quick responses: doors that "fail closed" in ways that trap people or cause panic are unsafe, not secure. Reliable access control is more than a feature; it's a daily responsibility the institution must maintain.

Inclusivity by Design

Access control is essential for genuine inclusion. Using a one-size-fits-all user model misses many needs: people with mobility aids, those with temporary injuries, neurodiverse users seeking predictability, staff moving carts, students with instruments, and community members needing accommodations. Inclusivity should be a default feature, not an afterthought.

This means that ADA-compliant and accessible routes must be reliable both physically and digitally. Door timing should be consistent and sufficient. Touchless solutions require thoughtful placement. Mobile-friendly interfaces should account for users without smartphones or with limited connectivity. Most crucially, accommodations should be normalized— not seen as special exceptions that demand repeated requests, public disclosure, or awkward negotiations at the entrance. A dignity-centered approach guarantees that people can enter, move around freely, and feel included without barriers that imply they are "outside the norm."

Strong Privacy Posture and Biometrics

Identity systems can quickly become sources of trust issues or liabilities, especially if biometrics are introduced, expanded, or even discussed without proper governance. Although biometric methods improve convenience and reduce credential sharing, they also raise valid concerns, such as the permanence of identifiers, potential misuse, function creep, and unequal impacts across populations. The perception of biometric overreach alone can trigger widespread resistance across campuses.

A dignity-first approach treats privacy as a core element of architecture and governance. When using biometrics, organizations should follow the highest levels of justification and restraint: clearly define purposes, opt for opt-in systems when possible, enforce strict data retention policies, ensure secure storage, implement strong access controls, and communicate transparently with the community. They should also be able to specify what they are not doing—such as no hidden facial recognition, no behavioral analysis, and no secondary uses beyond the original intent. Privacy is not just about compliance; it's about building legitimacy.

Guest Access and Temporary Credentials:

Universities are inherently open environments, designed to welcome prospective students, visiting scholars, event attendees, families, contractors, and community partners—often in large numbers. Guest and temporary access are essential aspects of their daily operations, not rare exceptions. When visitor procedures become complicated, excessively strict, or inconsistent, it unintentionally sends the message: "You are not welcome unless you're already on campus."

A dignity-centered access approach sets clear policies for guest entry to guarantee safety and maintain hospitality standards. This includes issuing credentials with time limits, assigning permissions according to roles, streamlining check-in procedures, managing access for events, and establishing contractor protocols to prevent unauthorized credential sharing. It also defines accountability, such as who authorizes a guest, which areas are appropriate, and the proper ways to revoke access. An effective guest access system reduces risks by discouraging workarounds.

Building Trust

In higher education, trust develops through small, daily experiences. Access control is one of the most common interactions between individuals and institutional systems. When it functions seamlessly, people hardly notice — which is a sign of success. However, when it fails, individuals feel monitored, hindered, or frustrated, making modernization emotionally and politically costly.

Designing access as a dignity system protects the institution's core goals: a safer, more humane campus. It ensures security improves the campus experience without creating obstacles. That digital identity supports learning, research, and community engagement—avoiding it from becoming a source of friction that hinders the mission.

Governing Smart Campus Data

A core duty of unified leadership involves managing the institution's Data Estate. While most higher education institutions already have some form of data governance—mainly focused on student records, institutional reports, compliance, and privacy—an AI-Enabled Sustainable Smart Campus significantly broadens the scope and importance of this effort. Traditional data governance systems were not built to handle the continuous, high-volume cyber-physical telemetry produced by modern OT systems, IoT sensors, energy platforms, space-use analytics, safety technologies, and digital twin ecosystems. Consequently, the smart campus initiative introduces new data categories, institutional risks, and expectations for stewardship that demand more extensive, integrated governance capabilities.

Unlike traditional enterprise data, smart campus data blurs the lines between digital systems, physical spaces, and human experiences. Data such as occupancy signals, IAQ measurements, equipment telemetry, energy profiles, access control logs, and predictive maintenance results carry operational, ethical, privacy, and safety considerations that extend beyond standard institutional datasets. Managing this expanded Data Estate requires cross-functional decision-making authority, clear policies on data collection and retention, transparent definitions of purpose, strong security measures, and accountability frameworks. It also needs active oversight to prevent "mission creep"—the unintended broadening of data use without community approval or strategic intent.

Furthermore, a modern Data Estate should be managed as a core enterprise asset rather than just a technology byproduct. It serves as the foundation for training AI models, building digital twins, automating operational decisions, and assessing sustainability and resilience. Without strong governance, institutions risk deploying systems that are technically advanced but lack operational trustworthiness—undermining confidence in analytics, leading to data silos, and increasing cyber-physical vulnerabilities on campus.

The governance structures must therefore evolve to encompass this expanded landscape: to set the policies, decision rights, security expectations, and ethical guardrails that enable the Data Estate to be used responsibly and at scale. Chapter 10 examines the architecture and enablement of the institutional Data Estate in greater detail. Still, its effective use begins here—with leadership recognizing that governing data is not an ancillary task, but a defining capability of the AI-enabled, integrated, and trustworthy smart campus.

Governing Digital Sovereignty

A strong, smart campus governance strategy needs a broader framework called Digital Sovereignty. This strategic focus requires an institution to retain full control over its entire intelligence ecosystem. As campuses become more 'sentient,' relying on a third-party vendor to manage core logic or access data risks outsourcing the institution's collective memory and central nervous system. Digital Sovereignty guarantees that, even if a vendor rebrands, goes bankrupt, or changes its pricing, the campus remains aware and functional.

The Pillars of a Digital Sovereignty Strategy

Data Ownership and Portability (The Foundation)
Institutions must move beyond simply "owning" their data in a legal sense to owning it in a technical sense. This means all data generated by campus IoT, building systems, and occupant interactions must be accessible in real-time via open APIs and stored in a vendor-neutral **Data Lake.**

For example, contracts must explicitly state that the institution has the right to extract its data in a structured, machine-readable format without egress fees or proprietary encryption that requires vendor-specific tools to decipher.

Open Standards and Ontologies
Digital sovereignty is maintained through the use of open-source semantic layers. By adopting standards such as Project Haystack or Brick Schema, the institution ensures that the "meaning" of its data—the relationships among sensors, spaces, and systems—is not trapped within a proprietary vendor dashboard. For example, if a vendor's visualization platform fails, the underlying structured data and its relationships (the ontology) can be migrated to a new platform with minimal remapping.

Decoupling the Layered Architecture
A sovereign digital twin distinguishes among the Data Layer (which stores information), the Intelligence Layer (where AI models analyze data), and the Presentation Layer (the user interface). Its purpose is to avoid "Vertical Monoliths," where a single vendor supplies sensors, the database, and the user interface. By separating these layers, institutions can replace individual components, such as an AI inference model, without overhauling the entire system.

Mitigating the "Blindness" Risk
The biggest risk to a smart campus is Digital Blindness, which occurs when a proprietary Digital Twin becomes inaccessible after the provider's cloud service ends. To avoid this, a Digital Sovereignty strategy requires:

- **Escrow Agreements**
  For essential proprietary software, the source code must be kept in escrow to ensure the university can maintain the system if the vendor ceases operations.
- **Edge Autonomy**
  Core operational logic, such as life-safety and fundamental HVAC scheduling, should be implemented on on-campus hardware ('edge') rather than relying solely on a vendor's remote cloud for essential functions.

By emphasizing Digital Sovereignty within governance, the institution guarantees its digital twin becomes a lasting institutional asset rather than a temporary vendor subscription. This approach shifts the smart campus from merely a "managed service" to a sovereign infrastructure that evolves alongside the university.

Chapter Summary

Chapter 9 emphasizes that governance is the critical factor determining whether an AI-enabled sustainable smart campus succeeds or fails. Today's campuses function as integrated cyber-physical ecosystems, where daily experiences, safety, sustainability, and credibility hinge on the coordinated actions of digital and physical systems. A lack of disciplined, shared governance can lead to fragmented pilots, shadow operational technology (OT), inconsistent security measures, and conflicting data interpretations—risks highlighted repeatedly throughout the chapter.

Effective governance is not just a bureaucratic layer but the framework of trust within an institution. It ensures that as systems become more interconnected, the campus remains not only vulnerable but also more predictable, secure, and aligned with its goals. The establishment of the Smart Campus Steering Committee (SCSC)—co-led by the CTO and CFAO—exemplifies this change. Its purpose is to unify decision-making authority, promote open standards, prevent shadow OT activities, and oversee the cyber-physical boundaries, making it the central coordinating body for the campus. This structure enables proactive governance: identifying risks early, guiding investments, and ensuring coherence across the institution's digital and physical aspects.

Governance allows the smart campus to grow safely, transparently, and intentionally. It integrates technology, facilities, data, and human elements into a unified, adaptive framework. By incorporating privacy, interoperability, security, and ethical use into the architecture, governance maintains institutional trust while fostering responsible innovation.

As the campus becomes a more intelligent and data-driven environment, governance plays a crucial role in ensuring that modernization aligns with the university's core mission—focusing on resilience, decarbonization, seamless user experience, and fostering a true "living laboratory" for education and research. In this context, governance is not the final goal of the smart campus initiative but an ongoing process that maintains coherence in progress and makes outcomes measurable. It reflects the institution's dedication to managing its cyber-physical future with integrity, transparency, and shared purpose.

Chapter 10

The Smart Campus Data Estate

The good governance of data is crucial for smart campuses because we insist on high reliability and perfect operational intelligence.

As OT systems converge with campus IT and analytics capabilities, the university begins to assemble something more consequential than a collection of sensors and dashboards. It begins building a **facilities data estate**—a continuously growing body of information about how the campus operates and how people experience the built environment. This estate can include sensor streams, equipment metadata, spatial context, work order history, access and usage signals, indoor air quality measures, energy and carbon performance benchmarks, and the operational "story" of buildings over time.

That estate is powerful because it converts the campus from a set of independent systems into a learning system. It can enable repeatable optimization, better capital planning, predictive maintenance, and—when readiness is earned—digital twin use cases that support simulation, scenario planning, and resilience modeling. In executive terms, the data estate is the foundation for moving from reactive operations to strategic, continuously improving stewardship of the physical campus.

Once a data estate is established, responsibilities grow. Data enables capabilities, which in turn create obligations. A campus that can monitor more may also risk overstepping bounds. As OT data is integrated with identity systems, spatial analytics, and behavioral patterns, information that once seemed purely technical can become socially sensitive. The institution's choices regarding privacy, access, ethical use, data retention, and transparency are as crucial as technological choices. These governance decisions are ultimately human choices—shaped by values, boundaries, and institutional identity.

A mature facilities transformation program views the data estate as a stewardship responsibility: a strategic asset managed responsibly, used for legitimate purposes, and safeguarded against misuse. This stewardship approach—more than any dashboard—shapes whether the campus community views modernization as helpful infrastructure or as a potentially suspicious tool of control. This chapter examines the types of data that support the smart campus and discusses the importance of ontologies.

Understanding Estate Data

As campuses transition into digitally enabled, cyber-physical ecosystems, data's role shifts from a mere operational byproduct to a vital institutional asset that drives intelligence, resilience, and strategic decisions. However, "data" is not a single entity. The modern smart campus relies on various categories of estate data, each with distinct purposes, levels of detail, governance needs, and operational risks. Recognizing these data types is crucial, as the quality and consistency of the data estate ultimately influence whether analytics, AI models, and digital twins provide reliable insights or instead generate noise, blind spots, and institutional risks.

Estate data reflects the full reality of the campus: the **spatial form** (buildings, rooms, utility networks), the **static truth** (asset inventories, configurations, commissioning records), and the **live operational signals** (temperature, occupancy, energy flow, equipment behavior) that reveal how the campus behaves moment to moment. When these dimensions remain fragmented—locked in departmental spreadsheets, vendor dashboards, or outdated building automation systems—the institution loses its ability to forecast failures, optimize energy, validate sustainability claims, or design spaces that truly serve students and faculty. But when these data types are integrated into a governed, interoperable architecture, they become the foundation for predictive maintenance, intelligent scheduling, carbon-aware operations, and a campus experience that is visibly responsive and trustworthy.

This section outlines the main categories of estate data and explains their importance for an AI-enabled sustainable smart campus. It clarifies the differences among spatial, static, and live data, and describes how integrating these domains enhances safety, reliability, sustainability, and mission coherence. By fostering a common understanding of estate data across IT, Facilities, Sustainability, Research, and Executive Leadership, institutions lay the semantic and operational groundwork to scale digital twins, automate insights, and manage the built environment with confidence and accountability.

Structured vs Unstructured Data

In the context of an AI-enabled sustainable smart campus, data is the foundational asset that fuels both operational efficiency and sustainability. While higher education has historically relied on **structured data**—information organized into well-defined formats such as spreadsheets, SQL databases, and registrar records—the true potential of a modern campus lies in harnessing **unstructured data**.

Structured Data is quantitative and follows a rigid, predefined schema. On a campus, this includes student ID numbers, GPS coordinates of assets, utility billing records, and temperature setpoints. Because it is highly organized, it is easily searchable and can be ingested directly by traditional analytics tools.

Unstructured Data does not have a pre-defined model or organization. It is often qualitative and heavy on context. In a campus environment, this manifests as video feeds from security cameras, natural-language maintenance logs, PDF floor plans, social media sentiment from the student body, and raw audio.

In a smart campus ecosystem, unstructured data acts as the "sensory intelligence" that fills the gaps left by traditional databases. While structured data provides the metrics (the *what*), unstructured data provides the context (the *why* and *how*).

Exhibit 9. Structured vs Unstructured Data Examples

Data Domain	Structured Data Examples (Quantitative/Schema-based)	Unstructured Data Examples (Qualitative/Context-based)
Buildings and Infrastructure	BACnet/Modbus sensor readings (temperature, CO_2), HVAC setpoints, electrical load metrics, and GIS coordinates of utility lines.	PDF floor plans, hand-annotated blueprints, photographs of mechanical equipment, LiDAR point clouds of structural "as-builts."
Campus Safety	Access control logs (time/location of badge swipes), fire alarm trigger history, and emergency blue light phone locations.	Live CCTV video feeds, audio recordings of security dispatches, witness statements in text reports, and social media safety alerts.

Data Domain	Structured Data Examples (Quantitative/Schema-based)	Unstructured Data Examples (Qualitative/Context-based)
Sustainability and Energy	Smart meter interval data (kWh), solar array output, water flow rates, greenhouse gas emissions calculations.	Environmental impact reports, thermal imaging photos of building envelopes, and narrative summaries of waste audits.
Space and Occupancy	Room reservation schedules, Wi-Fi device counts per AP, library seat booking data, and parking garage stall counts.	Heat maps derived from computer vision, student feedback on study space "vibe," and video footage of foot-traffic patterns in atria.
Maintenance and Operations	Work order numbers, equipment age/lifecycle dates, inventory SKU counts, and replacement part costs.	Technician "field notes" describing intermittent issues, voice memos from inspections, manufacturer manuals, and warranty PDFs.
Student Life and Academic	Registrar records (grades, enrollment), LMS login timestamps, financial aid status, and dining hall meal plan usage.	Discussion board posts, video recordings of lectures, student sentiment on campus forums, and open-ended survey responses.

Unstructured data represents the vast majority of information generated on a campus, yet it is often the least utilized. Its role is to provide:

- **Contextual Depth**
  Moving beyond simple sensor readings to understand the environment through natural language and visual cues.
- **Human-Centric Insights**
  Capturing the sentiment, behavior, and needs of students and faculty that cannot be distilled into a spreadsheet.
- **Historical Narrative**
  Leveraging decades of digitized documents, floor plans, and maintenance logs to inform future structural and operational decisions.

Key Use Cases of Unstructured Data in the Smart Campus

Predictive Maintenance and Asset Management

In addition to monitoring structured data from vibration or heat sensors, AI can interpret unstructured technician notes and work order descriptions. By recognizing patterns in the language used to describe equipment "quirks" over several years, the system can forecast potential failures before hardware alarms are triggered. Furthermore, computer vision applied to infrastructure photos can detect cracks or corrosion early, preventing critical issues.

Campus Safety and Intelligent Security

Modern security extends beyond just recording video; it includes real-time analysis of unstructured video feeds. AI systems can identify "anomalous behavior,' such as unexpected crowd gatherings or someone entering a restricted area, and notify staff promptly. Additionally, audio analytics can detect specific sounds, such as breaking glass or raised voices, enabling faster responses than manual monitoring.

Student Engagement and Sentiment Analysis

Universities can monitor campus life in real time by analyzing unstructured social media mentions, forum posts, and survey comments. Natural Language Processing (NLP) helps categorize students' overall sentiment about campus services, dining, or safety issues, enabling administrators to proactively address problems early.

Energy Optimization and Occupancy

Although thermostats provide organized temperature data, unstructured data from Wi-Fi traffic and optical sensors provide a more accurate picture of space usage. Knowing that a lounge is crowded every Tuesday at 2:00 PM—irrespective of the scheduled calendar—enables the building automation system to pre-cool the area, optimizing comfort and energy efficiency.

Institutional Knowledge Recovery

Many campuses face challenges with "dark data"—such as unorganized PDF blueprints, handwritten electrical schematics, and outdated contracts. Digitizing and indexing these using Optical Character Recognition (OCR) allows a smart campus to develop a searchable "knowledge graph." This ensures that a new facilities manager can quickly access the current "as-built" details of a building that's been there for 50 years.

Spatial Estate Data

Grasping the spatial aspect of the campus is essential for any AI-enabled, sustainable smart-campus plan. Spatial data goes beyond simple floor plans or architectural diagrams; it serves as the vital connector that unites people, assets, systems, and environments into a unified operational framework.

When well-managed and organized, spatial data shows how the campus truly operates: how air circulates, how people move, how energy is distributed, and how spaces are used, stressed, or underutilized. However, on most campuses, spatial information is fragmented, inconsistent, and stored in outdated formats, which reduces its strategic usefulness.

This section emphasizes the essential elements of spatial data and argues that a modern smart campus should see it not as a fixed record but as a dynamic, interoperable asset. This asset facilitates digital twins, operational intelligence, and human-centered design.

The three-dimensional and geographic attributes of the campus define the spatial data. It is often categorized into three primary layers:

- **Building Information Modeling (BIM)**
  High-fidelity 3D models that detail building geometry, spatial relationships, and utility infrastructure. Digital twins leverage BIM to visualize what lies behind walls, including HVAC ductwork, electrical conduits, and plumbing systems.
- **Geographic Information Systems (GIS)**
  Large-scale data detailing the broader campus layout, including utility mainlines, green spaces, topography, and environmental zones.
- **Indoor and Reality Capture**
  2D/3D CAD drawings, LiDAR scans (laser mapping), and photogrammetry. These provide "as-built" accuracy for indoor navigation and terrain analysis.

Spatial data does more than show where things are; it provides the visual and contextual framework for **Location-Based Intelligence**. By integrating spatial data with AI, organizations can move from static maps to dynamic operations:

- **Predictive Maintenance**
  Using AR to overlay "digital X-ray" views of infrastructure for technicians, showing exactly where a faulty valve is located behind a ceiling tile.
- **Resource Optimization**
  AI algorithms analyze spatial data to optimize routes for autonomous shuttles or to plan security patrols based on area vulnerability and line-of-sight.
- **Space Utilization**
  Real-time analysis of how physical spaces are actually used, enabling data-driven decisions on campus expansion or room scheduling.

By viewing the campus as a "Canvas," spatial estate data ensures that each sensor reading and operational event has an exact coordinate, transforming raw data into meaningful, contextual insights.

Static Estate Data

Static estate data underpins the campus's long-term digital memory. While spatial data indicates locations, static estate data describes the identity, specifications, constraints, and operational histories of assets. It includes essential records such as equipment metadata, asset registries, commissioning documents, warranty details, maintenance logs, and the unchanging characteristics of the built environment that influence campus operations over time.

On a smart campus, this data transcends mere background documentation; it acts as the definitive source of truth, supporting precise analytics, reliable digital twins, and practical operational insights. Proper curation of static estate data anchors the cyber-physical system, guaranteeing that all optimizations, predictions, and automated choices reflect the actual capabilities and constraints of the physical infrastructure.

Static data provides the factual and historical context required to understand asset behavior. It is primarily composed of three knowledge layers:

- **Technical Specifications and As-Builts**
  High-fidelity documentation, including Building Information Models (BIM) and CAD drawings. These define the "as-constructed" reality of architectural, structural, and MEP (Mechanical, Electrical, Plumbing) systems.
- **Asset Lifecycle Records**
  The historical "paper trail" of physical infrastructure. This includes equipment manuals, warranty details, operational playbooks, and comprehensive maintenance logs that track past performance and failures.
- **Governance and Organizational Context**
  The administrative framework of the estate. This includes organizational charts, departmental boundaries, policy documents, and historical records that define how the space is governed and intended to function.

Static data serves as the foundation for advancing from basic monitoring to Contextual Intelligence. By offering a reliable reference point, it enables AI to shift focus from merely describing "what is happening" to understanding "why it matters" through:

- **Informed Predictive Maintenance**
  AI algorithms analyze live sensor anomalies in conjunction with historical maintenance logs and manufacturer specifications. When a pump vibrates, the AI examines its "Static Data" — including warranty, age, and last repair — to decide whether it requires a straightforward tune-up or a more comprehensive intervention, such as replacement.

- **Automated Operational Protocols**
  When a fault occurs, the system can automatically access the appropriate digital manual or operational playbook, giving technicians instant, asset-specific guidance instructions.
- **Trend Analysis and Simulation**
  Digital twins incorporate historical sensor data and static building details to run "what-if" scenarios, such as predicting how a 40-year-old HVAC system would perform during an extreme heatwave versus its initial design specifications.

Real-time data shows the building's current "pulse," while static estate data offers the "medical history" necessary for an informed diagnosis.

Live and Dynamic Data

Live and dynamic data represent the real-time pulse of the campus. Unlike spatial or static estate data—which describe *where* assets are and *what* they are—dynamic data captures *how things are behaving right now* and *how conditions evolve moment-to-moment.* This includes streaming telemetry from HVAC and BAS controllers, sensor-based readings of occupancy, environmental quality, vibration, energy and water flows, security events, and any other signal that reflects the active state of the campus ecosystem.

In an AI-enabled sustainable smart campus, this data is indispensable: it fuels predictive maintenance, continuous commissioning, carbon-aware load management, adaptive environmental controls, responsive mobility services, and the digital twin's ability to mirror reality with precision and immediacy. Properly governed, live data transforms the campus from a static collection of systems into a sensing, learning, and continuously optimizing organism—one capable of rapid insight, anticipatory action, and measurable institutional improvement.

Dynamic data reflects the "now,' offering essential material for instant decisions and ongoing improvements. It consists of several rapid data streams:

- **IoT and Environmental Feeds**
  Real-time data from smart meters (energy/water), occupancy sensors, and air quality monitors. This also includes external environmental data—such as weather patterns, outdoor humidity, and solar load—that influence how a building should respond to its surroundings.
- **Building Management Systems (BMS)**
  The operational status of "heavy" infrastructure, such as HVAC setpoints, fan speeds, elevator positions, and lighting levels.
- **User and Behavioral Data**
  Aggregated, anonymized data showing how people navigate and engage with the campus. This includes Wi-Fi usage density, parking space availability, and EV charging station locations utilization.
- **IT and Security Logs**
  Real-time network health metrics, bandwidth usage, and security alerts from motion detectors or access control systems.

Live data acts as the "fuel" for the AI engine, facilitating the shift from manual oversight to Autonomous Operations.

- **Dynamic Optimization**
  Rather than operating the HVAC on a fixed schedule, AI examines real-time occupancy and weather data to adjust temperature setpoints — cooling a crowded lecture hall while switching to "eco-mode" for an empty office wing.

- **Anomaly Detection and Rapid Response**
  AI algorithms monitor data streams to detect hidden issues, such as a water leak indicated by a midnight flow spike or a cybersecurity threat revealed by atypical network activity.
- **Closed-Loop Simulations**
  The digital twin employs real-time data to perform ongoing "What-If" simulations. For instance, it can forecast how a sudden increase in outdoor temperature might affect energy expenses over the next four hours and proactively modify the building's thermal mass accordingly.
- **Enhanced User Experience**
  Real-time data feeds campus apps that guide students to available study spaces or provide commuters with live shuttle locations and parking availability.

Static data can be seen as the "medical history,' while live data represents the "vital signs." It offers the real-time feedback needed for a campus to operate in a responsive, efficient, and resilient manner.

Operational and Business Systems Data

Operational and business systems capture the core functions of the institution behind the physical campus, integrating administrative, academic, financial, and service activities that support daily operations. This includes ERP transactions, work-order data from the CMMS, student information and scheduling details, research compliance records, financial and procurement activities, and various digital signals from academic technologies, safety systems, and support services. Unlike live telemetry—which shows real-time environmental conditions—operational data indicates organizational plans: who is scheduled to teach where, space reservations, planned maintenance, budget allocations, and potential resource constraints or bottlenecks.

In a smart campus ecosystem, this data provides critical context. It supports AI models, enhances predictive analytics, assists in space optimization, and connects the cyber-physical environment to the institution's strategic, financial, and academic objectives. When well-governed and integrated into the broader Data Estate, data from operational and business systems enable the campus to function not just as a streamlined physical space but as a unified, mission-driven enterprise — aligning routine activities with long-term institutional aims through digital intelligence.

Operational and business data connect the physical assets to the organizational mission through three primary systems:

- **Facilities Management (FM) and Asset Logs**
  The detailed record of "work" encompasses maintenance schedules, work order histories, and asset depreciation. It serves as a connection between a sensor alert and the deployment of a technician.
- **Enterprise Resource Planning (ERP)**
  The essential financial and human infrastructure includes procurement records, vendor contracts, HR staffing levels, and budget allocations. It provides details not only about what an asset is but also about its cost and the parties responsible for it.
- **Student and Stakeholder Systems (CRM/LMS)**
  The engagement record comprises student enrollment information, alum interactions, and activity logs from campus services. This data indicates the "purpose" of a space at any specific moment.

This operational data enables the AI-powered estate to advance from purely engineering metrics to Business Intelligence by:

- **Workflow Efficiency**
  AI examines Facilities Management logs to pinpoint bottlenecks in repair processes. For instance, if

HVAC repairs regularly take longer in a particular building, the AI can link this to procurement delays for certain parts or staffing shortages.

- **Intelligent Procurement**
  By analyzing ERP and procurement data, AI can uncover opportunities for "Green Sourcing" or volume-based discounts, consolidating purchases across the campus to lower costs and reduce the carbon footprint.
- **Predictive Financial Modeling**
  Integrating energy usage (Live Data) with utility billing and budget data (Business Data) enables precise financial forecasting, supporting leadership in planning future capital expenditures or sustainability investments.
- **Service Personalization**
  CRM data enables the campus to be more "responsive" to its community. When student engagement signals an increase in the need for collaborative study spaces, the digital twin can recommend converting underused administrative offices to meet that demand.

If Live Data is the "pulse," Static Data is the "medical history," and Operational Data is the "lifestyle and insurance record," it describes the costs, effort, and the primary reason for the estate's existence.

Academic, Research, and Experiential Data

Academic, research, and experiential data are the institution's core intellectual assets. They capture the processes of learning, discovery, and engagement among students, faculty, and researchers within the campus ecosystem. This data includes course schedules, pedagogical interaction patterns, research instrument outputs, laboratory environmental conditions, experiential learning analytics, learning management system signals, and the operational details of specialized research facilities. Unlike operational data—which reflects administrative goals—and live telemetry—which monitors physical conditions—academic and research data illustrate how the campus fulfills its main mission: teaching, learning, and inquiry.

In a smart campus setting, this data gains new strategic value. It offers insights needed to enhance learning spaces, safeguard sensitive research, analyze usage trends in valuable labs, and design facilities and services that support changing instructional approaches. When incorporated into the broader Data Estate—managed with strict privacy, ethical standards, and clear purpose—academic and research data allow the campus to act as a "living laboratory." Here, students interact directly with cyber-physical systems, researchers utilize real-world datasets, and campus leaders align space, technology, and resources with pedagogical and scientific goals. With proper governance, this data turns the built environment into an innovation platform, helping the campus pursue its mission with clarity, integrity, and tangible results.

Academic, Research, and Experiential data capture the intellectual and social engagement of the campus community, often categorized into three pillars:

- **Learning and Engagement Metrics**
  Aggregated data from Learning Management Systems (LMS) and library platforms. This tracks course engagement, digital resource utilization, and peak academic activity cycles.
- **Student Success and Wellness Indicators**
  Anonymized patterns of interactions with student services, including academic advising, wellness centers, and career services.
- **The "Living Lab" Research Output**
  Data generated by campus-based research projects, particularly those using the smart campus infrastructure itself to study sustainability, energy behavior, or urban planning.

- **Sentiment and Feedback Loops**
  Qualitative data from campus climate surveys, service satisfaction forms, and sustainability awareness assessments.

When ethically integrated, academic data enables the estate to shift from a mere "facility" to a "Supportive Ecosystem" that proactively encourages student and faculty success.

- **Proactive Student Support**
  AI can detect anonymized patterns in LMS activity, such as decreased interaction with core resources, to notify advisors about students who could benefit from proactive outreach or academic support.
- **Dynamic Resource and Schedule Alignment**
  Instead of relying on traditional scheduling methods, AI can analyze library usage and LMS activity to identify demand trends and better allocate time among labs, study areas, and faculty offices.
- **Personalized Campus Experiences**
  By analyzing students' broader interests and resource usage, the system can offer tailored suggestions for campus events, research opportunities, or wellness services, helping students stay connected to relevant campus resources and the community.
- **The Continuous Innovation Loop**
  By considering the campus as a "Living Lab," research results on topics like "optimal study lighting" and "thermal comfort and cognitive performance" are integrated into the Building Management System (BMS) to enhance the physical environment.

Academic, research, and experiential data ensure that the "Smart" in Smart Campus is ultimately measured by its ability to improve the human mission of learning and discovery, not just the efficiency of its machines.

Exhibit 10. Data Domain Integration Table

Data Domain	Primary Question	AI Outcome
Spatial	*Where is it?*	Visualization: 3D navigation and "X-ray" views of infrastructure.
Static	*What is it?*	Context: Mapping relationships (e.g., this pipe feeds that chiller).
Live	*What's happening?*	Automation: Real-time optimization of HVAC and lighting.
Operational	*What does it cost?*	Efficiency: Streamlined work orders and procurement.
Academic	*Why are we here?*	Personalization: Tailoring the campus to the needs of students and researchers.

Each data domain requires a "Data Trustee" (Executive-level) and a "Data Steward" (Operational-level) to ensure quality and compliance.

Exhibit 11. Domain Trustees and Stewards

Data Domain	Primary Trustee	Primary Steward	Security Focus
Spatial	Chief Facilities Officer	GIS/BIM Manager	Physical Security: Protecting blueprints of sensitive infrastructure.
Static	VP of Operations	Maintenance Manager	Asset Integrity: Ensuring "as-built" records are updated after renovations.
Live (IoT)	Chief Info Officer (CTO)	OT/Network Lead	Cybersecurity: Preventing unauthorized control of HVAC or Access systems.
Operational	CFAO / Provost	Registrar / HR Dir	Data Privacy: Protecting financial and employee records.
Academic	Provost / Dean	LMS Admin	Ethics and Consent: Ensuring student data is anonymized and FERPA compliant.

The Data Estate as a Strategic Enabler

By effectively curating, integrating, and managing these diverse data streams within a cohesive Data Estate, institutions can consolidate their information assets, ensure high standards of data quality and integrity, significantly enhance data security and privacy compliance, and, most importantly, leverage this unified intelligence to support superior decision-making and transformative business insights. The Data Estate is thus fundamental to unlocking the full, multifaceted potential of an AI-enabled Sustainable Smart Campus. It drives the creation of a verifiable "single source of truth," fosters deeper institutional knowing, stimulates cross-disciplinary innovation, and becomes the engine for generating demonstrable returns on investment and continuous improvements in all aspects of the campus experience.

The strategic applications, powered by robust data analytics and AI operating on this rich Data Estate, are extensive:

- **Revolutionized Operational Efficiency**
  AI-driven analytics, drawing on integrated infrastructure and operational data, enable continuous monitoring and meticulous optimization of resource utilization—energy, water, and physical space—pinpointing inefficiencies invisible to traditional methods and enacting automated adjustments to achieve peak performance and reduce waste.
- **Proactive Condition-Based Predictive Maintenance**
  By feeding AI models with continuous streams of live asset performance data correlated with static maintenance histories and operational parameters, the campus can shift from reactive repairs to a highly proactive predictive maintenance posture. This foresight prevents costly unexpected breakdowns, significantly extends the lifespan of critical equipment, minimizes operational disruptions, and optimizes maintenance team deployment.
- **Elevated and Evidence-Based Decision-Making**
  Comprehensive, real-time, and historically contextualized data empowers university administrators with profound insights for strategic planning, effective resource allocation, and responsive operational adjustments. This leads to more agile, informed, and impactful governance across all facets of the institution.

- **Achieving Ambitious Sustainability Goals**
  The systematic monitoring and AI-powered analysis of environmental parameters—from granular energy consumption and campus-wide carbon emissions to water usage and waste generation—indispensably support the development of effective sustainability strategies, precise tracking of progress towards ambitious targets (such as carbon neutrality or zero waste), and the identification of new opportunities to reduce environmental impact.
- **Enhanced Student Engagement and Personalized Experiences**
  Ethically managed data analytics can illuminate pathways to enhance the student journey significantly. This includes AI-driven personalization of learning support, optimization of campus services based on real-time demand (such as dining, library access, or transportation), dynamic class scheduling, and proactive measures to enhance campus safety and well-being.
- **Fostering a "Living Lab" for Research and Innovation**
  A well-documented and accessible Data Estate transforms the smart campus into a dynamic "living laboratory". This provides an invaluable resource for researchers across disciplines—urban planning, environmental science, data science, AI, engineering, and social sciences—to conduct impactful research, simulate complex scenarios, and develop novel solutions that can benefit both the institution and society at large.

Essentially, the strategic development and management of a comprehensive Data Estate, built from the beginning to support advanced AI and data analytics capabilities, are not just technical requirements but the core features of a truly intelligent and adaptable Sustainable Smart Campus. It serves as the foundation for operational excellence, environmental leadership, and an exceptional campus experience.

Structured and Unstructured Data in the Campus Estate

On a smart campus, estate data includes all details about the physical environment—such as buildings, infrastructure, utilities, spaces, assets, and their usage and operation. This data is divided into two main types: structured and unstructured. While both are crucial, they serve different purposes in supporting smart campus applications, analytics, and decision-making.

Structured Data

Structured data forms the foundation of operational intelligence. It adheres to a defined schema or model, organized into consistent fields, tables, time-series, or graphs, making it naturally understandable by machines. In campus estate management, structured data records key operational details necessary for automation, optimization, and large-scale system management. High-quality structured data enhances value by simplifying root-cause analysis and optimization. Examples include:

- **Operational Telemetry**
  Temperatures, flows, pressures, vibration, energy consumption, power quality, and occupancy counts.
- **Asset Metadata**
  Equipment identifiers, locations, capacities, manufacturers, install dates, and maintenance histories.
- **Spatial and Relational Data**
  BIM/GIS entities, room hierarchies, system topologies, asset-to-space relationships, and how assets are working together.
- **Control States and Events**
  Setpoints, alarms, faults, schedules, and runtime hours.

- **Transactional Systems**
  CMMS/EAM work orders, meter readings, utility invoices, and space scheduling data.

Structured estate data is organized within the Institutional Data Estate, where it is standardized and semantically tagged. Ontologies such as spatial, asset, and system taxonomies enable this data to be reused across various applications, reducing vendor lock-in and semantic fragmentation. Structured data is what enables:

- Predictive maintenance and fault detection
- Energy and carbon optimization
- Digital twins and simulations
- AI-driven analytics and automation
- Audit-grade sustainability reporting
- Interoperability across vendors and systems

> ***You cannot scale intelligence without structure.***
> ***AI, digital twins, and smart automation depend on reliable, well-modeled data.***

Unstructured Data

Unstructured data offers contextual insights and human knowledge. It has no fixed schema and is mainly intended for human use rather than machine processing. In facilities and campus management, this data type encompasses experience, context, and institutional memory—often valuable but challenging to implement without extra effort. Examples of unstructured estate data include:

- PDF plans, as-built drawings, and specifications
- Commissioning reports and TAB documents
- O&M manuals and safety procedures
- Field technician notes and inspection logs
- Incident reports, emails, and root-cause narratives
- Photos, videos, and thermal images
- Regulatory documents and compliance attestations

Unstructured data is valuable because it:

- Captures **intent**, not just measurements
- Preserves **human judgment and expertise**
- Provides a **historical explanation** for why systems behave as they do
- Supports troubleshooting, training, and knowledge transfer

However, on its own, unstructured data **does not scale**. It cannot directly drive analytics, automation, or AI without transformation.

Making Unstructured Information Useful

A critical smart campus maturity move is **to systematically convert high-value unstructured information into structured references**, rather than trying to "analyze everything." Effective approaches include:

- Linking documents and media to **assets, systems, or spaces** in the digital twin

- Extracting key fields (e.g., capacity, sequence parameters, acceptance criteria) into structured records
- Using AI selectively to classify, summarize, and index documents—not as a substitute for governance, but as an accelerator
- Treating unstructured content as **contextual augmentation**, not primary operational input

In this model, structured data drives **detection, prediction, and optimization** while unstructured data provides **explanation, validation, and confidence.** The digital twin becomes the convergence point where both coexist.

Exhibit 12. Usage in Smart Campus Applications

Application	Role of Structured Data	Role of Unstructured Data
Predictive maintenance	Sensor telemetry, fault codes, asset history	Maintenance notes, OEM manuals
Energy & carbon management	Meter data, schedules, emissions factors	Energy audits, policy documents
Digital twin	Modeled assets, spatial graph, live signals	Drawings, commissioning records
Capital planning	Condition scores, utilization metrics	Master plans, feasibility studies
Risk & resilience	Alarm patterns, failure rates	Incident narratives, lessons learned

Structured data enables the campus to sense, learn, and act.
Unstructured data enables leaders and operators to understand, build trust, and improve towards greater outcomes.

Planning and Building The Data Estate

Creating a functional Data Estate involves going beyond just collecting data. It requires setting up a pipeline to ingest, clean, and add context to information. This process turns raw files and sensor pings into a machine-readable **Knowledge Graph**.

1. Spatial Data as the Backbone of a Predictive Campus

Spatial data acts as the backbone of a predictive digital twin, consisting of two closely connected subcategories: Geometry and Hierarchy.

Geometry captures the 2D and 3D spatial forms of key concepts such as buildings, floors, rooms, zones, utility corridors, outdoor areas, and assets. It defines where these elements are located and offers the necessary spatial context for visualization, simulation, navigation, and spatial analysis. Geometry allows the twin to assess proximity, adjacency, volume, capacity, line-of-sight, and physical constraints, shaping the visual and physical reality of the campus.

Hierarchy, represented as a spatial knowledge graph, clarifies what entities are and their relationships. It models the logical organization of the campus—showing how rooms are part of floors, which are within buildings, and how these buildings belong to districts. It also links these entities to systems, assets, schedules, ownership, policies, and mission outcomes. This semantic layer turns raw geometric data into meaningful information, enabling the twin to comprehend roles, dependencies, constraints, and operational goals.

Together, geometry and hierarchy allow the digital twin to function as more than a map or model. They enable **spatial reasoning**, in which physical form and institutional logic reinforce each other. A room is no longer just a polygon; it becomes a learning space with capacity limits, air-quality requirements, academic schedules, safety rules, energy targets, and accessibility constraints—each derived from its position in the hierarchy and its geometric properties.

As twins become predictive, this dual-layer spatial model becomes essential. Operational AI agents rely on geometry to act safely in the physical world and on hierarchy to act correctly within institutional rules. An HVAC-optimizing agent must understand a space's volume and orientation (geometry) while respecting its academic use, occupancy patterns, and ventilation standards (hierarchy). An emergency-response agent must reason about egress paths and distances (geometry) while honoring command structures, roles, and decision rights (hierarchy).

Methods of Collection

Spatial data is gathered differently depending on a building's stage in its lifecycle. Each method contributes a distinct layer of accuracy, context, and operational value to the digital twin.

New Construction (As-Designed)

For new buildings, spatial data originates in the design process itself. Architects and engineers create high-fidelity digital models using platforms such as Revit and other BIM tools. These models represent the *intentional* or "perfect" version of the building—capturing geometry, systems, and spatial relationships before a single brick is laid. As-designed models provide a precise starting point for lifecycle management, commissioning, and long-term operational planning.

Existing Buildings (As-Built Reality Capture)

For existing facilities—especially those where drawings are missing, incomplete, or outdated—spatial data is captured directly from the physical environment using reality-capture technologies. Two methods are most commonly used:

- **LiDAR (Laser Scanning)**
  LiDAR scanners emit millions of laser pulses per second to measure distance with extreme precision. The result is a point cloud—a dense, three-dimensional field of measured points that forms an exact digital "ghost" of the structure. With millimeter-level accuracy, point clouds reveal true dimensions, deviations from plans, and the physical condition of spaces and systems.

- **Photogrammetry**
  Photogrammetry uses high-resolution cameras—often mounted on drones or handheld devices—to capture thousands of overlapping images. AI-driven algorithms analyze and stitch these images together, producing a textured 3D mesh. While typically less dimensionally precise than LiDAR, photogrammetry adds critical visual realism, making it invaluable for contextual navigation, inspections, and stakeholder engagement.

From Geometry to Intelligence: Integration into the Digital Twin

On their own, 3D models, point clouds, and meshes remain static representations. They become operational assets only when integrated into the broader Digital Twin ecosystem. This transformation occurs through two essential integration steps.

Interoperability (IFC)

To ensure long-term usability across tools, vendors, and teams, spatial models are commonly exported using Industry Foundation Classes (IFC). IFC acts as a neutral, open standard for building data, preserving structural relationships, asset metadata, and system hierarchies regardless of the platform used downstream. This interoperability prevents vendor lock-in and enables spatial data to remain meaningful across design, construction, facilities, analytics, and AI platforms.

Geographic Anchoring (GIS)

To move from a *model* to a *true twin*, spatial data must be georeferenced. By integrating the 3D geometry with Geographic Information Systems (GIS), each building—along with its floors, rooms, and assets—is assigned real-world coordinates (latitude, longitude, and elevation). This grounding situates the building within its broader physical, environmental, and civic context, allowing it to interact with weather systems, utilities, transportation networks, and regional infrastructure.

The Integration Point: Where Insight Becomes Action

The real value of spatial data emerges at the integration point—where geometry, hierarchy, and location converge. When a user clicks on a virtual thermostat inside the Digital Twin:

- Spatial geometry identifies the exact location of the device within the building using precise x, y, z coordinates.
- Hierarchy and metadata define what the device is, which room it serves, and which systems and policies govern its behavior.
- GIS context situates the building within the campus, city, and region—informing decisions influenced by weather, energy pricing, or grid conditions.

II. Static Data: Creating the Institutional Brain

Static Data serves as the Digital Twin's brain and memory. It includes history, specifications, and protocols often hidden away in filing cabinets or nested folders. To make this data truly useful for a modern Digital Twin, we must go beyond simple storage and adopt Semantic Intelligence. Static Data covers everything from Operations and Maintenance (O&M) manuals and warranties to repair history and safety protocols. The aim is to turn static, text-based information into an active, searchable layer of intelligence.

Phase 1: Ingestion (From Paper to Structured Data)

The first hurdle is **Digitization**. Most institutional knowledge is locked in non-searchable formats, such as flattened PDFs or physical binders.

- **Advanced OCR and Document AI:**
  Standard Optical Character Recognition (OCR) merely turns images of letters into digital text. **Document AI** goes further—it uses computer vision to understand the *layout*. It recognizes that a specific string of numbers is a "Part ID" because it appears in a table, or that it identifies a "Warning" because it is near a specific icon.
- **Semantic Enrichment**
  During ingestion, metadata tags are applied. A component manual isn't just "Document_123"; it is tagged as "Chiller Maintenance Manual," "North Wing," and "Brand: York," creating a multi-layered map of the content.

Phase 2: Vectorization (The Language of AI)

To make this information "LLM-ready," it must be translated from English into a format a machine can calculate.

- **Chunking**
  The AI breaks long documents into smaller, logical "chunks" (e.g., a single paragraph about pump lubrication).

- **Embedding**
  Each chunk is passed through an embedding model that maps the text's meaning into a long sequence of numbers (a vector).
- **The Vector Database**
  These numbers are stored in a specialized database. In this mathematical space, similar concepts sit close together. For example, the vector for "repair" will be mathematically closer to "fix" than it is to "cafeteria."

Fortunately, the tools for ingesting static data from paper and digital representations have improved significantly. Today, it is possible to ingest data via automated, integrated scanning equipment or to upload it to an LLM for a RAG application.

Phase 3: The AI Loop (Retrieval-Augmented Generation)

This is where the "Brain" begins to speak. Instead of a technician searching a 500-page PDF, they engage in **RAG (Retrieval-Augmented Generation)**.

1. **The Query:** A technician asks, *"What is the maintenance interval for the North Wing Chiller?"*
2. **The Retrieval:** The system converts that question into a vector and searches the database for the most relevant "chunks" of data.
3. **The Context:** It pulls the exact page from the specific York Chiller manual and feeds it to the Large Language Model (LLM).
4. **The Generation:** The LLM reads the provided snippet and generates a natural language answer: *"According to the York YK Manual (page 42), the North Wing Chiller requires a bearing inspection every 8,000 hours."*

The Integration Point

By linking this "Brain" to the "Skeleton" (Spatial Data), a user can click on the physical chiller in the 3D model and immediately ask the AI for its specific maintenance history—bridging the gap between the physical asset and its digital documentation.

III Live Data: The Nervous System (APIs and Agents)

Live data is the real-time "nervous system" of the smart campus—continuously transmitting signals between the physical environment and the digital ecosystem. This constant exchange enables the institution to sense conditions, understand what is happening, and act quickly and intelligently.

- **Digital Agents as Translators and Messengers**
  Modern Building Management Systems (BMS) speak in OT-native protocols such as BACnet or Modbus. To make this data usable at the enterprise level, lightweight digital agents operate on local gateways. These agents continuously poll the BMS, convert its proprietary signals into modern, structured formats such as JSON, and securely publish them to the cloud via MQTT. This process turns isolated mechanical systems into live data sources that can fully participate in analytics, automation, and digital twin models.
- **API Harvesting for Distributed Sensors**
  Not all sensors—such as standalone air quality monitors or specialty lab devices—connect to a central BMS. For these, the integration layer uses RESTful APIs to pull data from each device's cloud service at regular intervals. API harvesting ensures that every sensor, regardless of vendor or location, contributes to a unified institutional data estate.
- **Edge Processing to Preserve Bandwidth and Efficiency**
  Raw data at the edge can be extremely high-volume. To avoid overloading the network, edge devices

perform intelligent filtering before sending anything upstream. For instance, a camera-based occupancy sensor does not transmit video; it sends only the occupancy count needed for decision-making. This preserves privacy, reduces bandwidth consumption, and ensures that only actionable information flows into analytics and digital-twin platforms.

IV. Operational and Business Data: Harvesting the Records

Operational and business data form the "institutional memory" of the smart campus. Unlike live sensor telemetry—which reflects the present moment—these records describe **how the campus has operated over time**, how work is performed, where resources are allocated, and how assets evolve across their lifecycle. Integrating this information into the Digital Twin transforms it from a purely real-time system into a context-aware **intelligence engine** capable of connecting faults to work orders, energy use to occupancy, and asset performance to capital planning.

This layer is harvested primarily through **ETL (Extract, Transform, Load)** pipelines that draw data from existing enterprise platforms, such as ERP, CMMS, space management systems, and scheduling tools. When stitched together through disciplined data architecture, these systems provide the longitudinal datasets needed for planning, compliance, optimization, and executive decision-making.

• Database Crawling:
Connecting the Digital Twin to System-of-Record Data

To ensure accuracy and reliability, the Data Estate connects directly to the authoritative databases that power institutional operations—such as Workday for finance and HR, or Maximo for asset management and maintenance history. Through secure connectors, the platform periodically extracts structured data such as:

- asset inventories and specifications
- historical work orders and repair logs
- procurement and lifecycle records
- space assignments, cost centers, and operational budgets

This process transforms operational systems from static record-keepers into **active contributors** to campus intelligence. It ensures the Digital Twin is grounded in the same authoritative "truth" that governs payroll, procurement, and regulatory compliance—allowing analytics, AI models, and scenario planning tools to operate with institutional legitimacy.

Trigger-Based Updates:
Real-Time Synchronization of Work and Workflow

While scheduled ETL pipelines maintain historical datasets, real-time accuracy requires event-driven updates. This is where **webhooks**—lightweight, trigger-based notifications—play a critical role. When a technician closes a work order in the CMMS, submits a meter reading, or updates an asset's status, a webhook instantly alerts the Digital Twin. The asset transitions from *"Under Repair"* to *"Operational"* without waiting for a batch process to complete. This eliminates data lag and ensures the twin always reflects the **true operational state** of the campus. This allows the institution to:

- verify fault detection and corrective action in near-real time
- support continuous commissioning workflows
- keep predictive models aligned with current field conditions

- link operational outcomes directly to sustainability and performance metrics

Why This Layer Matters

Operational and business data give the Digital Twin *memory*, enabling it to describe not just what is happening but **why, for how long,** and **with what cost or consequence.** It connects technical insight to governance, budgets, staffing, risk, and strategic planning. Without this layer:

- analytics lack context,
- capital planning lacks evidence,
- predictive maintenance lacks verification, and
- Sustainability metrics lack audit-grade integrity.

With it, the institution gains a living, continuously updated operational ledger—one capable of transforming physical infrastructure from a collection of assets into a measurable, optimizable, enterprise-grade ecosystem.

V. Synthesis: The Unified Namespace (UNS)

A smart campus produces data everywhere—meters, controllers, space systems, work orders, schedules, sensors, and vendor clouds. Without a disciplined structure, this data becomes fragmented, duplicated, and nearly impossible to use at scale. The **Unified Namespace (UNS)** solves this problem by creating a single, hierarchical, institution-wide directory for all campus data—spatial, static, and live.

Rather than scattering information across multiple databases and vendor portals, the UNS organizes every asset and data stream into a shared, intuitive path:

Campus / Building A / Floor 2 / Room 201 / VAV_Box_04 / Temperature_Sensor

This consistent naming convention becomes the backbone of the Digital Twin. It ensures that AI models, operators, and applications do not need to "guess" what a point represents or where it exists physically. A temperature reading in the live telemetry layer automatically links to:

- its precise physical location inside the 3D BIM model,
- its equipment manual and static metadata, and
- its work-order history inside the CMMS.

The Unified Namespace transforms a chaotic campus into a **coherent, navigable, machine-readable environment**—the foundation for analytics, automation, and institutional intelligence.

Why the Unified Namespace Matters

A well-constructed UNS turns the Data Estate from a storage repository into an **institutional operating system**:

- **AI becomes dramatically more accurate** because entities and relationships are consistently defined.
- **Digital Twins load faster and stay synchronized**, since each asset has one canonical identity.
- **Technicians can instantly locate equipment** and the documents tied to it.
- **Governance improves** because ownership, lineage, and access controls follow the namespace hierarchy.
- **New systems integrate more easily**, eliminating brittle point-to-point connections.

Without the UNS, the Data Estate becomes a patchwork of incompatible feeds. With it, the Smart Campus becomes a **composable, scalable architecture** that can grow for decades.

Chapter Summary

A Smart Campus cannot exist without a governed, unified Data Estate. The Unified Namespace is the essential architectural layer that transforms scattered telemetry, documents, and enterprise records into a coherent, machine-readable campus model. When paired with early ontology adoption, strong data ownership requirements, and latency-driven design, the Data Estate becomes far more than a data warehouse—it becomes the single source of truth powering strategic planning, operational automation, risk governance, digital-twin intelligence, and academic-research enablement. By using this naming convention across **Spatial, Static, and Live** data, the AI doesn't have to guess. It knows exactly where a sensor is located in the 3D BIM model and can instantly find the corresponding manual in the Static repository.

The following practices ensure that the Data Estate becomes a durable, governed, and enterprise-ready foundation rather than another set of disconnected databases.

Adopt an Ontology Early

Use an industry-standard building ontology, such as Project Haystack or Brick Schema, from Day 1. Retrofitting semantic tags *after* data ingestion is slow, costly, and error-prone.

2. Enforce Data and API Ownership

Every smart campus vendor contract should clearly state that the institution owns all data generated by its physical systems, that the institution controls API access and documentation, and that vendors cannot restrict, surcharge, or "hold hostage" operational data. This protects the university from vendor lock-in and ensures that every system can participate in the UNS, regardless of proprietary origins.

3. Prioritize Latency Based on Use-Case

Not all data needs the same speed. A disciplined Data Estate defines *latency classes*:

- **Sub-second:** fault protection, electrical anomalies, safety-critical systems
- **Seconds:** occupancy sensors, dynamic load management
- **Minutes:** HVAC temperatures, utility meters, IAQ
- **Hours or Daily:** ERP updates, asset inventories

Prioritizing latency ensures efficient networking, cost control, and reliable analytics. The rule is simple: *only collect speed where speed creates value.*

4. Govern Through Joint CTO–CDO Leadership

The Data Estate is not merely an IT project; it is an institutional asset that demands shared leadership from the CTO and CFAO to oversee.

- architecture and integration,
- semantics and ontologies,
- data quality and access controls,
- privacy and compliance,
- lifecycle governance.

CTOs and CDOs should collaborate to map, govern, and design the data estate. This joint approach guarantees that the Data Estate underpins all enterprise areas—from capital planning and research computing to physical operations. Additionally, the Data Estate serves as the single source of truth, facilitating strategic planning, operational automation, and research support.

Chapter 11
The Smart Campus Technology Stack

A unified, layered architecture that spans the technology stack facilitates secure connectivity, sustainable operations, integrated systems, and scalable intelligence.

A sustainable smart campus involves more than just smart devices; it creates an interconnected environment that senses, learns, and responds across buildings, utilities, transportation, academic spaces, and digital services. The main difference between a campus that is simply experimenting with new technology and one that genuinely becomes intelligent lies in its architecture: a thoughtfully designed technology framework that is scalable, seamlessly integrated, secure, and capable of evolving.

Most institutions consist of a patchwork of systems: legacy building controls with proprietary controllers, new IoT pilots managed by departments, cloud services used by academic units, and enterprise platforms overseen by central IT. This often results in fragmented data, inconsistent security, unclear ownership, and innovation that struggles with staff turnover or budgeting constraints. The goal of this chapter is to shift the focus from merely listing technologies ("what do we have?") to understanding how these technologies come together to form a cohesive institutional capability ("how do they work together?").

This section of the manifesto outlines a practical reference architecture for the AI-enabled Sustainable Smart Campus. It features a layered technology stack built on interoperable standards, shaped by OT realities, and overseen through cybersecurity and data stewardship. The target audience includes CTOs and CFaOs focused on ensuring reliability, safety, and continuity, rather than merely innovation.

Architecture serves as the bridge linking ambition to reliability. Many smart campus strategies fail not because of a flawed vision but because the foundational infrastructure cannot support them. While governance can define decision rights, ethics, and processes, it needs a reference architecture to guide its authority. Without it, convergent architecture turns into an unmanaged collection of tools.

A sustainable smart campus emerges where architecture, data, and governance intersect: it's a layered, interoperable framework built for resilience and antifragility, managed to build trust, and aimed at delivering specific institutional outcomes.

Layers of Technology

The AI-enabled Sustainable Smart Campus technology stack begins with the AI and cognitive layer. AI forms the core of the Smart Campus, as it is widely deployed and integrated throughout almost all applications, services, and interactive technologies. Most campus users engage with AI, either intentionally or unconsciously. This list of applications keeps evolving. Beneath these applications and system integrations lies crucial infrastructure that ensures reliable networking, computing resources, and data storage.

This infrastructure includes standards and protocols—essentially the "language" of technology—that enable communication between devices and systems. Together, these components make up the technological stack:

- AI The Cognitive Technology Layer
- Applications and Services Layer
- Infrastructure Layer
- Systems Integration Layer
- Standards
- Protocols

This overview of the AI-enabled Sustainable Smart Campus technology stack is extensive and intricate. Although every detail cannot be covered in the manifesto, it establishes a foundation for key principles. It emphasizes the need to intentionally design the technology to be 'composable' and to scale efficiently throughout the institution.

This can help guide an inventory and evaluation of existing applications, infrastructure, standards, and protocols to identify strengths, weaknesses, opportunities, and risks. It also encourages discussion on technological priorities for achieving the desired future state.

A smart campus is not a collection of devices. It is an institutional capability: a connected, governed, and continuously improving system that links the built environment, digital services, and human experience to measurable outcomes—sustainability, resilience, reliability, accessibility, and student success. The defining difference between "technology on campus" and a "smart campus" is integration with intent: common standards, shared governance, secure data flows, and operational workflows that turn signals into decisions and decisions into better daily performance.

> *A building you talk to is just concrete. A building that listens is intelligent. A building that talks back is smart.*

This section outlines a hierarchical stack that begins with foundational infrastructure and progresses through cyber-physical systems, data, and governance layers, ultimately reaching user-facing applications. For each layer, the emphasis is on its purpose—what it enables, its significance in higher education, and the consequences of its absence—rather than on specific features.

Artificial Intelligence: The Cognitive Layer

As an executive guiding the future of higher education, it is essential to recognize that Artificial Intelligence (AI) is more than just a standalone software tool — it forms the cognitive core of the institution. It should not be viewed as a collection of isolated pilot projects but as a managed, institutional capability.

To guide this transformation, leadership must recognize that AI is becoming embedded in all systems and aspects. Therefore, we start the conversation about technological stacks with a brief overview of AI, ensuring institutional leaders clearly understand how various types of AI and automation are implemented across the campus.

Types of Artificial Intelligence

Machine Learning (ML) – The Pattern Finder

ML algorithms process large volumes of historical and real-time data to detect patterns that humans might overlook. On a smart campus, ML powers predictive maintenance and sustainability initiatives. For example, it can notify Facilities about a chiller likely to fail next week by identifying subtle vibration signals or automatically optimize microgrid energy use to avoid peak utility rates.

Large Language Models (LLMs) – The Reasoning Engine

LLMs are the core technology for understanding and interpreting human language. They serve as translators, able to process complex information—like detailed vendor manuals or complicated institutional policies—and distill it into straightforward, conversational responses for staff and students.

Generative AI (GenAI) – The Creator:

Building on LLMs, Generative AI produces entirely new content. Instead of merely searching for answers, GenAI can quickly draft an emergency communication protocol, create a personalized study guide from a professor's lecture transcript, or assist your IT team in writing code for a new software integration.

RAG (Retrieval-Augmented Generation) – The Truth Tether

A primary concern regarding AI is "hallucination"—the system confidently fabricating false information. RAG addresses this by directing the AI to avoid general internet searches and instead query a secure, private repository containing your university's authentic data, such as the official student handbook, course catalogs, or live HVAC telemetry, before producing a response. This approach ensures the AI is accurate, contextually aware, and legally defensible.

Agentic AI – The Doer

This marks the transition from mere conversation to active implementation. Unlike traditional AI, which might notify a facility manager of a boiler's malfunction, Agentic AI has the authority to act. An AI "Agent" can independently log in to maintenance systems, create work orders, purchase replacement parts, and contact contractors—only halting at predetermined human-approval steps. In student services, Agentic AI can proactively assist students through complex administrative procedures, liaising with the Registrar and Financial Aid to serve as a personalized digital concierge.

AI as an Enabling Technology

AI's real physical influence becomes evident when combined with kinetic technology. Robotics and autonomous vehicles are at the forefront of this evolution, where digital intelligence engages physically with the campus environment.

Service Robots

The use of service robots is rapidly becoming part of everyday life. Today, universities employ autonomous delivery robots to streamline logistics and make food services more affordable. Additionally, autonomous floor scrubbers and groundskeeping robots handle repetitive tasks, allowing staff to focus on more complex maintenance. As this technology advances, we will likely see robotic assistive companions designed to help visually impaired students navigate campuses, opening up new possibilities for accessibility and management.

Autonomous Vehicles (AVs) and Mobility:

Campuses operate like small cities, serving as ideal testing grounds for AVs. AI is the key technology enabling autonomous shuttles to safely navigate mixed-traffic areas, offering dependable, accessible transportation that links distant parking areas or research parks to the main campus. Additionally, edge computing AI allows these vehicles to interact directly with smart intersections (Vehicle-to-Infrastructure or V2I), efficiently adjusting traffic flow and significantly enhancing pedestrian safety during class transitions.

Computer Vision and Spatial Intelligence

AI redefines standard optical cameras, turning them from passive recorders into dynamic, multi-purpose IoT sensors. Instead of human security guards monitoring screens, computer vision algorithms analyze video feeds locally to detect specific events—such as spotting a slip-and-fall in a remote stairwell, tracking queue length in a dining hall to allocate staff accordingly, or enabling seamless parking through license plate recognition. In a governed smart campus, this approach uses anonymized spatial tracking (like counting "shapes" and flow) instead of intrusive facial recognition.

The Dynamic Digital Twin

A traditional Digital Twin is merely a static 3D architectural (BIM) model. AI acts as the catalyst that transforms it into a predictive simulation engine. By continuously integrating real-time data from the Digital Integration Fabric, AI enables Facilities leaders to perform "what-if" analyses. For example, an executive might ask the Digital Twin, "If a severe heatwave occurs next Tuesday and our primary chiller fails, how soon will the research labs reach unsafe temperatures?" AI then models the buildings' thermal behavior, providing an accurate timeline to support emergency response planning.

Algorithmic Energy Management and Microgrids

Renewable energy assets such as solar panels and batteries are highly volatile. AI is a crucial technology for managing Distributed Energy Resource Management Systems (DERMS) at the institutional level. AI algorithms forecast future campus energy needs by analyzing class schedules, past usage data, and weather forecasts. Based on these predictions, the system automatically determines whether to store solar power in campus batteries, pre-cool large lecture halls before peak pricing begins, or sell excess energy back to the regional utility grid through Automated Demand Response (OpenADR).

Cyber-Physical Security and Anomaly Detection

As the campus incorporates thousands of physical devices into the network, traditional rule-based firewalls are no longer enough. AI facilitates large-scale behavioral cybersecurity by monitoring the normal activity of each connected device. For example, if a standard thermostat in a residence hall—only transmitting small temperature data packets—starts attempting to download a large file or ping the university's financial server, the AI detects this behavioral anomaly and immediately disconnects the device's network port without human input.

The Smart Call Center

AI enables the automated call center by shifting it from scripted responses to genuine interactions and actions. Talking *at* a system—leaving voicemails, navigating phone trees, or waiting on hold—adds little value. When the system can listen, it becomes intelligent: understanding intent, emotion, urgency, and context across multi-turn conversations. And when it can talk back, it becomes truly smart—resolving issues, scheduling appointments, routing requests, and confirming outcomes instantly.

Platforms like **OpenCall.ai** embody this shift by acting as a voice-first AI workforce that answers inbound and outbound calls, manages SMS, and books appointments 24/7. Its conversational AI agents go beyond basic chatbots, handling nuanced conversations and integrating directly with CRMs, EHRs, calendars, and appointment-booking systems. Increasingly, systems are HIPAA-compliant and composable, lowering operational costs and reducing potential technical debt. AI transforms the call center from a bottleneck into an always-on, human-capable service layer.

Ambient Accessibility and Natural Language Processing (NLP)

AI fosters a more inclusive campus by removing physical and language barriers instantly. In classrooms, NLP provides real-time, precise transcription and translation of lectures, displaying captions on students' devices in their native languages. On a physical level, AI-powered mapping apps integrate GIS data with live operational feeds to help mobility-impaired students navigate around broken elevators, construction zones, or crowded hallways.

AI Summary

AI should be viewed as an enabling technology rather than merely an application. As robotics and Agentic AI expand, institutions will encounter a new governance challenge. You are no longer just overseeing software in a data center; you're managing autonomous machines and agents functioning within human environments. Success depends on strong cyber-physical governance to ensure that the intelligence enhancing your campus upholds the safety, privacy, and dignity of its inhabitants.

Applications and Services Layer

Facilities, Energy, and Work Management

Facilities, energy, and work management form the operational engine room of the smart campus—where the physical plant, utility systems, and daily maintenance practices converge to shape the institution's reliability, sustainability, and lived experience. This domain is where strategic intent meets operational reality: boilers and chillers must run, energy must be produced and consumed efficiently, and thousands of assets must be monitored, maintained, and optimized in real time. As campuses become increasingly instrumented and data-rich, the traditional boundaries between maintenance, energy operations, and facilities management dissolve, giving way to an integrated, cyber-physical workflow driven by telemetry, analytics, and intelligent automation.

Each paragraph is expanded for clarity, narrative flow, and executive readability while remaining concise and consistent with the manifesto's style.

Energy Management Information Systems (EMIS)

Energy Management Information Systems (EMIS) consolidate and standardize energy data across buildings, transforming raw consumption information into actionable insight. By benchmarking performance, surfacing anomalies, and tracking trends over time, EMIS platforms make energy efficiency legible to both operators and strategic leaders. Their true value lies in turning a scattered utility landscape into a coherent operational narrative—empowering data-driven energy planning, continuous performance improvement, and more-informed investment decisions.

Fault Detection and Diagnostics (FDD) and Continuous Commissioning

Fault Detection and Diagnostics (FDD) systems continuously scan building operations for hidden inefficiencies—such as simultaneous heating and cooling, stuck dampers, and improper equipment staging—and elevate them before they become costly failures. When combined with continuous commissioning, FDD ensures that the performance gains achieved through modernization persist long after construction teams leave. This is where smart infrastructure begins to compound value: ongoing optimization becomes routine, savings become repeatable, and buildings evolve from static assets into continuously tuned systems.

Carbon Accounting and Climate Reporting Platforms

Carbon accounting and climate reporting platforms translate operational energy, utility, and equipment data into clear emissions metrics that support internal planning and external scrutiny. They provide the transparency needed to validate climate commitments, meet regulatory requirements, and inform capital investment strategies. By converting complex operational signals into auditable carbon insights, these platforms strengthen institutional credibility and enable leaders to monitor progress, prioritize interventions, and communicate impact with confidence.

CMMS/EAM (Work Orders, Preventive Maintenance, Lifecycle)

Computerized Maintenance Management Systems (CMMS) and Enterprise Asset Management (EAM) platforms serve as the operational record of truth for maintenance activity, asset condition, preventive schedules, and lifecycle planning. When connected to sensors, analytics, and digital-twin insights, these systems transform from administrative tools into engines of predictive maintenance. By coordinating work orders with real-time diagnostics, CMMS/EAM platforms enable teams to allocate labor more effectively, reduce downtime, and extend the useful life of critical infrastructure.

Mobile Field Tools

Mobile field tools equip technicians with real-time, context-rich information on demand, including diagnostics, location data, schematics, and required parts. These tools close the operational loop by enabling fast documentation, evidence capture, and verification of completed work. In a smart campus environment, mobile solutions become the connective tissue between analytics and action—ensuring insights reach the field and that field outcomes flow back into the institutional data ecosystem.

Contractor Management Platforms

Contractor management platforms support safer, more consistent, and more accountable vendor operations by centralizing documentation, tracking compliance, managing access permissions, and recording performance. As campuses rely on increasingly complex vendor ecosystems for controls, HVAC, electrical, security, and specialized systems, these platforms ensure that external work aligns with institutional standards. They reduce risk, improve transparency, and reinforce operational continuity across a diverse network of service partners.

Parts Inventory and Procurement Systems

Parts inventory and procurement systems maintain real-time visibility into stock levels, lead times, and supply-chain dependencies, reducing downtime caused by missing components or delayed orders. By aligning procurement with preventive maintenance needs and predictive fault insights, these systems help avoid operational interruptions, control costs, and enable technicians to respond quickly and effectively. They transform materials management from a reactive function into a strategic enabler of reliability.

Condition Monitoring (Vibration, Thermal, Power Quality)

Condition monitoring technologies—such as vibration analysis, thermal imaging, and power-quality sensing—provide early warning indicators for mechanical and electrical systems. By detecting subtle changes long before equipment fails, these tools help prevent major outages, optimize maintenance timing, and extend asset lifespan. In a mature smart-campus environment, condition monitoring forms the sensory layer that underpins predictive maintenance and transforms facilities from reactive to anticipatory operations.

Water Management Systems

Modern water management systems integrate leak detection, submetering, smart irrigation, and quality monitoring to reduce waste, lower costs, and protect buildings from water damage. As climate pressures and infrastructure aging intensify water-related risk, these systems play an increasingly critical role in both resilience and sustainability strategies. They allow campuses to detect issues early, adjust consumption intelligently, and safeguard one of the institution's most essential—and most vulnerable—resources.

Waste Analytics and Smart Routing

Waste analytics and smart routing technologies use sensor-equipped bins and dynamic collection schedules to reduce labor, lower emissions, and improve service reliability. By providing insight into waste streams, contamination patterns, and disposal behavior, these platforms help institutions enhance operational efficiency while advancing sustainability education and diversion goals. More advanced programs integrate with broader ESG initiatives, turning waste management into a measurable component of institutional stewardship.

The Campus Digital Twin

The way facility management and campus operations are understood is changing, shifting from reactive, static maintenance to more dynamic, predictive, and autonomous systems. A key driver of this change is the digital twin technology, which is proving to be a groundbreaking tool for enhancing the built environment.

Digital Twins represent the cutting edge by integrating data to generate actionable insights about facility performance. Specifically, they serve as highly detailed virtual replicas of physical assets, systems, and processes, enabling continuous, bidirectional data flow that mirrors the campus's actual condition. Digital Twins help campuses meet increasing sustainability goals, reduce lifecycle costs, and improve efficiency by enhancing processes and skilled trades work.

Grasping the nuances of this technology—covering data engineering for AI and organizational best practices—is crucial for any stakeholder involved in modernizing large-scale built environments.11 The manifest discusses the fundamental data ontology, interfaces, and protocol constructs of the digital twin in this chapter, with an in-depth look at Digital Twins implementation in Chapter 20.

Safety, Security, and Emergency Operations

A smart campus safety plan is not merely about having 'more cameras." Instead, it involves a coordinated system that enhances awareness and response, all while maintaining privacy and trust. Safety, security, and emergency management are fundamental components of an AI-powered, sustainable smart campus. In a time when physical infrastructure, digital systems, and human activities are intricately linked, the institution's duty of care goes beyond traditional life-safety measures. Modern campuses must predict threats, detect anomalies, coordinate responses, and ensure operational continuity amid a rapidly changing landscape of cyber-physical risks. Safety is no longer just a collection of separate functions; it is a unified operational capability driven by data, guided by clear policies, and executed by tightly coordinated teams.

As building systems, access controls, sensing technologies, and emergency communications integrate, institutions now enjoy unprecedented real-time visibility and operational flexibility. However, this advancement also introduces increased complexity and potential vulnerabilities. Developing a mature safety and security infrastructure demands clear governance, coordination across IT, Facilities, and Campus Safety, and careful integration of intelligent systems that enhance—not hinder—response efforts. The primary goal remains: to protect individuals, maintain trust, and keep the campus a resilient and predictable environment, whether under normal conditions or in times of crisis.

Access Control and Visitor Management

Modern access control systems—including digital ID cards, mobile credentials, cloud-based permissions, and role-specific policies—are essential for maintaining a secure and accessible campus. Combining these with thoughtful visitor management strikes a balance between welcoming guests and ensuring security, allowing guests, contractors, and event participants to be received properly while safeguarding institutional spaces. In busy venues and large events, this integrated approach enhances safety and service by providing clear records of who is present, their purpose, and their movement within sensitive zones.

Video Management Systems (VMS) and Analytics

Video Management Systems (VMS) unify camera feeds, storage, and monitoring into a single operational interface, enabling safety teams to identify incidents and coordinate their responses swiftly. When enhanced with analytics—such as motion detection, behavioral cues, or automated triage—VMS platforms can greatly improve situational awareness. However, since some analytical models might reveal sensitive personal information, their deployment must adhere to strict policies, transparency standards, and privacy commitments to ensure ethical use keeps pace with technological capabilities.

Intrusion Detection and Perimeter Monitoring

Intrusion detection systems—such as sensors, alarms, and perimeter monitoring—protect buildings and restricted areas, particularly after hours or in places with sensitive research or valuable assets. When combined with overall safety procedures, these signals enable quick coordination among security teams, dispatch, and responders. However, this system must be carefully managed to prevent unwarranted alerts, false alarms, and operational exhaustion; the aim is to ensure responses are accurate, measured, and based on defined thresholds and policies.

Emergency Communications and Mass Notification

Emergency communication systems—including blue-light phones, PA systems, SMS and email alerts, digital signage, and app notifications—offer vital infrastructure for quick, campus-wide responses during emergencies. When used thoughtfully, these tools enhance message accuracy, minimize confusion, and facilitate quicker responses by linking situational information with the most appropriate communication channels. A well-developed system prioritizes clarity, redundancy, and accessibility to ensure that every community member receives precise guidance at critical moments.

Panic Apps and Escort Services

Panic apps, mobile safety tools, and campus escort services provide quick, respectful ways for individuals to summon help, enhancing safety and speeding response times. Their design should prioritize accessibility and empowerment for users over surveillance, ensuring that students, faculty, staff, and visitors feel supported—not monitored. When effectively deployed, these services foster trust by offering the community straightforward, dependable methods to connect with safety teams during moments of vulnerability.

Emergency Management and EOC Platforms

Emergency Operations Center (EOC) platforms organize institutional response by coordinating roles, tasks, communication channels, and documentation during incidents. In higher education, which encompasses residential life, research, public events, and community engagement, these platforms are crucial for ensuring coordinated action across various teams. Additionally, they maintain institutional memory through after-action reports and lessons learned, supporting ongoing improvements in emergency preparedness and resilience.

Lockdown Coordination Capabilities

Lockdown coordination systems enable institutions to swiftly secure buildings or zones in the event of extreme threats. However, these capabilities should be governed by clear decision rights, proportionality measures, and transparent escalation procedures. Automation can aid responses by speeding up notifications or securing predefined areas, but final authority should always remain with humans to ensure actions are appropriate, context-aware, and accountable. The goal is to improve protection while maintaining trust, safety, and operational integrity.

Student Experience and Campus Services

The true value of a smart campus is realized when it minimizes friction and enhances everyday life. The student experience stands as the ultimate test of a smart campus's effectiveness. All investments in digital infrastructure, operational technology, data governance, and intelligent services ultimately impact students' daily lives—how they learn, live, move, and feel a sense of belonging. In an AI-enabled sustainable smart campus, the student journey becomes a connected ecosystem, with adaptable spaces, intuitive services, and support systems that proactively address needs rather than merely react to problems. This approach redefines campus services from isolated transactions to a comprehensive experience supported by an integrated digital and physical infrastructure.

As expectations grow for reliability, personalization, accessibility, and well-being, campuses must shift from just providing services to actively shaping experiences. Areas like dining, housing, mobility, safety, advising, recreation, and student life increasingly rely on shared data, interoperable platforms, and real-time insights. When these systems are thoughtfully integrated, they reduce friction, promote equity, and foster an environment where students feel supported academically, socially, and emotionally. In this setting, student experience is not merely a "soft" outcome but a key strategic factor influencing recruitment, retention, and lifelong ties. A smart campus makes this experience measurable, adaptable, and aligned with institutional goals.

Campus mobile app and digital service portal

A "front door" application can integrate various services such as digital ID, digital key (for building access), ticketing (for events), campus maps and navigation, service requests, notifications, telehealth access, and more. These services are increasingly combined into a single digital companion designed to be the always-on guide for student success. When implemented effectively, it serves as the experience layer, ensuring digital services are consistent and easy to access.

Digital Signage and Content Management

Digital signage has become a vital communication infrastructure for the modern campus, facilitating movement, fostering community engagement, and acting as a key emergency communication tool. When properly managed—with clear content standards, accessibility compliance, and seamless integration with event feeds and alert systems—

digital signage transcends mere display screens, becoming a dynamic, real-time reflection of campus life. Its true value comes from consistency and purposefulness: providing timely information, enabling inclusive wayfinding, enhancing community identity, and improving situational awareness for diverse audiences.

Service Queue Management and Virtual Waiting Rooms

Service queue management platforms and virtual waiting rooms became mainstream during the COVID pandemic. They remain relevant because they address a key barrier to student services: administrative friction. By enabling students to join queues remotely, view estimated wait times, and access services asynchronously, these tools enhance predictability and fairness in support delivery. They assist institutions in transitioning from a first-come, first-served approach—often benefiting those with flexible schedules—to a system based on fairness, transparency, and reduced stress. In a smart campus, these systems are part of a comprehensive strategy to improve access to services, allowing students to engage with support without sacrificing their time, privacy, or academic progress.

Case Management for Student Services

Structured case management systems unify advising, financial aid, disability services, student success centers, and other support areas into a seamless operational workflow. They provide continuity at various touchpoints, preventing students from having to recount their stories or navigate institutional complexity on their own. These platforms boost accountability by tracking interventions, recording follow-ups, and facilitating cross-functional teamwork—especially crucial as students engage through in-person, virtual, and hybrid channels. In a smart campus setting, case management fosters a coordinated ecosystem of care, where data-driven, proactive outreach ensures that student needs are consistently and clearly addressed.

Kiosks and Service Points

Kiosks and intelligent service points offer instant, location-based support for navigation, information access, and essential services. They play a key role in fostering inclusion on a smart campus by helping visitors, new students, and those unfamiliar with campus protocols confidently navigate without prior insider knowledge. When linked to schedules, events, maps, and service directories, kiosks enhance the digital campus experience by connecting online resources to the physical environment, enabling a smooth, integrated journey for every user.

AI Chatbots and Service Agents

AI chatbots and virtual support agents provide frontline assistance across IT help desks, student services, enrollment, and campus activities. When implemented with robust governance—covering accuracy, clear escalation processes, defined boundaries, and privacy protections—these tools greatly enhance service hours, cut down wait times, and efficiently manage routine inquiries. Their main benefit isn't replacing human experts but enhancing their capabilities: allowing staff to focus on more complex, impactful tasks, while students get quick, dependable support anytime and anywhere.

Unified communications and contact centers

In an AI-enabled smart campus, the unified communications and contact center platform ceases to be a legacy telephone system. It becomes the critical human interface of the **Digital Integration Fabric**. By dismantling siloed departmental help desks in favor of a single omnichannel architecture, the institution delivers frictionless, context-aware support, with a user's history traveling seamlessly across IT, Facilities, and academic services without requiring them to repeat their story. During a crisis, this layer acts as the central nervous system for the Emergency Operations Center, providing location-aware E911 routing, unified first-responder communications, and micro-targeted mass notifications to ensure absolute situational awareness. Furthermore, by empowering support staff with AI copilots that offer real-time knowledge surfacing, sentiment analysis, and proactive call deflection driven by live campus telemetry—such as automatically alerting residents to a localized Wi-Fi outage before they call—the university transforms reactive customer service into an empathetic, highly efficient, and scalable operational capability.

Academics and Learning

A smart campus extends beyond buildings to include the learning environment and to ensure academic continuity. The core mission of a smart campus focuses on the academic and learning realms. While operational technologies such as HVAC systems and microgrids provide the physical infrastructure, academic technologies serve as the intellectual core. An effective smart campus not only enhances energy efficiency but also minimizes barriers to learning, safeguards research activities, and leverages the physical setting as a teaching resource.

Smart Classrooms and HyFlex Environments

Spaces designed with integrated AV systems, automated lecture capture, and high-density Wi-Fi support both in-person and remote learning (HyFlex). In a mature smart campus, these systems integrate with building automation systems. When the AV system starts a lecture, the room's lighting dims automatically for presentations, and HVAC adjusts for high occupancy. These integrated AV systems, lecture capture, and HyFlex features minimize friction and enhance instructional resilience. Proactive monitoring, condition-based maintenance, and remote support help reduce class disruptions.

Dynamic Collaboration spaces

Spaces on campus are reserved or available for spontaneous use by the campus community. They offer AV integration and serve as gathering places where people can leverage smart technologies to promote learning, collaboration, and sharing.

Learning Analytics and LMS Integration

The Learning Management System (LMS) functions as the central digital classroom. Smart integration involves collecting real-time engagement data—such as login frequency, assignment completion rates, and forum participation—and sending it to advising dashboards. This enables faculty and student success teams to detect students who are struggling weeks before midterm grades are released.

Connected Lab Management and Instrumentation

High-value laboratory environments demand strict environmental management. IoT sensors track conditions in ultra-low-temperature freezers, fume hoods, and clean rooms. When a compressor shows signs of failure, the system sends an alert before any biological research or sensitive materials are compromised, safeguarding the institution's research funding and reputation.

AI-Powered Adaptive Learning and Tutors

Advancing from static textbooks to intelligent tutoring systems, which utilize AI to assess a student's current performance, modify question difficulty dynamically, recommend tailored reading resources, and target specific knowledge gaps. Generative AI "copilots," carefully trained only on the professor's syllabus and approved materials, offer round-the-clock, precise tutoring without hallucinating external information.

Predictive Student Success and Intervention Engines

Integrating academic data (such as LMS engagement and grades) with behavioral and operational information (like swipe access in the dining hall and Wi-Fi usage at the library) to develop a comprehensive, predictive model of student wellbeing. When the AI observes a sudden decline in a student's usual campus interactions, along with missed assignments, it automatically initiates a workflow to contact an academic advisor or wellness counselor. (Note: This process necessitates strict data privacy governance and ethical AI oversight.

Immersive XR (Extended Reality)

Leveraging AR (Augmented Reality) and VR (Virtual Reality) for experiential learning that is too dangerous, expensive, or impossible to replicate physically. Nursing students practice triage in high-stress VR emergency rooms; architecture students walk through a 1:1 Digital Twin of a building before it is constructed. This relies heavily on the campus's high-bandwidth, low-latency infrastructure (Wi-Fi 7 or Private 5G).

Smart Proctoring and Credentialing (Blockchain/Verifiable Credentials)

The use of decentralized, cryptographically secure ledgers to issue diplomas, micro-credentials, and skills badges. This allows learners to own their academic records portably and securely. At the same time, smart proctoring environments use edge-AI to ensure academic integrity during digital assessments by verifying identity and monitoring testing environments for anomalies.

The Smart Library Information System

The library is transforming from a static repository into an active, sensory environment. IoT-enabled "smart shelves" use RFID to continuously monitor inventory and instantly notify staff if a book is misshelved. Digital Twins of the library provide 3D, turn-by-turn indoor wayfinding on student mobile apps to locate specific resources. Furthermore, environmental sensors strictly monitor temperature and humidity around rare archives, directly linking physical preservation to the building's HVAC automation.

AI-Assisted Collaboration and Content Generation

Multi-application collaboration platforms provide essential tools like word processors, spreadsheets, presentation builders, and note-taking features. They allow students to access course materials, collaborate with peers, and get feedback on their projects. However, they lack direct integration of enterprise AI agents, such as Microsoft© 365 Copilot or Google© Gemini, into students' and faculty's workflows. These AI assistants act as collaborative partners, supporting activities like analyzing large datasets in Excel/Sheets, quickly summarizing recorded Teams/Meet or lecture transcripts into study guides, or drafting outlines for group presentations. The key advantage is their governance: these tools operate within university standards, securely managing data to prevent leaks of intellectual property or student information to public AI models.

Attendance Monitoring

Collecting attendance data is a tedious and time-consuming process. Smart campuses leverage technologies that automate attendance. Attendance data is no longer used merely for grading; it becomes a critical diagnostic signal for student success.

Proactive intervention

The AI Analytics Engine analyzes attendance and academic performance data, LMS engagement, and historical course success models. If the system detects that a typically engaged student has missed consecutive classes, stopped co-authoring in their Google© Workspace group project, or is performing poorly on quizzes and exams, it autonomously flags an academic advisor to initiate a wellness check—shifting the institution from reactive remediation to prediction-based success interventions.

The Smart Syllabus

In a Smart Campus learning ecosystem, the Syllabot serves as a personalized interface connecting the institution's data—such as policies, schedules, and objectives—with students. It plays a key role in establishing a cognitive infrastructure that fosters an integrated environment, where students avoid navigating fragmented platforms such as LMS, portals, and email. Instead, they engage with a unified system where the syllabus proactively inquires about their goals and adapts to their individual learning pace and style. For instance, Arizona State University has implemented nearly 1,000 *Syllabots*, enabling continuous support for large classes. The system not only answers questions but also detects patterns of student engagement and confusion, helping faculty improve course design on the fly. By automating routine inquiries—such as late policy or deadline issues—it relieves administrative burdens, allowing faculty to focus more on teaching and mentorship. The Syllabot is built with a "guardrail" to prevent it from generating false answers; when uncertain, it directs students back to instructors, reinforcing trust—vital for large-scale AI use in institutions. It acts as the connective tissue within a broader TechEd architecture, linking the Academic Record of Capabilities (ARC)—which tracks what students know—with PraxisAI, where they apply their knowledge. This creates a closed-loop system that actively supports learners' agentic development.[9]

AI-Assisted Curriculum Design

Artificial intelligence on a smart campus enhances curriculum development by converting institutional data into valuable academic insights. It examines trends in enrollment, learning outcomes, labor market signals, course usage, and student feedback to help academic leaders pinpoint misaligned, redundant, or slow-to-adapt curricula that fall short of job market demands. Digital mapping of academic programs allows simulation of the effects of new courses, delivery methods, or prerequisite changes before implementation, minimizing risk and shortening approval processes. AI-powered tools also assist faculty by quickly mapping competencies to outcomes, ensuring courses meet accreditation standards, and identifying evidence-based improvements by comparing peer institutions and internal data. This approach shifts from one-off, committee-led curriculum updates to an ongoing, data-driven process—enabling institutions to adapt programs more quickly, use instructional resources effectively, and keep curricula relevant to students, employers, and society.

Research Computing and Edge Data Services

Connecting High-Performance Computing (HPC) and localized edge servers directly to research labs enables the processing of large-scale, high-velocity data from scientific instruments (such as electron microscopes or genomic sequencers) locally before transferring it to the cloud, thereby reducing latency and network load. HPC supports campus researchers in combining operational data with academic research models, fostering innovative interdisciplinary scholarship.

- **Sustainability & Climate Science**
  High-resolution climate modeling, campus microclimate simulations, and carbon forecasting at building and portfolio levels.
- **Engineering & Applied Sciences**
  Modeling of energy systems, microgrids, materials performance, and embodied carbon using real campus assets as data sources.

- **AI, Data Science, and Systems Research**
  Training and validating machine learning models on real OT data—fault detection, predictive maintenance, anomaly detection, and optimization algorithms.
- **Social Sciences & Public Policy**
  Studying space utilization, mobility patterns, equity of building conditions, and behavioral responses to environmental changes using anonymized data.

HPC supports research that would be computationally infeasible on regular infrastructure, while reflecting real-world campus conditions.

The Campus as a "Living Laboratory"

The Living Laboratory exemplifies the smart campus at its best as an academic institution. It is not just a location for learning but also a proactive player in discovery, experimentation, and applied research. In this setup, the university deliberately and securely grants access to parts of its Operational Technology (OT) ecosystem, including microgrid telemetry, HVAC system performance, indoor air quality data, and anonymized spatial and energy-use information, to students, faculty, and researchers.

Instead of using abstract datasets or simulated environments, learners work with real systems functioning at an institutional level. Engineering students examine live chiller performance and thermal dynamics in thermodynamics and controls courses. Computer science and data science students develop machine learning models based on actual campus energy use and occupancy data. Sustainability and public policy researchers assess decarbonization strategies using real-world emissions data directly linked to campus operations. Ultimately, the campus itself serves as a constantly evolving research platform.

In this environment, theory and practice come together. Research progresses quickly as hypotheses are tested against real-world conditions, at actual scale and complexity. Teaching shifts from textbook examples to systems thinking, helping students understand the tradeoffs among comfort, energy, carbon, resilience, and cost. Facilities and Technology teams also gain, as academic insights inform better operations, optimization, and innovation pilots.

The Living Laboratory transforms the physical campus from mere infrastructure into a shared, institutional cognitive system for learning and research. Buildings are seen as datasets, utilities as models, and operations as case studies. This creates a virtuous cycle in which education, research, sustainability, and campus performance mutually enhance one another—making the university a place that not only investigates the future but also actively manages it.

Health and Wellness Services

Smart AI-enabled campuses go beyond clinical services by integrating health support into students' everyday lives. The upcoming sections explain how ethically and transparently managed intelligent systems broaden access, enhance prevention, and foster a campus environment that actively promotes physical, emotional, and social well-being.

Smart Wellness Platforms and Behavioral Insights (Opt-In)

Smart wellness platforms enable students to voluntarily monitor stress levels, sleep quality, study routines, and well-being markers through privacy-conscious analytics. They offer tailored advice, gentle prompts, and curated resources, helping students recognize patterns they might miss otherwise. When managed with strict consent and data minimization, these platforms support student empowerment rather than surveillance. Over time, anonymized and aggregated data can reveal systemic issues such as academic hurdles or seasonal declines in well-being.

This allows institutions to refine policies, adapt services, and promote equity—all while safeguarding individual autonomy.

AI-Enabled Clinic Scheduling, Triage, and Care Navigation

AI-powered triage tools improve clinic access by helping students describe their symptoms, connecting them with suitable providers, and finding available appointment slots without long waits or complex procedures. These tools can direct urgent issues to immediate care and provide self-service options for routine concerns. Integrated care navigation guides students to the right services—whether counseling, medical care, disability support, or community referrals—reducing confusion and drop-off. This creates a more efficient clinical environment, optimizing resource use, decreasing wait times, and ensuring students get timely assistance tailored to their needs.

Virtual Care, Telehealth, and Remote Monitoring

Telehealth has become an essential part of modern campus wellness, expanding access to mental health experts, medical advice, and specialists who may not be on campus in person. AI-powered tools aid in scheduling, sending follow-up reminders, and preparing students before visits, enabling care from residence halls, remote campuses, or while traveling. Remote monitoring—done with student consent and appropriate medical needs can help students manage chronic illnesses, recover from surgery, or address ongoing mental health challenges. This hybrid approach offers convenience, consistent care, and improved access, allowing students to receive support on their terms.

Environmental Health Dashboards

Environmental health dashboards display indoor air quality (IAQ), temperature, humidity, CO_2, and other environmental indicators in a clear and accessible way for the campus community. These dashboards foster trust by openly sharing information that affects cognitive function, comfort, and health. They offer real-time data on space performance, reinforce commitments to wellness and safety, and enable targeted operational actions—such as proactive ventilation and quick responses to anomalies.

Smart Environments that Support Wellbeing

The smart campus environment promotes well-being by leveraging AI to enhance indoor air quality, lighting, acoustics, and thermal comfort. Classrooms and dorms can adjust ventilation based on occupancy, identify poor air quality that could worsen respiratory issues, or adjust lighting to match natural circadian cycles. These features go beyond mere conveniences; they are key health factors affecting sleep, thinking, and emotional health. By integrating wellbeing into the campus design, stress is minimized, and conditions are improved for better academic performance.

Crisis Intervention, Mental Health Support, and AI-Assisted Routing

AI-enhanced mental health platforms offer immediate, confidential, and culturally aware support through chat-based agents that detect crisis language and direct students to appropriate human support. These systems do not diagnose or replace clinicians; rather, they lower barriers for students reluctant to seek help and facilitate faster connections with trained professionals when warning signs appear. When combined with campus safety measures, counseling centers, and after-hours services—operating under strict oversight—they help ensure students in crisis receive prompt, compassionate, and stigma-free support.

Smart Disability and Accessibility Supports

AI-powered accessibility tools—like real-time transcription, adaptive learning aids, mobility-focused campus navigation, and personalized guidance—boost inclusion for students with disabilities. Smart buildings can provide automated door access for designated users, locate accessible seating instantly, or modify acoustics and lighting for sensory-sensitive individuals. When integrated with student services, these tools simplify accommodation procedures, lessen administrative workload, and ensure consistent support across academic, residential, and experiential settings.

Preventive Analytics and Campuswide Wellbeing Signals

While privacy constraints must always govern individual health data, aggregated and anonymized patterns can help institutions identify systemic needs or emerging pressures. Sudden increases in counseling requests, longer wait times, or geographic clustering of issues in residence halls may indicate environmental, scheduling, or policy-related stressors. With appropriate safeguards, predictive analytics can aid in staffing, proactive outreach, and institutional measures that enhance community wellbeing—without profiling individual users.

Smart Fitness, Recreation, and Engagement Tools

Wearable integrations, smart gym scheduling, and participation analytics enable students to pursue physical wellbeing flexibly. Intelligent fitness systems can suggest classes aligned with individual goals, monitor equipment availability in real time, and create wellness challenges that foster community. These tools transform recreation from a static amenity into an inclusive, personalized wellbeing partner throughout the semester.

Information Analytics and Space Optimization

As campuses develop into complex, data-rich ecosystems, analytics, and AI serve as the cognitive layer that provides meaning, coordination, and foresight to the physical environment. This section examines how intelligence—used ethically and intentionally—transforms raw data into operational capabilities, student-focused services, and evidence-based planning. These technologies do not replace human judgment; instead, they enhance it by making the unseen visible, simplifying the complex, and turning the systemic into actionable insights. When properly managed, AI-driven analytics enable the modern campus to function as a continuously learning entity—anticipating needs, adjusting conditions, and aligning resources with the institution's mission and community well-being.

AI Analytics and Optimization Engine

The AI Analytics and Optimization Engine functions as the core intelligence for the smart campus. It continuously learns from high-frequency OT telemetry, space-use patterns, energy flows, and environmental data. The engine forecasts loads, detects anomalies, finds inefficiencies, and suggests interventions—tasks that would be difficult to do manually. Whether optimizing HVAC to lower carbon emissions, predicting equipment failures weeks in advance, adjusting mobility patterns, or alerting operators to cyber-physical issues, it converts real-time data into actionable insights. This engine underpins the goals of reliability, resilience, and sustainability on a large scale.

AI Applications and Agent Services

AI Applications and Agent Services deliver intelligence directly to students, staff, faculty, and operators via user-friendly tools. These include student success concierges, proactive advising agents, service desk copilots, operator copilots in facilities control rooms, classroom support agents, and other knowledge-based assistants. Built on governed, high-quality data and aligned with institutional privacy standards, these agents enhance response times, lessen administrative workload, and make support more intuitive. Their purpose is not automation alone but augmentation—enabling campus experts to focus on more complex tasks while ensuring routine support is quick, consistent, and fair.

Scheduling and Utilization Systems

Scheduling and utilization systems combine room booking, occupancy visualization, and space analytics into a unified operational toolset. They aid academic planning, streamline processes for students and instructors, and optimize the use of limited space. When privacy is maintained through strong governance—like de-identification, purpose limitation, and transparency—these analytics facilitate smarter scheduling, dynamic room allocation, and better access to spaces. Overall, they transform space planning from guesswork into a data-driven approach.

Reservations and Occupancy Displays

In hybrid learning and flexible work settings, reservations and live occupancy displays are essential for minimizing obstacles. They allow students, staff, and faculty to view real-time availability of study rooms, collaboration spaces, and desks, thus decreasing frustration and promoting fair access. These tools assist campuses in optimizing space investments and adapting operations by showing how spaces are used throughout the day. Reliable access to spaces builds community trust and improves the overall campus experience.

Utilization Analytics

Utilization analytics collect deidentified, purpose-limited data on space usage, aiding evidence-based decisions in scheduling, planning, and capital allocation. Proper governance is crucial: institutions must ensure data is aggregated responsibly, retained minimally, and communicated transparently. When managed appropriately, these analytics help prioritize renovations, prevent unnecessary new construction, and allocate high-demand spaces to maximize academic and student benefits.

Comfort Analytics and Complaint-to-Fix Workflows

Comfort analytics connects occupant feedback directly with environmental and operational data, establishing a closed-loop improvement process. By linking comfort complaints to HVAC telemetry, occupancy, and equipment performance, facilities teams can identify issues faster, reduce repetitive calls, and demonstrate operational accountability. This method turns comfort from a subjective concern into actionable information—enabling the campus to maintain environments that foster cognition, well-being, and inclusivity.

Experience Measurement and Correlation

Experience measurement tools, such as surveys, pulse feedback, sentiment indicators, and environmental correlations, help the campus understand how operational conditions affect academic success and well-being. When properly governed, these signals uncover patterns without incurring the risk of invasive monitoring. Institutions can identify when environmental discomfort aligns with student outcomes or when service bottlenecks disrupt academic workflows, enabling ongoing improvements based on concrete data.

Institutional Research Data Portals

Institutional Research (IR) data portals offer controlled access to aggregated insights across academic, operational, and experiential areas. Designed with a privacy-first approach, these portals foster accountability, facilitate shared governance, and equip faculty, analysts, and administrators with the data they need to assess progress and inform policy decisions. They enhance transparency and allow the university community to actively engage with institutional performance metrics, all while safeguarding sensitive individual information.

Data Dictionary Services

A smart campus creates a "data estate" where governance dictates what data is collected, how it is used, who has access, and how long it is retained. A Data dictionary serves as the central repository for metadata, detailing their security classifications and how governance and system engineers utilize them. It also supplies metadata relevant to prior governance actions, data element risk classification, and audit practices, among others. When sensing involves human experience, transparency is crucial. Communication norms—clarifying what is sensed, why, and what is not sensed—help reduce suspicion and foster legitimacy.

Lab Management and Instrumentation Monitoring

Lab management and instrumentation monitoring systems oversee equipment uptime, environmental factors, access, and maintenance within research settings. These systems help prevent costly disruptions, support compliance, and maintain continuity of funded research by forecasting failures and notifying staff before conditions become unsafe or violate protocols. Integrated into the smart campus network, they ensure the research infrastructure stays dependable, resilient, and aligned with the institution's scientific goals.

Mobility, Logistics, and Auxiliary Enterprises

This is the movement layer. Mobility is both a sustainability and a user experience domain. It forms the circulatory system of the modern campus, moving people, goods, and services through complex physical and digital environments. As universities expand multimodal transportation, manage rising delivery volumes, and operate large auxiliary systems like housing, dining, parking, and retail, the efficiency of these flows influences daily life as much as classroom or building systems do. In an AI-enabled, sustainable smart campus, mobility, and logistics transition from separate operational functions to an integrated, data-driven network that can anticipate demand, optimize movement, reduce emissions, and enhance safety and accessibility. Auxiliary services, often major financial drivers for universities, also benefit by gaining tools to provide responsive, equitable, and high-performing services aligned with institutional missions and sustainability targets. This section discusses how intelligence, sensing, and cyber-physical integration transform mobility and auxiliary operations into strategic assets, boosting campus reliability, resilience, and overall experience.

Shuttle Tracking and Route Optimization

A smart shuttle system integrates GPS telemetry, historical ridership data, class schedules, and event calendars to predict demand and dynamically manage routes. Riders receive precise ETAs and occupancy updates, lowering uncertainty and wait times. Dispatchers benefit from heat maps of origin–destination movements, allowing them to rebalance vehicles proactively. When integrated into the campus energy and sustainability scorecard, the system reduces deadhead miles, idling, and peak emissions, while offering auditable KPIs such as passenger miles per gallon-equivalent, trip cost, and on-time rates. Linking with major events and weather alerts enables adaptive routing, maintaining service reliability during surges and disruptions.

Bike/Scooter Integration and Trip Planning

Multimodal trip planning integrates bike lanes, scooter options, micro-mobility hubs, and pedestrian routes into a unified experience through the campus app and public kiosks. Policy-aware routing can automatically favor safer, more accessible, and low-carbon paths. Data from dock/charge telemetry, geofenced slow zones, and parking compliance help minimize clutter and avoid conflicts with pedestrians and ADA paths. Analytics guide infrastructure investments, such as adding racks, lighting, or protected lanes, encouraging short trips to shift from shuttles and cars to active modes that promote health, equitable access, and lower carbon emissions.

Smart Parking Systems

A smart parking platform combines wayfinding, dynamic guidance to open spaces, digital permits, and seamless entry/exit using ALPR or credentials. Its strategic benefit goes beyond convenience, focusing on demand shaping. Applying variable policies, such as zones, pricing, time limits, and EV prioritization, helps smooth peak demand near popular facilities. It encourages mode shifts to transit or micromobility options. When connected with event schedules and class timetables, the system predicts surges, pre-allocates parking inventory, and communicates proactively to minimize circling, congestion, and emissions. It also provides transparent, board-ready data on utilization, turnover, and revenue.

Curb Management and Delivery Zoning

Curbspace is a limited infrastructure. Using digital curb permits, geofenced loading windows, and designated ride-share pickup and drop-off zones (PUDO) transforms chaotic edges into predictable, safer operations—especially near residence halls, labs, and stadiums. By incorporating safety cameras, pedestrian counts, and event schedules, the system dynamically adapts curb regulations, such as switching lanes to prioritize pedestrians during peak times. This leads to fewer conflicts among delivery vehicles, micro-mobility users, and pedestrians, as well as fewer instances of double-parking and near-misses. It also enables faster service for vendors by enforcing clear rules.

Fleet Telematics and Electrified Fleet Optimization

Facilities, safety, and service fleets achieve a comprehensive view of operations through telematics, which track location, duty cycle, idle time, driver behavior, and diagnostic trouble codes. AI models help determine the appropriate size and power of vehicles, identify routes suitable for electrification, and schedule maintenance based on actual condition rather than fixed schedules. When combined with energy price and carbon intensity data, this information enables dispatch decisions to prioritize low-carbon operating periods. In the event of disruptions, the fleet increases resilience by pre-positioning vehicles, batteries, or mobile chargers where they can make the greatest operational difference.

EV Charging Management

EV charging bridges mobility, energy, and user experience. A centralized platform manages signals such as prices, carbon levels, building demand, and reservations to avoid overload and stress on transformers, while ensuring a consistent user experience. It can prioritize essential uses such as public safety vehicles, ADA spaces, and fleet servicing, and align load management with renewable energy, microgrid limits, or demand-response needs. Fully integrated with parking, permits, and payments, it offers clear information on availability and equitable access, along with detailed reports on carbon impact and costs.

Cashless Payments and POS Analytics

Modern POS systems in dining, retail, events, and transit simplify transactions through tap-to-pay and mobile wallets, while providing valuable operational insights. Anonymized and regulated analytics can identify demand patterns by venue, time, and event type; help optimize staffing; forecast inventory needs; and align shuttle and event schedules to minimize queue spillover. When combined with accessibility and affordability initiatives, payment systems can grant entitlements discreetly, ensuring privacy and adhering to data-minimization standards.

Inventory and Waste Reduction Systems

IoT scales, smart labels, and POS-linked forecasting synchronize purchasing with actual consumption, minimizing waste in dining and concessions. Predictive models tailor preparation quantities based on weather, events, and academic calendars; back-of-house dashboards identify overproduction risks in time for corrective action. Visibility also covers packaging and compost streams, facilitating targeted behavior changes and supplier adjustments. These efforts lead to lower costs of goods sold, decreased waste hauling, and tangible progress on scope 3 emissions and sustainability goals.

Venue Operations Systems

For arenas, theaters, and large lecture venues, an integrated operations system manages ticketing, access, staffing, concessions, safety procedures, and egress planning. Real-time occupancy and entry analytics help optimize gate staffing and concession throughput; environmental controls prepare spaces based on occupancy patterns; and post-event data support ongoing improvements. With unified views for Facilities, Safety, and Events, venues enhance guest experiences, improve safety, and boost revenue—all while reducing energy use and waste.

IT Operations and Cybersecurity Services

In a sustainable, AI-enabled smart campus, IT Operations and Cybersecurity Services serve as the digital foundation supporting all academic, administrative, and physical systems. As the campus integrates IT and operational technology (OT) into a unified cyber-physical framework, traditional distinctions between network performance, application reliability, data governance, and campus safety blur. This results in a critical operations layer tasked with ensuring that every connected device—from enterprise systems to building controllers—operates reliably, securely, and in line with institutional policies.

Consequently, modern IT must shift from reactive troubleshooting to proactive approaches like engineering, observability, automation, and ongoing resilience. Simultaneously, cybersecurity expands to encompass campus-wide safety, safeguarding data, identities, and physical systems that support research, housing, utilities, and learning environments. A mature IT and cybersecurity organization fosters trust, mitigates risks, and establishes a stable digital foundation essential for scaling smart-campus intelligence.

IT Operations Services

Infrastructure and Platform Operations

Infrastructure and Platform Operations serve as the digital backbone of the AI-enabled sustainable smart campus, ensuring that essential systems remain reliable, secure, and scalable. This area covers the management of core compute, storage, and platform services—whether on-premises, in the cloud, or in a hybrid environment—that support a wide range of functions, including enterprise applications, analytics, AI workloads, and digital twins.

- Server and compute management (on-prem, cloud, hybrid)
- Storage and backup services (including DR/BCP readiness)
- Virtualization and container platform operations
- Database administration (maintenance, tuning, patching)
- Middleware and integration platform operations (API gateways, iPaaS)

By treating these platforms as resilient, standardized services rather than custom systems, institutions reduce operational risk, enhance recovery preparedness, and establish a stable foundation for safe growth in innovation, interoperability, and campus-wide intelligence.

Network Operations

Network Operations provides the connective tissue of the AI-enabled sustainable smart campus, ensuring that digital services and physical systems communicate reliably, securely, and at scale. This domain encompasses the end-to-end operation of wired and wireless networks that support academic, administrative, research, and operational technology (OT) environments.

- Core network management (routing, switching)
- Wireless/Wi-Fi operations
- Load balancing, DNS/DHCP, IPAM services
- Network performance monitoring and optimization
- Circuit and ISP management

By treating the network as a strategic shared utility—rather than invisible plumbing—institutions ensure predictable performance, resilient connectivity, and the capacity to support emerging demands such as IoT sensor grids, real-time analytics, cloud integration, and digital twins. Well-governed network operations are essential to campus reliability, user experience, cybersecurity posture, and the safe convergence of IT and OT systems.

Cloud and Identity Services

Cloud and Identity Services establish the trusted digital foundation of the AI-enabled, sustainable, smart campus by providing secure, scalable, and policy-driven access to systems, data, and services. This domain encompasses the day-to-day operation of cloud platforms and the institutional identity fabric that governs how people, devices, applications, and vendors interact with campus resources.

- Cloud platform operations (Azure/AWS/GCP)
- Identity and Access Management (IAM), including SSO, MFA, RBAC, privileged access workflows
- Directory services (Azure AD/Entra, AD DS)

By treating cloud and identity as core enterprise utilities—designed once and reused everywhere—institutions enable innovation at scale while maintaining strong security, auditability, and operational consistency across IT and OT environments. Well-run cloud and identity services are essential to zero-trust security, digital-twin scalability, vendor accountability, and a seamless user experience.

Endpoint and Device Management

Endpoint and Device Management is a foundational operational discipline that ensures every device accessing campus systems—desktops, laptops, tablets, kiosks, and mobile phones—is secure, reliable, and consistently managed throughout its lifecycle. On an AI-enabled sustainable smart campus, endpoints are no longer peripheral; they are control surfaces for academic work, facilities operations, safety systems, and digital twin interaction.

- Endpoint configuration and patch management
- Software deployment and license management
- Mobile device management (MDM/UEM)

By standardizing configuration, automating patching, and governing software and mobile device use, institutions reduce cyber risk, improve the user experience, and maintain operational continuity across IT and OT environments. Well-run endpoint management enables Zero Trust security, supports a distributed workforce, and ensures that edge innovation does not become unmanaged risk at scale.

Service Management and User Support

Service Management and User Support is the connective tissue between smart campus strategy and daily lived experience. As technology becomes embedded in learning spaces, facilities operations, safety systems, and digital services, the quality of service delivery increasingly defines trust in the institution. Effective service management ensures that issues are resolved quickly, changes are controlled, and knowledge is shared—while continuously measuring what users actually experience, not just whether systems are "up."

- IT Service Desk (Tier 1–3)
- ITIL/ITSM processes (incident, problem, change, request)
- Knowledge management and self-service portals
- Experience Level Management (XLAs)

On an AI-enabled sustainable smart campus, service management evolves from reactive ticket handling to a proactive, experience-driven discipline that supports reliability, adoption, and continuous improvement across both IT and OT environments.

Observability and Reliability Engineering

Observability and Reliability Engineering provides the situational awareness and operational discipline required to run an always-on, AI-enabled smart campus with confidence. As campuses evolve into distributed cyber-physical systems—spanning cloud platforms, edge devices, operational technology, and safety-critical services—leaders must be able to see system behavior in real time, understand how failures emerge, and design systems that degrade gracefully under stress.

- Monitoring, alerting, and observability platforms
- Log and event management
- Resilience engineering and SRE practices
- Capacity planning and performance engineering

Observability and reliability engineering shift operations from reactive troubleshooting to predictive insight, enabling institutions to detect weak signals early, manage complexity at scale, and continuously improve performance. In the context of an antifragile campus, this discipline ensures that disruptions generate learning, resilience is engineered by design, and both IT and OT services remain trustworthy as complexity increases.

Application Operations

Application Operations ensures that the institution's core digital systems—academic, administrative, and operational—run reliably, evolve safely, and integrate seamlessly across the smart campus ecosystem. As ERP, SIS, and LMS platforms increasingly intersect with analytics, AI, digital twins, and operational workflows, application operations become a strategic capability rather than a back-office function. ERP/SIS/LMS operations

- Application hosting, versioning, and lifecycle support
- Release and deployment automation (CI/CD)
- Integration platform support (ETL, event streaming)

Disciplined application operations enable continuous improvement without disruption, protect data integrity, and ensure that enterprise systems can scale with institutional demand while remaining interoperable with both IT and OT platforms.

Cybersecurity Services

Security Operations (SecOps)

Security Operations (SecOps) is the institutional nervous system for detecting, responding to, and learning from cyber threats across a converged IT/OT campus. As smart campuses expand their digital and cyber-physical footprint—linking cloud platforms, endpoints, building systems, research environments, and vendor access—security can no longer be episodic or perimeter-based.

- Security Operations Center (SOC)
- Threat detection and response (SIEM, SOAR, XDR)
- Malware/endpoint protection
- Vulnerability scanning and remediation tracking
- Threat intelligence and threat-hunting

SecOps provides continuous vigilance, rapid response, and institutional learning, enabling the campus not just to withstand attacks but to improve its security posture over time. In an AI-enabled, sustainable smart campus, effective SecOps is foundational to trust, safety, operational continuity, and the responsible scaling of digital innovation.

Identity, Access, and Privileged Access Management

Identity, Access, and Privileged Access Management is the foundation of trust in a modern smart campus. As institutions converge IT and Operational Technology (OT), identity becomes the primary control plane for security, accountability, and governance—replacing implicit network trust with explicit, verifiable authorization. MFA, SSO, certificate, and credential management

- Privileged Access Management (PAM)
- Vendor access control and session monitoring
- Least-privilege enforcement and access review workflows

A disciplined IAM and PAM strategy ensures that every human, device, application, and vendor interaction is authenticated, time-bound, least-privileged, and auditable. Done well, this capability protects life safety systems and critical data, enables zero-trust security across cyber-physical environments, and allows the campus to safely scale digital twins, cloud platforms, and partner ecosystems—without slowing operations or compromising uptime.

Network and OT Security

Network and OT Security is the structural safeguard that protects the smart campus as a unified cyber-physical system. As buildings, utilities, safety systems, and IoT devices connect to enterprise networks and cloud platforms, traditional perimeter security is no longer sufficient. Network and OT security establishes clear trust boundaries, constrains blast radius, and enforces safe interaction between IT and operational environments without disrupting uptime or life-safety obligations.

- Zero Trust segmentation (IT and OT)
- Firewall management and secure remote access
- OT-specific protections (passive monitoring, anomaly detection)
- Digital Certificate Management
- "Zones and conduits" architecture for IoT/OT systems

By combining Zero Trust principles, OT-aware protections, and standards-based segmentation, institutions can securely scale connectivity, enable analytics and digital twins, and support remote operations—while ensuring that cyber incidents cannot cascade into physical disruptions or safety events.

Governance, Risk, and Compliance

Governance, Risk, and Compliance (GRC) provides the institutional backbone that allows a smart campus to scale innovation without eroding trust, safety, or accountability. As IT and Operational Technology converge—and campuses rely more heavily on cloud platforms, data sharing, AI, and external partners—governance and risk management must evolve from static policy oversight to continuous, enterprise-wide discipline.

- Policy development and standards enforcement
- Risk assessments (IT, OT, cyber-physical)
- Compliance frameworks (NIST CSF, ISO 27001, SOC2, FERPA, HIPAA)
- Vendor security assessments and supply chain risk management

A mature GRC function aligns technology decisions with institutional values, regulatory obligations, and risk tolerance, ensuring that cyber-physical systems, data use, and vendor relationships are managed deliberately rather than reactively. When embedded into everyday operations, GRC enables confident decision-making, protects institutional credibility, and ensures that modernization strengthens—not undermines—resilience, safety, and public trust.

Security Architecture and Engineering

Security Architecture and Engineering ensures that security is deliberately designed into the smart campus rather than bolted on after systems are deployed. As universities adopt cloud platforms, AI, digital twins, and tightly integrated IT/OT environments, architectural weaknesses can quickly become systemic risks that span cyber, physical, and safety domains. Security architecture and engineering provides the discipline to evaluate new systems holistically—across applications, infrastructure, data flows, and identities—before they operate at scale. In this context, cybersecurity programs will implement:

- Secure-by-design reviews for new systems,
- Application security reviews (DevSecOps),
- Cloud security configuration and posture management (CSPM),
- Encryption, key management, and tokenization.

By embedding secure-by-design principles, proactive application security, strong cryptographic controls, and continuous cloud posture management, institutions reduce technical debt, prevent fragile architectures, and enable innovation that is resilient, auditable, and aligned with institutional risk tolerance.

Incident Response and Digital Forensics

Incident Response and Digital Forensics is the institution's last—and most critical—line of defense when prevention fails, and disruption occurs. In a smart campus, cyber incidents are no longer confined to data systems; they can directly affect buildings, safety systems, research environments, and residential life. Effective incident response and forensics enable detection, containment, investigation, and recovery from events across IT and OT domains without creating secondary safety or operational risks.

- Cyber incident response planning and execution
- Forensic investigations

- Disaster recovery security integration
- Tabletop exercises and playbook management

By integrating cyber response with disaster recovery, exercising cross-functional playbooks, and preserving forensic integrity, institutions move from ad hoc crisis management to disciplined cyber-physical resilience—ensuring that incidents become learning events rather than cascading failures.

Data Security and Privacy

Data security and privacy are foundational to trust, legality, and ethical stewardship on an AI-enabled smart campus. As institutions integrate IT, OT, analytics platforms, and digital twins, sensitive academic, operational, and human data increasingly flows across systems and organizational boundaries.

- Data classification and governance
- DLP (Data Loss Prevention)
- Privacy impact assessments
- Logging/monitoring of sensitive data access
- OT/data lifecycle governance

A disciplined data-security and privacy practice ensures that this information is classified appropriately, protected throughout its lifecycle, and used only for clearly defined institutional purposes—enabling innovation and insight while preserving confidentiality, integrity, compliance, and public trust. In a smart campus, strong data governance is not a constraint on progress; it is the condition that allows scalable, responsible use of analytics and AI across both digital and physical domains.

Combined IT Ops + Cybersecurity "Converged Services."

As campuses evolve into tightly integrated cyber-physical environments, the historical separation between IT operations and cybersecurity becomes a liability rather than a safeguard. AI-enabled, sustainable smart campuses increasingly require **converged services** that unify operational reliability, security assurance, and governance across both IT and OT domains, such as:

- IT/OT security architecture
- Microsegmentation and identity-driven network design
- Operational telemetry + security analytics fusion
- Change enablement tied to cybersecurity safeguards
- Digital twin integrity and data governance controls
- Unified incident command for cyber-physical systems
- IT Service Management / Service and Incident management.

By aligning IT Ops and SecOps into a shared operating model, institutions gain faster detection, safer change, clearer accountability, and coordinated response across digital and physical systems. Converged services ensure that innovation—digital twins, automation, and AI—is delivered with built-in security, operational discipline, and mission continuity rather than added risk.

IT service management (ITSM) and change enablement

IT Service Management (ITSM) is a strategic, comprehensive approach that enables an organization to design, provide, operate, and continuously enhance its IT services to support its goals. Instead of seeing technology merely as a collection of systems, ITSM considers IT as a service portfolio—a range of capabilities that must be dependable, secure, user-focused, and aligned with organizational objectives.

ITSM offers the governance, processes, and operating model necessary for consistent, effective, and sustainable delivery of technology services. Frameworks like ITIL direct the functions, including incident response, change management, asset control, service request handling, configuration management, and ongoing service enhancement.

ITSM implements structured processes such as incident, change, and problem management to minimize outages, improve response times, and ensure stable service delivery. This consistency is crucial for modern organizations that rely on digital systems for teaching, research, administration, and safety.

By organizing services around user needs, service catalogs, self-service portals, and well-defined SLAs/XLAs, ITSM improves the interaction of students, faculty, staff, and operators with IT. It guarantees that services are dependable, transparent, and straightforward to use.

ITSM also offers a governance layer that designates ownership, standardizes approval processes, and maintains auditable records of changes and incidents. This transparency fosters trust throughout the institution and mitigates risks associated with ad hoc practices.

Effective management of IT services enhances security. Implementing controlled change procedures, maintaining configuration discipline, keeping asset inventories, and managing the service lifecycle help reduce the attack surface and boost cyber-physical resilience.

Finally, IT Service Management acts as the operating system for the IT organization. It shifts IT from reactive firefighting to a proactive, accountable, user-focused enterprise service. By guaranteeing reliability, security, and strategic alignment, ITSM becomes essential in supporting the institution's mission within an increasingly digital and cyber-physical landscape.[10]

Housing and Residential Life

This is the "always-on" campus layer. Residence halls are 24/7 environments where reliability and dignity matter deeply. Campus housing is more than a place to sleep—it is a living environment that shapes wellbeing, belonging, academic success, and daily life for thousands of students. In an AI-enabled, sustainable smart campus, residential life becomes an intelligent ecosystem where thoughtfully governed digital and physical systems support comfort, safety, logistics, and community experience.

Smart housing does not mean intrusive sensing or automated control for its own sake. Instead, it means using technology to elevate dignity, reduce friction, enhance comfort, ensure safety, and create reliably supportive environments for residents. When combined with strong privacy standards, transparent governance, and human-centered design, the residential environment evolves into a responsive, equitable, and sustainable living platform—one that meaningfully contributes to student success and institutional mission.

Smart Controls and Resident Comfort

Smart controls enable residence halls to offer consistent comfort while using energy efficiently. Their full benefits come when guided by a resident-first approach. Features such as modern thermostats, occupancy-aware HVAC scheduling, adaptive lighting, and comfort analytics can significantly reduce complaints about hot or cold spots, helping students focus on learning and their well-being. These systems must operate within strict privacy boundaries—collecting only essential data, avoiding personally identifiable occupancy data, and clearly explaining how automation functions. When built with transparency and user control, smart controls promote healthier sleep, better indoor environmental quality, and an intuitive feeling of comfort, enhancing student satisfaction and retention.

Common-Area IAQ, Safety Systems,

Shared spaces in residence halls—such as lounges, study rooms, and social areas—are enhanced by real-time IAQ monitoring, safety alerts, and digitally augmented services that support daily living. Indoor air quality dashboards help maintain healthy CO_2 levels, proper ventilation, and optimal humidity without exposing individual behaviors. Collectively, these systems foster safer, healthier, and more accessible communal environments while prioritizing privacy by default.

Laundry and Smart Food Stores

Smart laundry systems offer machine availability, cycle alerts, and energy-efficient scheduling, helping to minimize frustration and avoid unnecessary trips. Small smart food stores and micro-markets allow 24/7 access to essentials and fresh products through frictionless checkout, while also improving inventory management and reducing waste.

Laundry, Packages, and Logistics Services

Students regularly use laundry rooms, package centers, and logistics systems. Smart campus technology enhances these interactions by decreasing wait times, uncertainty, and administrative hurdles. Features such as digital package alerts, smart lockers, and real-time pickup scheduling help prevent long lines and missed deliveries. Laundry platforms provide visibility into machine use, cycle notifications, and predictive maintenance to keep equipment operational. Overall, these systems enhance student experience by saving time, reducing stress, and making frequent tasks more efficient—small yet impactful improvements that contribute to a more supportive campus environment.

Guest Access Workflows

Residence halls serve as private homes and lively social hubs, needing a careful balance of hospitality and security. Smart guest access systems use accountable, time-limited digital credentials that simplify check-in while protecting the community. These passes, sent via mobile or email, can be linked to access controls, event sign-ups, or resident approvals. Coupled with visitor analytics and strict privacy policies, these systems help campuses handle more visitors, reduce tailgating, and keep safety intact without fostering hostility. This creates a secure yet inviting living environment based on trust and openness.

Resident Work Order and Communications Platforms

In modern residential living, prompt communication and transparency in service are just as important as the repairs themselves. Resident work order systems offer students easy ways to report issues, track progress, receive updates, and give feedback—making maintenance processes more transparent and predictable. When combined with facilities workflows, analytics, and mobile tools for fieldwork, these platforms help prioritize urgent matters, reduce recurring problems, and speed up repairs. Together with real-time communication tools and community notices, they build trust, lessen uncertainty, and demonstrate the institution's dedication to maintaining safe, well-kept living spaces.

Institutional Business and Resource Management Services and Resources Management

An AI-powered, sustainable smart campus does more than optimize buildings and infrastructure; it also modernizes the core systems that keep the university functioning. Business and resource management systems form the foundation of every academic, administrative, and operational activity. When these systems become intelligent, integrated, and data-driven, they enhance overall efficiency, minimize friction for students and staff, and bolster financial stability. This area represents the "enterprise operations" layer of the smart campus—where analytics, automation, and workflow intelligence come together to enable quicker decision-making, better service delivery, and resource alignment with the institution's mission and strategy.

Dynamic Scheduling and Spatial Utilization

Traditional scheduling systems assume a reserved room is occupied, which is not accurate in the hybrid era. Modern campuses utilize real-time, privacy-compliant telemetry from Wi-Fi density, IoT occupancy sensors, and digital scheduling platforms. These systems compare actual space usage with registrar bookings, uncovering underutilized classrooms, empty meeting areas, and discrepancies between scheduled and actual use. This data helps avoid the "ghost town" phenomenon, where rooms seem full but are empty, allowing institutions to reallocate space efficiently, increase study zones during busy times, and cut energy waste by shutting down HVAC and lights in unused areas. It also facilitates flexible work arrangements by supporting on-demand booking of offices or meeting rooms for hybrid employees who need short-term, private spaces on campus.

HR Systems

Human Resources systems serve as a core operational layer, supporting recruitment, hiring, onboarding, labor compliance, and workforce development. In a smart campus, HR tools become more predictive and interconnected—linking staffing trends to demand, identifying signs of attrition or burnout, and automating routine tasks such as position approvals, benefits updates, and training compliance. When integrated with the wider campus data ecosystem, HR systems help institutions match the right talent with dynamic cyber-physical operations, sustainability goals, and strategic priorities.

ERP (Enterprise Resource Planning)

The ERP system serves as the core of the institution's financial and administrative functions. In a smart campus ecosystem, ERP platforms transition from simple record-keeping tools to strategic sources of intelligence. When combined with data from facilities, energy, academic, and student services, ERP enables leadership to identify cost drivers, assess financial risks, project resource needs, and ensure budgets support mission goals. By integrating ERP with IoT telemetry—such as connecting utility billing to actual building performance—it provides real-time transparency, minimizes waste, fosters more flexible financial decision-making, and contributes to institutional antifragility.

Fiscal Systems

Fiscal management platforms enable budgeting, forecasting, financial reporting, and strategic planning. On a smart campus, they leverage real-time data streams to reveal asset performance, resource utilization, and investment returns. Predictive analytics help model the financial effects of energy savings, reduced deferred maintenance, staffing changes, and enrollment variations, allowing institutions to allocate funds more effectively and identify risks proactively.

Procurement Systems

Smart procurement platforms enhance vendor selection, contract management, purchasing workflows, and compliance. When connected with facilities, cybersecurity, and IT standards, these systems prevent the emergence of "shadow OT" by ensuring all equipment and services adhere to interoperability, cybersecurity, and data governance requirements. They can automate bidding processes, leverage AI to expedite contract analysis, analyze spending habits, and enforce lifecycle standards—assisting institutions in avoiding vendor lock-in and lowering overall ownership costs.

Workflow Automation and E-Signature Systems

Workflow automation and digital signature platforms enhance administrative efficiency, covering everything from hiring approvals to the documentation of capital projects, student petitions, and financial transactions. When scaled effectively, these tools significantly reduce process bottlenecks, speed up completion times, maintain consistent policies, and relieve staff from mundane manual tasks. On a smart campus, workflow automation leverages agentic AI, serving as the crucial link that transforms data into actionable insights—helping decisions to be made swiftly, transparently, and in line with governance standards.

Policy Automation and Compliance Engines

Tools that automate policy enforcement—including accessibility standards, procurement rules, export controls, cybersecurity initiatives, and sustainability goals—use agentic AI to minimize human mistakes and promote uniformity across decentralized units.

Grants and Research Administration Systems

Smart campuses can connect grant management systems with data on space utilization, energy consumption, and compliance. This integration lessens the administrative workload for researchers and enhances the institution's ability to oversee research-related expenses and risks.

Student Information and Lifecycle Management

Although often considered in isolation, SIS platforms are integral to the broader institutional resource ecosystem. When combined with real-time operational data and enhanced by agentic AI, they enhance retention analytics, proactively detect student needs, and promote equitable service delivery.

Records Management and Digital Archives

Automated retention, secure document storage, and digital retrieval systems help lower risk, support accreditation, and decrease administrative workload.

Institutional Performance Dashboards

Executive dashboards consolidate fiscal, operational, academic, and campus experience data into a unified view to support mission-driven decision-making. They evolve into powerful tools that can reveal trends or new relationships when integrated with AI technologies.

Facilities Management

In a smart campus ecosystem, Facilities Management evolves from a labor-intensive task into a coordinated, intelligence-enabled service layer. The use of robotics, automation, sensing, and data-driven workflows boosts workforce capacity, enhances safety and reliability, and ensures consistent service across extensive, complex campus networks. These technologies do not replace human judgment but supplement it, freeing teams to concentrate on high-value activities while smart systems handle routine, risky, or precise tasks. As a result, a facilities organization emerges that can operate at an institutional scale, deliver reliable quality, and continually improve through data-driven feedback loops.

Cleaning Robots and Grounds Automation

Autonomous cleaning systems, such as robotic scrubbers, vacuums, and lawn-care robots, provide reliable, repeatable service in buildings and outdoor areas while greatly reducing physical demands on staff. They are especially effective for high-frequency, low-complexity tasks and can automatically adapt their routes and schedules to occupancy, weather, or events. Automating routine tasks frees up facilities teams to focus on skilled work, enhancing overall quality and team wellbeing.

Warehouse Automation and Automated Storage and Retrieval Systems (ASRS)

Automation in warehouses, libraries, and central stores significantly enhances retrieval speed, space utilization, and inventory accuracy. ASRS systems replace manual stacking and aisle use with space-saving vertical storage and automated retrieval robots. For facilities with large archives, research materials, or parts inventories, these systems minimize space requirements, improve operational efficiency, and shorten fulfillment times. When connected with procurement and inventory management systems, ASRS serves as a strategic tool for controlling costs and boosting operational flexibility.

Inspection Drones

Drones equipped with high-resolution visual, thermal, and multispectral cameras make it easier to inspect roofs, solar panels, façades, storm-damaged zones, and remote infrastructure. They enable facilities teams to carry out assessments more safely and more frequently, minimizing the risks associated with scaffolding, ladders, and lifts. Automated flight routes and AI-powered defect detection improve documentation accuracy and help identify developing problems early—boosting resilience and lowering long-term maintenance costs.

Autonomous Lawn and Field Maintenance

Robotic mowing and smart irrigation tools enable continuous ground maintenance while drastically reducing water usage, fuel costs, and manual labor. They can adjust based on weather data, soil moisture readings, and scheduled field use, ensuring high-quality landscapes with minimal manual input. For large campuses with athletic fields, greenways, and lawns, autonomous systems enhance sustainability and dependability.

Predictive Pest Management

IoT-enabled traps, acoustic sensors, and digital monitoring platforms detect pest activity early and decrease reliance on chemical treatments. AI-driven analytics analyze behavioral patterns, environmental factors, and probable entry points—enabling teams to prevent infestations before they cause health or operational issues. Predictive pest management supports wellness standards and minimizes environmental impact.

Smart Snow and Ice Management

In cold-weather areas, embedded surface-temperature sensors, connected plows, and automated de-icing systems enhance winter response. These technologies enable staff to monitor walkway and roadway conditions in real time, helping prevent slip hazards and minimizing chemical usage. Smart routing directs crews to prioritize critical areas, boosting safety and lowering expenses.

Digital Lockout/Tagout (LOTO) and Safety Systems

Digitizing lockout/tagout procedures helps technicians follow proper safety protocols when working on electrical, mechanical, or pressurized systems. Digital LOTO systems offer step-by-step guidance, ensure compliance, log timestamps, and enable remote verification—reducing risks and enhancing regulatory compliance. As equipment becomes more connected, LOTO workflows incorporate greater intelligence by leveraging telemetry, alarms, and fault conditions.

Connected Tools and Asset-Tracking Systems

BLE, RFID, and GPS-enabled tagging systems enable facilities teams to quickly locate tools, equipment, and mobile assets. This helps minimize downtime, prevent loss, and guarantee that essential gear is maintained and accessible. Additionally, connected tool ecosystems can automatically record usage hours, supporting maintenance planning and reducing the risk of equipment failures.

Autonomous Indoor Delivery Robots

Indoor delivery robots are ideal for large research institutions or dispersed campuses, as they transport mail, lab samples, supplies, and documents across buildings. These robots reduce the workload of human staff by handling routine transport tasks and help ensure consistent logistics. When connected with building access and scheduling systems, they can navigate autonomously while maintaining secure delivery processes.

Smart Waste Collection and Routing

Sensor-enabled waste bins, compactors, and recycling systems monitor fill levels and optimize collection routes, lowering labor costs, emissions, and unnecessary trips. Advanced waste analytics platforms help identify contamination sources, track diversion rates, and offer educational insights to support sustainability efforts. When integrated with campus engagement initiatives, these tools strengthen a culture of environmental stewardship.

Dining, Retail, and Auxiliary Enterprises

Dining, retail, and auxiliary services are central to the campus's commercial life, connecting student experience, operational efficiency, safety, and financial health. In an AI-powered, sustainable smart campus, these functions shift from manual, time-sensitive tasks to intelligent, data-driven systems that predict needs, minimize waste, boost safety, and increase customer satisfaction. Through IoT sensors, predictive analytics, automation, and digital commerce, campuses develop environments where food safety is ongoing, retail is optimized in real time, and every transaction—whether physical or digital—fosters a smooth, student-focused experience. These smart auxiliary services not only enhance operational dependability but also serve as strategic resources supporting wellness, sustainability, and institutional resilience.

Cold-Chain Monitoring and Kitchen Equipment Monitoring

Cold-chain monitoring systems offer real-time oversight of refrigeration and freezer units, food storage temperatures, and other essential kitchen equipment. IoT sensors continuously track temperature, humidity, door-open events, and equipment performance, automatically identifying deviations that could threaten food safety or quality. These systems help prevent spoilage, cut food waste, and avoid costly service interruptions, all while supporting regulatory compliance. When connected with work-order systems or digital twin platforms, cold-chain alerts can prompt immediate maintenance actions, keeping kitchens operational, efficient, and safe. This intelligence enables dining operations to shift from reactive crisis handling to proactive, predictable performance.

E-Commerce and Point-of-Sale (POS) Systems

Modern e-commerce and POS platforms serve as the digital foundation for campus retail, dining, bookstores, and additional services. Cloud-based POS systems update menus, prices, promotions, and inventory instantly across multiple sites. They enable seamless payment methods such as mobile wallets, tap-to-pay, online ordering, automated lockers, and self-checkout kiosks, minimizing wait times and improving service accessibility for students and visitors. Integrated analytics provide valuable insights into peak times, product diversity, staffing, inventory turnover, and supply chain trends. E-commerce ecosystems also support loyalty programs, campus-wide currency, student meal plans, and targeted marketing efforts. When combined with AI forecast tools, they help institutions streamline labor, prevent stockouts, reduce waste, and improve financial outcomes.

Smart Vending and Micro-Markets

AI-enabled micro-markets and smart vending machines provide autonomous convenience on campuses by integrating sensors, computer vision, connected payments, and real-time inventory management. These systems enable students and staff to access fresh meals, snacks, and essentials 24/7 without the need for staffed counters. Features such as embedded weight sensors, shelf cameras, and RFID tags automatically monitor what customers take, update stock levels in real time, and initiate restocking workflows, reducing stockouts and labor costs. The systems support seamless payments through mobile wallets, tap-to-pay, and stored campus credentials, ensuring a smooth user experience. For institutions with staffing shortages or longer service hours, smart vending significantly increases access while ensuring consistent and efficient operations.

Digital Menu Boards and Demand Forecasting

Digital menu boards leverage real-time data, dynamic content management, and AI-driven demand forecasting to support responsive, sustainable dining operations. These boards update menus based on ingredient stock, crowd levels, dietary needs, and sustainability targets, such as reducing food waste or carbon emissions. Forecasting models consider historical traffic, academic calendars, weather, and mealtimes to predict demand and improve efficiency. Consequently, kitchens minimize overproduction, labor is managed more effectively, and students face shorter lines with more dependable options. When integrated with procurement and inventory systems, digital menu boards become smart interfaces that mirror operational conditions and the institution's commitments to health and sustainability.

Waste Reduction and Food Recovery Analytics

Smart waste analytics platforms use IoT-enabled scales, image recognition, and AI-driven pattern analysis to detect where and when food waste occurs—whether from overproduction, prep waste, plate waste, or poor purchasing decisions. By gathering detailed data throughout different stages of the food service process, these tools reveal structural inefficiencies that traditional reports often overlook. Dining teams can leverage these insights to identify menu items that frequently generate excess, adjust portions at specific stations, and improve demand forecasting. When combined with food recovery initiatives, these analytics bolster sustainability efforts, reduce food waste and disposal costs, and enable institutions to clearly showcase progress in responsible resource management.

Smart Locker and Pickup Systems

Smart pickup lockers—temperature-controlled, IoT-connected, and fully automated—expand dining and retail access across campus by offering secure, contactless, and convenient pickup options. They accommodate hot, chilled, frozen, or ambient storage, enabling students and staff with busy or unpredictable schedules to pick up meals, packages, or retail items at their convenience without waiting in line. Connected to mobile ordering apps and

POS systems, these lockers boost fulfillment accuracy, ease congestion in dining areas, and facilitate late-night service without extra staffing. Additionally, they improve accessibility for students managing work, studies, and caregiving duties.

Queue Analytics and Crowd-Flow Monitoring

Using computer vision, Wi-Fi density mapping, and crowd-flow analytics, dining facilities gain real-time insights into wait times, seating availability, and traffic patterns across locations. These technologies enable operational teams to predict busy periods, optimize staffing, modify food production, and guide customers to less crowded areas. Over time, analytics identify patterns related to academic calendars, weather, events, and menu options—leading to more accurate forecasting and an improved dining experience. When combined with digital signage or mobile apps, queue intelligence enables students to make more informed decisions and reduces friction during peak times.

Mobile Ordering and Food Delivery Robots

Mobile ordering platforms make campus dining more efficient by allowing students and staff to browse menus, customize orders, choose pickup or delivery times, and pay via their phones. When linked with kitchen display systems, scheduling software, and autonomous delivery robots, these platforms create a smooth entire process—improving prep timing, minimizing delays, and enabling quick last-mile delivery across large or spread-out campuses. Delivery robots improve access, especially for students with mobility issues or busy schedules, and help decrease vehicle congestion and staffing needs. Overall, mobile ordering combined with automation boosts convenience and operational reliability.

Sustainability Dashboards for Dining

Dining sustainability dashboards compile real-time data on energy consumption, carbon footprint, water use, sourcing transparency, and waste reduction into a single, easy-to-access interface. These dashboards enable institutions to track progress toward sustainability targets, detect inefficiencies, and share performance updates with students, leaders, and external partners. When integrated with procurement, menu systems, and waste analytics, they provide a dynamic overview of how everyday dining activities affect the institution's climate goals. Additionally, they support educational initiatives by revealing previously unseen environmental impacts and encouraging the community to adopt sustainable dining practices.

Personal Productivity, Communications, and Collaboration

A successful smart campus depends on its community—students, faculty, staff, and operational teams—working together with clarity and confidence. This part of the technology infrastructure emphasizes tools that enhance human productivity, communication, and coordination. When properly integrated with the larger data ecosystem and responsibly governed, these systems revolutionize institutional collaboration: speeding up decision-making, minimizing friction, supporting hybrid work and learning, and enabling insights to flow smoothly across organizational boundaries. In an AI-enabled setting, these applications support—rather than replace—human judgment, making daily tasks easier, quicker, and more connected.

Unified Communication and Collaboration Platforms

Modern collaboration platforms bring messaging, video conferencing, shared documents, and cross-team workspaces into a single ecosystem. They create real-time channels for academic departments, administrative offices, IT,

Facilities, Safety, and executive leadership—reducing siloed communication and enabling faster, better-informed decisions across the institution.

AI-Enhanced Productivity Tools and Copilots

AI-powered assistants amplify human capability by summarizing meetings, generating drafts, analyzing data, and guiding operational staff through complex tasks. When tied to institutional data and governance, these copilots become trusted partners that enhance productivity while maintaining accuracy, privacy, and accountability.

Workflow Automation and Task Coordination Systems

These tools streamline processes by routing approvals, automating routine administrative tasks, and coordinating work across departments. Integrated with ITSM, CMMS, and campus service platforms, they improve transparency and ensure issues reach the right people quickly—thereby reducing delays and operational friction.

Campuswide Notification and Communication Systems

Emergency alerts, targeted announcements, and context-aware messaging help ensure that essential information reaches the community quickly and appropriately. When integrated with digital signage, mobile apps, and scheduling systems, these tools support coordinated response and clear communication during both daily operations and emergencies.

Knowledge Management and Institutional Memory Systems

Searchable repositories, standardized templates, runbooks, and AI-enhanced knowledge bases preserve institutional expertise and reduce reliance on "tribal knowledge." These systems enable continuity when personnel changes and ensure that procedures, policies, and historical decisions remain accessible and consistent.

Project and Program Management Tools

Large-scale initiatives such as OT modernization, digital twins, and AI deployment require structured coordination. Portfolio and project management tools provide shared roadmaps, task boards, and dashboards that help cross-functional teams plan, track, and deliver work with discipline and alignment.

Smart Meeting Spaces and Collaborative Classrooms

These environments combine intelligent AV, occupancy-aware controls, and hybrid collaboration tools to create responsive, intuitive spaces that support teaching, research, and administrative coordination. They ensure that meetings and classes run smoothly, whether participants are on campus or remote.

Mobile Productivity and Campus Apps

Mobile platforms serve as the unified entry point for communication, scheduling, access, service requests, and information. Integrated with identity systems and campus services, they simplify daily tasks and ensure that the smart-campus experience extends to every user, everywhere.

Digital Accessibility and Inclusive Collaboration Tools

AI-driven accessibility features—such as captioning, reading aids, language translation, and interface personalization—ensure the campus experience is equitable. These tools help institutions meet accessibility commitments while enabling all community members to participate fully in academic and operational life.

The Infrastructure Layer

The Infrastructure Layer is the crucial foundation of the AI-Enabled Sustainable Smart Campus, providing a stable and resilient backbone that supports all advanced features. Although applications, analytics, and digital intelligence often draw attention, they rely on stable, secure, and scalable infrastructure. This layer includes networks, computing resources, storage options, edge devices, and connectivity, which together integrate the physical and digital elements of the campus into a cohesive operational system.

For executive leaders, the Infrastructure Layer represents more than mere technical infrastructure; it signifies the institution's long-term resilience strategy. A contemporary smart campus requires infrastructure designed not only for present demands but also to accommodate future expansion in sensing, automation, AI orchestration, and digital-twin features. When implemented successfully, this layer becomes a valuable institutional asset—adaptable enough to promote innovation, secure enough to protect essential activities, and reliable enough to sustain teaching, research, and daily operations seamlessly.

Connectivity and Edge Infrastructure

Connectivity and edge infrastructure represent the campus nervous system. Smart campuses start with dependable connectivity—not just more Wi-Fi—but with a secure, segmented, and observable network infrastructure. This backbone safely supports both human traffic (students and faculty) and machine traffic (building and operational systems). It also ensures that advanced analytics and automation can be implemented without sacrificing reliability or security.

The Campus fiber backbone and core network

The fiber backbone and core switching/routing infrastructure serve as the campus's circulatory system. They link buildings, data centers, and essential services with high-bandwidth, low-latency connections. For a smart campus, the core infrastructure must enable segmentation, redundancy, and consistent performance—since OT telemetry, safety services, and critical research rely on uninterrupted connectivity.

SD-WAN and resilient WAN design

SD-WAN enhances resilience and operational management across multiple locations, such as satellite campuses, clinics, and research farms, by intelligently directing traffic and ensuring service continuity even when circuits degrade. In the context of smart campuses, SD-WAN allows for consistent policy enforcement and optimal performance for both enterprise applications and connected physical systems.

Wi-Fi 6/6E/7 and indoor coverage engineering

Wi-Fi serves as the primary user experience layer and a secondary connection point for many devices. On a smart campus, the goal extends beyond coverage to include capacity, reliability, and location-aware performance, especially in crowded areas such as lecture halls, residence halls, and arenas. As Wi-Fi technology advances, it more effectively enables low-latency learning experiences and supports sensor-rich environments that provide operational insights.

Private LTE/5G, DAS, and small cells

Private cellular networks and distributed antenna systems address coverage gaps and ensure dedicated connectivity for mission-critical services, especially in mobility systems, safety communications, and outdoor areas. On a smart campus, private cellular can also offer a more reliable network path for specific IoT/OT devices where Wi-Fi may be unstable.

Low-power IoT networks (LoRaWAN and similar)

Low-power wide-area networks enable thousands of small sensors, such as temperature sensors, leak detectors, and occupancy counters, to operate with long battery life and minimal network overhead. They are designed to expand sensing capabilities economically and reliably, especially in areas where wiring or high-speed networks would be too costly.

Network Time synchronization (NTP/PTP) services

Many smart campus features—like security logs, automation processes, and event correlation—depend on accurate timing. Time synchronization relies on NTP (Network Time Protocol) and PTP (Precision Time Protocol). NTP provides millisecond precision across extensive networks, whereas PTP offers microsecond-to-nanosecond accuracy for specialized, high-performance, and local use. Reliable DNS/DHCP services also ensure that devices stay consistently addressable and manageable, even at scale.

Network segmentation, NAC, and microsegmentation

Segmentation is a key principle for safety and resilience. A smart campus isolates IT, OT, research networks, guest access, and public services to minimize the impact of a breach and prevent lateral movement. Network Access Control (NAC) manages who and what can access the network, while microsegmentation allows detailed policy enforcement for high-risk systems.

Edge compute nodes and building gateways.

Edge computing moves processing nearer to data sources—inside buildings, at plants, or near sensors. This minimizes latency, ensures local operation during outages, and facilitates OT-safe analytics such as fault and anomaly detection without continuous cloud access. Building gateways act as protocol converters and security checkpoints, enabling legacy OT systems to exchange data securely.

Device provisioning, certificates, and secure onboarding

At scale, device onboarding becomes a matter of governance. Secure provisioning, certificate-based authentication, and standardized onboarding workflows prevent "shadow devices" and reduce long-term risk. This is where smart campus programs quietly succeed or fail: unmanaged devices and inconsistent onboarding undermine everything else.

Digital Certificate Management

Digital certificate management is a core infrastructure element that ensures trusted, encrypted interactions within the smart campus ecosystem. By centrally overseeing the issuance, renewal, rotation, and revocation of certificates for devices, applications, controllers, APIs, and cloud services, the institution creates a robust trust network. This network supports Zero Trust security, secure OT/IT integration, and dependable machine-to-machine communication. This crucial function underpins every aspect of the smart campus by offering a stable, authenticated, and encrypted base that modern digital services rely on.

The Intelligent Microgrid

Initially, campus electrification viewed EV charging mainly as a "risk"—an unpredictable load that threatened grid stability and increased demand charges. Yet advanced energy management now treats the thousands of parked vehicles as a Mobile Microgrid rather than liabilities. By evolving from basic Grid-to-Vehicle (G2V) charging to a Vehicle-to-Everything (V2X) system, the university can convert its parking areas into a distributed battery energy storage system (BESS).

The Hierarchy of Bidirectional Energy Flow

The "Integrative Architect" must design for multiple layers of bidirectional energy exchange, moving beyond the simple "plug and charge" model:

- **Vehicle-to-Grid (V2G)**
  During peak utility demand periods, the campus can aggregate the energy from connected EVs and sell it back to the utility, generating revenue and stabilizing the regional grid.
- **Vehicle-to-Building (V2B)**
  Parked EVs can discharge power directly into campus facilities during "Demand Charge" events. By "shaving" the building's peak load with vehicle batteries, the institution can save thousands in monthly capacity costs without affecting occupant comfort.
- **Vehicle-to-Everything (V2X)**
  This broader ecosystem includes **Vehicle-to-Infrastructure (V2I)**, where EVs communicate with smart signals to optimize traffic flow, further reducing the campus's overall energy footprint.

EVs as Emergency Resilience Assets

Many campus disaster recovery plans overlook the potential of EVs as mobile power sources. In the event of a complete grid failure, V2X-enabled electric vehicles—particularly large-battery models like shuttle buses—can isolate and power critical research labs or emergency shelters, providing 48–72 hours of backup power without the noise and emissions typical of diesel generators. An AI-driven Data Estate can forecast which vehicles are likely to stay on campus the longest, based on class schedules or faculty routines, and assign those batteries for deep-cycle support. This approach guarantees that short-term commuters always depart with sufficient charge.

Overcoming the "Battery Anxiety" Barrier

The main challenge to adopting V2X technology is worries about battery degradation. AI can be used to track the State of Health (SoH) of each connected vehicle, ensuring discharge cycles stay within optimal parameters and could actually prolong battery life by avoiding calendar aging. To promote community involvement, the university could introduce "Energy Credits" that enable students and staff to exchange stored battery energy for discounts on parking fees or tuition.

By adopting V2X, the CTO transforms the campus from a mere "Passive Consumer" of energy into an "Active Orchestrator," utilizing the community's assets to build a more resilient, sustainable, and financially stable institution.

Building Systems and Operational Technology

The cyber-physical layer is where outcomes are won or lost. This is the layer where smart campus ambitions meet physical reality. Buildings and utilities are long-lived assets; they must operate safely under constraints. The smart campus does not replace OT—it modernizes OT into a measurable, governable, interoperable system.

Building Automation System (BAS/BMS)

The BAS manages HVAC, ventilation, and various mechanical subsystems. It serves as the core of building performance, influencing comfort, air quality, and energy consumption. Modern smart campus projects typically start with stabilizing BAS fundamentals, including standardizing naming conventions, implementing trending, rationalizing alarms, ensuring reliable operational sequences, and establishing secure remote access.

Lighting controls and connected lighting networks

Lighting operates as both an energy and infrastructure sector. Networked lighting reduces energy use by utilizing occupancy sensors and daylight harvesting. In some designs, the lighting system also acts as a physical backbone for sensors and wireless networks. However, this dual use must be carefully managed to avoid creating fragile dependencies that serve multiple purposes.

Advanced metering and submetering

Electric, thermal, gas, and water metering turn sustainability from a goal into measurable results. Submetering promotes accountability at the building, zone, or system level, which is essential for benchmarking, ongoing commissioning, and reliable carbon reporting.

Central plant controls and optimization

Central plants like chillers, boilers, CHP, and thermal storage usually consume the most energy and pose significant operational risks. Implementing smart campus plant optimization improves efficiency and reliability by coordinating equipment staging, setpoints, and demand management, while maintaining redundancy and safety.

Indoor Air Quality (IAQ) sensing

IAQ sensors that measure CO_2, particulates, VOCs, humidity, and temperature improve health, comfort, and public trust. Their primary functions are to optimize ventilation management and provide clear performance data when required. In higher education settings, IAQ is becoming more associated with institutional credibility, especially in busy learning and residential spaces.

Conveyance monitoring (elevators/escalators)

Conveyance condition monitoring and predictive maintenance help minimize outages and enhance accessibility. On a campus, elevator reliability is not only an operational metric but also a matter of dignity—failures particularly affect users with mobility limitations.

Fire and life-safety systems integration

Integration can support quicker situational awareness where permitted and appropriate, but it requires strict governance and compliance. Life-safety systems are subject to regulated constraints and should never be considered merely as "just another dataset."

Cold chain and critical environment monitoring

Monitoring and escalation for freezers, research storage, and specialized environments are crucial because failures can destroy research assets and disrupt mission-critical work. This represents one of the most valuable smart campus use cases, as early detection helps prevent catastrophic losses.

Building envelope monitoring

Leak detection, humidity intrusion monitoring, and thermal performance data help protect assets and minimize long-term deferred maintenance costs. On many campuses, envelope failures lead to hidden expenses, making early detection a key aspect of resilience and stewardship.

Access/egress infrastructure health monitoring

In addition to access control, the condition of door hardware and the effectiveness of egress influence safety, accessibility, and daily operation. Regular monitoring can help prevent frequent small failures that diminish trust and cause inconvenience.

Demand response and grid-interactive building controls

These features enable campuses to adjust their load based on grid signals, pricing, or internal resilience requirements. As the grid moves towards decarbonization, load flexibility emerges as a key strategic advantage.

Microgrids, storage, and islanding controls

Microgrids and battery storage enhance resilience and facilitate greater integration of renewable energy. The smart campus remains operational by prioritizing critical loads, ensuring continuity, and managing transitions during grid disruptions.

Renewable generation monitoring

PV and other renewable sources need to be continuously monitored, forecasted, and maintained. Their effective integration—particularly in storage and load management—transforms renewables from mere symbolic assets into a robust and resilient energy infrastructure.

Identity, Security, and IT Operations Backbone

A smart campus relies on a trusted digital foundation that verifies connections, safeguards operations, and maintains resilience for mission-critical services. This backbone comprises the enterprise identity system that manages access across thousands of systems, the cybersecurity framework that protects both digital and physical assets, and operational practices that ensure the reliability of networks, applications, and services at scale. As campuses transform into dense cyber-physical ecosystems—where people, devices, sensors, and AI agents interact constantly—this backbone guarantees that each connection is authenticated, every action is accountable, and services are consistently delivered with transparency and trust. It is the unseen infrastructure that ensures a safe, interoperable, and cohesive campus environment, fostering innovation without sacrificing security or reliability.

Identity and Access Management (IAM)

Identity and Access Management (IAM) acts as the central trust framework on a smart campus. It offers seamless login options like SSO and MFA, but its importance extends beyond user convenience. IAM creates a reliable identity infrastructure that manages how students, faculty, staff, contractors, devices, and AI agents access both digital and physical systems. In cyber-physical settings—where logins can control data, building systems, sensors, or safety-critical infrastructure—identity becomes a key element of risk mitigation. IAM guarantees that all interactions are traceable, auditable, and governed by consistent rules, helping the institution maintain security while supporting openness across countless applications and interconnected operational technology systems.

Privileged Access Management (PAM)

Privileged Access Management (PAM) is the most critical tool for managing cyber-physical risks in an institution. It controls how administrative access is granted, tracks configuration changes, and monitors vendor sessions. PAM removes uncontrolled or shared accounts, enforces session recordings and time-limited access, and blocks privilege escalation—the usual route to serious breaches. For a smart campus, PAM isn't just a good IT practice; it's essential for safety. It guarantees that those adjusting HVAC systems, power settings, or security panels do so within strictly managed, fully traceable procedures. As a result, PAM underpins resilience, provides clarity for investigations, and enhances institutional accountability.

Security Operations (SOC), SIEM/SOAR, and OT-Aware Monitoring

Security operations serve as the 24/7 "mission control" for cyber-physical security. A modern SOC combines SIEM—which collects logs and signals from thousands of devices—and SOAR—which automates triage and response—into a unified command center. In smart campuses, the SOC also needs to interpret OT telemetry, vendor protocols, and safety constraints, since operational systems differ from typical IT endpoints. OT-aware monitoring establishes behavioral baselines for equipment like chillers, air handlers, laboratory hoods, access control, and utility plants. This approach ensures responses maintain safety and operational uptime, preventing disruptions such as shutting down ventilation in active chemistry labs or interrupting district energy flow. Ultimately, the SOC acts as a steward of the cyber-physical environment, extending beyond traditional cybersecurity roles.

Endpoint and Server Security (EDR/XDR)

EDR/XDR safeguards the compute ecosystem powering the smart campus, from edge servers hosting analytics and digital twin services to operator workstations used by Facilities and Campus Safety. These tools identify malicious activities, isolate compromised systems, and prevent lateral movement within networks. As campuses implement distributed intelligence, AI workloads, and micro-segmented OT networks, endpoint protection becomes a crucial strategic component: it protects the compute layers that enable smart capabilities. XDR's cross-domain analytics help link anomalies across IT and OT environments, enhancing situational awareness and reducing the risk of silent breaches that could disrupt operations.

Vulnerability Management with OT Constraints

Vulnerability management in a smart campus environment demands subtlety and discipline. Unlike IT systems that can be patched weekly or monthly, OT assets operate within limited maintenance windows and cannot be taken offline without risking safety, research, or operational disruptions. Successful vulnerability management combines risk prioritization, compensating controls, segmentation, vendor communication, and a formal exception process. It focuses on containment and continuous monitoring when immediate patching is unsafe or not feasible. This strategy safeguards system uptime while minimizing risk exposure, ensuring cyber-physical resilience.

Backup, Restore, and Disaster Recovery

A resilient smart campus relies not only on backups but also on thoroughly tested recovery procedures and well-defined fallback strategies. Disaster recovery should encompass OT configurations, control logic, campus integration systems, and the entire data estate—not only servers and applications. Institutions must be able to swiftly restore building systems to safe operational states following cyberattacks, natural disasters, or equipment breakdowns. Achieving this requires collaborative testing with Facilities, comprehensive runbooks, offline configuration backups, and established "safe mode" procedures. In a cyber-physical setting, recovery isn't solely about data — it's about ensuring the continued operation of heating, cooling, research environments, safety systems, and essential utilities.

Research Computing and Data Services

Research computing plays a crucial role in enhancing an institution's competitiveness. Technologies such as High Performance Computing (HPC), secure data environments, scalable storage, and compliant platforms support a wide range of fields, including engineering, biology, climate science, and the humanities. As campuses develop into AI-enabled and sustainable environments, these services increasingly connect with energy planning, digital twin technology, and sensor-rich infrastructure. Smart campus leaders need to ensure that research computing is supported by resilient infrastructure, energy-conscious scheduling, secure data management, and integration with the overall data ecosystem. This coordination boosts grant success, speeds up discovery, and fosters a unified innovation ecosystem across the institution.

The Systems Integration Layer

This layer guarantees that the smart campus is both composable and scalable. It avoids isolated integrations and promotes reuse across different domains. The systems integration layer is based on standards that, when applied enterprise-wide, transform a collection of connected projects into a reusable institutional platform. It distinguishes between a one-time system integration and the development of a capability that can be safely expanded and updated over many years.

Without a properly designed integration layer, smart campus initiatives often encounter predictable setbacks: each project develops its own integration methods, data definitions, dashboards, and vendor dependencies. While this "one-off" method may be suitable for pilot programs, it becomes problematic when scaled across a portfolio. It results in increased lifecycle costs, inconsistent cybersecurity measures, fragile operations, and a disjointed array of tools that operators and leaders cannot depend on consistently.

Composable Systems

A robust systems integration layer enables a campus to be highly modular—new use cases can be built using reliable, pre-existing components. In a composable campus, capabilities are designed to be assembled and expanded without recreating fundamental parts. For instance:

- If the institution has a standard way to ingest BAS data, then adding a new building is primarily onboarding—not a custom integration effort.
- If it has a common event stream for alarms and work orders, then connecting a new analytics tool or dispatch workflow becomes straightforward.
- If it has a shared semantic layer for spaces, assets, and equipment tags, then energy optimization logic developed for one building can be replicated across many.

Composability enables affordable and manageable innovation. It generates a "platform effect": each deployment enhances the institution's capabilities instead of creating a separate technology silo.

Scalable Systems

Scalable systems allow successful usage patterns to be replicated across buildings, departments, and vendors without overhauling everything each time. Scalability isn't just about increasing volume; it's about consistent, disciplined repetition. A scalable campus intelligence foundation offers:

- **Standard integration patterns** (APIs, connectors, gateways, event streams) that vendors and internal teams must follow.
- **Consistent identity and access controls** so data and control pathways are governed across the estate.
- **Data quality and metadata standards** so leaders can trust comparisons across buildings and time.
- **Operational workflows** that turn insight into action—fault → triage → work order → fix → verify—without requiring heroics.

When this is present, "scaling" a smart campus involves replicating established patterns rather than improvising.

Digital Integration Fabric and Data Services

A modern smart campus functions as an interconnected ecosystem, where data, events, and decisions flow smoothly across physical infrastructure, enterprise systems, and human-centered services. The Digital Integration Fabric is the key architectural layer enabling this connectivity. It facilitates the efficient, secure, and meaningful transfer of information from sensors and operational systems into analytics, AI models, applications, and institutional workflows. This layer is more than just technical infrastructure; it serves as the engine of interoperability, supports institutional intelligence, and helps campus leaders translate fragmented digital signals into tangible outcomes. By integrating API services, event brokers, data platforms, semantic modeling, and master data management, it creates a unified operational backbone that provides the consistency, reliability, and scalability essential for an AI-enabled, sustainable smart campus.

API Gateways and Services

API gateways act as the controlled access point into the campus's digital environment, ensuring consistent data exchange and workflow coordination. They facilitate secure, traceable, and governed interactions among enterprise applications, operational technology platforms, cloud services, and AI tools. For leaders, the strategic importance of API gateways is in preventing vendor lock-in, boosting innovation, and ensuring interoperability across infrastructure spanning many years. By mandating that vendors connect via institution-defined APIs rather than proprietary methods, the university retains long-term control over data and system behavior, making it easier to expand analytics and digital twins without overhauling integrations every time a system is updated.

Agents

Agents are lightweight, task-focused software components positioned near data sources, devices, or workflows that execute specific integration or processing tasks. In a smart campus setting, they may normalize sensor data, enforce security measures, trigger automated responses, or monitor conditions in real time. They connect the physical and digital worlds by applying localized logic while adhering to enterprise standards. Agents support scalability—thousands can operate across buildings, systems, and devices—and remain under an institutional governance framework to ensure consistency, safety, and auditability as intelligence spreads.

Message Brokers

Message brokers serve as the event-driven foundation of the integrated campus, routing real-time signals—such as sensor data, alarms, occupancy updates, or system states—between operational systems and analytics platforms.

In a smart campus, they facilitate high-speed data streaming without overloading core systems, allowing applications and AI models to subscribe only to the signals they need. This decoupling minimizes integration issues, enhances resilience, and enables the addition of new features like digital twins or predictive maintenance without overhauling the architecture. For executives, message brokers mark a transition from poll-based, point-to-point integrations toward a scalable, flexible platform aligned with long-term modernization goals.

Central Data Platform (Lakehouse / Warehouse)

The central data platform serves as the institutional repository for operational, spatial, environmental, academic, financial, and experiential data. The lakehouse/warehouse architecture facilitates both structured analytics—such as reporting, dashboards, and compliance—while handling unstructured or semi-structured data, such as IoT telemetry, logs, and streaming events. Its strategic significance is considerable: it converts raw, isolated data into governed, decision-ready intelligence that supports sustainability reports, digital twin models, AI decision-making systems, and executive performance metrics. By unifying data in a shared, secure platform, the campus establishes a single source of truth that drives cross-functional insights and diminishes dependence on inconsistent departmental datasets.

Master Data Management (MDM)

Master Data Management (MDM) guarantees that key institutional data—such as buildings, rooms, assets, organizations, people, and schedules—remains consistent across all systems. Without MDM, a classroom might be listed under different names in the registrar system, BAS, energy dashboards, and space analytics, which diminishes the usefulness of analytics and digital twins. MDM sets authoritative definitions, resolves data conflicts, and ensures accurate, standardized information is shared throughout the organization. For leaders, MDM is essential for reliability and governance, supporting transparent reporting, accountable operations, and uniform decision-making across both digital and physical enterprise environments.

Semantic Layer and Campus Ontology

The semantic layer captures the "meaning" of data, including how rooms connect to systems, how assets relate to telemetry, and how operational events link to campus functions. A formal campus ontology serves as a blueprint for digital twins, advanced analytics, and AI models, enabling consistent interpretation and efficient reuse of data from various systems. This abstraction layer transforms the smart campus from merely a collection of dashboards into a cohesive intelligence system. For leaders, the semantic layer is a vital infrastructure that facilitates interoperability, preserves institutional knowledge, supports scalability, and ensures that technological investments deliver consistent, reliable results over time. For a more complete discussion of Ontology, see Chapter 10.

The Data Dictionary

As previously discussed, the Data Dictionary service enhances the semantic layer. Although no single universal standard protocol for data dictionaries exists, established best practices, industry-specific standards, and structural guidelines help maintain consistency. A well-developed data dictionary is integrated into technological sources and governance, making it a valuable tool for managing metadata and linking to auditable cybersecurity practices.

Building the Digital Integration Fabric

In the absence of a unified architecture, smart campus projects inevitably default to "one-off integrations." In this reactive model, each project team independently reinvents the wheel: custom-linking data sources, manually translating vendor-specific names and units, creating isolated dashboards, and granting temporary vendor access.

These ad-hoc, repeated choices compound into massive institutional technical debt. This debt manifests as hidden operational costs, brittle point-to-point connections that break during routine software updates, inconsistent reporting across departments, and a dangerous reliance on undocumented "tribal knowledge" held by a few key individuals.

To break this cycle, the Integrative Architect must establish a **Digital Integration Fabric**—a set of shared, governed enterprise services that every new project must utilize. This fabric replaces chaotic, point-to-point wiring with a managed, scalable architecture built on four key pillars, ultimately culminating in true spatial intelligence.

The Integration Backbone (Moving the Data)

Rather than hardwiring systems together, the campus must use decoupled integration middleware (iPaaS) to connect systems through repeatable, secure patterns.

- **API Gateways and Service Catalogs**
  APIs make institutional services and Operational Technology (OT) data reusable. An API Gateway manages this traffic, providing critical functions like authentication, throttling, and logging. This mechanism allows OT data to be safely exposed to analytics engines, student apps, and external partners without compromising the core system.
- **Event Streaming and Message Brokers:**
  While APIs handle data requests, real-time campus operations depend on *events*—alarms, occupancy shifts, work order updates, and security triggers. An event streaming architecture utilizes a publish/subscribe model, allowing a sensor to broadcast an event once while multiple systems react simultaneously, eliminating fragile point-to-point dependencies.

The Centralized Data Platform (Housing the Intelligence)

Data must flow out of vendor silos and into a centralized—yet federated—data platform, such as a Lakehouse or Data Warehouse. The goal here is not centralization for its own sake; it is about creating a secure, consistent environment where operational telemetry and institutional data can be normalized and synthesized at scale. This platform is the engine room for cross-domain insights, supporting heavy AI workloads, longitudinal benchmarking, and the rigorous performance verification required for sustainability reporting.

Governance, Identity, and Security (Protecting the Trust)

Security and governance cannot be bolted on to the smart campus after the fact; they are the absolute prerequisites for scaling connectivity without simultaneously increasing enterprise risk.

- **Identity Controls**
  The smart campus requires consistent authentication and authorization across all systems. The same rigorous identity standards that protect the ERP must be applied to vendor access and privileged OT operations.
- **Data Governance**
  The institution must be able to govern the lifecycle of its intelligence. Governance ensures the university can confidently answer: Who owns this data? Where did it originate? How reliable is it? How long is it retained? In the context of a "sensory campus," privacy protections and civil liberties become enforceable architectural mandates rather than aspirational goals.

The Culmination: The Digital Twin

When these four pillars operate in harmony, they enable the ultimate operational tool: the Digital Twin. A digital twin is only valuable when it integrates the real-time operational telemetry from the Data Lakehouse with the exact spatial context of Geographic Information Systems (GIS) and Building Information Modeling (BIM).

By fusing the integration backbone, semantic meaning, centralized data, and rigorous governance, the Digital Twin provides a dynamic map of assets and spaces. It transforms raw, disconnected building data into navigable, actionable intelligence that drives the AI-enabled sustainable campus.

The Platform Dividend: Strategic Reuse Across the Enterprise

Once the Digital Integration Fabric is fully deployed, it shifts from being merely an IT project to an enterprise-scale enabler. Its real value—financial and operational—comes from its ability to be reused across various initiatives. Instead of investing in separate, costly "mini-platforms" for each new departmental project, the organization capitalizes on a unified, governed data ecosystem that supports multiple strategic goals at once.

Because the data is normalized, secured, and semantically tagged in the fabric, a single sensor's output can serve five different masters at once:

- **Facilities Operations**
  Raw equipment telemetry is processed into Fault Detection and Diagnostics (FDD). This intelligence automatically triggers work orders in the CMMS and, crucially, verifies that the technician's corrective action actually resolved the mechanical anomaly.
- **Sustainability and ESG**
  That same telemetry—combined with metered energy and Building Automation System (BAS) data—is routed to the sustainability office. It fuels automated carbon reporting and enables continuous algorithm-driven commissioning to prevent energy drift.
- **Space and Capital Planning**
  Occupancy sensors and Wi-Fi flow data, rigorously stripped of Personally Identifiable Information (PII) by the governance layer, generate real-time utilization heat maps. This empowers the Provost and Facilities planners to make multi-million-dollar capital decisions based on how spaces are actually used, rather than how they were scheduled.
- **Campus Safety and Security**
  Anomalous environmental alerts and situational signals are instantly pushed through the event-streaming backbone into unified incident-management workflows, accelerating emergency response and physical-security protocols.
- **The Student Experience**
  Real-time operational status of the campus is exposed via secure APIs to digital signage, service portals, and the unified student mobile app, creating the frictionless, responsive environment that digital-native students expect.

Evolving AI Integration Fabric and Standards

Just as BACnet standardized how physical building systems communicate, a new generation of technical standards is rapidly evolving to govern how AI models are built, deployed, and interconnected.

For the Integrative Architect, these standards are the ultimate defense against "AI vendor lock-in." If a university buys a proprietary AI system that cannot share its models or insights with the rest of the **Digital Integration Fabric**, it has simply purchased an expensive, intelligent silo.

To future-proof the AI-enabled smart campus, CTOs and CFaO must begin baking these emerging standards and interoperability frameworks into their procurement language today. Here are the critical evolving standards supporting AI future-state development:

1. Model Exchange and Framework Interoperability (The Engine Room)

AI models are built using various programming frameworks (such as PyTorch or TensorFlow). Historically, moving a model from one vendor's environment to another was nearly impossible.

- **ONNX (Open Neural Network Exchange):**
  This is the gold standard for AI interoperability. ONNX acts as an open-source translator for machine learning models. If a university research team develops a brilliant energy-optimization model in one framework, ONNX allows the IT team to seamlessly deploy that exact model into the campus's edge computing gateways, regardless of the hardware vendor.

2. Semantic Data Standards (The AI's Ground Truth)

AI is entirely blind without context. An algorithm cannot optimize a room if it doesn't know what the data point "TMP_VAV_2" means. The interoperability of AI is directly dependent on the physical campus's semantic standards.

- **Brick Schema and Project Haystack**
  While we discussed these as data standards, they are fundamentally *AI-enabling* standards. They provide the standardized ontology (the digital dictionary) that allows an AI model trained on a building in Arizona to be instantly deployed to a building in Massachusetts without retraining on a new set of data tags.
- **RealEstateCore (REC):** An emerging, cloud-native ontology specifically designed to map the physical built environment into a format that AI and Digital Twins can natively understand and manipulate.

3. AI Governance, Risk, and Trust (The Guardrails)

As AI moves from providing recommendations to taking autonomous actions (Agentic AI), Boards of Trustees require certifiable proof that the systems are safe, unbiased, and transparent.

- **NIST AI RMF (Artificial Intelligence Risk Management Framework):** Developed by the U.S. government, this is rapidly becoming the enterprise standard for mapping, measuring, and managing the risks of AI. It provides a shared lexicon for the CISO, CTO, and CFaO to govern the deployment of AI in cyber-physical environments.
- **ISO/IEC 42001:** This is the world's first AI management system standard. Just as ISO 27001 certifies cybersecurity, ISO 42001 provides a certifiable framework to ensure that an organization develops and uses AI safely, transparently, and traceably.

4. API and Agentic Interoperability (The Connectors)

As the campus deploys multiple AI agents (e.g., an AI managing the HVAC, an AI managing the IT help desk, and an AI managing student advising), these agents must be able to communicate securely with one another to resolve complex, cross-domain issues.

- **OpenAPI Specification (OAS):** The foundational standard for how software APIs are described. For an AI agent to act (e.g., submitting a work order in the CMMS), it must read the OpenAPI documentation to understand exactly how to interact with that software.

- **Emerging Agent-to-Agent Protocols:** Consortia like the **Coalition for Secure AI (CoSAI)** and the **IEEE** are actively developing the next generation of standards that will dictate how autonomous AI agents authenticate each other, share context, and delegate tasks without human intervention.

The future state of the smart campus will not rely on a single, monolithic "God AI" that runs everything. Instead, it will be an ecosystem of specialized, localized AI models—one optimizing the chiller plant, another managing cybersecurity, and another answering student financial aid questions.

The Standards and Protocol Layer

Interoperability standards define how different IT/OT systems communicate. The protocol layer, serving as their language, determines whether interoperability is achieved or lost. By implementing and enforcing these standards across the university's IT/OT infrastructure, the institution ensures that its ecosystem of applications and services remains flexible, secure, and fully managed.

Over the past 40 years of experience in higher education technology, a key lesson has been that interoperability determines the success or failure of innovation. Repeatedly, proprietary silos have turned promising campus solutions into costly technical debt. In today's AI-enabled, sustainable smart campus, we must avoid these isolated systems.

OT systems are often highly proprietary and based on unique standards and protocols, suggesting that sticking to a single vendor's solution promotes easier interoperability and scalability across the campus. However, CFaOs face major hurdles with "vendor lock-in" when attempting to combine solutions from multiple vendors, making the process costly and challenging. This situation also limits the ability to obtain competitive bids for projects. Consequently, innovation across the various layers of smart campus applications is constrained, often resulting in isolated silos.

Today, higher education offers a diverse range of applications and a strong push to leverage data within our smart campus ecosystem. This highlights the importance of adopting "open" standards and protocols that are openly shared among technology vendors, fostering innovation within a highly integrated technological infrastructure.

The CTO/CFaO should intentionally design the smart campus architecture around clear institutional standards, definitions, policies, processes, and procurement language that promote a resilient, vendor-neutral ecosystem. This includes developing a "Digital Constitution," a shared set of languages, reference frameworks, and enforceable rules—allowing, for example, a central plant chiller to securely connect with a cloud-based AI optimizer without costly translation layers.

Reference Architectures You Can Govern

A campus reference architecture cannot be purely conceptual; it must align directly with decision rights, procurement language, network segmentation strategy, and operational accountability. Three "reference frames" consistently help institutions reduce complexity and strengthen security.

The Purdue / ISA-95 Mental Model: Separating Enterprise from Control

Campuses often face challenges because enterprise IT, building automation, and utility Operational Technology (OT) are combined without clear distinctions. ISA-95 outlines levels from enterprise planning to physical control systems. Although strict adherence isn't required, it's essential to follow its core principle: the nearer to real-time physical control, the more attention you must give to determinism, safety, and containment. Conversely, as it moves closer to enterprise functions, the emphasis shifts to scalability, analytics, and broad accessibility.

IEC 62443 Zones and Conduits: Cyber-Architecture for Converged IT/OT

When IT and OT integrate, "flat networks" pose an existential risk to the enterprise. IEC 62443 offers a framework that segments systems into zones based on trust and criticality and manages data flows between these zones through controlled conduits. A sustainable smart campus must typically define zones such as:

- Enterprise IT zone
- Academic and residential user zones
- OT control zones (Building Management Systems, central plant, life safety)
- IoT device zones (tiered by device trustworthiness)
- Data/analytics zone (where data aggregation and AI operate)

This segmentation represents the clearest common language between Facilities and IT. It guarantees that a vulnerability in a smart vending machine cannot be exploited to access the Registrar's records or disable a research lab.

NIST ICS Guidance: Aligning Security with OT Realities

Industrial control environments have specific constraints: limited patching windows, strict uptime requirements, and outdated protocols. NIST's ICS security guidance addresses these challenges by aligning security measures with actual operational needs. For smart campuses, the goal is to adopt modern security practices—such as identity management, ongoing monitoring, and Zero Trust segmentation—through compensating controls and phased upgrades that account for the critical safety aspects of the physical infrastructure.

The Domain-by-Domain Matrix

Vision without enforceable contractual provisions is merely an illusion. To prevent vendor lock-in, this matrix segments the smart campus into operational areas, emphasizes the preferred standards, and provides specific procurement language for use in every RFP and vendor agreement.

A smart campus protocol strategy focuses on governance rather than always pursuing the latest technology. By rigorously applying these standards during procurement, the Integrative Architect guarantees that field protocols provide control, MQTT messaging scales telemetry, semantic schemas ensure data reusability, and security segmentation safeguards the entire system. This disciplined approach removes previous technical debt and creates a solid foundation that will benefit the university for many years.

Exhibit 13. Contract Language by Operational Domain

Operational Domain	Preferred Standard(s)	Procurement-Ready Contract Language
Building Automation Systems (BAS / HVAC)	BACnet/SC (Secure Connect)	*All provided controllers and supervisory devices must support native BACnet/SC (Standard 135) communication. The use of proprietary protocols, middleware translators, or gateway devices to achieve BACnet compliance is strictly prohibited. The system must support TLS 1.3 encryption for all network traffic.*
Utility Metering (Energy/Water)	Modbus TCP/IP or OPC UA	*All IP-enabled metering devices must support native Modbus TCP/IP or OPC UA. Devices must be configurable to operate entirely within an isolated, non-routable OT network segment (IEC 62443 Zone). They must not require outbound internet access to vendor cloud environments for core functionality.*

Operational Domain	Preferred Standard(s)	Procurement-Ready Contract Language
Smart Lighting	DALI-2 with MQTT or BACnet/IP uplink	*Luminaire-level control networks must utilize the open DALI-2 standard. The lighting control gateway must expose granular telemetry data (occupancy, ambient light, energy draw) natively via BACnet/IP or MQTT publish/subscribe protocols directly to the enterprise data broker, free of proprietary APIs."*
IoT Sensor Mesh (Occupancy/Environmental)	MQTT (Sparkplug B payload)	*"All IoT sensor arrays and associated edge gateways must support MQTT publish/subscribe messaging protocols (Sparkplug B payload standard preferred) to enable decoupled, many-to-many data streaming. Devices must support certificate-based authentication and TLS encryption.*
Electric Vehicle (EV) Charging Infrastructure	OCPP 2.0.1+ (Open Charge Point Protocol)	*All EV charging hardware and associated network management software must be fully, natively compliant with OCPP version 2.0.1 or higher. The university retains the right to provision the hardware to a different OCPP-compliant software management platform at any time, without modification or penalty.*
Microgrid and Energy Management (DERMs)	OpenADR 2.0b and OPC UA	*Distributed Energy Resource Management Systems (DERMS) must be certified compliant with OpenADR 2.0b to ensure automated, interoperable demand-response capabilities with the regional utility grid. Industrial plant telemetry must be structured and shared via OPC UA.*
Life Safety and Access Control	Hardwired / BACnet (Read-Only) + SAML 2.0	*Life safety systems must operate autonomously on physically or logically isolated networks. Any required monitoring integration to enterprise dashboards must be strictly unidirectional (read-only). Administrative access to the management console must integrate natively with the university's enterprise Identity Provider via SAML 2.0 / SSO.*
Semantic Data Governance (Campus-Wide)	Project Haystack or Brick Schema	*The vendor is required to comprehensively tag all data points, equipment, and spatial relationships using [Insert Institution's Choice: Project Haystack 4 or Brick Schema]. A complete, validated semantic model file must be turned over to the university before final commissioning and payment.*

This approach shifts your technology from merely having "cool gadgets" to becoming a Strategic Asset that is defensible on the board, secure, and prepared for the AI-driven future. To turn these concepts into an institutional standard, your reference architecture should explicitly declare:

- **Semantic Requirements:** No data enters the estate without proper tagging.
- **Eventing Standards:** How alarms are routed and audited.
- **Privileged Access Patterns:** How vendors get "time-bound" access to systems.
- **Operational Acceptance Criteria:** What "done" looks like for a new project—naming, sequences, and documentation.

Exhibit 14. Architectural Layers and Examples

Architecture Layer	Examples
Physical + OT layer:	sensors, meters, controllers, PLCs/HMIs, BMS

Edge layer:	gateways, local compute, buffering, local control logic
Connectivity layer:	fiber + Wi-Fi 6/6E + private cellular/CBRS + LoRaWAN (as appropriate), with segmentation and device identity
Integration layer:	protocol translation (BACnet/Modbus/etc), MQTT/event broker, API gateway, integration middleware
Data estate:	time-series + asset master + spatial/static repositories + metadata/lineage + RBAC/privacy controls
Analytics/AI:	anomaly detection, forecasting, optimization, predictive maintenance, model governance, and monitoring
Applications:	dashboards, work management, command center, mobile experiences, digital twin interface
Security everywhere:	Zero Trust identity, zone/conduit segmentation, monitoring (SIEM/OT visibility), and incident readiness

Chapter Summary

For Cabinet leaders and Trustees, the key difference isn't choosing one technology over another, but rather between achieving isolated digital successes and establishing a sustainable institutional framework. Smart campus projects tend to produce eye-catching pilots—such as a renovated engineering building, an AI-based dashboard, or a smart entry system in residence halls. While these demos are useful, they lack scalability. When built on fragile, one-off integrations, pilots turn into maintenance challenges that often fail due to complexity.

Chapter 11 emphasizes that authentic modernization goes beyond just implementing smart tools. It involves creating a Digital Integration Fabric—an enterprise-level infrastructure that turns isolated technology initiatives into a lasting institutional capability. This foundation allows the university to pursue ongoing modernization without system fragmentation, security risks, or unmanageable technical debt.

With this architectural core established, the smart campus is both easily configurable—allowing for simple assembly, expansion, and upgrades with interoperable components—and scalable across millions of square feet. It transforms the institution from episodic innovation into a sustainable, long-term operating model that withstands technological and leadership changes.

A smart campus application stack is not just a top-down set of tools; instead, it functions as a layered ecosystem where each level relies on the stability of the one below.

- Without connectivity and edge infrastructure, data cannot be trusted.
- Without modernized OT systems, analytics remain shallow and fragile.
- Without governance, identity, and cybersecurity, scale erodes trust and increases risk.
- Without integration platforms and a governed data estate, every pilot becomes custom—and therefore unsustainable.
- Without work management and operator enablement, insights never translate into measurable outcomes.
- Without a human-centered experience, adoption fails—even when the technology is sound.

The executive insight is clear yet transformative: Successful smart campus initiatives rely less on obtaining the right technologies and more on building the appropriate institutional framework—one that is secure, interoperable, governed, and tailored for those who operate, maintain, and experience the campus daily. This architectural discipline transforms modernization from merely impressive experiments into a lasting strategic advantage.

Chapter 12
Human and Cultural Factors

The success or failure of smart campus modernization depends on the lived experiences of operators, occupants, and institutional stewards.

Facilities transformation is often discussed in technical language, covering sensors, controls, analytics, digital twins, microgrids, and security platforms. Yet the crucial element is seldom the technology alone. Moreover, it hinges on the human system—how individuals interpret, trust, use, resist, and ultimately adopt it on a broad scale.

On a campus, the human system is particularly intricate. Universities are not monolithic; they function as communities with shared governance, distinct professional identities, varied stakeholder expectations, and a broad array of building types—from residence halls and laboratories to arenas and clinics. OT modernization impacts all aspects of this system. Consequently, facilities transformation goes beyond engineering; it represents a fundamental institutional change visible in the physical structures.

This chapter focuses on the human realities that determine whether an AI-enabled, sensor-rich campus becomes a more resilient and supportive environment—or a source of friction, distrust, and operational fatigue.

The Human Side of Transformation

Convergence encourages the development of new collaboration approaches. Facilities staff may need to implement updated access procedures, analytics-driven maintenance plans, or revised vendor management strategies. Meanwhile, IT personnel might have to accept longer change windows, stricter stability requirements, and new escalation procedures related to LIFEST technology; they typically resist uncertainty, especially when system reliability is at risk.

Structured change frameworks are especially useful here. The Prosci ADKAR model—covering Awareness, Desire, Knowledge, Ability, and Reinforcement—offers CTOs a valuable perspective by emphasizing individual adoption alongside organizational objectives. (Prosci) During convergence efforts, ADKAR can be tailored into a set of defined institutional steps.

- **Awareness**
  Explain why convergence is necessary (risk, resilience, sustainability, service continuity).
- **Desire**
  Show what improves for each group (fewer emergencies, better diagnostics, safer access).
- **Knowledge**
  Provide role-based training and operational playbooks.
- **Ability**
  Run pilots with hands-on coaching and protected learning spaces.
- **Reinforcement**
  Publish outcomes, reward adoption, and lock improvements into governance.

Learning Via Conferencing and Peer Networks

The acceleration of smart campus transformation relies heavily on external benchmarks and credible patterns. Universities tend to progress more rapidly when they can confidently state, "This is backed by peers, established standards, and emerging models," rather than relying on speculation. Consequently, external learning is a core component of strategy, not just an add-on. Conferences and peer networks are crucial for the "three CXOs"—technology, finance, and sustainability/operations—as they provide a neutral environment to develop a common language. When leaders engage in external learning, they usually come back with fewer internal debates about possibilities and more focus on what suits their institution best.

From the CTO/CFaO viewpoint, conferences are essential, serving as an extension of governance. They introduce leaders to reference architectures, procurement strategies, cybersecurity trends, and integration methods that minimize reinventing the wheel. These events also highlight vendor directions and distinguish between hype and maturity. They offer insights from practitioners about the actual work involved—such as change management, data governance, commissioning, and training—beyond just technology. Most importantly, they enable institutions to benchmark themselves, not to imitate, but to adjust their ambitions and planning accordingly.

Events organized by higher education thought leadership groups like EDUCAUSE, NACUBO, CUPA, and APPA - Leadership in Educational Facilities foster cross-disciplinary learning that reflects the internal convergence universities need to achieve. Simultaneously, gatherings that focus on smart cities and infrastructure—such as Smart Cities Connect, IEEE communities, BOMI, and regional summits—highlight a vital point: campuses are increasingly functioning like small cities. They encounter similar cyber-physical risks, data architecture challenges, AI governance issues, and sustainability reporting requirements. Exposure to these broader ecosystems enables universities to recognize that a "smart campus" is not just a niche project; it's part of a larger shift toward integrated digital-physical operating systems.

The most effective leaders leverage external learning as an ongoing process rather than a one-time event. They deploy cross-functional teams with defined learning objectives, conduct structured debriefs after their trips, turn insights into revised standards and roadmaps, and cultivate peer relationships that function as "call-a-friend" networks during implementation. Over time, these peer networks offer what consultants and vendors often cannot: reliable, real-world experience that helps institutions avoid common mistakes and adopt proven patterns tested in actual campus settings.

Transformation Is Culture Change With Infrastructure Consequences

OT modernization offers higher education a unique chance—a once-in-a-generation opportunity—to reconfigure the built environment as a strategic platform for accomplishing its mission. When executed effectively, facilities transformation can enhance buildings, making them safer, healthier, more efficient, and more resilient. It helps lower waste and emissions while boosting comfort and dependability. Additionally, it can serve as a dynamic learning environment—fostering applied research, workforce development, and leadership in sustainability, resilience, and responsible innovation.

But the defining truth is this: the transformation is not delivered by technology alone. It is delivered by people—operators, engineers, supervisors, IT security leaders, campus safety teams, and vendors—who understand the mission, trust the systems, and have the skills and authority to use them well. In complex institutions, infrastructure is never "just infrastructure." It shapes work, behavior, expectations, and accountability. Every new sensor, dashboard, automation rule, and access control change carries cultural meaning: *Who is in control? Who is responsible? What is being measured? What happens when it fails?*

That is why facilities transformation is fundamentally a culture change—with infrastructure consequences.

The Opportunity: From Buildings as Costs to Buildings as Mission Platforms

Most institutions have traditionally viewed buildings primarily as cost centers — assets requiring upkeep, utility expenses, and deferred maintenance management. OT modernization shifts this perspective, allowing campuses to see buildings as mission-critical platforms. These platforms foster student success, faculty productivity, research continuity, accessibility, well-being, and community trust. It also makes sustainability a practical, measurable, and continuously improving effort, rather than just a lofty goal. Furthermore, it establishes a foundation for resilience amid climate volatility, grid instability, and increasing operational complexity.

This opportunity is focused on dependability rather than just "smartness." It involves a campus that functions reliably, recovers swiftly, and continues to improve with time.

The Delivery Mechanism: People, Not Platforms

Technology offers visibility, insights, and automation, but it cannot replace judgment, legitimacy, or coordination, which are solely human skills. When experienced operators lack trust in a system, they tend to bypass it, risking safety and operational continuity. Unclear governance leads to an increase in pilot proliferation, causing scale issues and fragmentation. Vague privacy boundaries foster suspicion, hindering adoption rates. Without rehearsed safety protocols, integrated sensing can generate noise instead of effective responses.

In short, modern OT boosts institutional capability only when the human system is designed with the same discipline as the technical system. A campus becomes "smart" when it achieves coherence—both technically and socially.

Human Factors as First-Class Design Constraints

A mature program considers human factors as fundamental design constraints rather than afterthoughts, influencing architecture, procurement, workflows, and governance from the outset. The key elements are as follows:

- **Operator UX**
  Tools that reduce cognitive load, support real workflows, and build trust through reliability—not dashboards that perform for leadership but frustrate the people on call.
- **Trust and transparency**
  Clear explanations of what is measured, why it is measured, and what protections exist—so modernization feels supportive rather than surveillant.
- **Clear decision rights**
  Defined authority for change, automation, access, escalation, and override—so people know who owns what and how risk is governed.
- **Thoughtful safety protocols**
  Human-in-the-loop response models, triage logic, and rehearsal—so sensing and analytics compress response time without triggering overreaction.
- **Sustained training and enablement**
  Role-based pathways, scenario practice, embedded job aids, and continuous improvement loops—so capability grows with the system.
- **Phased adoption**
  Progressive intelligence that earns trust through repeated wins—visibility first, optimization next, automation only when readiness is proven.

When these factors are honored, modernization becomes socially affordable and operationally durable. People adopt what they trust. They trust what works. And what works in facilities is what performs under stress.

The Measure of Success: A Campus That Is Dependable

The smart campus is not defined by connectivity. It is defined by dependability: stable performance, safer operations, healthier spaces, lower emissions, faster recovery, and a community that experiences modernization as mission-aligned rather than intrusive. That is the true measure of transformation in the built environment: **a campus that works better for the people it exists to serve—consistently, transparently, and with dignity.**

Cultural Factors

Psychological Safety and Change Fatigue

As institutions invest millions in the Digital Integration Fabric, IoT sensor meshes, and AI-driven analytics, there is a tendency to treat campus modernization as a purely technical challenge. However, the most sophisticated cyber-physical architecture will inevitably fail if it is deployed into a workforce paralyzed by change fatigue and fear.

The frontline operators—the mechanics, control technicians, network engineers, and dispatchers—are the actual engine of the smart campus. They already carry a high physical and cognitive load just keeping the legacy campus running. When leaders introduce radical technological shifts without addressing the human element, they are not building capability; they are engineering resentment.

To successfully transition to an AI-enabled sustainable smart campus, the CTO/CFaO must recognize that **psychological safety is an operational prerequisite for transformation, not an HR nicety.**

The Emotional Risk of Modernization

For a veteran facilities operator or a mid-level IT administrator, the shift to a smart campus does not feel like an exciting "digital transformation"—it feels like profound emotional and professional risk. Leaders must directly address the three primary anxieties driving change fatigue:

- **The Fear of Automation (Obsolescence)**
  When executives tout "AI-driven efficiency" and "autonomous fault detection," frontline workers often hear "layoffs." If an algorithm can diagnose a failing chiller and automatically dispatch a part, the mechanic wonders if their hard-earned diagnostic intuition is no longer valued.
- **The Fear of Exposure (Surveillance)**
  For decades, building operations have been somewhat opaque. A smart campus suddenly shines a blinding light of data on every action. Operators fear that granular telemetry will be used punitively—exposing inefficiencies, highlighting delayed maintenance, or proving that a long-held operational habit was actually wasting energy.
- **Anxiety Over Unfamiliar Systems (Competence Loss)**
  As we noted in our discussion of workforce constraints, the smart campus demands hybrid skills. A master electrician or HVAC technician who has spent thirty years perfecting their physical craft is suddenly confronted with IP subnetting, BACnet/SC certificates, and complex dashboard interfaces. This creates intense imposter syndrome; their deeply rooted competence is suddenly thrown into question by a poorly designed User Experience (UX) that increases their cognitive load.

The Evolution From Trade to Cyber-Physical

The "Cyber-Physical Technician" is a hybrid professional—a master of the physical world who is equally fluent in the digital layer that controls it. This is not about teaching an electrician to be a software engineer; it is about empowering them with the digital tools to see through walls and predict the future of their equipment.

The Plumber's Pivot: From Reactive to Predictive

- **The Traditional Role:** Responding to a "wet floor" report, hunting for a leak behind drywall, and manually checking water meters.
- **The Cyber-Physical Skillset:** Utilizing ultrasonic leak detection and flow-rate analytics via the Digital Twin.
- **Upskilling Focus:** Training in **Acoustic Sensor Analytics** and **Smart Metering Infrastructure.** The plumber now spends less time digging and more time analyzing flow anomalies to prevent bursts before they occur.

The Electrician's Expansion: From Wiring to Logic

- **The Traditional Role:** Pulling copper, testing circuits with a multimeter, and replacing breakers.
- **The Cyber-Physical Skillset:** Managing Power-over-Ethernet (PoE) lighting, variable frequency drives (VFDs), and smart grid integration.
- **Upskilling Focus:** Proficiency in **Network Protocols (IP-based building controls)** and **Energy Management Systems (EMS).** The electrician becomes a "Power Architect," optimizing load balancing and renewable energy storage across the campus microgrid.

The HVAC Tech's Transformation: From Mechanics to Thermodynamics AI

- **The Traditional Role:** Checking refrigerant levels and manually adjusting dampers based on seasonal calendars.
- **The Cyber-Physical Skillset:** Interrogating AI-driven occupancy models to automate air changes and utilizing thermal imaging to identify envelope leaks.
- **Upskilling Focus:** Understanding **Building Automation Logic** and **Data Visualization.** The technician shifts from "fixing the air" to "curating the environment," using the Digital Twin to simulate the impact of a 500-person event on the atrium's cooling load.

The Upskilling Roadmap: A Three-Phase Approach

To move the workforce through this transition, the institution must provide a structured path:

- **Phase I: Digital Literacy and Data Access.**
  Provide every technician with a mobile interface to the Data Estate. Success in this phase is simply moving from paperwork orders to interacting with the Digital Twin's "Meaning Layer" in the field.
- **Phase II: Diagnostic Fluency.**
  Shift the training focus from "How to fix" to "How to interpret." Technicians learn to trust and interrogate the AI's predictive alerts, using data as their primary diagnostic tool.
- **Phase III: Systemic Optimization.**
  In this final phase, the Cyber-Physical Technician contributes to the "Institutional Intellect." They are no longer just maintaining assets; they are providing feedback to the AI models to refine the campus's operational logic.

Reinvesting the "Savings"

Automating 30% of a technician's time spent on manual inspections and travel between buildings saves the institution hundreds of hours. These saved hours are redirected toward addressing deferred maintenance and strategic infrastructure upgrades, tasks that were previously unfeasible due to staffing limitations. This approach embodies the concept of "People Saving'—protecting the workforce from obsolescence and safeguarding the institution's physical future.

Why Psychological Safety Matters to the Architecture

If these fears are not mitigated through psychological safety, they do not simply disappear; they manifest as operational friction. Fear breeds resistance, slow adoption, and most dangerously, **silent workaround cultures.**

If a mechanic does not feel safe admitting they do not understand the new AI-driven HVAC dashboard, they will not ask for help. Instead, they will quietly walk down to the mechanical room, physically override the actuator, and put the system into manual mode—instantly neutralizing a million-dollar investment in enterprise software. The smart campus becomes a facade, with the actual work happening in the shadows via legacy habits.

Psychological safety means an employee knows they will not be punished, humiliated, or marginalized for speaking up with ideas, questions, concerns, or mistakes. In a cyber-physical environment where a single keystroke can shut down a research lab, an operator's ability to quickly and safely say, "I made a mistake," or "I don't understand how this controller works," is a critical risk-management capability.

Leadership Guidance: Engineering the Safe Space

The CTO/CFaO cannot mandate trust; they must engineer it into the operational culture just as deliberately as they engineer the network. Here is how leaders create the safe spaces required for modernization:

- **Fund the "Sandbox" (Experimentation Without Consequence):** Before a new control system or AI agent is deployed to a live building, operators must have a digital and physical sandbox. This is a safe, offline environment where they can push buttons, trigger alarms, and deliberately break the system to see how it reacts, without the fear of impacting live academic or research environments.
- **Normalize the Gap (Admit Executive Ignorance):** Psychological safety starts at the top. The CTO/CFaO must publicly acknowledge the difficulty of this convergence. When a leader says, "This new semantic tagging ontology is confusing to me, too. Let's figure it out together," it instantly removes the stigma for the frontline technician who is struggling to adapt.
- **Reward Problem Identification over Blind Compliance:** When an operator points out that the new AI algorithm is making a dangerous recommendation—for example, trying to close an outside air damper in a chemistry lab—they should be publicly praised, not dismissed as "resistant to tech." Validating their physical expertise proves that the AI is meant to augment their judgment, not replace it.
- **Decouple Telemetry from Performance Reviews:** Leadership must make an ironclad, public commitment that the initial wave of data generated by the smart campus will be used to fix *systems*, not to punish *people*. If an operator's first experience with the new data platform is being reprimanded for a metric they didn't know was being tracked, trust is permanently broken.

By prioritizing psychological safety, leaders ensure that the workforce views the new Digital Integration Fabric not as a threat to their livelihood but as a powerful new toolset that elevates their craft.

Merging the Wrench and the Keyboard

The convergence of IT and Operational Technology (OT) on a smart campus is not merely a technical integration of networks and sensors; it is a profound collision of two distinct professional cultures. Historically, the CTO and the CFaO have led entirely separate tribes, each with its own language, values, and deep-seated professional pride.

When modernization forces these two worlds to share the same Digital Integration Fabric, leaders who ignore the human element will quickly encounter fierce cultural defensiveness. If the transition to an AI-enabled campus strips away an operator's professional identity, the technology will be rejected, regardless of its operational brilliance.

The Tale of Two Tribes: Craft vs. Structure

To lead this convergence, it requires a quantum shift in leadership perspective—recognizing that the IT and Facilities domains are no longer binary, but deeply interconnected states of campus operation. Leaders must first understand the identities they are merging:

- **Facilities (The Tacit Craft Knowledge)**
  The Facilities identity is rooted in the tangible, physical world. It is a culture of *craft.* A master HVAC mechanic or central plant operator possesses deep, tacit knowledge—an intuitive understanding of thermodynamics, fluid dynamics, and the unique "personality" of a thirty-year-old boiler. Their pride comes from keeping the campus physically alive, often battling failing infrastructure in the middle of the night with a wrench in hand. To them, the building is a physical asset that must be protected from abstract digital interference.
- **Information Technology (The Structured Governance)**
  The IT identity is rooted in the digital and the abstract. It is a culture of *structured systems and governance.* Network architects and systems engineers pride themselves on standardization, scalability, and security. Their tools are code, APIs, and the keyboard. To IT, physical devices are simply endpoints on a network that must be managed, patched, and secured in accordance with rigorous, enterprise-wide policies.

The Risk of Ignoring Professional Identity

When these identities clash without leadership intervention, the smart campus stalls. If IT professionals treat Facilities staff like legacy hardware jockeys who need to "learn how to use a computer," Facilities will retreat into silent resistance. Conversely, if Facilities views IT as ivory-tower bureaucrats who do not understand how a chiller actually works, they will circumvent the network by buying insecure "Shadow OT" to get their jobs done.

Ignoring these distinct professional identities fosters a culture of "us versus them," in which both sides protect their turf at the expense of the institution's sustainability and resilience goals.

Strategies for Honoring Both Identities

To build a sustainable smart campus, the CTO/CFaO must actively architect a culture that honors the legacy of the physical craft while embracing the rigor of digital governance. Respecting the craft elevates morale and accelerates the ownership of new AI and digital tools.

- **Co-Design Workshops (Building Together)**
  Never hand a Facilities operator a completed dashboard designed exclusively by an IT software developer. Bring the mechanic and the UI designer into the same room. The mechanic explains the physical sequence of operations; the developer builds the digital interface to match it. When the mechanic sees their tacit knowledge reflected in the software, the dashboard stops being an "IT mandate" and becomes *their* tool.

- **Recognition Rituals (Celebrating the Save)**
  Universities are excellent at celebrating academic achievements, but operational victories are often invisible unless something breaks. Leadership must establish rituals that publicly recognize cross-functional success. When a major winter storm hits, and the campus remains operational, publicly celebrate both the mechanic who physically bypassed the boiler *and* the network engineer who maintained the OT telemetry under stress.

- **Elevating the "Wrench + Keyboard" Hybrid**
  The highest evolution of this cultural merger is the celebration of the new **Cyber-Physical Professional.** Leaders must actively rewrite job descriptions and career pathways to reward those who master both domains. When a plumber learns to securely configure an IP-enabled smart valve, or a network engineer earns an ASHRAE certification, it must be celebrated as the pinnacle of smart campus craft. By elevating the "wrench and keyboard" hybrid role, you signal that modernization is not erasing their professional identity; it is upgrading it for the next century.[11]

Connecting Daily Tasks to Mission Outcomes

In the daily grind of campus operations, a dangerous disconnect often forms between the boardroom's strategic vision and the mechanical room's reality. This is the "Meaning Gap." Executives look at dashboards and celebrate carbon neutrality milestones, while frontline technicians stare at a backlog of 400 routine work orders in the CMMS. When new smart campus technologies—like AI analytics or mobile field tools—are introduced to the frontline simply as "a faster way to close tickets," they are inevitably met with apathy.

To drive true cultural transformation and technology adoption, leaders must connect the operator's daily, physical tasks directly to the university's existential mission.

Translating the Work Order into Mission Delivery

Operators do not lack work ethic; they often lack context. If a technician does not understand *why* a new system matters, they will default to the legacy processes they already trust. The Integrative Architect must actively translate routine maintenance into its strategic equivalent, proving that the frontline worker is the ultimate guarantor of the university's success.

Leaders must systematically bridge the gap between the wrench and the mission across four critical domains:

- **Sustainability and Decarbonization**
  A technician responding to an AI-generated Fault Detection and Diagnostics (FDD) alert for a stuck damper is not just performing mechanical triage. They must understand that fixing a simultaneously heating-and-cooling air handler is a direct intervention in the university's climate action plan. When operators see how their specific repair visibly drops the energy consumption curve on the building's dashboard, they transform from mechanics into active decarbonization agents.

- **Student Wellbeing and Learning Environment**
  Recalibrating a CO_2 sensor or tuning an automated lighting schedule in a lecture hall is not merely a matter of comfort. Poor indoor air quality directly degrades cognitive function, while optimal lighting improves focus and reduces fatigue. The frontline staff must know that they are literally engineering the atmospheric conditions required for student success and academic excellence.

- **Research Integrity**
  When a central plant operator responds to a predictive vibration alert on a chiller, they aren't just saving a compressor from burning out; they are ensuring that the ultra-low-temperature freezers in the biology building do not fail over a holiday weekend. By connecting the chiller's uptime to the preservation of multi-million-dollar grants and decades of irreplaceable scientific research, the operator's role is elevated to that of the silent guardian of the research enterprise.

- **Risk Prevention and Resilience**
  When an IT network engineer and a Facilities control technician collaborate to patch a vulnerable BACnet controller or segment an OT network, they are not simply executing an IT compliance checklist. They are actively defending the physical safety of the students and the operational continuity of the campus against crippling cyber-physical attacks. They are the frontline defenders of institutional trust.

Why Meaning Drives Mastery

Humans are meaning-making machines. If an operator believes a new smart campus dashboard is just a surveillance tool for management to track their productivity, they will resist it, criticize its flaws, and find workarounds. However, if they understand that the same dashboard empowers them to protect the campus ecosystem and directly support the students they see walking across the quad every day, they will master it.

When leadership takes the time to close the Meaning Gap, it fundamentally alters the cultural dynamic. The workforce stops asking, "Why do I have to use this new software?" and begins asking, "How can I use this data to make my building perform better?" Data must flow up to the executive to inform strategy, but *purpose* must flow down to the operator to inspire execution. By explicitly connecting the daily task to the institutional mission, leadership ensures the smart campus is powered not just by electricity and AI, but by profound professional purpose.

Building Trust in Data, Sensors, Analytics, and AI

In the legacy campus, trust was inherently physical. A master mechanic trusted their own senses—the high-pitched whine of a failing bearing, the smell of an overheating motor, or the physical resistance of a manual valve. The AI-enabled smart campus asks that same operator to transfer their trust away from their physical senses and place it into a digital dashboard, an IoT sensor mesh, and an invisible algorithm.

If that transfer of trust fails, the entire smart campus capability collapses. When leaders deploy predictive tools without intentionally building trust in the data, the system becomes a very expensive digital paperweight.

Epistemic Trust and the "Bad Sensor" Reality

To successfully deploy analytics and AI, leadership must understand the concept of **epistemic trust**—the confidence a user has in *how* the system knows what it claims to know.

Frontline operators have a deeply ingrained, historically justified skepticism of sensor accuracy. They know from years of experience that a $50 temperature sensor installed in a return air duct can easily drift out of calibration, or that a newly installed bookshelf can block an occupancy sensor. If a new AI-driven dashboard triggers an urgent alert that a lecture hall is freezing, but the technician arrives with a calibrated handheld thermometer and reads a perfect 72 degrees, epistemic trust is instantly shattered.

When operators are burned by bad data or subjected to relentless "alarm fatigue"—where poorly tuned systems flood their screens with constant false positives—they adapt by ignoring the system altogether. They lose faith not just in that specific sensor, but in the entire modernization initiative.

Explaining the Black Box: "Why Does the Algorithm Want This?"

The fastest way to destroy credibility on a smart campus is to deploy a "black box" AI. If a predictive maintenance algorithm outputs a mandate— *"Replace compressor on Chiller 3 immediately"*—without providing context, the operator will view it as an arrogant, unjustified demand.

To build trust, AI-driven recommendations must be transparent and explainable. The interface must translate algorithmic logic into mechanical reality. The system should instead communicate: *"Recommending inspection of Chiller 3 compressor. The current vibration signature matches historical failure patterns with 89% confidence, and the secondary sensor confirms an abnormal 12-degree spike in discharge temperature over the last 48 hours.* "When the AI shows its work, it stops being a robotic dictator and becomes a highly capable digital assistant.

Strategies for Leaders to Engineer Trust

The CTO/CFaO must actively architect this trust through governance and operational design:

- **Validate the Data (Ground-Truthing):** Before turning on an AI optimization engine, leadership must fund a rigorous "ground-truthing" phase. Technicians should be dispatched to verify the digital readings physically. When frontline staff help calibrate the system and sign off on the sensor's accuracy, they co-own the integrity of the data.
- **Show Data Lineage and Confidence:** Interfaces should transparently display data health. If a sensor hasn't reported in 24 hours, the dashboard should flag the resulting analytics with a "low confidence" warning. Operators respect a system that is honest about its own blind spots.
- **Empower the Human Override:** AI algorithms lack human context. An energy-saving algorithm might recommend shutting down the HVAC in a seemingly empty pavilion, entirely unaware that the President is hosting a donor dinner there in an hour. Operators must be explicitly empowered to override algorithmic recommendations without fear of reprimand.
- **Create a Learning Loop, Not a Disciplinary One:** When an operator executes an override, the system should ask "Why?" (e.g., *"Special event," "Safety concern," "Sensor physically broken"*). This feedback loop trains the AI to become smarter while reinforcing that human judgment remains the ultimate authority on the smart campus.

When operators trust the sensors, understand the algorithm's logic, and feel empowered to apply their own contextual judgment, they will stop fighting the technology. They will lean into predictive analytics, utilizing the AI to catch cascading failures days before they happen, ultimately delivering the resilience and efficiency the smart campus promises.

Bridging the IT and Facilities Divide

We have spent considerable time detailing the architecture of the Digital Integration Fabric—the APIs, the message brokers, and the network topology required to connect the smart campus. However, the harshest reality of enterprise modernization is this: **technical integration will inevitably fail without social integration.** You cannot successfully hardwire an IP network to a central plant chiller if the network engineer and the plant mechanic refuse to speak to one another.

Historically, IT and Facilities have operated in deeply entrenched silos. When an IT server goes down, it is a digital crisis handled in the Network Operations Center (NOC). When a pipe bursts, it is a physical crisis handled in the mechanical room. But in the AI-enabled smart campus, these domains converge. A failed network switch now means the building automation system loses visibility of a critical lab environment. When a cyber-physical failure occurs, the lack of cross-domain social capital instantly devolves into tribal finger-pointing: IT blames the legacy hardware, Facilities blames the firewall, and the campus suffers the downtime.

To fulfill the "bridging worlds" mandate of the Integrative Architect, the CTO/CFaO must actively manufacture interpersonal trust. Social capital must be engineered just as deliberately as the network infrastructure.

The Architecture of Interpersonal Trust

Trust between a software developer and a master electrician is not organic; they do not naturally share the same operational priorities or vocabulary. Leadership must implement structured strategies to force collaboration and build mutual respect.

- **Cross-Shadowing and Joint Walk-Throughs**
  You cannot protect an environment you do not understand. Network architects should be required to wear a hard hat and shadow a mechanic during a preventive maintenance route through the central plant. Conversely, building control technicians should sit in the Security Operations Center (SOC) to watch how anomalous traffic is flagged. These physical job swaps demystify "the other side" and build profound empathy for the constraints each team faces.
- **Shared Language Training**
  IT speaks in bandwidth, latency, and subnet masks. Facilities speak in BTUs, static pressure, and sequences of operation. The institution must develop a shared glossary. When both teams understand that an "SLA" (Service Level Agreement) in IT equates directly to an "uptime requirement" in Facilities, they stop talking past one another and start collaborating on solutions.
- **Joint Tabletop Exercises**
  Trust is forged in the crucible of simulated crises. Leadership should mandate quarterly joint tabletop exercises that simulate cyber-physical disasters. *Scenario: A ransomware attack locks the Active Directory, simultaneously shutting down credentialed door access and disabling the campus microgrid.* By forcing IT and Facilities to work through the incident response together in a low-stakes conference room, they build the muscle memory and interpersonal trust required to execute when a real crisis hits.
- **Rituals and Touchpoints**
  Break down the silos in the weekly calendar. Establish a joint Friday-morning huddle for the IT Service Desk manager and the Facilities Dispatch manager to review the week's critical tickets. Require joint walk-throughs for the commissioning of any new smart building. These consistent rituals prevent relationships from reverting to transactional email threads.
- **Shared KPIs and Shared Wins**
  As long as IT is bonused solely on network uptime and Facilities is bonused solely on energy reduction, they will remain misaligned. The CTO/CFaO must establish shared Key Performance Indicators (KPIs). If a new AI-driven HVAC optimization routine successfully reduces a building's carbon footprint without generating network security alerts, the victory belongs to the joint task force. Celebrate the network engineer and the mechanic on the same stage.

Why Social Capital is the Ultimate Antifragile Asset

When an institution invests in building cross-domain social capital, it transforms its operational culture. The relationship shifts from adversarial ("IT is blocking my vendor's access") to collaborative ("Can IT help me secure this new sensor array?").

In the smart campus era, the most resilient, antifragile asset you have is not a redundant data center or a backup generator. It is the social capital that allows a systems administrator and a master mechanic to look at the same malfunctioning piece of equipment, inherently trust each other's expertise, and solve the problem together.

Leadership-Behaviors That Shape Culture

We often mistakenly believe that organizational culture is transformed through sweeping mandates, multi-million dollar software launches, or annual all-hands meetings. In reality, the culture of a smart campus is built—or destroyed—in the aggregate of hundreds of daily micro-signals sent by the CTO and the Chief Facilities Officer (CFaO).

In the high-stakes, cyber-physical environment of an AI-enabled campus, the leader must embrace a new identity: the **Leader-as-Culture-Architect**. Every raised eyebrow in a budget meeting, every email sent at midnight, and every reaction to a failed pilot project acts as a structural blueprint for how the rest of the organization will behave.

To successfully lead the human element of this transformation, executives must consciously engineer their micro-behaviors to signal safety, curiosity, and profound respect for the frontline operator.

The Framework of Cultural Micro-Signals

To build an antifragile, highly integrated campus culture, leadership must intentionally practice the following micro-behaviors:

- **Modeling Curiosity Rather Than Certainty**
  The convergence of IT, Operational Technology (OT), and AI is at the forefront; no one possesses all the answers. If an executive pretends to understand every nuance of a new machine learning optimization algorithm, they signal to the organization that ignorance is a weakness to be hidden. Conversely, when the CTO openly asks a frontline boiler technician, "I don't fully understand how this thermal lag affects the sequence of operations—can you teach me?", it radically alters the power dynamic. Modeling curiosity gives the entire organization permission to ask questions, admit knowledge gaps, and learn collaboratively.
- **The Physical "Gemba Walk" (Visiting the Field)**
  You cannot lead a cyber-physical environment exclusively from an executive conference room. The CFaO and CTO must regularly visit the "gemba"—the real place where the work happens. When a CTO puts on Personal Protective Equipment (PPE) to walk through a noisy central plant to see a Wi-Fi dead zone firsthand, or when the CFaO sits in the IT Network Operations Center to watch a firewall deployment, it sends a deafening cultural signal. It proves that the executive values the physical reality of the work as much as the dashboard's abstraction.
- **Praising the Early Adopters (and the Constructive Skeptics)**
  How leadership responds to the initial awkward phases of a technology rollout determines its long-term adoption. When a veteran mechanic uses the new mobile work-order app and makes a mistake, the leader must publicly praise the *attempt* rather than penalize the error. Furthermore, when a constructive skeptic points out a legitimate flaw in a new AI scheduling tool, the leader must thank them for protecting the institution, rather than dismissing them as a roadblock. You must actively reward the behaviors you want to scale.
- **Enforcing Vendor Accountability for Operator Training**
  A subtle yet devastating micro-behavior allows a vendor to install a complex, million-dollar system, hand over a dense manual, and walk away. This signals to the operators that leadership cares more about the procurement press release than the staff's ability to succeed. The Culture Architect ensures that every vendor contract mandates rigorous, hands-on training for the frontline staff. By refusing to sign off on a project until the operators feel confident operating it, the leader proves that the workforce's competence and comfort are non-negotiable priorities.

Why the Micro-Signal Matters

Transformation depends entirely on these micro-signals because culture is simply the shadow of leadership. If the CTO dismisses a broken building sensor as "just a Facilities problem," the entire IT department will adopt that same dismissive posture. If the CFaO complains that "IT security is just slowing us down," the Facilities team will view every cybersecurity protocol as an enemy to be routed around.

However, when leaders consciously align their daily micro-behaviors with the vision of an integrated, respectful, and curious institution, the silos begin to dissolve. The "Leader-as-Culture-Architect" understands that they are not just installing new software; they are actively designing the psychological environment required for the smart campus to thrive.

Cultural Models for Scaling Adoption

Many institutions address the human aspect of implementing a smart campus mainly via individual change management models. This manifesto highlights the importance of adopting a structured change management approach. Chapter 16 focuses on applying ADKAR. Although ADKAR is highly effective for guiding individuals through transitions, it falls short in addressing the complex, systemic social dynamics of a university's operations division.

To effectively expand adoption across hundreds of buildings and various operational teams, the Integrative Architect needs to look beyond individual and team levels and understand how new behaviors propagate through social networks.

The Diffusion of Innovations

Leaders must recognize that adoption will never be uniform. Everett Rogers' *Diffusion of Innovations* theory provides a vital framework for understanding this reality.[12] The campus workforce will naturally stratify into five categories when introduced to new cyber-physical tools:

- **Innovators (2.5%) and Early Adopters (13.5%)**
  These are your technology enthusiasts. They will actively seek out the new mobile CMMS app or the AI-driven fault detection dashboard. They are forgiving of early software bugs because they love the tool's potential.
- **The Early Majority (34%) and Late Majority (34%)**
  This is the campus's operational core. They are pragmatists. They do not care about "AI" or "digital transformation"; they only care if the tool makes their shift easier or harder. They will not adopt the technology until it is proven reliable.
- **Laggards (16%)**
  The staunch traditionalists will actively resist the new system until legacy alternatives are entirely removed.

Scaling Strategies for the Pragmatic Majority

If a smart campus initiative fails, it usually dies in the chasm between the Early Adopters and the Early Majority. To bridge this gap, leaders must deploy specific cultural strategies:

Identify and Empower "High-Trust Influencers"

Every facility and IT department has informal leaders. This is the master mechanic or senior network engineer whom everyone else calls when things go wrong, regardless of their official title. If leadership forces a new tool from the top down without consulting these influencers, the Early Majority will reject it. Leaders must co-opt these high-trust individuals early and bring them into the pilot phase. When the most respected mechanic in the central plant says, "This new dashboard actually caught a bearing failure before it happened," the Early Majority will instantly follow.

Match Adoption Cadence to Support Cadence

You cannot scale software faster than you can scale human support. If you turn on thousands of IoT sensors across ten buildings at once, the resulting flood of data will overwhelm the frontline staff. This induces immediate "alarm fatigue," causing the Early Majority to abandon the system. Rollouts must be phased so that support teams can tune the algorithms and eliminate false positives before the broader workforce interacts with them.

Engineer the "Early Win"

Momentum is the most valuable currency in change management. Do not start a smart campus rollout by tackling the most complex, politically fraught system. Find a highly visible, universally despised operational headache—like the manual process for tracking down temporary chiller keys—and solve it flawlessly with the new technology. Early, undeniable wins build the institutional capital needed to tackle more complex integrations later.

The Politics of Space, Territory and Control

We frequently treat the modernization of the campus as a purely technical endeavor, focusing on API gateways and network segmentation. In reality, a smart campus disrupts established power dynamics. When you digitize the physical environment, you are fundamentally renegotiating invisible boundaries, shifting control, and redefining ownership. Many of the fiercest conflicts in a smart campus deployment are not technical; they are deeply political.

If the CTO/CFaO does not proactively manage the politics of space and territory, the integration will stall in endless committee debates and passive-aggressive turf wars.

The Battlegrounds of Modernization

As disparate systems converge onto the Digital Integration Fabric, leaders must navigate several highly territorial flashpoints:

Ownership of Naming Conventions

When integrating systems, IT often wants to rename physical assets to fit a structured, database-friendly ontology (such as Brick Schema). Facilities will strongly object, as they have used a specific legacy naming convention in their blueprints for 40 years. It is a battle over whose mental model of the campus is the "correct" one.

Control of Setpoints and Schedules

Historically, the building occupant (e.g., the Dean of the Business School) or the local building manager controlled the thermostat. When a centralized AI optimization engine takes over HVAC scheduling to shed load during peak pricing, occupants experience a significant loss of autonomy. The technology shifts power from the local occupant to the central algorithm.

Defining Alarm Thresholds

Who decides what constitutes an "emergency"? If IT configures the integration middleware to page the on-call Facilities technician whenever a VAV box loses network connectivity for 5 minutes, Facilities will view IT as dictating their labor priorities.

Principles for negotiating territory

To resolve these political conflicts, leadership must transition the culture from "domain ownership" to "enterprise stewardship":

Establish Clear Decision Rights Before Turning the System On

Do not wait for a conflict to occur. Use a RACI matrix (Responsible, Accountable, Consulted, Informed) to explicitly define who owns the algorithm, who owns the sensor, and who has the authority to override the system.

Create a Shared Data Governance Council

The rules of the smart campus cannot be dictated exclusively by IT. A cross-functional council—including representation from IT, Facilities, Sustainability, and Academic leadership—must collaboratively agree on naming conventions and data standards.

Design for Graceful Overrides

The system must respect human context. If a professor needs to extend the HVAC schedule for a late-night study session, the smart campus app should allow a frictionless, temporary override. Providing localized, temporary control prevents the system from feeling like an oppressive surveillance state.

Inter-Generational Workforce Dynamics

If there is a single, silent existential risk facing the operations of higher education institutions today, it is not a cyberattack or a grid failure—it is demographics. We are witnessing an unprecedented generational change of the guard within campus facilities and IT departments. If leadership fails to actively manage this demographic collision, the smart campus will stall not from a lack of technology but from a catastrophic loss of institutional memory and a failure to attract new talent.

Over four decades of observing higher education technology and facilities operations, a consistent truth emerges: the campus is held together by human memory just as much as by concrete and fiber optics. We must explicitly architect our culture to bridge the gap between the outgoing masters of the physical craft and the incoming masters of the digital realm.

The "Silver Tsunami" and the crisis of tacit knowledge

Higher education is currently in the midst of the "Silver Tsunami"—a massive wave of retirements among baby boomer facilities directors, central plant operators, and master mechanics. These individuals are the custodians of the campus's tacit, tribal knowledge.

They know that the blueprints for the 1970s science building are wrong, and they know exactly which unmarked valve in the sub-basement actually shuts off the steam. They understand the unique, undocumented physical quirks of every legacy air handler. When these veterans retire, they take the mental map of the physical campus out the door with them. If that knowledge is not captured, the AI optimization algorithms will ultimately fail because they operate on flawed baseline assumptions.

Capturing the craft: The digital twin as institutional memory

The smart campus architecture provides the exact vessel needed to save this institutional memory, but leadership must force the structured handover. The CFaO and CTO must treat the retiring workforce as the most valuable data source on campus.

Digitizing the tribal knowledge

Before a master mechanic retires, they must be paired with the data engineering team to map the semantic layer (the Campus Ontology). The veteran translates their mental map into the metadata.

The Digital Twin as the Hei

The Digital Twin ceases to be just a 3D architectural model; it becomes the permanent repository of this tacit knowledge. When the veteran explains that a specific chiller always runs hot in July due to the sun's angle on the courtyard, that constraint is permanently coded into the predictive maintenance algorithm. The machine inherits the human's wisdom.

The Expectations of the Digital Native

Simultaneously, the institution is desperately trying to recruit the next generation of technicians, operators, and engineers. This incoming cohort consists of digital natives. Their expectations of how work should be accomplished are fundamentally different, and they have zero tolerance for technological friction.

Consumer-Grade UX is Non-Negotiable

If you hand a 24-year-old technician a clipboard with triplicate paperwork or force them to navigate a clunky, text-based legacy Building Automation interface from the 1990s, they will view the institution as an archaic dead end. They will leave.

The Demand for Augmented Workflows

Younger staff expect mobile-first workflows. They expect to stand in front of a boiler, pull out their phone, scan a QR code, and instantly see the real-time IoT telemetry, the last five years of maintenance history, and a link to the digital manual. They expect AI copilots to assist with data entry and automation to handle repetitive administrative tasks. For this generation, consumer-grade User Experience (UX) is not a luxury; it is a baseline requirement for retention.

Bi-Directional Mentorship: The Hybrid Engine

To survive this demographic shift, leadership must abandon the traditional, one-way training model (where the older worker simply dictates instructions to the younger worker). Instead, institutions must formalize **bi-directional mentorship models.**

Blending the Wrench and the Dashboard

Pair the generations together deliberately. The senior mechanic mentors the junior technician on the profound physical realities of the central plant—teaching them the physics, the sounds, and the smells of the equipment that no sensor can replicate. In return, the junior technician mentors the senior mechanic in navigating the new mobile CMMS, building a custom data visualization on the analytics dashboard, and interpreting the AI's fault-detection logic.

By structuring this reciprocal relationship, the institution achieves two critical cultural victories: it honors the lifelong legacy and profound craft of its veteran workforce, while simultaneously empowering the new generation to lead the digital operations of the future. The result is the ultimate operational asset: a unified, cyber-physical workforce.

Ethics of Automation, Surveillance, and Human Dignity

As the campus is blanketed with IoT sensors, high-definition cameras, and AI-driven analytics engines, the physical environment itself becomes a data-gathering machine. While this sensory network is critical for optimizing energy and automating maintenance, it generates a massive, continuous "data exhaust" of human behavior. If leadership manages this data purely as an operational asset while ignoring its human impact, the smart campus will quickly devolve into a panopticon.

For the Integrative Architect, the ultimate test of a smart campus is not its technological sophistication, but its ethical restraint. If modernization erodes academic freedom, creates a culture of paranoia among the workforce, or strips the dignity from the student experience, the technology has failed the institution.

The Threat of the "Surveillance Culture"

There is a profound difference between measuring a space and surveilling a person. When operators and students feel that every keystroke, badge swipe, and hallway conversation is being tracked, logged, and analyzed by an algorithm, it breeds a toxic "surveillance culture." This culture actively destroys psychological safety. Faculty will alter their routines, students will feel stripped of their autonomy, and facilities staff will assume the new analytics dashboard is simply a tool to monitor their break times.

To prevent this, the CTO/CFaO must establish absolute clarity: **the purpose of the smart campus is to monitor infrastructure health, not to police community behavior.**

Privacy by Default: The Ethical Use of Spatial Data

The foundation of ethical analytics is "Privacy by Design." Privacy cannot be an afterthought patched onto a system via an opt-out checkbox; it must be the default architectural state of the Digital Integration Fabric.

Aggregated and Anonymized by Default

When deploying spatial analytics to understand how a new active-learning classroom is being used, the AI should be strictly governed to count *shapes* and *flow*, not identities. The institution needs to know that 45 people are in the room to adjust the HVAC and lighting; it does not need to know their names.

Purpose Limitation

Data gathered for one operational purpose should never be repurposed for disciplinary action without explicit governance in place. If Wi-Fi triangulation data is collected to optimize the shuttle bus schedule, it must not be handed over to the Dean of Students to verify if a specific student attended class. When data crosses these boundaries, institutional trust is permanently broken.

The Un-Automatable: What Should Never Be Handed to an Algorithm

In the rush to deploy Agentic AI and algorithmic workflows, leaders must explicitly define the ethical boundaries of automation. Automation excels at scale, speed, and pattern recognition, but it possesses zero empathy, moral context, or institutional wisdom. Leadership must draw a hard line around processes that require human judgment — what should never be automated?

Decisions Impacting Employment or Dignity

An AI can flag that a technician is taking 30% longer to complete preventative maintenance routes than the baseline average. However, the AI cannot know that the technician is spending that extra time mentoring a new apprentice or recovering from a physical injury. Disciplinary actions, performance reviews, and resource allocations must always remain in the hands of a human leader who can apply context.

Life-Safety Overrides and Escalations

While an algorithm can autonomously shut down a localized HVAC zone to save energy, it must never be given unilateral authority to override fire-alarm pressurization protocols or dictate active-shooter lockdown sequences without a "human-in-the-loop" to verify the situational context.

Academic and Behavioral Sanctions

Predictive analytics might flag a student as "at-risk" based on dining hall swipes and library exit times. Still, an algorithm must never autonomously issue academic warnings or initiate wellness interventions. The AI's role is to notify a trained human counselor quietly; the human delivers the care.

Keeping Dignity at the Center of Automation

Ultimately, the ethics of the smart campus come down to the preservation of human dignity. Automation should be deployed to elevate human work, not to demean the worker.

When robotics and AI are used to handle the dull, dirty, and dangerous tasks—like an autonomous drone inspecting a hazardous roof or an AI script instantly reconciling 5,000 utility bills—it frees the human workforce to engage in higher-order problem-solving, mentorship, and creative engineering.

The smart campus is an awe-inspiring technological achievement, but its true power lies in its restraint. By fiercely protecting privacy, setting strict boundaries on what can be automated, and deliberately designing systems that augment rather than replace human capabilities, leaders ensure that technology serves the dignity of the people rather than forcing people to serve the technology.

Feedback Loops and Listening

The deployment of a new smart campus platform—whether an AI-driven fault-detection engine or a mobile work-order application—is never flawless on day one. Interfaces will be clunky, workflows will misalign with physical realities, and certain algorithms will generate false positives. If the CTO/CFaO treats deployment as the finish line, they are guaranteeing the system's failure.

When frontline operators encounter friction in a new digital tool, they will initially complain. If those complaints disappear into an institutional black hole, the operators will stop complaining—and they will stop using the tool. To prevent this silent rejection, leadership must build a **Listening Architecture**: a highly structured, transparent approach to collecting, organizing, and responding to operator feedback.

The Architecture of Listening

A true listening architecture moves beyond the annual, anonymous employee satisfaction survey. It embeds feedback mechanisms directly into the rhythm of daily operations, treating frontline operators as co-developers of the smart campus.

Field Listening Sessions

Executives and system architects cannot rely solely on IT help-desk ticket data to understand user friction. They must regularly conduct field listening sessions—standing in the mechanical room or the IT operations center, watching an operator attempt to use the new dashboard on their mobile device. When a leader observes a technician clicking six times to bypass a poorly designed drop-down menu, they witness the exact moment where adoption dies.

The "Voice of the Technician" Council

Establish a formalized, rotating council composed of end-users—electricians, HVAC techs, network admins, and dispatchers. Before any major UX update is pushed to the enterprise, or before a new AI vendor is selected, this council must test the workflow. Giving the frontline a formalized veto or approval right ensures that the technology serves the mission, rather than forcing the workforce to serve the software.

Closing the Loop: Transparency and Trust

Listening is only half of the equation; the psychological safety of the workforce is forged in the response. If operators provide critical feedback but never see it materialized, they will assume the listening sessions are merely executive theater.

The Transparent UX Backlog

When a mechanic submits a request to change how an alarm is prioritized on their screen, that request should not be swallowed by an opaque IT ticketing queue. The smart campus team should maintain a publicly visible Agile backlog. An operator should be able to log in and see: *"Requested feature: Group VAV alarms by floor. Status: In Development, expected Q3."* This transparency validates their frustration and proves that help is on the way.

Publishing the "Because You Said" Updates

The single most powerful cultural micro-signal a leader can send is publicly changing a system in response to frontline input. When a software update is pushed, the release notes should not just list technical bug fixes. They should explicitly state: *"Because the overnight shift noted that the lighting schedule was overriding manual lab controls, we have permanently adjusted the algorithm."*

Why Co-Creation Matters

A robust listening architecture fundamentally shifts the cultural dynamic of modernization. It transforms the smart campus from a "digital mandate" imposed from the top down into a "co-created capability" built from the ground up.

When an institution operationalizes feedback loops, it continuously reinforces psychological safety. It proves to the workforce that their tacit knowledge is respected, their daily frustrations are taken seriously, and their voices have the power to shape the technology that governs their environment. Ultimately, the institution that listens best adapts fastest.

Human-Centered Commissioning and Decommissioning

In the traditional construction and IT realms, "commissioning" is a strictly technical checklist. The project engineer verifies that the physical valves actuate, the API endpoints connect, and the network packets route correctly. Once the technical boxes are checked, the keys are handed over to the operations team, and the project is considered a success.

However, in the cyber-physical reality of an AI-enabled smart campus, technical commissioning is vastly insufficient. If a multi-million-dollar AI optimization engine functions perfectly but the frontline operators do not know how to read its dashboard, the system is operationally broken. The Integrative Architect must enforce **Human-Centered Commissioning**—a process that verifies not just the technical health of the hardware, but the operational readiness of the humans who must govern it.

The Commissioning Phase: Engineering Operational Readiness

Human-centered commissioning treats the operator's experience as a mission-critical deliverable, requiring the same rigorous testing as a fire alarm sequence. The CTO/CFaO must mandate that vendor contracts and internal project plans include four non-negotiable human milestones:

Change-Impact Assessment

Long before a new system goes live, leaders must map exactly how it will alter the daily routine of the workforce. If a new IoT sensor array shifts a technician's job from scheduled preventative maintenance to dynamic, predictive dispatch, that is a massive workflow shock. Anticipating this impact allows leadership to prepare the team mentally and structurally.

Usability Verification

Do not accept a vendor's promise that a dashboard is "intuitive." Usability verification requires putting the new tool in the hands of a veteran mechanic and observing them navigate a high-stress scenario, such as locating a critical VAV fault. If it takes them ten minutes to find the data they previously accessed in ten seconds, the interface fails commissioning.

Hands-on, Contextual Training

Dense, 500-page PDF manuals and generic vendor slide decks are unacceptable. Training must be kinetic. Operators must be trained on *their* actual campus equipment, using *their* specific naming conventions, while standing in front of the physical asset.

The Operator Sign-Off

This is the ultimate shift in the project power dynamic. A smart campus integration should not be legally or financially "completed" until the frontline operator signs a document stating: *"I have been trained on this system, I understand how to operate it safely, and I am confident in my ability to maintain it."* When the operator holds the final signature authority, vendors are forced to prioritize usability and knowledge transfer.

The Decommissioning Phase: The Empathy of Retirement

Just as we must carefully onboard the new, we must respectfully retire the old. Decommissioning legacy technology is not merely unplugging a server; it is an emotional and psychological transition for the workforce.

When an institution retires a 20-year-old "green screen" Building Automation System or a legacy ticketing platform, they are not just removing outdated software. They are stripping away the operator's **confidence and comfort**. That legacy system represents decades of hard-won mastery. The operator knows all its quirks, shortcuts, and blind spots.

Honoring Nostalgia and Mastery

Leaders must acknowledge the anxiety of returning to "beginner" status. Dismissing the old system as "garbage" insults the professionals who used it to keep the campus running for decades. Acknowledge its service and validate the operator's mastery of it.

The Safety Net of Parallel Operations

Never execute a "hard cutover" where the legacy system is unplugged on a Friday and the new AI platform is the only option on Monday. Decommissioning must involve a period of parallel operations. Allow the operators to rely on the legacy system as a safety net while they build their "muscle memory" and trust in the new platform.

By centering the human in both the birth and the death of campus technologies, the institution ensures a smooth, resilient transition. Human-centered commissioning proves to the workforce that they are the ultimate stewards of the smart campus, transforming technical handoffs from moments of profound anxiety into moments of operational empowerment.

The Executive Field Guide for Cultural Transformation

The success of the AI-Enabled Sustainable Smart Campus relies more on the dignity and capabilities it provides to its users than on the complexity of its algorithms. To effectively address human and cultural factors, we need to integrate these twelve dynamic elements into a practical framework. For the executive—referred to as the Integrative Architect—leading this effort, it is crucial to recognize that the smart campus is not merely a hierarchy of hardware but a connected, evolving ecosystem. The cyber, physical, and human components are deeply intertwined. Managing this environment calls for a departure from traditional, siloed approaches toward a holistic, quantum leadership style in which every technological update influences human culture broadly.

Phase 1
Architecting Trust and Safety (The Foundation)

Before a single algorithm is deployed, leadership must engineer the psychological and ethical foundation to support it.

- **Guarantee Psychological Safety**
  Treat change fatigue and the fear of automation as operational risks. Fund "sandbox" environments for risk-free experimentation, and explicitly reward operators who identify system flaws.
- **Engineer Epistemic Trust**
  Never deploy a "black box." Ensure AI recommendations are explainable and that operators are empowered to override the system without fear of reprimand. Ground-truth all sensor data before linking it to analytics.
- **Establish Ethical Guardrails**
  Declare firmly that the network of sensors exists to monitor infrastructure health, not to surveil human behavior. Mandate "privacy by design" and draw hard lines around what decisions—especially those impacting employment or student discipline—will never be fully automated.

Phase 2
Honoring Identity and Purpose (The Motivation)

A smart campus scales only when the workforce adopts it as their own. This requires respecting their history and connecting their daily work to the institution's future.

- **Close the Meaning Gap**
  Do not sell new tools simply as "efficiency drivers." Connect the operator's daily physical tasks directly to the university's existential mission—whether that is decarbonization, student wellbeing, or protecting the research enterprise.
- **Honor the Craft**
  Respect the profound, tacit knowledge of the Facilities professionals. Co-design digital interfaces with frontline mechanics so the software reflects the physical reality of the central plant.
- **Bridge the Generational Divide**
  Manage the "Silver Tsunami" by formalizing bi-directional mentorship. Pair veteran mechanics with digitally native junior technicians to blend tribal knowledge with modern data fluency, capturing institutional memory in the Digital Twin before it retires.

Phase 3
Navigating Convergence and Politics (The Alignment)

The friction of modernization is rarely technical; it is almost always territorial.

- **Manufacture Social Capital**
  Force the collision of IT and Facilities cultures. Mandate cross-shadowing, establish shared language training, and conduct joint tabletop exercises to build the interpersonal trust required to survive a cyber-physical crisis.
- **Pre-Negotiate Territory**
  Use cross-functional governance councils to define naming conventions, alarm thresholds, and setpoint controls before systems go live. Shift the culture from siloed "domain ownership" to collaborative "enterprise stewardship."

Phase 4
Scaling and Sustaining the Change (The Execution)

Culture is the shadow of leadership. The executive must actively manage the cadence and rollout of the transformation.

- **Govern via Micro-Behaviors**
  Recognize that every executive action sends a cultural signal. Model curiosity, physically visit the field teams, and praise the early adopters.
- **Scale Pragmatically**
  Understand the *Diffusion of Innovations*. Do not force-feed the Early Majority. Co-opt high-trust frontline influencers to champion the new tools and deliberately engineer high-visibility "early wins" to build momentum.
- **Build a Listening Architecture**
  Establish formal "Voice of the Technician" councils and transparent UX backlogs. Prove to the workforce that modernization is a co-created, iterative process by publicly publishing system updates based directly on their feedback.
- **Enforce Human-Centered Commissioning**
  Never sign off on a smart campus project until frontline operators have received hands-on, kinetic training and formally verified the system's usability. When decommissioning legacy tools, do so with empathy for the comfort and confidence the workforce is leaving behind.

Figure 9. Phases of Cultural Transformation

Chapter Summary

The success of an AI-enabled, sustainable smart campus ultimately depends on human and cultural factors, not on technology alone. Although infrastructure, integrations, and analytics represent the visible aspects of modernization, it is the unseen framework of trust, culture, identity, communication, and competence that decides if the institution becomes more resilient or more fragile during change.

Throughout the chapter, the manifesto highlights that transforming a smart campus primarily focuses on people. Successful modernization depends on staff understanding the reasons for change, feeling psychologically secure when using new tools, and viewing technology as an enhancer rather than a threat. This involves deliberate investments in training, user support, role-specific workflows, and open communication, all aimed at making digital insights a help rather than a hurdle for operators, technicians, faculty, and students.

The chapter emphasizes the need to bridge the cultural gap between IT and Facilities, which have traditionally been separate areas with distinct timelines, values, and risk approaches. Successful transformation depends on developing a shared language, mutual respect, and cross-domain understanding, allowing CTOs and Facilities leaders to work together as co-architects of the cyber-physical campus. Only through this collaboration can institutions build the governance, decision-making authority, and operational confidence necessary to responsibly expand AI, digital twins, and predictive maintenance.

Equally crucial is structured change enablement, guided by models such as ADKAR, which help individuals progress from awareness to lasting adoption. Without addressing human psychology—such as fear of automation, perceived loss of skills, or fatigue from ongoing change—even the most sophisticated systems can end up unused. Conversely, by supporting change with clear communication, phased implementation, practical training, and continuous feedback, modernization becomes a positive evolution rather than a disruptive, forced change.

The manifesto states that genuine transformation isn't about introducing new technology, but about enhancing institutional capabilities. This includes fostering greater collaboration, improving operator experience, building trust in data, lowering cognitive load, and sharing ownership of results. A campus truly becomes "smart" only when people—regardless of their roles, ages, or skills—feel valued, supported, and connected to the mission.

The chapter ends with a decisive statement: culture functions as the operating system of the smart campus. While technology can highlight potential opportunities, it is human alignment and cultural preparedness that transform these into lasting, mission-driven realities.

Chapter 13
Academic and Research Alignment

The smart campus becomes a living laboratory that strengthens relationships, knowledge, competitiveness, innovation, and learning.

Even with advanced elements like the Digital Integration Fabric, the robust Data Estate, sophisticated AI controls, and digital management of the central plant, a smart campus's value extends beyond operational efficiency. Unlike commercial office parks, higher education institutions aim to generate knowledge, safeguard research, and develop future leaders. If the CTO or CFaO builds a highly optimized cyber-physical infrastructure without aligning it with academic and research missions, they end up creating a costly utility rather than a lasting institutional asset.

Moreover, we have thoroughly outlined the architecture, governance, and technology needed for the AI-Enabled Sustainable Smart Campus. However, any Chief Technology Officer or Chief Facilities Officer will face a simple operational truth: even with the latest digital twin technology and a high-end sensor network, hiring the right personnel to operate it remains a significant challenge.

Professionals who perfectly embody this integration—the Integrative Architects adept in both TCP/IP networks and chilled water systems—are extremely scarce. The industry suffers from a significant shortage of hybrid experts capable of designing, constructing, managing, and securing cyber-physical ecosystems. Meanwhile, we are upgrading our infrastructure at a pace that outstrips the development of our workforce.

This scarcity offers a rare and significant opportunity for higher education. To thrive and grow, the smart campus must be closely aligned with the university's fundamental mission: teaching, research, and fostering human potential. If the necessary workforce to run these environments is lacking, it is the university's responsibility to develop the curriculum that will produce it.

To fully realize the potential of the AI-Enabled Sustainable Smart Campus, the CTO/CFaO needs to shift from just managing the campus to becoming an academic collaborator. They should design the physical and digital spaces to function as premier educational and research platforms. Collaboration between the CTO/CFaO is essential to ensure the smart campus supports the institution's academic and research objectives.

The Campus as a Living Laboratory

Historically, campus operations have been hidden behind locked doors and "Authorized Personnel Only" signs. The mechanical plant and the IT data center were black boxes. In the smart campus era, the CTO/CFaO must safely open these environments, transforming the campus itself into a "Living Laboratory."

This requires a governed architecture in which the campus's operational telemetry—anonymized, secured, and strictly partitioned—is made available to the academic enterprise.

- **Data Science and Computer Science:** Instead of training machine learning models on stale, public datasets, students can pull live, read-only API feeds from the campus microgrid or HVAC telemetry to build predictive energy models.
- **Engineering and Architecture:** Civil and mechanical engineering students can access the university's Building Information Models (BIM) and real-time Digital Twins to study structural thermal loads, fluid dynamics, and automated fault detection in a live environment.
- **The Partnership Mandate:** The CTO/CFaO must establish the secure data enclaves and integration middleware that allow this academic access without exposing the core Operational Technology (OT) network to student hacking or accidental disruption.

Developing the Next Generation of Cyber-Physical Leaders

Higher education facilities and IT departments are facing a demographic crisis. The "silver tsunami" of retiring facilities directors and central plant operators is taking decades of mechanical tribal knowledge with it. Simultaneously, traditional IT graduates often lack any understanding of thermodynamics, fluid mechanics, or industrial control systems (ICS).

The institution must cultivate a new breed of professional: **The Cyber-Physical Engineer**. The CTO/CFaO has a strategic imperative to partner with academic Deans to build this talent pipeline right on their own campus.

- **Curriculum Co-Creation:** The CTO/CFaO should serve on academic advisory boards to help design interdisciplinary programs that merge computer science, cybersecurity, and mechanical engineering.
- **Apprenticeship in the Smart Plant:** The campus itself becomes the ultimate internship. Students learn BACnet/SC protocols, IEC 62443 zone segmentation, and edge-AI deployment by working alongside the campus network architects and automation controls directors. This not only prepares students for a rapidly growing job market but also provides the university with a continuous pipeline of highly trained, institutionally aware talent to run its own complex systems.

Catalyzing Applied Research and Securing Funding

A smart campus is a massive competitive advantage for securing federal research funding (e.g., NSF, DOE, DOD grants). Researchers need scalable, real-world environments to prove their hypotheses. The CTO/CFaO must architect the campus to support ongoing, applied research in domains directly relevant to the physical plant.

- **Materials Science and Civil Engineering:** The CFaO can designate specific new construction or retrofits as testbeds for advanced materials. Researchers can deploy and monitor phase-change materials for passive thermal regulation, self-healing concrete, or next-generation photovoltaic glass on actual campus buildings, using the Digital Integration Fabric to harvest the performance data.
- **Energy Transition and Edge AI:** The CTO can provide edge-compute nodes within the campus power substations for researchers developing advanced Distributed Energy Resource Management Systems (DERMS). The campus microgrid serves as a proving ground for **Edge AI**-driven load-shedding, battery-islanding, and bidirectional EV-charging algorithms.

Protecting the Research Enterprise (The Core Partnership)

Research continuity is the lifeblood of a Tier 1 institution. A single power fluctuation or a compromised HVAC controller can destroy decades of biological research or ruin millions of dollars of delicate instrumentation.

The alignment between the CTO/CFaO is most critical in protecting these high-stakes environments:

- **Cyber-Physical Security of Lab OT:** Research equipment (electron microscopes, genomic sequencers, ultra-low temperature freezers) often runs on outdated, unpatchable operating systems. The CTO/CFaO must partner to mandate strict OT microsegmentation, ensuring these vulnerable assets are isolated from the broader network while still allowing the facilities team to monitor their environmental conditions continuously.
- **Predictive Environmental Stability:** Using the AI Analytics Engine, the facilities team moves from reactive alarms to predictive stability. By analyzing vibration and thermal telemetry, the system alerts the CFaO that a compressor on a critical clean-room HVAC unit will fail in three days, allowing for a scheduled, non-disruptive replacement rather than a catastrophic midnight failure.

Data Ethics, Privacy, and Humanistic Alignment

Finally, the smart campus generates massive amounts of behavioral and spatial telemetry. As the campus becomes a sensory environment, the CTO/CFaO must recognize that they are not just managing hardware; they are managing human privacy.

This creates a vital opportunity to partner with the humanities, law, and social sciences:

- **The Ethical AI Governance Board:** The CTO/CFaO should bring faculty from the law school, sociology, and philosophy departments onto the AI Governance Council.
- **Societal Research:** The campus provides a live environment to study the social and technological impacts of ambient computing. How does constant spatial optimization affect student anxiety? How do we design an AI-driven digital concierge that respects neurodiversity? By embedding academic researchers into the governance of the smart campus, the institution ensures that its technological advancement remains tethered to its humanistic values and civil liberties.

When the CTO/CFaO and the architect design the smart campus in alignment with the academic mission, the physical estate ceases to be a cost center. It transforms into an active participant in the university's core mission—a dynamic engine for student success, a defensible fortress for the research enterprise, and a living testament to the institution's commitment to innovation and sustainability.

Part V

Executing Vision, Strategy, and Plans

Chapter 14
Smart Campus Implementation

Execution is the moment the campus shifts from modernization to momentum scaling vision, planning, governance, and assessment, beyond just an idea and into lived reality.

A smart campus strategy cannot function as a shadow IT project or an isolated facilities initiative. For long-term institutional impact, the digital strategy must be integrated into the institution's strategic plan. This chapter presents a high-level implementation framework, from discovery through ongoing optimization.

Strategic planning in higher education involves a disciplined effort to align various aspects of a complex institution—its mission, academic priorities, finances, personnel, facilities, and technology—toward a common developmental goal. Unlike many businesses, universities operate under shared governance, decentralized decision-making, long asset lifecycles, and multiple key areas, including academic units, research, student services, enrollment, auxiliaries, athletics, and health systems.

As a result, institutional plans typically consist of a set of interconnected strategies rather than a single, simple document. These include a broad institutional strategy supported by specific plans for academics, research, enrollment, finance, facilities, sustainability, and digital transformation. The planning process is usually iterative and involves consultation, blending leadership guidance with input from various stakeholders to ensure legitimacy and long-term commitment.

Within this planning ecosystem, the Chief Technology Officer (CTO) and the Chief Facilities Officer (CFaO) hold distinct yet increasingly interconnected roles. The CTO generally oversees planning related to the digital enterprise, including cyber risk and resilience, data strategy, enterprise architecture, academic and administrative systems, service management, and platforms that support learning, research, and operations. Their focus is often on viewing information as a strategic asset, technology as a service backbone, and cybersecurity and governance as essential for the institution's agility.

In contrast, the CFaO oversees planning for the enterprise's physical infrastructure, including campus utilities, space management, reducing deferred maintenance, and capital renewal sequencing. They also set construction standards, focus on operational reliability, and manage the physical environments that influence student experience, faculty productivity, research continuity, and safety. Their perspective prioritizes lifecycle management, real-world performance, and the long-term sustainability and resilience of physical assets.

Traditionally, these two planning lanes operated independently: the technology plan focused on systems and services, while the facilities plan focused on buildings, infrastructure, and space. However, with the rise of smart campus initiatives, decarbonization efforts, and resilience strategies, institutions are realizing that successful planning depends on shared assumptions and coordinated investments. Building systems now generate data and are digitally managed; cybersecurity has shifted to cyber-physical security; and sustainability objectives require unified operational insights across energy, carbon, mobility, and space.

This convergence is most evident in the university's traditional 10-year campus master plan. This plan serves as the long-term blueprint for the campus's physical development, covering land use, academic and residential spaces, circulation, utilities, landscape, sustainability, and the sequencing of major projects. Many institutions adopt a structured methodology: forecasting enrollment and growth, assessing space usage and conditions, establishing development principles, setting capital priorities and phases, and turning this into a detailed capital and funding strategy. Typically, the master plan is updated every 5–10 years, with periodic revisions reflecting changing conditions. Despite methodological differences, the goal remains consistent: to produce a clear, practical roadmap that safeguards institutional flexibility, minimizes long-term risks, and guides the campus's growth in line with the institution's mission, budget, and stakeholder needs.

This chapter outlines the key phases, from initial inventory to autonomous maturity, needed to develop an AI-Enabled Sustainable Smart Campus. Strategies, priorities, and timelines will differ among institutions and likely change over time. Grants or capital allocations can alter emphasis and priorities, accelerating some initiatives and setting rigorous deadlines. Funding shortages, procurement issues, and technological issues can slow initiatives or stop them completely. In short, life happens. So we need a consistent iterative approach to planning and implementation that helps the institution grow its smart campus. The Manifesto establishes five phases of implementation:

1. Governance and Discovery
2. The Digital Blueprint and Master Plan Integration
3. Foundation Integration and Exploratory Deployments
4. Scaling, AI Orchestration, and The Digital Twin
5. Continuous Optimization and Antifragility

The implementation phases and maturity levels are linked to the successful development of a smart campus. They establish the context of the OODA cycle (Observe, Orient, Decide, and Act) cycle that instills discipline without stifling flexibility and creativity. They acknowledge that a smart campus is not simply deployed but cultivated over time.

In a smart campus, maturity is measured not by how quickly new capabilities are added, but by how well the institution can pause, reassess, and then move forward with greater clarity, resilience, and purpose.

Each phase clearly defines its objectives, decision-making authority, and architectural readiness before introducing more complexity, avoiding common pitfalls such as pilot overload, fragile integrations, and layering automation on unstable operations. As institutions progress through these phases, they develop practical experience: leaders learn to govern convergence, operators gain trust in data, and plans are executed predictably well. With experience, campuses evolve from just complying to confidently orchestrating their operations—progressing not solely through ambition but through accumulated operational insight.

Importantly, these phase gates are not linear checkpoints to be once passed and forgotten. Disruptions like leadership changes, climate events, cyber incidents, enrollment shifts, or capital projects often require revisiting earlier phases to reassess assumptions, validate data, and realign resources through governance. This cyclical revisiting isn't regression; it reflects the natural development of maturity. In a smart campus, maturity is measured not by how quickly new capabilities are added, but by how well the institution can pause, reassess, and then move forward with greater clarity, resilience, and purpose.

Phase Gate / Level I: Governance and Discovery

The Objective: Establish governance and the undeniable ground truth of the campus's current physical, digital, and cultural state before a single vendor contract is signed or a single piece of software is purchased. The biggest temptation at the start of a smart campus project is to rush into procurement. Executives, eager to showcase modernization to the Board of Trustees, often fall for vendor demos of predictive dashboards and AI copilots. However, implementing advanced AI on a disorganized, undocumented, and resistant organizational foundation is a mistake. The first year should be dedicated to thoroughly understanding the current situation rather than designing the future. To reach this point, the Integrative Architect must guide the institution through four essential baseline assessments.

Exercising Governance

Establishing and utilizing the Smart Campus Steering Committee (SCSC) is a crucial first step in Phase I, as discovery without governance leads to confusion rather than clarity. Before inventorying systems, mapping OT assets, interviewing stakeholders, or assessing digital readiness, an institution must define authority, decision-making processes, and standards.

Chapter 9 extensively covers how governance is formed and highlights that the greatest risk to smart campus modernization isn't technical problems, but fragmentation, pilot sprawl, and inconsistent decision-making across IT, Facilities, Safety, Sustainability, and academic departments. Without a centralized governing body with clear authority, discovery efforts often become isolated observations rather than part of an integrated enterprise baseline.

The SCSC serves as the institution's main coordination body, ensuring that all discovery efforts align with a common mission, vocabulary, and architectural standards. As noted in the governance chapter, campuses often experience "Shadow OT," in which departments independently acquire smart systems, leading to incompatible data models, unregulated vendor integrations, and security vulnerabilities.

Launching discovery without the SCSC could entrench these silos rather than break them down. The committee offers the authoritative perspective that will shape all interpretations of the findings, establishing the criteria for interoperability, cybersecurity standards, data ownership, and defining what constitutes success for a "smart campus."

Establishing the SCSC from the start reinforces executive credibility and institutional legitimacy for the discovery process. The governance section emphasizes that the SCSC should be co-led by the CTO and the Chief Facilities Officer, with support from the CISO, data leadership, sustainability, procurement, and planning stakeholders. Initiating discovery within this framework sends a clear message: this is not merely an IT project, a facilities upgrade, or an isolated innovation effort. It signifies an enterprise-wide transformation. This early executive framing is crucial for gaining cooperation from various units, ensuring steady participation, and preventing political resistance, which often arises when discovery is seen as evaluative rather than collaborative.

In conclusion, the SCSC plays a vital role because governance sets the scope, which in turn defines what "discovery" entails. The governance section highlights the importance of institutions thoroughly cataloging cyber-physical systems—including OT devices, IoT sensors, vendor cloud connections, BAS controllers, spatial data, and integration points. Without the committee first clarifying which systems fall under smart-campus governance, discovery teams may either over-scope by attempting to map every device or under-scope by omitting key systems such as access control, utility meters, or research environments. The SCSC establishes the definitive boundary: determining what is in scope, what is out of scope, and what needs to be discovered now versus in future phases.

The Cyber-Physical Asset Inventory: Illuminating the Shadows

You cannot optimize, secure, or govern what you cannot see. Historically, Information Technology and Facilities Management have maintained entirely separate ledgers of reality. IT tracked servers, switches, and IP addresses; Facilities tracked chillers, air handlers, and thermostats. In the smart campus era, those lists collide.

This stage involves quickly forming a joint IT and Facilities team tasked with a thorough, comprehensive audit of all Operational Technology (OT). It entails a challenging yet essential journey through the campus's mechanical rooms and network closets. The team must physically and digitally document legacy Building Automation Systems (BAS), identify old proprietary controllers, and locate unsecured "Shadow IoT" devices—such as consumer smart plugs or unpatched environmental sensors—that run quietly on the network. At the same time, network engineers should review the core infrastructure, evaluating the fiber backbone's capacity, Wi-Fi coverage density, and the presence of edge-compute nodes needed for local data processing. The result of this process is a unified cyber-physical map of the entire institution.

Conducting the Inventory

To move from a conceptual "collision of lists" to a unified Cyber-Physical Asset Inventory, you need a standardized intake process. Below is a structured approach to data collection, along with a sample audit form.

Suggested Steps for Data Collection

Building a unified map requires a synchronized "boots-on-the-ground" and "eyes-on-the-wire" approach.

1. **Form the Hybrid Strike Team**
   Pair a Network Engineer with a Facilities Mechanic. The engineer identifies the MAC/IP address and port, while the mechanic identifies the physical make and model and the mechanical function.
2. **The "Sweep and Sniff"**
   Use network scanning tools (like Nmap or specialized OT discovery software) to "sniff" for active devices, then physically locate them to verify their environmental context.[13]
3. **Identify the "Shadows":** Specifically hunt for non-standard gateways—the "consumer-grade" smart plugs or unmanaged weather stations often plugged in by individual departments without IT oversight.
4. **Assess the "Vascular System"**
   While auditing devices, document the physical infrastructure supporting them. Is the fiber backbone at 40% or 90% capacity? Does the mechanical room have enough Wi-Fi density for a technician's tablet?
5. **Unified Tagging**
   Assign a single Global Unique Identifier (GUID) that links the asset in the IT database (e.g., ServiceNow) to the asset in the Facilities database (e.g., Maximo).

The Cyber-Physical Asset Audit Form

It is important to gather audit data through electronic methods that incorporate institutional nomenclature and coding systems. The Appendix includes a straightforward example of an audit form for recording assets during an asset audit. When accurate and up-to-date, data can often be imported from Fiscal or ERP systems containing information about IT/OT assets. However, exercise caution regarding data accuracy, and diligently test and clean the data as needed.

The Integration Point

The goal of this inventory is to move away from "Siloed Reality." By merging these data points, you create a **Functional Dependency Map.** If a specific network switch fails, the Digital Twin can immediately alert you that the "North Wing Chiller" has lost its "intelligence," even if the mechanical motor is still spinning.

The Knowledge Asset Inventory: Institutional Memory Ingestion

Implementation starts not with sensors, integrations, or analytics, but with truth. Before a digital twin can analyze the campus, it must first acquire the institution's memory. The Knowledge Asset Inventory serves as the formal process for this. Properly rephrased, it is not just cataloging or cleaning up. Still, it is Institutional Memory Ingestion—the intentional capture, organization, and integration of the institution's existing knowledge into the digital twin environment.

Each campus already holds a vast amount of operational knowledge. Some of it is documented in formal files such as drawings, specifications, commissioning reports, control sequences, and maintenance logs. Much of this knowledge, however, resides elsewhere: in shared drives, vendor portals, personal notebooks, email chains, whiteboards, and, importantly, as tacit knowledge among experienced operators, technicians, and engineers. Failures in implementation often occur when this knowledge remains scattered, hard to access, or reliant on specific individuals. Institutional Memory Ingestion ensures that this knowledge becomes durable, searchable, and actionable, preventing it from being tribal or fleeting.

From Documentation to Operational Memory

Traditional projects view documentation as a final deliverable to be stored after construction or deployment. In contrast, the digital twin approach treats documentation as a crucial input rather than an output. During onboarding, the Knowledge Asset Inventory systematically identifies, verifies, and incorporates materials that define operational context—covering existing assets, their expected behavior, failure modes, and safety procedures learned over time. This includes, at a minimum:

- As-built drawings, BIM models, and GIS layers
- Sequences of operation and control narratives
- Commissioning, retro-commissioning, and TAB reports
- Equipment submittals, O&M manuals, and warranty data
- CMMS/EAM histories, work orders, and failure notes
- Energy studies, audits, and prior optimization efforts
- Safety procedures, emergency runbooks, and escalation paths

In a mature implementation, these artifacts are more than just static files attached to assets; they are embedded within the digital twin, connected to spaces, systems, sensors, and workflows. This setup creates a baseline for assessing live telemetry, AI models, and future decisions.

Capturing Tacit Knowledge Before It Disappears

Equally important, but even more fragile, is tacit operational knowledge. Many campuses are nearing a demographic inflection point, risking decades of valuable experience as retiring staff or departing vendors leave. The Knowledge Asset Inventory considers this a strategic risk rather than just a human resources concern. During digital twin onboarding, structured interviews, shadowing sessions, and facilitated workshops are used to surface answers to questions such as:

- "What always goes wrong first during a heat wave?"
- "Which alarms can be ignored, and which can't?"
- "What workarounds exist because the system never behaved as designed?"
- "Which vendor changes require extra scrutiny?"

These insights are integrated into the digital twin as annotations, rules, constraints, and operational context. This process helps the institution turn experience into institutional memory, reducing reliance on heroics and speeding up onboarding for new staff. As a result, the digital twin evolves from merely a model of assets to a repository of operational knowledge.

Institutional Memory as a Readiness Gate

Institutional Memory Ingestion is essential and integral, serving as a prerequisite for advanced digital twin features. AI, predictive maintenance, and prescriptive optimization depend heavily on a solid knowledge base. Without documenting sequences, clarifying point mappings, or understanding past failures, the twins' reasoning will be flawed—regardless of the sophistication of the analytics. Therefore, the Knowledge Asset Inventory is performed during onboarding rather than later phases. This approach reveals gaps early, questions assumptions, and clearly marks unknowns. It creates what the manifesto calls operational truth: a common, evidence-based understanding of how the campus truly operates today, not how it was designed to function in the past.

A Living Asset, Not a One-Time Task

The Knowledge Asset Inventory is not static after ingestion; it is a dynamic, evolving institutional asset. As assets are replaced, sequences change, lessons are learned, and incidents happen, the digital twin's memory is continuously updated. Over time, knowledge accumulates, making the institution easier to manage, more resilient to disruptions, and less prone to knowledge loss. In this way, Institutional Memory Ingestion transforms implementation from a technology deployment into an act of stewardship. The digital twin becomes the keeper of what the institution knows—past, present, and evolving—ensuring that intelligence compounds rather than resets with each project, vendor, or leadership transition.

Phase Gate / Level 2: The Digital Blueprint and Master Plan Integration

The Objective: Formally weave the digital architecture into the institution's capital planning and academic mission, permanently shifting technology from an operational afterthought to a foundational architectural requirement. Once the Smart Campus Steering Committee (SCSC) has established the ground truth in Phase 1, the institution faces its most critical strategic pivot. Historically, universities have planned their physical estates and their digital networks on entirely separate tracks. A building is funded, designed, and constructed; only at the very end is Information Technology invited in to "pull the wire." In Phase 2, the Integrative Architect must permanently break this cycle. The digital strategy cannot continue to exist as a separate "IT Strategic Plan." It must be grafted directly into the DNA of the university's traditional 10-Year Physical Master Plan.

1. Capital Project Interception: Mandating the Digital Foundation

To change how a campus operates, it's essential to modify its procurement and construction processes. The CTO and CFAO need to implement a "Capital Project Interception," taking a role at the very beginning of the physical master planning. Starting from Year 1, every new building, major renovation, or central plant upgrade over the next decade must include strict digital requirements in its initial architectural Request for Proposal (RFP).

Merely submitting a 3D CAD file and physical plan is no longer sufficient. The RFP should demand a Digital Twin and a connected Sensor Mesh. Vendors must be contractually obligated to provide open data access, use standardized semantic tagging (like Brick Schema), and support IP-enabled operational technology. By engaging early in the capital process, the institution can prevent constructing "dumb" legacy buildings that would require costly digital retrofits later.

2. Designing the Digital Integration Fabric: The Enterprise Nervous System

While the physical buildings are being planned, the IT architecture team must design the "middleware" of the smart campus. If you connect thousands of new IoT sensors directly to legacy databases using point-to-point custom code, you will create an unmanageable, fragile spaghetti network.

Instead, the CTO must architect the **Digital Integration Fabric**. This is the secure central nervous system of the campus. It requires designing scalable API gateways, message brokers (such as MQTT or Kafka) to handle millions of sensor telemetry pings per second, and data lakes to normalize incoming OT data. Crucially, this fabric must be designed with a Zero-Trust security posture from day one, ensuring that a compromised smart thermostat in a residence hall cannot be used to pivot into the university's financial or research networks.

3. Academic Alignment: The Living Lab Blueprint

A smart campus built solely for operational efficiency is a missed opportunity. To secure widespread faculty and Board support, the blueprint must explicitly serve the core mission of the university: teaching and research.

During Phase 2, the CTO/CFaO must partner with the Provost and academic Deans to draft the "Living Lab Blueprint." This identifies which upcoming physical infrastructure projects will double as active research testbeds. For example, a planned business school renovation might be designated as the testbed for AI-driven spatial analytics. At the same time, a new engineering annex might be wired to grant students read-only API access to the building's structural health and microgrid telemetry. By actively aligning the physical plant with the curriculum, the operations division transforms from a cost center into a direct contributor to student success and research grant acquisition.

4. Developing the AI Governance Charter: The Ethical Rulebook

Before any algorithmic control is handed over to a machine, the institution must define the rules of engagement. Technology moves faster than policy, and if the institution waits until an AI is making decisions to decide *how* it should make them, it will face severe cultural and legal backlash. The SCSC must draft and ratify the **AI Governance Charter**. This is the ethical rulebook for the smart campus. It must explicitly define:

- **Privacy by Default:** How spatial and behavioral data will be anonymized and aggregated to prevent the creation of a "surveillance culture."
- **The Boundaries of Automation:** Clear, non-negotiable lines dictating what an algorithm is *never* allowed to do. For example, while AI can optimize a chiller, the charter must state that it cannot autonomously initiate employee disciplinary workflows or execute student academic sanctions without a human-in-the-loop.
- **Algorithm Auditing:** A mandated schedule for testing predictive models for data drift and bias, ensuring the cognitive layer of the campus remains accurate and equitable.

By the end of this phase, the institution has achieved something important: it has locked the digital vision into the concrete of the university's capital plan, engineered the integration architecture, and established the ethical boundaries required to preserve human dignity.

Phase Gate / Level 3: Foundation Integration and Exploratory Deployments

The Objective: Construct the foundational digital plumbing, secure the cyber-physical perimeter, and engineer early, high-visibility operational wins to build unstoppable institutional momentum. After auditing, architectural design, and governance framing, Phase 3 is where the digital blueprint collides with the physical reality of the campus. This is often the most dangerous phase of the 10-year master plan. It is the period where the initial excitement fades, the unglamorous integration work begins, and the institution risks falling into a "trough of disillusionment."

To survive this phase and maintain the support of the Board of Trustees, the CTO/CFaO must balance long-term foundational upgrades with immediate, highly visible exploratory pilot successes. Although quick wins are important and helpful, a long-term strategy is favorable towards seizing opportunities for progress when conditions are right. As a result, specific timelines are often avoided. It is essential to continuously align the strategic implementation with campus governance, technological capabilities, and the ongoing development of culture and human factors throughout the journey toward an AI-enabled, sustainable smart campus.

1. Network Modernization and OT Microsegmentation: The Highway and the Fortress

Before deploying advanced AI algorithms or thousands of new IoT sensors, the underlying network must be radically upgraded and secured. First, IT must build the "highway." This involves deploying next-generation connectivity—such as Wi-Fi 7, private 5G networks, and distributed edge-compute nodes—that can handle the massive influx of real-time telemetry from the physical plant.

Simultaneously, the CISO and the network architects must build the "fortress." The cyber-physical asset inventory from Phase 1 likely revealed terrifying vulnerabilities: unpatchable legacy HVAC controllers, 15-year-old lighting panels, and sensitive research equipment directly exposed to the enterprise network. In Phase 3, these vulnerable Operational Technology (OT) assets must undergo **Microsegmentation**. By placing these legacy devices into strictly controlled, isolated network segments, the institution ensures that a threat actor cannot use a compromised smart valve to pivot into the university's financial systems or student records. This dual approach guarantees that the campus network is both fast enough to support AI and secure enough to survive a cyber-physical attack.

2. Semantic Tagging and Data Normalization: The Ontological Heavy Lifting

Artificial Intelligence cannot optimize a building if it does not understand the data the building is producing. Phase 3 requires the arduous, unglamorous work of data normalization. The joint IT and Facilities task force must begin translating the chaotic, proprietary data points discovered in Phase 1 into the unified campus ontology (e.g., Brick Schema or Project Haystack). A legacy data point cryptically labeled "AHU2_Z4_T" must be semantically tagged. Hence, the future AI inherently understands that it is an *Air Handling Unit*, located on the *second floor*, serving *Zone 4*, and measuring *Temperature*.

This is the ontological heavy lifting of the smart campus. It is often tedious, requiring veteran mechanics to work alongside data engineers to map physical reality onto the digital twin. However, if leadership skips this step and feeds raw, untagged legacy data into a machine learning algorithm, the AI will confidently generate operational garbage. Clean, standardized data is the absolute prerequisite for the cognitive layer.

3. The "Lighthouse" Pilot: Engineering the Early Win

You cannot boil the ocean, and you cannot modernize a 150-building campus simultaneously. To build institutional momentum and secure ongoing funding from the Cabinet, the Integrative Architect must engineer a flawless

"Early Win." This is achieved through the **Lighthouse Pilot**. Leadership must select a single, high-impact building—ideally a highly visible new student union, or a notoriously energy-intensive research laboratory. In this single building, the CTO/CFaO deploys the full smart campus stack: the dense IoT sensor mesh, the mobile operator workflows, and the early AI-driven Fault Detection and Diagnostics (FDD).

The Lighthouse building becomes the proving ground. It is where the integration middleware is stress-tested, and where the first automated work orders are generated. Crucially, the Lighthouse pilot is designed to prove immediate Return on Investment (ROI). When the CFaO can stand before the Board in Year 3 and demonstrate how the pilot building reduced HVAC energy consumption by 18% while simultaneously reducing hot/cold work orders by 40%, the smart campus ceases to be a theoretical concept and becomes a proven financial capability.

Phase Gate / Level 4: Scaling, AI Orchestration, and The Digital Twin

The Objective: Expand the digital capabilities from isolated pilot projects to a campus-wide ecosystem, fundamentally shifting operations from reactive monitoring to predictive, autonomous optimization. While Phase 3 demonstrated the concept within a single "Lighthouse" building, Phase 4 seeks to extend that success across the entire organization. Scaling up is the most difficult part of the transformation. The core infrastructure—including the network, cybersecurity microsegmentation, and semantic data tagging—is now largely in place. With this foundation in place, the CTO/CFaO is prepared to implement the cognitive layer at a broader scale. In this phase, the campus shifts from a set of isolated, basic assets to a cohesive, intelligent system.

1. Scaling the Digital Twin: From Model to Simulation Engine

In the pilot phase, the Digital Twin was likely a highly detailed, real-time dashboard of a single facility. In Phase 4, the Integrative Architect scales this capability across the entire campus map. The Digital Twin ceases to be just a 3D architectural rendering; it becomes a dynamic, predictive simulation engine. By continuously ingesting live telemetry from the Digital Integration Fabric, the twin allows executives to run complex "what-if" scenarios without risking physical assets.

- **Space and Utilization**
  Academic planners can simulate how changing a Monday/Wednesday/Friday lecture schedule will impact pedestrian traffic bottlenecks and dining hall peak loads.
- **Resilience Planning**
  Facilities leaders can simulate a Category 4 hurricane or a regional grid failure, watching the Digital Twin model to see exactly how long backup generators will sustain critical research freezers based on current ambient temperatures and battery degradation. The campus is no longer guessing its resilience; it is mathematically proving it.

2. Deploying the AI Control Plane: The Predictive Shift

With clean, semantically tagged data flowing freely, the institution flips the switch on the AI Control Plane. This is the moment the operational posture permanently shifts from reactive to predictive. Machine Learning (ML) algorithms take over the relentless, impossible task of monitoring millions of daily data points from the Building Automation Systems (BAS). Instead of a human operator staring at a screen, waiting for a red alarm to flash—indicating that a piece of equipment has already failed—the AI identifies subtle, invisible anomalies that precede failure.

- **Automated Workflows**
  When the AI detects a 3% degradation in the vibration signature of a primary chilled water pump, it doesn't just send an email. It autonomously generates a work order in the CMMS, orders the required bearing from the supplier via the supplier API, and schedules the repair for the following Tuesday, when the building's thermal load is lowest. The AI handles the diagnosis and the administration, allowing the mechanic to focus entirely on the physical craft of the repair.

3. Microgrid and Energy Algorithmic Control: The Decarbonization Engine

As the campus expands its renewable energy footprint—by adding solar arrays, battery storage, and EV charging stations—the complexity of managing these Distributed Energy Resources (DERs) outstrips human capabilities. In Phase 4, the AI is integrated directly into the campus microgrid. The algorithm continuously ingests external data (regional weather forecasts, live utility pricing, and class schedules) to optimize the campus's energy posture dynamically.

- **Automated Load Shedding**
  If the AI predicts a severe heatwave tomorrow afternoon that will spike utility peak-demand charges, it automatically instructs the central plant to over-chill the massive campus water tanks overnight when energy is cheap and abundant. The next afternoon, it uses that stored thermal energy to cool the campus, shedding massive electrical load and saving the university tens of thousands of dollars in a single day—all without human intervention.

4. Agentic AI and Ambient Experiences: The Human Interface

Up until this point, the smart campus has mostly been invisible to the students and faculty, operating quietly in mechanical rooms and data centers. In Phase 4, the intelligence surfaces to the human interface layer through Agentic AI and ambient computing.

- **For the Workforce**
  Generative AI copilots are embedded in the tools operators use every day. A technician standing on a ladder can ask their mobile device, "What is the exact sequence to bypass the actuator on this specific 1998 VAV model?" and the AI instantly synthesizes the answer from the digitized legacy manuals and the institution's specific maintenance history.

- **For the Student**
  The campus experience becomes frictionless and ambient. The smart campus application doesn't just show a map; it acts as a personalized concierge. If a student's calendar shows a group project meeting, the AI checks real-time IoT occupancy sensors across the library and student union, predicts which collaboration rooms will be empty in ten minutes, and autonomously reserves the ideal space, guiding the student there via an accessible, indoor-mapped route.

By the end of phase 4, the AI-Enabled Sustainable Smart Campus has achieved critical mass. It is saving energy, preventing catastrophic equipment failures, and elevating the daily experience of every human who interacts with it.

Phase Gate / Level 5: Continuous Optimization and Antifragility

The objective of Phase Gate / Level 5 is to elevate the campus from a state of managed resilience to true antifragility, in which the cyber-physical ecosystem autonomously learns, adapts, and improves its structure under operational stress and environmental volatility. As the institution enters the final stages of a plan, the heavy lifting of construction and integration is complete. The network is secure, the data is normalized, and the algorithms are actively managing the physical plant. However, the Integrative Architect understands that a smart campus never reaches a "finished" state. The goal of Phase 5 is not maintenance; it is evolution. In this phase, the CTO/CFaO transitions from building the system to governing an autonomous ecosystem. The campus ceases to survive shocks merely; it begins to use them as fuel for continuous improvement.

1. Self-Healing Infrastructure: The Antifragile Core

The highest evolutionary state of the AI-Enabled Sustainable Smart Campus is **antifragility**. While a resilient campus can absorb a shock and bounce back to its baseline, an antifragile campus learns from the shock and physically reorganizes itself to become stronger. In Phase 5, the Digital Integration Fabric achieves this through self-healing infrastructure.

- **Cyber-Physical Immune Systems**
  If a zero-day ransomware attack attempts to breach an IoT gateway in the engineering building, the network does not wait for a human security analyst to intervene. The AI instantly detects the behavioral anomaly, autonomously severs the compromised port, quarantines the subnet, and rewrites the firewall rules across the entire enterprise to inoculate the rest of the campus.
- **Mechanical Self-Correction**
  Similarly, if a severe winter storm causes an unexpected drop in primary steam pressure, the AI-driven microgrid and HVAC controllers instantly negotiate a new operational state. They autonomously shed non-critical thermal loads, prioritize research labs, and rewrite the predictive algorithms so the campus is mathematically prepared to handle that exact weather anomaly even better the next time. The system improves *because* it was stressed.

2. Kinetic Integration: Robotics and Autonomous Mobility

In the mature smart campus, intelligence is no longer confined to static buildings; it becomes kinetic. Robotics and Autonomous Vehicles (AVs) are fully integrated into the institution's daily operations and physical right-of-way.

- **The Autonomous Fleet:** Groundskeeping drones perform thermal roof inspections autonomously, while robotic floor scrubbers and intelligent delivery rovers navigate the pedestrian layers. Autonomous electric shuttles provide reliable, accessible transit, connecting distant parking lots to the campus core.
- **Vehicle-to-Infrastructure (V2I) Orchestration:** These machines are not operating as rogue, isolated devices. They are deeply tethered to the Digital Integration Fabric. An autonomous shuttle communicates directly with the smart intersections and pedestrian sensors, dynamically adjusting its speed during class changes. If a fire alarm triggers in a residence hall, the central AI instantly commands all delivery robots to clear the ADA ramps and egress routes, ensuring that kinetic technology never impedes life-safety operations.

3. The Iterative Listening Architecture: The Human Pulse

The paradox of extreme automation is that it requires profound human oversight. As the campus becomes more autonomous, the risk of alienating the workforce and the student body increases. If the AI becomes an opaque "black box" that dictates operations without human context, the culture will fracture. To prevent this, Phase 5 relies heavily on the **Iterative Listening Architecture to ensure that the technician's voice is heard**. The cross-functional councils established in earlier phases are now permanent institutional fixtures. As the AI begins to make more complex autonomous decisions, veteran mechanics and network engineers continuously review the AI's logic. They manage the transparent UX backlogs, ensuring the mobile dashboards and Agentic AI copilots evolve alongside the changing needs of the workforce. The operators remain the ultimate masters of the technology, continuously co-creating the environment.

4. Continuous Algorithmic Auditing: Governing the Cognitive Layer

Over a multi-year horizon, the physical and behavioral realities of a campus change. New buildings are erected, enrollment shifts, and climate patterns alter. If an AI algorithm was trained on data from Year 1, by Year 4 it will begin to suffer from "model drift"—making inaccurate, inefficient, or biased decisions based on outdated assumptions. AI governance is a continuous operational discipline. The Smart Campus Steering Committee (SCSC) mandates regular, rigorous audits of predictive models. They test the energy algorithms for efficiency drift and audit the spatial analytics to ensure no biases have crept into the digital twin, BMS or scheduling systems.

This continuous auditing ensures that the cognitive layer of the campus remains firmly tethered to institutional truth and human dignity. It proves to the Board of Trustees and the academic community that the university is not blindly surrendering control to a machine but is actively governing its intelligence and actively improving systems and services.

Systemic Continuous Improvement

Leading a smart campus means managing a dynamic, evolving system. Static strategic plans no longer suffice because the environment—made up of countless interconnected sensors and human actions—shifts every millisecond. For today's IT and Facilities leaders, Continuous Improvement isn't just a periodic review; it's an ongoing, rapid cycle of decision-making. The core structure for this responsiveness is the **OODA** Loop. Created by military strategist John Boyd to gain an edge over opponents in critical situations, the OODA Loop has been repurposed as the 'nervous system' of the smart campus. It guarantees that leadership and automated systems operate not only quickly but also accurately, guided by validated data.[14]

The Four Original Stages of the OODA Loop

The strength of OODA comes from its straightforward approach and emphasis on the "Feedback Loop"—the capacity to learn from the outcome of an action and quickly incorporate that insight into the next cycle.

1. Observe (The Raw Intake)

Observation encompasses the "Sensing" and "Connecting" phases. In a smart campus, this involves collecting raw telemetry data from the physical environment. It requires aggregating information from various isolated sources—such as temperature sensors, PIRs for occupancy, badge readers, and smart meters. To observe effectively, leaders must ensure the network infrastructure is strong enough to transmit data seamlessly from edge devices to the central system without delay. The overall aim is to digitize physical events, like "The library is at 90% capacity."

2. Orient (The Contextual Filter)

Orientation is a crucial and challenging phase that focuses on "Normalizing" and "Analyzing" observations. Raw data is meaningless without context. Orientation removes noise, such as data errors or anomalies, and adds semantic meaning by asking: What does this data signify in relation to our institutional goals? For example, the Smart Campus Goal ensures that "75 degrees" in a server room prompts a different response than "75 degrees" in a student lounge.

3. Decide (The Orchestration)

After the data is organized, a course of action is chosen. In manual settings, this decision is made by leadership; in a smart campus, it often involves the Orchestration Layer. The system or leader assesses the analyzed patterns using pre-set rules, AI inference, or institutional policies to decide the optimal response. The goal of a Smart Campus is to formulate an appropriate response, such as "Since the lab is empty for the next three hours, shift the HVAC to Economy Mode."

4. Act (The Control Layer)

Action represents the physical realization of a digital decision, completing the digital-to-physical cycle. The decision is transmitted via the network to an actuator—such as a smart lock, lighting controller, or VAV box—to alter the campus environment. The Smart Campus Goal focuses on closing this loop by implementing changes, such as dimming lights or adjusting airflow.

The Advanced Framework: A 7-Step Evolution

While the four-step OODA loop offers a strategic foundation, the complexities of Operational Technology (OT) often demand a more detailed approach to maintain data integrity. Extending from 4 to 7 steps is not a rejection of Boyd's model but a technical "unbundling" driven by modern data complexities. Originally, in the military OODA loop, "Observe" and "Orient" were cognitive processes handled by humans. In a Smart Campus, these functions are replaced by automated technical layers. In the advanced smart campus model, we expand OODA into a **7-step iterative cycle:**

1. **Sense** (Physical Layer)
2. **Connect** (Network Layer)
3. **Normalize** (Data Layer)
4. **Analyze** (Intelligence Layer)
5. **Decide** (Orchestration Layer)
6. **Act** (Control Layer)
7. **Learn** (Optimization Layer)

This expanded approach divides "Observe" and "Orient" into technical milestones—Sense, Connect, Normalize—to guarantee that, before a decision is made, the data is clean, tagged, and reliable. Most importantly, it introduces the Learn phase, where Machine Learning and human leaders evaluate the results of the "Act" phase to improve the next cycle—helping the institution shift from a Reactive to a Predictive state.

The last steps of the 7-step model—Analyze, Decide, Act, Learn—expand the Feedback Loop, a fundamental element of Boyd's OODA and the PDCA cycle. Adding "Learn" as a separate seventh step highlights the influence of Machine Learning and Predictive Analytics, where the system's "memory" enhances future iterations.

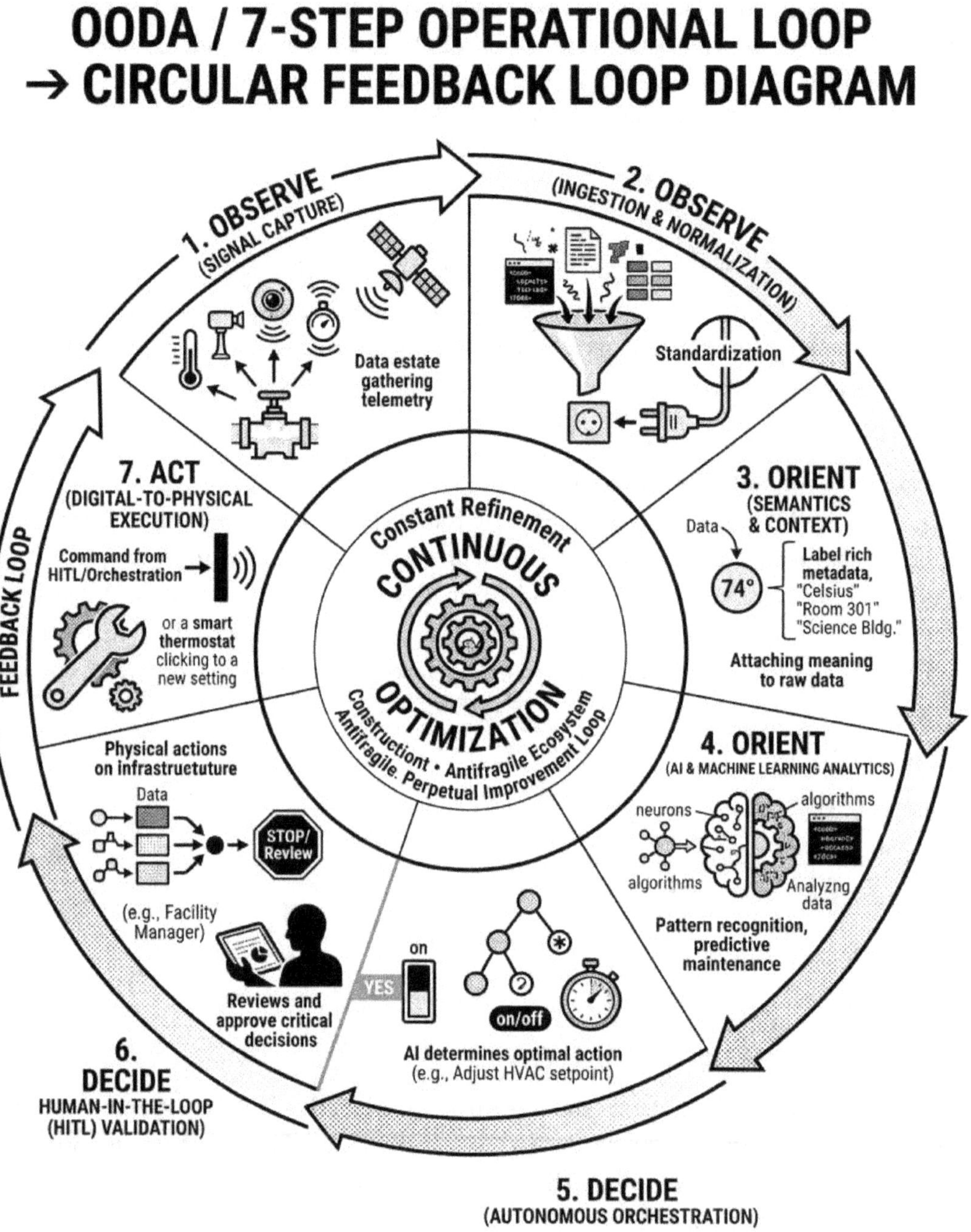

Figure 10. OODA 7 Step Operational Loop

Institutions may use various methods or systems to drive ongoing improvements. For leaders, it's crucial to understand that smart campus capabilities are typically iterative, evolving as more data is collected and analyzed or as feedback from the campus community is incorporated. To become truly smart, these applications must be enabled to learn under the supervision of administrators, allowing them to build accurate knowledge of different scenarios and outcomes. This way, their capabilities can lead to better operations and enhanced experiences.

Chapter Summary

This chapter clearly emphasizes that creating an AI-enabled, sustainable smart campus is more than just installing technology; it represents an institutional transformation. Achieving success depends on disciplined governance, a unified digital-physical framework, and a human-centered approach that ensures modernization enhances, rather than disrupts, the university's mission.

A smart campus can't be implemented with isolated pilots or vendor-led projects. Instead, it develops through a structured, multi-phase roadmap—starting with governance and discovery, moving to digital blueprinting and core integration, and ending with AI orchestration, scaling, and ongoing optimization. Each phase strengthens the institution by establishing shared standards, upgrading OT security, ensuring data is clean and semantically tagged, maintaining stable integrations, and enabling more predictive and automated operations.

Five Phases of Smart Campus Implementation

1. **Phase Gate/ Level I: Governance and Discovery**
   Governance and Discovery form the foundation, as evidenced by the establishment of key organizational components, including the **Smart Campus Steering Committee (SCC)** and an **Ethical Data Framework**. It emphasizes a thorough **Comprehensive Campus Audit** to understand existing assets before moving forward.

2. **Phase Gate / Level II: Digital Blueprint and Master Plan Integration**
   Digital blueprint and master plan integration moves the institution towards integrated strategic planning. This phase highlights the digital blueprint and ensures seamless integration with the physical campus plan, while also initiating vital technology vendor and partner evaluations.

3. **Phase Gate / Level III: Foundation Integration and Pilot Deployments**
   Integration and pilot deployments mark the beginning of hands-on implementation. Here, the focus shifts to creating the **Central Smart Campus Data Hub**, installing a campus-wide IoT Network, and launching targeted pilot projects in areas such as energy management and occupancy tracking.

4. **Phase Gate / Level IV: Scaling, AI Orchestration, OT Applications, and the Digital Twin**
   The scaling of AI and the evolution of OT applications such as the Digital Twin reflect the expansion and integration of advanced technologies. Success from earlier phases leads to scaling across campus. This is where artificial intelligence takes center stage in active AI orchestration, and the innovative concept of a campus Digital Twin is introduced to optimize real-time operations through simulations.

5. **Phase Gate / Level V: Continuous Optimization and Antifragility**
   Continuous optimization and the journey towards antifragility reflect the initiative's mature state. The diagram shows an ecosystem focused on constant refinement using data and AI. This phase emphasizes moving beyond mere resilience to achieve Antifragility—where the campus actually improves in response to stress—and institutionalizes predictive maintenance and community feedback for long-term success.

See the Appendix for a diagram depicting these phases.

The chapter also emphasizes a vital truth: technology's success depends on people. Human adoption, operator training, and psychological safety are essential, not just secondary concerns—they directly influence implementation success. Without effective communication, role-specific training, reliable workflows, and human-in-the-loop governance, even advanced technology will revert to workarounds, overrides, and "Shadow OT." Integrating change management models like ADKAR throughout every stage of implementation is crucial to ensure skills develop alongside increasing complexity.

Risk management—covering cyber, operational, ethical, and environmental aspects—is an ongoing aspect throughout implementation. Chapter 14 highlights that today's risk is cyber-physical: a misconfigured controller can be just as hazardous as a cybersecurity flaw. To address this, OT segmentation, privileged access controls, vendor governance, and continuous monitoring are integrated not merely as IT responsibilities but as essential to campus continuity.

The chapter ends by viewing implementation as an investment in long-term institutional resilience. Each stage enhances the university's capacity to foresee failures, handle energy, and carbon issues, improve learning environments, safeguard research, and provide consistent, transparent experiences for students and faculty. The aim is not merely to deploy devices or dashboards; it is to develop an operating model that supports ongoing learning, adaptation, and antifragility—making the institution stronger through volatility rather than fragile because of it.

Chapter 15
Inventory and Readiness Assessment

Institutions must first understand their current systems, risks, constraints, data, and capabilities before embarking on large-scale innovation and modernization.

The vision of an AI-powered, sustainable smart campus is admirable but difficult to realize. This isn't because the necessary technologies—such as sensors, analytics platforms, digital twins, and advanced building controls—are unavailable or advancing rapidly. Rather, campuses are intricate, living systems composed of legacy infrastructure, ingrained operational habits, distributed decision-making authority, and data that, despite being plentiful, can often be unreliable. Success in this environment depends less on launching impressive pilot projects and more on honest assessments and clear-eyed truth-telling from the start.

Truth-telling manifests as a joint assessment: a collaborative, evidence-driven evaluation of the current situation and readiness. This is where aspirations turn into responsible action. Before installing sensors, developing a dashboard, or implementing a Digital Twin, or carving out a line for "smart campus" in the budget, the institution must confidently answer a key question: what do we actually possess, what can we dependably incorporate, and what can we maintain over time?

Inventory as Operational Truth

The institution doesn't need a catalog solely for its own record-keeping. Instead, it requires operational truth—critical information that supports decision-making, budgeting, contracts, security, and measurable outcomes. This truth begins with a clear understanding of assets and systems. For example, which buildings have modern BAS controllers, and which still rely on legacy or proprietary controls? Where is detailed metering available, and where is it only accessible through monthly utility bills? Which buildings utilize occupancy sensors, and can this data be utilized beyond local subsystems? Additionally, what is the current state of the network infrastructure capable of supporting thousands more endpoints? Finally, where are the often-hidden boundaries between OT and IT, presenting potential risks?

Operational truth also requires data integrity. As buildings and systems are now represented in data models for analytics platforms to interpret consistently across vendors and device types, many organizations adopt common semantic frameworks, such as Brick Schema, and tag-based methods, such as Project Haystack. These approaches aim to reduce ambiguity and prevent reinventing point-naming and data-structure definitions for each building. While they do not eliminate integration challenges, they make integration more manageable and help ensure the institution develops a cohesive platform instead of a patchwork.

This is where the inventory and readiness phase quietly becomes transformative. It prompts the establishment of naming conventions, metadata discipline, and governance questions that many campuses tend to postpone until they face costly consequences of remediation.

> ***Readiness Begins with a Simple Commitment:***
> ***Replace Assumptions with Evidence***

Campuses often believe they understand their infrastructure, but their knowledge is usually fragmented: some drawings, a spreadsheet of assets, a BAS head-end that "sort of works," a CMMS filled with work orders, a network map that is more theoretical than accurate, a sustainability plan missing interval data, and a security environment that has evolved gradually over time.

Smart Campus Phase I Inventory Checklist

This list organizes what *should be included in the Phase I Current-State Inventory* to support a Smart Campus modernization program.

1. Architectural and Data Readiness Inventory

These items establish the "physical truth" and "data truth" needed before any blueprinting.

- **Complete OT asset inventory:** BAS controllers, HVAC equipment, submeters, lighting controls, access control, security devices, research/lab systems, utility plant equipment.
- **OT network architecture:** VLANs, subnets, firewalls, gateways, edge devices, bandwidth constraints.
- **IoT sensors and telemetry sources:** occupancy, IAQ, environmental, vibration/acoustic, metering sensors.
- **Integration points and interfaces:** APIs, drivers, gateways, middleware, vendor cloud services.
- **Semantic tagging state:** Any use of Project Haystack, Brick Schema, or other ontologies.
- **Spatial and static data sources:** BIM, GIS, as-builts, asset records, space inventory databases.
- **Data storage and data estate state of health:** data lakes, historians, time-series databases, retention rules.

2. Cybersecurity and Risk Posture Inventory

Necessary to establish a cyber-physical baseline for safe modernization.

- **Segmentation model review:** Actual vs. intended OT micro-segmentation, ISA/IEC 62443 alignment.
- **Privileged access and remote access posture:** Vendor tunnels, VPN paths, session logging, MFA.
- **Incident response readiness:** Current OT-aware IR playbooks, SOC involvement, and tabletop performance.
- **Shadow IoT and unmanaged systems:** Department-purchased sensors, rogue smart devices, vendor cloud pathways.
- **Policy and compliance mapping:** NIST CSF 2.0, NIST SP 800-82, IEC 62443 compliance status.

3. Operational and Workflow Readiness Inventory

Ensures discovery captures how the campus actually functions—not just the equipment list.

- **Maintenance workflows:** Preventive, reactive, and condition-based maintenance processes; CMMS usage.
- **Alarm/alert structures:** Building alarms, thresholds, noise levels, response expectations; current filtering logic.
- **Dispatch and work order integration:** How BAS events become work orders (or fail to).
- **Operator experience (OX):** Current interfaces, mobile tools, usability barriers, training gaps.
- **Deferred maintenance profile:** Critical risks, failing controls, drifted sequences, commissioning gaps.

4. Cultural and Change Enablement Readiness Inventory

The appendix emphasizes that modernization succeeds or fails based on people.

- **Cross-functional trust and collaboration status:** IT–Facilities–Safety–Sustainability working patterns.
- **Change management maturity:** ADKAR readiness, training cadence, communication channels.

- **Operator capability and psychological safety**: Comfort with digital systems, fear points (automation, monitoring).
- **User/stakeholder experience feedback**: Current challenges from faculty, students, and researchers.
- **Living Lab participation**: Academic partnerships, research access to telemetry, pilot engagement.

5. Governance and Strategic Alignment Inventory

This section maps current decision-making structures and readiness for the SCSC model.

- **Existing governance bodies**: Committees for IT, Facilities, Data, Risk, AI, and Sustainability.
- **Decision rights mapping**: Who currently approves OT purchases, integrations, network changes, and data sharing?
- **Procurement controls**: Presence or absence of "No Shadow OT" safeguards, contract standards, open-protocol requirements.
- **Capital alignment**: Whether digital twin readiness, cyber-physical modernization, and lifecycle planning appear in capital plans.
- **Ethical/AI governance**: AI oversight, privacy rules, acceptable use, risk registers.

6. Energy, Sustainability, and Utilities Inventory

Supports decarbonization, optimization, and digital twin scenarios.

- **Energy plant systems**: Chillers, boilers, generators, thermal storage, BESS, renewables integration.
- **Metering and submetering**: Electric, gas, thermal, water, steam; interval data availability and telemetry quality.
- **Demand response and grid-interactive capability**: Current automation, tariffs, utility signals, shed/load routines.
- **Carbon and ESG reporting structures**: Data sources, methodology, and verification credibility.

7. Space, Experience, and User Environment Inventory

Required for digital twin baselining and dynamic space optimization.

- **Occupancy data sources**: Wi-Fi analytics, cameras, sensors, scheduling systems.
- **Environmental quality systems**: IAQ sensors, ventilation controls, comfort complaints.
- **Smart classroom tech and AV systems**: Monitoring, control interfaces, support workflows.
- **Accessibility and inclusion touchpoints**: Digital signage, wayfinding, app integrations.

8. Mobility, Safety, and Security Inventory

Required for cyber-physical resilience and unified campus experience.

- **Access control systems**: Credentials, standards, vendor ecosystems, and segmentation state.
- **Video security and analytics**: Storage, retention, AI usage, privacy boundaries.
- **Emergency management tech**: Blue lights, panic systems, alarm integration, dispatch workflows.
- **Mobility infrastructure**: Parking systems, LPR, transit optimization, EV charging network maturity.

Readiness Assessment Is Not Optional

The readiness assessment involves systematically organizing fragments into a clear, coherent picture. It transitions the mindset from "we believe" to "we can prove," ensuring each aspect is supported by evidence.

This is not an academic task; it involves financial stewardship, risk management, and operational realism. Additionally, it increasingly becomes a reputational obligation, as smart campus initiatives are now linked to sustainability commitments and public reporting. You cannot convincingly report progress on energy, carbon reduction, resilience, or student experience without establishing a solid baseline and using a transparent measurement method.

As operational technology (OT) advances, the campus threat surface grows wider. This is a real concern, not just a theoretical one. A connected Building Automation System (BAS) isn't just a convenience—it can serve as an entry point for actual disruptions. Unlike typical IT failures, OT failures can directly affect safety and cause operational downtime, resulting in immediate physical effects.

That's why the readiness assessment needs to incorporate an OT-informed cybersecurity review, rather than just applying an IT checklist designed for a different setting. OT systems often face uptime constraints, contain legacy devices that can't be patched as frequently as IT endpoints, require vendor remote access, and have long replacement cycles, making quick "rip and replace" approaches impractical.

A practical foundation for this work is NIST guidance on securing industrial control and OT environments, which highlights the need to adapt controls to the specific constraints and risks of control systems rather than assuming IT standards are directly applicable. Additionally, many organizations and vendors follow the ISA/IEC 62443 standards, which provide a structured approach to security capabilities and levels in industrial automation and control environments.

A readiness assessment isn't about perfection. Instead, it aims to pinpoint areas where current conditions hinder safe or scalable future integration, such as porous segmentation, weak authentication, uncontrolled vendor access, lack of monitoring in OT networks, and incident response plans that consider only digital events rather than cyber-physical ones.

Readiness Is Ultimately Human Skills, Governance, and the Capacity to Sustain Change

A comprehensive technical inventory isn't sufficient if the institution lacks an operating model that enables ongoing transformation. Smart campus initiatives go beyond installations; they embody a continuous lifecycle evolution. This encompasses procurement practices, commissioning standards, control sequence management, cybersecurity exceptions, data governance, vendor accountability, and ongoing training.

This is where readiness really turns into a cultural issue. Does the institution have solid, cross-functional routines to ensure IT, Facilities, Sustainability, Public Safety, Finance, and Academic leadership stay aligned? Are decision-making rights clearly assigned, or do they become ambiguous when issues involve multiple areas? Can the campus effectively oversee a multi-year modernization effort without ending up with siloed budgets, conflicting priorities, or change fatigue?

The readiness phase is the stage where these questions can be answered truthfully. Not based on opinions, but supported by evidence such as skills inventories, vacancy and retention patterns, project post-mortems, collaboration telemetry, adoption data from previous technology changes, and the maturity of governance practices.

Benchmarks Matter Because They Turn Internal Claims into External Credibility

One reason why readiness assessments are gaining popularity with boards and executive leadership is that they can be based on well-known frameworks. Sustainability reporting and building performance are now more often assessed from external perspectives—sometimes voluntarily, sometimes under market pressure, and increasingly due to regulatory requirements.

Frameworks like AASHE STARS, LEED, and WELL don't prescribe specific strategies but offer credible benchmarks, common terminology, and external comparability. For building controls modernization, operational best practices such as ASHRAE Guideline 36 serve as reference points for what constitutes "high-performance sequences" as systems transition from simple automation to fully optimized HVAC control.

The key focus is not just compliance but building confidence. Benchmarks assist the institution in clearly stating, with clear justification, "this is our starting point, this is our goal, and this is how we will evaluate our progress."

The Readiness Report:

A readiness assessment becomes truly valuable when it results in a readiness report that serves as a decision-making tool rather than just a document. Its purpose is to convert raw inventory data into a shared story of reality and a preliminary plan of action.

At a minimum, that report should produce:

- an executive storyline that connects readiness to mission, risk, and financial stewardship.
- baseline metrics that can anchor ROI, carbon claims, and operational improvement.
- a clear identification of the few gaps that could stall everything (often governance, identity, segmentation, data quality, or skills).
- a sequenced action roadmap that distinguishes critical-path remediation from early pilots and enabling foundations.
- explicit ownership—who leads what, who funds what, and how success will be measured.

Readiness becomes momentum when leadership recognizes a clear diagnosis and a feasible sequence. This allows the institution to focus funding on the essential foundations first, rather than spreading resources thin across unrelated initiatives.

Chapter Summary

The most successful campuses start with clarity rather than the latest technology. They prioritize a common understanding of both the physical and data infrastructure. Cybersecurity is considered a fundamental requirement from the start, not an add-on for future development. They address governance and skill gaps early on, instead of waiting until pilot projects fail to expand.

In this way, the readiness assessment isn't just a preliminary step before transformation. It actually marks the transformation itself—when the campus shifts from viewing IT and Facilities as separate entities to functioning as a cohesive, cyber-physical institution.

Once the campus establishes that unity—when it can confidently say "this is our current state, and what we can support"—every future investment is then seen as credible, secure, measurable, and more likely to have a lasting impact.

Chapter 16

Change Management

Structured enablement, communication, and adoption practices ensure technology becomes an institutional capability rather than an abandoned idea.

Viewing IT/OT convergence on a smart campus only as a technical issue—focusing on protocols, API gateways, network segmentation, device inventories, and data lakes—misses the bigger picture. Although these are crucial architectural components, they are not the primary obstacles to modernization. The real difficulty is how convergence fundamentally changes human operations within the institution. It impacts who can access physical systems, how life-safety uptime is preserved, redefines data ownership, and alters the daily routines of frontline staff.

In higher education, this tension is intensified by extensive decentralization and varied professional cultures. Facilities teams have traditionally focused on managing physical risks, safeguarding life safety, and providing uninterrupted service. Meanwhile, Information Technology teams specialize in identity management, network security, application deployment, and responding to rapidly changing cyber threats. Each culture is finely tuned to its respective legacy responsibilities.

Convergence, however, requires a unified enterprise capability. It necessitates an environment that upholds the strict safety and reliability limits of Operational Technology (OT) while also maintaining the rigorous cybersecurity and data governance standards of modern IT.

Since this transformation spans multiple historical boundaries, it cannot be assigned to a single project manager. Managing change for IT/OT convergence requires a collaborative, coordinated effort with clear visibility, involving both the CTO and the Chief Facilities Officer (CFaO). If executive leaders are not aligned, the campus risks splitting along its old fault lines.

Why Transformation Feels Hard

When IT and OT come together, both sides are concerned about taking on each other's risks. Facilities fears that IT's fast-paced culture of quick updates and regular patching could destabilize sensitive mechanical systems. Meanwhile, IT is concerned that unmanaged OT connectivity could introduce significant cyber vulnerabilities and create blind spots.

This tension is not primarily a communications problem; it is a matter of design. The CTO/CFaO is tasked with developing a change strategy that accounts for OT limitations while gradually advancing the institution's digital maturity. As per NIST, OT security needs to address "unique performance, reliability, and safety requirements," highlighting why change management cannot just be replicated from traditional IT software implementations.

The Human Architecture of Convergence: Enablement, Risk, and Antifragility

Smart Campus initiatives are often—yet mistakenly—presented as mere technology upgrades. We focus on new edge controllers, segmented networks, sleek dashboards, and predictive analytics. However, at a university, the real transformation is rooted in people. Modernization changes routines, shifts decision-making authority, and permanently redefines how professional expertise is demonstrated.

If an institution only installs software without enabling the human workforce, it does not achieve modernization; it achieves expensive disruption. To reach a state of true operational **antifragility**—where the campus ecosystem actually learns and grows stronger from stress—leadership must treat human change management with the same engineering rigor applied to the network architecture.

The Change Curve and the Generational Gap

Adoption usually takes time, and technical problems rarely drive frontline resistance. In settings where life safety, research continuity, and critical reliability are essential, such resistance is a logical response to uncertainty.

Facilities organizations hold valuable, practical tribal knowledge that is often undocumented. A seasoned mechanic can sense when a building is off-key or identify which legacy air handler struggles during shoulder seasons. When modernization focuses solely on a "digital twin-first" approach, experienced staff might feel their judgment is being replaced by remote telemetry. Meanwhile, newer, digitally native staff may readily adopt the digital twin but lack the physical intuition needed to stabilize a building when sensors fail.

If the institution does not actively address this experiential gap, it risks undermining its operating culture, leading to modern tools being bypassed through manual workarounds as soon as a crisis occurs. Enablement serves as the link between installation and adoption. Without it, modernization risks becoming unused shelfware, fostering an override culture, and causing significant disappointment.

The Five Pillars of Operator Enablement

Enablement should be viewed as a funded, strategically planned, and ongoing component of the project, rather than just a one-day training session scheduled alongside the "go-live" date. To shift the workforce from reactive responses to proactive stewardship, the CTO/CFaO needs to establish five key pillars of enablement.

1. Role-Based Training Paths

Different roles perceive the smart campus in distinct ways. Training should be tailored to individual job responsibilities and decision-making authority rather than standard vendor feature demonstrations.

- **Operators** need fast pattern recognition and alarm response discipline.
- **Supervisors** need situational awareness, staffing coordination, and escalation clarity.
- **Engineers** need deep configuration understanding and root-cause analytics.
- **Security Partners** need to understand how building automation interacts with lockdown protocols.

2. Scenario Training: Rehearsing the "Moments that Matter."

Smart campuses prove their resilience during moments of extreme stress, such as a rapid temperature drop in a key lab, a cyber-attack on the BAS system, or a sudden, large-scale surge in events. Scenario training helps build cross-functional muscle memory by safely simulating these crises. It reveals operational weaknesses—like unclear decision-making authority, missing access credentials, or vague alert definitions—allowing institutions to refine their protocols before facing real emergencies.

3. Embedded Job Aids: Reducing Cognitive Load

The system must support the operator by providing contextual guidance directly within dashboards and alarm workflows. When a fault occurs, thc intcrfacc should suggest next steps, include links to pertinent graphics, and offer straightforward escalation checklists. This form of "ambient guidance" embeds the standard operating procedures where they are needed most, helping to reduce anxiety and cognitive load during rotating shifts or stressful situations.

4. The Feedback Loop: Operators as Co-Designers

The initial version of a dashboard is seldom accurate when scaled. True enablement occurs when frontline staff have structured tools, such as daily huddles, ticket tags, and post-incident reviews, to refine graphics, alarm thresholds, and sequences based on their experiences. Involving operators as co-designers sends a clear cultural signal: modernization is a collaborative effort built with them, not imposed on them.

5. Reinforcement Mechanisms: Preventing Relapse

The biggest challenge in modernization is the tendency to revert to old habits gradually. Achieving lasting change involves continuous reinforcement, such as coaching supervisors, recognizing outstanding operational discipline, and using metrics to measure concrete results—like quicker diagnoses or fewer nuisance alarms. This reinforcement should be presented as empowering professionals rather than as a form of monitoring.

As institutions move toward AI-enabled sustainable smart campuses, the clear priority is that human readiness must keep pace with modernization. Successful antifragility happens only when both technical and human systems develop in tandem, creating an ecosystem that benefits from stress rather than being hindered by it.

This is precisely why the next phase of transformation must turn from architecture to adoption. Having established the human foundations of convergence—enablement, risk capability, psychological safety, and adaptive culture—the institution must now translate these principles into a structured, repeatable change-management discipline. for sustained behavior change.

Fostering Organizational Change with ADKAR

For the executive leadership team—the Chief Technology Officer, the Chief Facilities Officer, and the Cabinet—the mandate of building an AI-Enabled Sustainable Smart Campus is ultimately a test of human orchestration. Procuring the technology is a capital exercise; getting a deeply entrenched, multi-generational workforce to adopt it is a cultural crucible.

When the physical domain of wrenches and chillers collides with the digital domain of algorithms and network segments, resistance is guaranteed. This resistance is not malicious; it is a rational defense of institutional stability. To navigate this, the Integrative Architect must move beyond traditional "train-the-trainer" mandates and adopt a structured, psychological approach to human change.

Unlike change models that focus solely on organizational milestones (such as Kotter's 8-Step Process, which is excellent for creating executive urgency), ADKAR focuses on the *individual* building blocks of change. So, while leaders should understand complementary models—such as Rogers' Diffusion of Innovations — to identify early adopters and scale momentum, ADKAR provides the most practical, human-centric playbook for the CTO and

CTO to govern the convergence journey. Unlike top-down directives that often fail in the decentralized culture of higher education, ADKAR treats transformation as a sequential, individual journey scaled to the enterprise. It recognizes that a university only transforms when one mechanic, one network engineer, and one academic dean successfully transition.

The Foundational Premise

Before applying the ADKAR framework, the CTO/CFaO must agree on a foundational design principle: **Modernization must improve the operator's day.**

OT modernization lives or dies in the daily routines of building operators. If a new AI-enabled platform creates more nuisance alarms, inconsistent interfaces, or fragile integrations that fail during a winter storm, operators will do exactly what their duty requires: override the system, create manual workarounds, and abandon the technology. This is not stubborn resistance; it is professional judgment. Therefore, Operator User Experience (UX) is not aesthetic polish—it is a critical reliability infrastructure. The CTO/CFaO must ensure the system delivers:

- **Actionable Alerting, Not Noise:** Alarm fatigue destroys trust. Alerts must be curated, prioritized by consequence (life-safety vs. efficiency), and presented with context and recommended next steps.
- **Consistent Naming and Tagging:** Operators should not have to translate ten different vendor dialects mentally. The CFaO must standardize on a semantic ontology (e.g., Brick Schema) so that a "chilled water valve" means the same thing across every building dashboard.
- **Workflow-Aligned Tools:** Analytics must map to how work actually happens—rounds, service calls, and shift handoffs. If tools do not align with physical workflows, they become shelfware.
- **Visible Cause-and-Effect for Automation:** Operators must be able to see *why* an algorithm made a change and have a safe, explicit method to override it if physical safety or research continuity demands it.

ADKAR

ADKAR is a change enablement framework that organizes adoption around five conditions that must be true for people and organizations to change successfully: **Awareness** (of why), **Desire** (to participate), **Knowledge** (how to change), **Ability** (to implement in practice), and **Reinforcement** (to sustain). In IT/OT convergence, ADKAR is useful because it treats adoption as a managed sequence—not a single "go-live"—and it makes the human work of modernization as explicit as the technical work. To operationalize this cultural shift, the CTO/CFaO must jointly execute an ADKAR-aligned strategy.

A — Awareness: Establish a Shared Definition and "Why"

The leadership team must first clearly define what "IT/OT convergence" means for their campus. Without a common understanding, vendors and separate departments will fill the gap with their own assumptions. The CTO/CFaO should collaboratively craft a clear, mission-focused story that connects both areas: it should assure safety and reliability for Facilities and uphold the identity and monitoring needs of IT. Awareness is reached when stakeholders across the campus can explain the change using the same words.

D — Desire: Create a Joint Operating Rhythm

Desire develops from mutual trust rather than executive persuasion. If IT and Facilities only collaborate during crises, the desire will never grow. The goal is to create a consistent, cooperative operational rhythm. The CTO/CFaO needs to require shared roadmap reviews, aligned maintenance schedules, and joint incident reviews.

This demonstrates mutual respect: Facilities has final say over the physical space, while IT manages the connected systems at an enterprise level.

K — Knowledge: Train for Responsibilities, Not Features

OT modernization fundamentally transforms campus operations by requiring hybrid skills that combine hands-on physical craftsmanship with digital fluency. The Knowledge phase moves away from generic vendor feature tours, emphasizing instead tailored, scenario-based training. The CTO must ensure that training is role-specific and focuses on real-world situations. Operators should develop quick pattern recognition for alarms; engineers need Root Cause Analysis expertise; and IT security teams must understand how server patching could affect fire-pressurization processes. Focus on practicing the "moments that matter"—simulating cyber-physical disaster scenarios—so cross-functional teams develop muscle memory before an actual crisis occurs.

A — Ability: Show, Don't Tell (The Sandbox)

Ability refers to demonstrated performance under pressure. Operators require psychological safety to experiment with new systems without fearing campus disruptions. The goal is to facilitate hands-on engagement using demonstration buildings and "living lab" pilots. Offer operators digital sandboxes to activate alarms, test overrides, and safely verify system responses. Establish a feedback loop that involves operators as co-designers, allowing their input to influence dashboard designs and alarm settings directly.

R — Reinforcement: Institutionalize the Capability

The primary risk is not the initial adoption but the potential regression into manual overrides and isolated "Shadow OT" procurement practices. The CTO and CTO should integrate convergence into the institution's core functions. This involves rewriting procurement standards to require interoperability, revising onboarding checklists for new employees, and developing shared IT and Facilities scorecards that track digital security compliance and physical energy efficiency.

Linking Change Management to the Implementation

The ADKAR strategy cannot exist in a vacuum; it must be tightly coupled to the implementation playbook (as detailed in Chapter 16). Change management must pace the technical rollout to ensure the institution is never pushed past its capacity to absorb disruption.

Phase 1 (Discovery and Governance)

Focus only on **Awareness and Desire**. The CTO and CTO align the Cabinet, build the cross-functional Steering Committee, and map the legacy environment to prove they understand the baseline before changing it.

Phase 2 and 3 (Digital Blueprints and Pilots)

Focus on **Knowledge and Ability.** As the plan is made, people are aware and learn the skills necessary to engage. The initial "Lighthouse" pilots. Hands-on scenario training begins. The pilot proves that modernization reduces chaos and improves the operator's day.

Phases 4 and 5 (Scaling and Continuous Optimization)

Focus on **Reinforcement**. As the AI control plane and Digital Twins scale campus-wide, the new operational behaviors are locked in through shared KPIs, strict procurement gates, and continuous iterative listening and optimization loops.

A sustainable smart campus is characterized not by cutting-edge technology but by strong governance and a capable workforce. When the CTO/CFaO collaborate to guide this human-centered transformation, convergence shifts from a merely disruptive IT initiative to a new, resilient model of university operations.

If you cannot confidently check the boxes in the *Awareness*, *Desire*, and *Knowledge* sections, do not authorize the deployment. Scaling technology ahead of human readiness does not create a smart campus; it creates expensive, fragile shelfware.

Sources for ADKAR Training

EDUCAUSE has teamed up with Prosci to boost change management capabilities at colleges and universities. Members receive discounted access to PROSCI's comprehensive role-based training and certification programs. The page highlights that higher education institutions face increasing pressures—such as information security, privacy, digital learning, and technology upgrades—yet only a small percentage are implementing structured change management to tackle these challenges. Prosci, with over 100,000 certified practitioners globally, offers a proven methodology and popular tools such as the ADKAR model to help institutions manage change more effectively.

The partnership offers EDUCAUSE members virtual programs and access to a wide range of offerings, such as the Practitioner Certification Program, leadership-focused courses, ADKAR mastery levels, and enterprise change management bootcamps. Additionally, EDUCAUSE provides complementary Learning Labs led by Prosci-certified instructors, forming a unified professional development pathway for individuals and teams.[15]

ADKAR Checklist

See ADKAR Matrix for IT/OT Convergence in the Appendix

Transitioning from People to Process: The Mechanics of Execution

If Change Management represents the psychology of transformation, then Project Management embodies its physics. To create a cyber-physical ecosystem, the Integrative Architect must effectively combine human enablement with precise, disciplined execution.

We have confirmed that the success of the AI-Enabled Sustainable Smart Campus depends on a thorough change management approach. It's essential to build awareness, desire, and knowledge among the human workforce before implementing any algorithms. Nevertheless, cultural readiness is only part of the challenge. Enthusiasm alone doesn't operate fiber optic cables, set up Zero-Trust API gateways, or arrange complex chiller valves. You may foster an eager workforce for modernization, but if sensors arrive six months late, budgets run out, or software fails to run, the entire transformation could fail under its own weight.

Change Management vs. Project Management

In many organizations, these two disciplines are often dangerously mixed up. A leader might give a Project Manager responsibility for a large smart campus deployment and believe that simply updating the Gantt chart will ensure the workforce adopts the new tool. This reflects a deep misunderstanding of both fields.

Change Management (The Human Capability) concentrates on guiding people from their existing state to a future one. It addresses resistance, fosters psychological safety, and ensures that, after technology implementation, it is used effectively. The results are Adoption and Proficiency.

Project Management (Technical Delivery) focuses on the practical aspects of delivering the solution. It oversees the scope, schedule, budget, quality, and risk associated with the technical and physical implementation. The goal: to deliver the agreed system successfully, on time, and within budget.

To succeed, the CTO/CFaO must view these as interlocking gears. Project Management builds the digital twin; Change Management ensures the mechanics actually log into it.

Disciplined Project Management is Non-Negotiable

A smart campus initiative differs from typical IT software deployments or facilities construction projects; it is a complex blend of both. Effective project management is essential for three key reasons.

1. **The "Hybrid Waterfall-Agile" Reality**
   Physical infrastructure—such as pulling copper wire, installing VAV boxes, and pouring concrete—is naturally linear and inflexible, demanding traditional, rigid "Waterfall" planning. In contrast, digital infrastructure—like tuning AI algorithms, designing mobile app UIs, and configuring cloud dashboards—is inherently iterative, necessitating an "Agile" approach. The Project Manager plays a crucial role as the synchronization engine, ensuring that rapid software sprints seamlessly align with the slower, physical construction schedules.
2. **Cyber-Physical Risk Containment**
   In a smart campus, a missed dependency isn't merely a 404 error on a webpage; it signifies a compromised physical security boundary or a frozen pipe in a high-value research lab. Strict project management upholds the Readiness Gates outlined earlier, guaranteeing that no system is activated until cybersecurity microsegmentation is completely in place and verified.
3. **Vendor Ecosystem Orchestration**
   A smart campus depends on collaboration among multiple vendors, including legacy HVAC suppliers, digital twin startups, local electrical unions, and enterprise IT providers. Without a dedicated and disciplined project management team, these vendors tend to work independently, leading to disconnected systems and costly change orders that can derail the project.

Rigorous project management must be engaged at the point of **Capital Interception**—the moment the Smart Campus Steering Committee approves a proposal and funding is allocated. Once activated, project management guides the entire lifecycle:

- **Initiation**
  Defining the project charter, aligning the OKRs and KPIs, and formally assigning the cross-functional team.
- **Planning**
  Drafting the precise technical architecture, baselining the budget, and sequencing the IT/OT dependencies.
- **Execution**
  Managing the daily friction of vendor deployment, hardware installation, and API integration.
- **Monitoring and Controlling**
  Tracking the "burn rate" of the budget, managing scope creep (the "Shadow IT" requests that inevitably arise), and verifying quality.
- **Closing and Transition**
  Handing the completed, documented system over to the daily operational teams alongside the ITIL-based service desk protocols.

The Smart Campus Project Manager: Certifications and Skills

The individual tasked with managing IT/OT convergence cannot be a traditional IT software manager who has never worn a hard hat, nor can they be a traditional construction manager who does not understand cloud APIs. They must be a bilingual translator.

When staffing this role, executive leadership should look for the following certifications and competencies:

Core Certifications

- **PMP (Project Management Professional)**
  The gold standard from the Project Management Institute (PMI). This proves the manager understands how to handle massive capital budgets, complex risk registers, and strict, linear physical dependencies (Waterfall).
- **Agile/Scrum Certifications (CSM, PMI-ACP)**
  Essential for managing the software integration, digital twin development, and UX feedback loops that require rapid, iterative sprints.
- **ITIL 4 Foundation (Information Technology Infrastructure Library)**
  While technically a service management framework, an ITIL-certified project manager understands how to design a project so that it smoothly transitions into daily operations, ensuring help-desk routing and incident management protocols are built into the smart campus tool from day one.

The Indispensable "Soft" Skills

- **Bilingual Translation:** The ability to translate the strict, life-safety requirements of the central plant mechanic into the rapid, sprint-based language of the cloud software developer.
- **Matrix Leadership:** The authority to lead cross-functional teams (IT network engineers, facilities electricians, academic researchers) who do not directly report to them.
- **Vendor Negotiation:** The operational grit to hold disparate, sometimes competing, third-party vendors accountable to a unified integration standard.

You simply cannot innovate your way out of poor execution. By pairing the cultural empathy of ADKAR with the operational ruthlessness of disciplined Project Management, the institution ensures that its strategic vision survives the journey from vision to physical reality.

Chapter Summary

Chapter 16 describes change management as the essential human infrastructure needed to ensure the success of IT/OT convergence in an AI-enabled, sustainable smart campus. It emphasizes that modernization often fails not because of a lack of technology but because institutions overlook the cultural, behavioral, and psychological changes necessary to operate within a cyber-physical environment.

The chapter fundamentally highlights that convergence is primarily a human transformation rather than a purely technical process. It explains how deeply rooted differences between IT and Facilities cultures—such as divergent risk perceptions, varying time horizons, and distinct work identities—lead to predictable resistance. This resistance can only be overcome through deliberate enablement rather than mandates. Change management is presented as the essential discipline that bridges this divide by fostering shared understanding, competence, and trust.

A key framework introduced in this chapter is the ADKAR change management methodology. ADKAR provides the structure to guide change within individuals and teams, helping the institution manage the pressures of change and achieve lasting adoption.

The manifesto underscores the Human Architecture of Convergence, emphasizing psychological safety, trust, and cross-domain learning as essential to adoption. It points out risks such as workarounds, operator disengagement, "shelfware," and heightened operational fragility stemming from poor change management. To address these, it recommends strategies such as transparency, user-focused UX, co-design, and structured engagement.

Finally, Chapter 16 links change management with governance through stakeholder councils, communication rhythms, role clarity, risk registers, and feedback loops. It presents change management not as an isolated task but as a core element of institutional modernization—supporting operational excellence, sustainability, and a better campus experience. In this context, change management is the key to a campus becoming adaptable, resilient, and ultimately antifragile.

Chapter 17
Smart Campus Risks

Cyber-and physical risk management requires unified security, segmentation, operational safeguards, and antifragility planning.

Risk in the AI-enabled Sustainable Smart Campus is now integrated into every stage of the design process, rather than being solely the concern of compliance, audit, or cybersecurity teams after the initial design. It influences architectural choices, procurement agreements, data collection—both what we gather and what we choose to avoid—the models we deploy, and the operational practices across IT, facilities, and public safety.

The smart campus expands the institution's risk profile in two key ways. Firstly, it adds digital complexity through increased endpoints, integrations, data flows, third-party involvement, and automation. Secondly, it heightens physical risks because these digital systems increasingly influence real-world functions such as doors, air, water, power, alarms, elevators, labs, and learning or living environments. Merging IT and OT does not just add more technology; it introduces a new cyber-physical risk that can disrupt operations, endanger safety, and erode trust.

Modern risk management focuses on building institutional resilience rather than eliminating uncertainty. This resilience includes the ability to predict, endure, adapt to, and recover from risks, all while preserving the institution's core mission, values, and community ties.

Identifying Risks and Implementing Risk Management.

A sustainable smart campus evolves from reactive controls to proactive governance. This begins with a change in perspective: consider risk as a core part of strategic design rather than merely a defensive checklist. Consequently, risk management must be an ongoing process throughout the campus lifecycle.

- **Before** systems are procured: risk is shaped by requirements, standards, and contracts.
- **During** implementation: risk is shaped by integration quality, segmentation, identity, logging, and operational readiness.
- **After** go-live: risk is shaped by monitoring, model drift, patching, training, incident response, and continuous improvement.

Within a smart campus, a practical strategy for grasping overall risk—rather than viewing it in isolated pockets—is to analyze it across the same five domains used in planning: social, economic, environmental, governance, and technology/data. This method helps prevent over-securing systems, which could hinder access and user experience, while also avoiding sustainability initiatives that might unintentionally cause safety, privacy, or reliability concerns.

Identifying Social Risk

As campuses implement sensors, analytics, and AI-driven decision-making, social risks often arise subtly—through concerns about surveillance, unequal effects, or opaque automation that diminishes human control. An occupancy program designed to boost energy efficiency may face community opposition if it lacks transparency, defined limits, and a clear purpose. Similarly, a safety analytics system can undermine trust if it appears to target specific groups or areas.

Effective social risk management on a smart campus begins with human-centered governance. This means openly communicating about what data is gathered, why it is collected, and how it's protected, and allowing the community to question the results. Additionally, systems should be designed with dignity, minimizing the collection of personally identifiable information unless needed, opting for data aggregation and de-identification whenever possible, and establishing clear boundaries to prevent scope creep.

This is also where the risk posed by AI shifts into human risk. When models influence advising, access, or resource allocation, the institution assumes responsibilities for explainability, contestability, and fairness. The most "advanced" campus isn't necessarily the one with the highest level of automation; rather, it's the one that can ethically, legally, and socially defend its decisions.

Identifying Economic Risk

Smart campus initiatives require long-term financial commitments. The risks go beyond project delays and include lifecycle challenges like proprietary lock-in, integration debt, hidden ongoing expenses, and dependence on vendors whose prices and strategies could change.

Operational risk emerges when IT's rapid iteration unintentionally destabilizes deterministic OT environments. The CTO/CFaO must treat operational continuity as a hard design constraint. Network scans, patching, and data pipelines must be engineered to align with maintenance windows and operational dependencies to prevent "random" equipment cycling or lab downtime.

Without proper oversight, institutions may implement isolated pilot projects that evolve into full-scale operations, leading to a fragmented landscape of vendor-specific platforms, duplicate contracts, incompatible data models, shadow IT/OT, and accumulating technical debt. An unrecognized pilot that cannot be scaled properly is a warning sign, not a success. The CTO/CFaO needs to ensure that pilots are officially acknowledged through governance, follow formal procurement procedures, and adopt standardized integration patterns and data ontologies from the start.

Economic risk is best managed through procurement as a governance tool. Contracts should reflect campus realities, such as multi-year planning, interoperability, data portability, security commitments, audit rights, and clearly defined "value." Performance claims, especially those about "avoided costs" or energy savings, should be treated as decision-support estimates rather than guarantees, because variables such as weather, energy market prices, occupant behavior, and deferred maintenance can significantly affect actual savings.

An effective campus that consistently manages economic risk focuses on three main actions: precisely pricing the full lifecycle, protecting the institution's exit strategies, and treating integration and operations as fundamental costs rather than secondary ones.

Identifying Environmental Risk

water, and waste. However, from a risk management perspective, an important additional dimension arises. As climate variability and infrastructure vulnerability increase, campuses need to plan not just for decarbonization but also to remain operational during disruptions.

The smart campus functions as a resilience platform through sensor networks, integrated operations, and digital twins, enhancing situational awareness and accelerating responses. However, this reliance also creates risks: network failures, identity system problems, or inadequate OT segmentation can result in a loss of essential visibility and control when it is most needed.

The key lesson isn't about reducing the speed of innovation but about embedding resilience into it. Wildfire detection exemplifies this: AI-powered camera systems can enhance early fire detection and speed up responses, but they can't influence environmental factors that worsen fires. Organizations must prepare for the gap between detection and containment. Recent reports on UC San Diego-led ALERTCalifornia demonstrate how AI cameras can rapidly identify fires.[16] However, they also emphasize that early detection is just part of a comprehensive resilience strategy.

Identifying Governance Risk

Smart campus risk is often ignored when roles aren't well-defined. Combining IT, OT, and campus operations creates shared responsibilities, but without clear governance, accountability may be overlooked. As telemetry increases, so does the concern about a 'surveillance state.' To reduce trust risks, strict governance is essential: openly sharing what data is collected, why it is collected, who can access it, and its retention period. Analytics should focus firmly on human-centered goals—safety, comfort, and sustainability—without infringing on unnecessary behavioral monitoring.

Therefore, effective governance risk management requires more than merely having a committee; it necessitates a clearly defined decision-making process model.

- Who sets standards for IT/OT integration and segmentation?
- Who approves sensor deployments and defines acceptable use boundaries?
- Who is accountable for model performance, drift, bias testing, and appeal pathways?
- Who owns incident response when an event spans IT security, facilities operations, and campus safety?
- Who adjudicates tradeoffs when sustainability optimization collides with comfort, safety, or research requirements?

This is where the institution benefits from anchoring governance within established frameworks, not because these frameworks solve problems, but because they create a common language across various fields. For example, the NIST Cybersecurity Framework 2.0 offers an updated structure for managing governance, risk, and cybersecurity operations across an organization. Likewise, the NIST AI Risk Management Framework helps organizations to formulate their processes for identifying, assessing, managing, and communicating AI-related risks. (ISA Global Cybersecurity Alliance)

Regulatory and Legal Risk: Compliance as a Component of Trust

Smart campus compliance now covers a wide array of obligations beyond traditional ones, including privacy laws, security standards, accessibility requirements, environmental disclosures, and emerging AI accountability norms.

The main approach is to view compliance as a core element of the design process rather than as a final checkpoint. Incorporating privacy, retention, and access controls into the architecture and data governance early on ensures more resilient, less prone-to-issues compliance. On the other hand, retrofitting compliance after the fact can lead to fragile systems, inefficient exception handling, and a persistent feeling of "risk debt."

Vendor and Partner Risks

Vendor and Partner risks run a wide gamut of issues, including vendor lock-in, remote access exposure, proprietary systems, unmanaged tunnels, and supply-chain vulnerabilities. These risks emerge not just from technology but from business models, incentives, lifecycle obligations, and misaligned operational practices across a diverse ecosystem of contractors, platform providers, and integrators. For example, a $50 sensor with a hidden vulnerability can lead to a $5M ransomware event. **Supply Chain Security** is not a procurement hurdle; it is financial insurance. By controlling the pedigree of our technology, we reduce the institution's "Technical Debt" and "Risk Debt".

Divergent Incentive Structures and Misaligned Success Criteria

Most vendor relationships in OT emerged from traditional facilities contracting models or were precipitated from decisions made by general contractors during new building construction. These contracts often focus on uptime and the convenience of service access, leading to long-term embedded support contracts. As OT becomes part of the cyber-physical enterprise, these incentives may conflict with institutional standards and technical requirements for Zero Trust, identity governance, segmentation, and data portability. For example, vendors may prefer persistent access, closed architectures, or proprietary APIs that make troubleshooting easier for their teams but increase the institution's cyber-physical exposure. The institution must recognize that **vendor goals are not automatically aligned with enterprise security, lifecycle transparency, or governance mandates**.

Variable Maturity Across Vendor Ecosystems

Smart campus modernization brings together mechanical contractors, controls integrators, SaaS providers, IoT startups, ESCOs, and cloud platforms—each with vastly different levels of cybersecurity maturity, documentation quality, and IT/OT governance. A campus may have one highly mature BAS integrator but five smaller subcontractors plugging in IoT devices with no secure onboarding, no patch cadence, and no semantic consistency. These maturity gaps create **weak links** that undermine network segmentation, telemetry reliability, and data estate accuracy.

Documentation and Knowledge-Transfer Risk

Many legacy vendors hold decades of "tribal knowledge" about system configurations, control logic, and equipment quirks. However, these records often exist in private vendor repositories rather than campus-governed systems. When staff turnover or contract transitions occur, institutions may lose access to system logic, API keys, commissioning documents, and controller maps—thereby increasing operational fragility and dependence on specific partners. The manifesto already emphasizes documentation as an operational backbone; **vendor control of that documentation creates strategic vulnerability**.

Short-Term Project Logic vs. Long-Term Institutional Lifecycle

Vendors typically engage on a project basis, with success measured by installation completion rather than long-term operability, data quality, or maintainability.
This conflicts with the campus requirement for **30–50-year mechanical lifecycles and multi-decade digital governance**. A vendor may deliver a functioning system today that becomes a technical debt burden tomorrow—because data models were inconsistent, commissioning was incomplete, or integrations were brittle. The result is **accelerated semantic decay**, incomplete digital twins, or non-portable data.

Multi-Vendor Interdependency and Cascading Failures

Modern OT ecosystems are interdependent—lighting depends on occupancy sensors, HVAC depends on BAS points, digital twins depend on structured telemetry, and access control may tie into campus identity systems. When one vendor updates firmware or cloud APIs without coordinating with others, integration failures cascade across systems. This risk is amplified when proprietary gateways or cloud connectors act as "single points of vendor failure." Institutions require governance that compels vendors to follow change management protocols and interoperability standards to avoid chain-reaction **outages**.

End-of-Life (EOL) and Business-Continuity

Vendors merge, go bankrupt, discontinue products, or alter licensing models. In a campus with 30–50-year-old physical assets, a five-year-old cloud vendor or IoT startup is a continuity risk. EOL decisions can instantly orphan thousands of deployed sensors or controllers, creating security gaps, integration failures, loss of patch support, and downtime in critical spaces. A sustainable smart campus architecture must plan for vendor exit scenarios and ensure that devices continue to function safely under institutional control.

Remote Support and "Invisible" Supply-Chain Dependencies

In a Sentient Campus, our security is only as strong as the most obscure component in our network. A smart campus is a massive assembly of global parts—sensors from one continent, controllers from another, and code from a third. party vendor. When a primary vendor is trustworthy, their subcontractors, offshore development teams, or cloud providers may not adhere to equivalent controls. Institutions often underestimate who actually has remote diagnostic access, where cloud data is processed, how many third parties have API keys, and which sub-vendors handle firmware, drivers, or analytics. This creates a multilayer supply-chain exposure that traditional contract language does not fully address. As such, we manage the "pedigree" of our technology long before it is bolted to a wall or plugged into a switch.

The SBOM Mandate: Knowing the "Ingredients"

Modern OT devices are rarely built from scratch; they are "assembled" using third-party software libraries and open-source code. NIST emphasizes the need for a **Software Bill of Materials (SBOM)**—an inventory of all components in a device's software. We require all vendors to provide an SBOM for any device entering the campus ecosystem. By knowing the "ingredients" of our controllers, we can determine whether a newly discovered global software vulnerability (like Log4j) affects our campus HVAC or lighting systems, enabling surgical patching rather than blind guessing.

Anti-Counterfeit and Hardware Authenticity

The physical provenance of hardware is just as critical as the code. Counterfeit or "grey market" hardware often lacks rigorous quality control and can contain "hardware backdoors" that bypass traditional firewalls. We align with NIST SR-11 by mandating that critical infrastructure components—PLCs, routers, and power controllers—are purchased only through authorized, audited channels. We prioritize vendors who provide hardware "roots of trust"—cryptographic chips that verify the device is genuine and that the firmware has not been tampered with during shipping.

The Managed Lifecycle: From Cradle to Grave

Security readiness includes the "Exit Strategy." A device that is no longer supported by its manufacturer is a permanent open window into the campus. We maintain a central registry of "End-of-Life" (EOL) dates for all OT hardware. We do not wait for a device to fail; we plan its retirement based on the expiration of its security support. Following NIST SR-12, any device being removed from the campus must undergo "Digital Sanitization." We ensure that Wi-Fi keys, IP addresses, and building maps are physically and logically wiped before the hardware is recycled or discarded.

Vendor Governance: Security as a Partnership

IT/OT leaders are shifting from "transactional" relationships with vendors to more "collaborative" partnerships. They now demand that vendors provide a minimum number of security updates within a set timeframe.

Mitigating Vendor and Partner Risk

Managing vendor ecosystems in IT/OT convergence presents a significant vulnerability. The CTO is experienced with enterprise software vendors that have strong cybersecurity measures. Meanwhile, the CFaO oversees a fragmented network of mechanical contractors, electrical integrators, and Energy Service Companies (ESCOs), many of whom vary in digital maturity and require seamless remote access to troubleshoot installed equipment.

To mitigate vendor and partner risk, the CFaO should implement IT standards, including Identity and Access Management (IAM), such as Multi-Factor Authentication (MFA), Role-Based Access Control (RBAC), and robust governance of service accounts for all external contractors. When remote access is necessary for maintenance, vendors must use the institution's Secure Gateway, employ multi-factor authentication, and ensure that all activities are continuously logged. Additionally, establishing backdoors for outside contractors is strictly forbidden.

Meanwhile, the CTO needs to grasp the operational specifics of these mechanical vendors to develop secure privileged access methods that do not hinder emergency maintenance response times. Both executives must work together on vendor risk management, rewriting contracts to clearly specify data ownership, define patching duties, and set up controls for third-party access.

Risk Transfer and the Insurance Advantage

The cyber-insurance market has shifted from a "soft" market to one defined by extreme scrutiny of OT assets. Carriers no longer ask if you have a firewall; they ask if you have a **NIST-aligned OT Security Program.** By documenting our adherence to the NIST RMF and demonstrating automated anomaly detection, we position the institution as a "Low Risk" entity. This leads to lower premiums and higher coverage limits for cyber-liability policies. Having a unified "Meaning Layer" that documents the security of every sensor makes the annual auditing process a data-driven exercise rather than a manual, labor-intensive scavenger hunt.

Cyber Risks

Technological risks naturally emerge as campuses integrate legacy building systems, IoT devices, vendor cloud platforms, and enterprise IT into a unified operational ecosystem. These risks encompass unmanaged or "Shadow OT" devices, inconsistent vendor practices, flat or inadequately segmented networks, weak identity and access controls, and unreliable or poorly governed data streams. Collectively, they broaden the cyber-physical attack surface and heighten the chance that digital vulnerabilities could lead to real-world operational failures—impacting safety, uptime, sustainability, and institutional trust.

These technological risks directly influence how the campus must defend, monitor, and govern its connected infrastructure. Consequently, this creates a critical field: Cybersecurity for OT. The following section covers the essential security architecture, controls, and governance frameworks needed to protect the integrated IT/OT environment and ensure resilient, reliable, and secure campus operations.

Cybersecurity for OT

We must recognize a serious reality: an attack on our Operational Technology (OT) poses more than just a threat to our data; it directly threatens our students' safety. For instance, a compromised HVAC system in cold winter conditions or a hacked laboratory ventilation system creates an immediate risk to human safety. The leadership's directive starts with the understanding that OT security is fundamentally about safeguarding life safety.

For many years, the Chief Technology Officer (CTO) and the Facilities Officer (CFaO) worked in separate areas. The CTO focused on data security issues, including data breaches, while the CFaO handled physical problems, such as pipe bursts. In the Sentient Campus, these concerns are integrated into a single, comprehensive Cyber-Physical Risk Profile. Traditional IT security emphasizes the CIA Triad—Confidentiality, Integrity, and Availability—when developing cybersecurity measures. However, in environments with steam plants, chillers, and power grids, these priorities need to be reversed. The Smart Campus operates according to the SRA Triad:

- **Safety:** Above all, the system must not harm the occupants. If a cyber-conflict occurs, the default "fail-state" must be a safe physical state.
- **Reliability:** The campus must function. A "secure" building that is uninhabitable because the security software crashed, and the boiler has failed.
- **Availability:** Data must be present and actionable.

This cultural shift requires the CTO to understand the terminology of mechanical engineering and the Facilities Officer to learn the details of network packet inspection. This merging of skills is more than just technical; it's a leadership challenge. We are no longer just managing "computers" and "buildings" — we are now overseeing a unified cyber-physical system.

The traditional cybersecurity approach relies heavily on compliance, fostering a "checkbox" mindset focused on satisfying auditors and meeting minimal regulatory standards. In the past, when systems were isolated, compliance alone was enough to ensure security. However, in today's interconnected environment, this method exposes vulnerabilities. As IT and OT converge, the focus shifts from simply protecting data to defending lives, research, and the institution's physical safety.

Standards for OT Cybersecurity

As the digital and physical realms merge, the smart campus requires more than just overarching policies; it calls for a thorough, multi-dimensional framework. Therefore, the manifesto presents the "Security Trinity"—a combination of three globally acknowledged standards that collectively transform our cybersecurity perspective from reactive expenses to a strategic, resilient asset.

For the CTO and CFAO, understanding how these three engines connect is essential to creating a robust, technically secure, and effectively managed campus.

Exhibit 15. The Security Trinity: How the Standards Integrate

Standard	Role	Focus
The Purdue Model	The Structural Blueprint	Segmentation: Defines "where" assets live and how to isolate them into functional zones.
NIST SP 800-82r3	The OT Implementation Guide	Specifics: Defines the "what"—the specific technical controls and "overlays" needed to secure physical hardware.
NIST RMF (800-37)	The Governance Lifecycle	Process: Defines "how" to manage risk through a continuous seven-step lifecycle of accountability.

The Purdue Model for Industrial Control Systems: The Structural Blueprint

The Purdue Model for Industrial Control Systems (ICS) guides our architecture. On a smart campus, it helps organize network segmentation across Levels 0–5, originating from ISA-99 standards. This model classifies equipment by function, from physical sensors (Level 0) and controllers (Level 1) up to the corporate network (Level 5). It also defines a hierarchy of Zones and Conduits that logically organizes the campus by criticality. ICS represents the guiding architectural vision.

Exhibit 16. ICS Layers (Purdue Model)

Level	Name	Description
Level 5	Enterprise Network	Business systems such as ERP, finance, HR, institutional apps, and cloud services support enterprise planning and analytics.
Level 4	Site Business Planning and Logistics	Campus-level operations: scheduling, reporting, energy dashboards, sustainability platforms; interfaces between enterprise systems and OT.
Level 3	Operations and Control (OT Management Layer)	Supervisory systems, including BAS head ends, SCADA servers, OT data historians, FDD/analytics engines, and digital twin integrations. Separates enterprise IT from lower-level controls.
Level 2	Area Supervisory Control	Localized automation servers, gateway controllers, and operator workstations that oversee building zones, plants, or subsystem groups.
Level 1	Basic Control Layer	PLCs, DDC controllers, field controllers, and other devices execute logic, sequences of operation, safety interlocks, and real-time control.
Level 0	Physical Process Layer	Sensors, actuators, meters, valves, VFDs, pumps, dampers, and all physical equipment that directly interact with the built environment.

In a smart campus environment, depending on the "One Big Network" myth is inadequate. Instead, using a Zones and Conduits strategy prevents a breach in a student dorm Wi-Fi router from spreading to the campus electrical substation.

Level 0/1: Protecting the Invisible Foundation

The main vulnerability in the Smart Campus lies at Levels 0 (Field Devices) and 1 (Direct Control). These include valves, actuators, and sensors that physically operate the campus. As noted in NIST SP 800-82r3, many legacy devices lack the processing capacity to support current encryption or authentication methods.

- **The Mitigation Strategy**
  Where we cannot secure the device, we must secure the environment. We implement physical tamper-evidence, port security, and "shrouding" techniques to prevent unauthorized physical or logical access to these "blind" endpoints.
- **The Electronic Security Perimeter (ESP)**
  We define a "circle of trust" around critical OT assets. Access to the ESP requires multi-factor authentication (MFA) and is logged with the same rigor as access to a financial database.

Level 2/3: The Intelligence Hub and Local Operations

This is the level where Building Automation Systems (BAS) and Supervisory Control and Data Acquisition (SCADA) systems are housed. In the Smart Campus, Level 3 serves as the link between local mechanical logic and the overall campus strategy.

- **Converged Monitoring**
  We are transitioning to a Unified Operations Center (UOC). The UOC eliminates the traditional separation between the NOC (Network Operations) and the FOC (Facilities Operations), offering a "single pane of glass" view that displays both network health and boiler status pressure.
- **Local Resilience**
  Although the Digital Twin resides in the cloud, Level 3 architecture must enable autonomous operation at the local level. If the campus loses its connection to the outside world, the building controllers need sufficient "embedded intelligence" to ensure life safety and environmental stability, independent of external instructions.

Level 4/5: The Cloud and the Data Estate

The top of the pyramid is where the **Digital Twin** and **AI Analytics** reside. This is the "brain" that optimizes energy based on the Meaning Layer.

- **The Secure Gateway**
  Whenever feasible, all data transferred from the campus (OT) to the cloud (IT) should go through a secure, unidirectional gateway, known as a "data diode." This setup ensures that a breach of the cloud-based AI cannot subsequently influence or manipulate physical system hardware.
- **Encryption at Rest and in Transit**
  Although the sensors are basic, the data they generate is a valuable strategic asset. We use end-to-end encryption for all data transmitted from the gateway to the Sentient Data Lake, safeguarding the privacy of our "Meaning Layer" insights.

By segmenting the campus into levels—from the physical sensors in a boiler room (Level 0) to the AI-driven Digital Twin in the cloud (Level 5)—we ensure that a breach in a student dormitory Wi-Fi cannot propagate physically to the campus electrical substation. **This approach results in** structural resilience. If one zone is compromised, the "Conduits" allow us to isolate the threat instantly, ensuring the rest of the campus remains operational.

IST SP 800-82r3: The Operational Technology (OT) Hardening Standard

The NIST SP 800-82r3 serves as the detailed technical guide, akin to a law enforcement manual for physical security. It acknowledges that in a Smart Campus, security extends beyond server data to include high-voltage switchgear, laboratory chemical sensors, and life-safety systems in crowded stadiums. For the CIO/CFaO, 800-82r3 acts as a bridge, translating IT security principles into the terminology of mechanical engineering and campus management.

The SRA Priority: Safety, Reliability, and Availability

In traditional IT, security is built on the **CIA Triad** (Confidentiality, Integrity, Availability). NIST 800-82r3 shifts the paradigm for the Sentient Campus, emphasizing that in OT, the priorities must be **SRA**:

- **Safety:** The system must never fail in a way that endangers students, faculty, or staff.
- **Reliability:** Systems must operate predictably and consistently under all conditions.
- **Availability:** Critical building functions (power, water, air) must be present when needed.

- **Institutional Importance:** This shift ensures that our security measures don't accidentally "brick" a building or shut down a research experiment to protect a non-critical password.

The OT Security Overlay: Tailored Controls

NIST 800-82r3 provides a specialized "overlay" for the standard NIST 800-53 security controls. It acknowledges that "standard" IT solutions—like forced weekly reboots or complex encryption—can be catastrophic for OT devices.

- **Compensating Controls**
  For a 20-year-old boiler controller that cannot support modern antivirus software, 800-82r3 recommends implementing "Compensating Controls," such as network isolation or physical safeguards, such as shrouding.
- **Institutional Importance**
  It enables us to maintain a diverse, multi-generational campus without the need to replace millions of dollars in legacy mechanical systems immediately.

Defense-in-Depth Architecture

The standard calls for a layered defense. It isn't enough to have a firewall at the campus perimeter; security must be integrated at all levels of the Purdue Model.

- **Zones and Conduits**
  800-82r3 specifies the technical standards for "Conduits"—the secure pipes enabling our AI Digital Twin to communicate with sensors. It guarantees that, even if a student's laptop is compromised on campus Wi-Fi, a hacker cannot "jump" the conduit to access lighting or HVAC systems controls.
- **Unidirectional Gateways**
  In high-risk areas such as a campus power plant, the standard suggests using data diodes—hardware devices that enable data to be sent to the AI for monitoring while physically blocking any signals from entering the controller.

Incident Response for the Physical World

When an IT system is attacked, we "unplug" it. NIST SP 800-82r3 emphasizes that, unlike IT systems, OT systems cannot always be "rebooted" without physical risk. For example, when a smart building is attacked, we cannot simply unplug the fire suppression system or the laboratory ventilation. We must plan for the **Graceful Degradation** of campus services. Graceful Degradation obliges us to prepare for system failures and develop procedures for "reverting to manual control." This ensures that Facilities staff are trained to operate the campus as usual during a cyber-event, maintaining safety and climate control even if the digital network goes down.

- **The Manual Override Mandate:**
  We require that all critical life-safety and utility systems include a "Manual Mode." In the event of a network failure, Facilities staff must be trained and ready to manually access mechanical rooms and operate chillers, pumps, and breakers directly.
- **Defining the Fail-Safe**
  Our architecture guarantees that if the "brain" (the Digital Twin) is disconnected, the "limbs" (building controllers) automatically switch to a safety mode—keeping doors secure, fire suppression functional, and vital ventilation operational.

Performance and Resource Constraints

Many smart campus devices, such as IoT sensors and smart LED controllers, have limited processing power. 800-82r3 describes methods for securing these "resource-constrained" devices without introducing latency that could cause system faults. It guarantees that our sustainability goals, which rely on these small sensors, are not compromised by overly demanding security protocols that exceed their hardware capabilities.

NIST Risk Management Framework (RMF): The Governance Engine

While the Purdue Model provides the architecture and NIST SP 800-82r3 provides the technical controls, the **NIST RMF (SP 800-37)** provides the **intellectual governance** necessary to sustain a smart campus. In the context of an AI-enabled institution, the RMF is what prevents technology from outpacing our ability to control it.

Moreover, the NIST RMF (SP 800-37) serves as the central framework ensuring the effectiveness of the other two standards. It provides the CFaO with a clear, continuous record of institutional accountability and explains how to manage risk consistently over time. The process follows a seven-step cycle: Prepare, Categorize, Select, Implement, Assess, Authorize, and Monitor, emphasizing that security is an ongoing effort rather than a one-time task. It also states that systems should be classified based on their physical impact on the mission, not on the likelihood of data loss. This approach leads to Operational Governance supported by Continuous Monitoring, offering real-time assurance that our structural and technical defenses remain strong throughout the year.

For the CIO/CFaO, the RMF ensures that every AI model, occupancy sensor, and automated HVAC valve is linked to a clear line of institutional accountability and a measurable risk threshold.

Preparation: Cultivating the "Convergence" Culture

The initial step of the RMF, often neglected, is Prepare. For a sustainable smart campus, this means forming a cross-functional risk council comprising leaders from Facilities, IT, Sustainability, and Academics. Addressing these issues breaks down the siloed "not my problem" mentality. Preparation ensures that before a single sensor is installed, the campus has defined its "Risk Appetite"—deciding, for instance, how much autonomous control we are willing to give an AI over life-safety systems in exchange for energy savings.

Categorization: Mapping Risk to the Academic Mission

In traditional IT, categorization centers on data privacy, but in the RMF for a smart campus, we classify systems based on their physical impact. For example, a breach of a central utility plant's cooling system poses a higher risk than a breach of digital signage. By prioritizing systems according to their mission—such as safeguarding multimillion-dollar medical research or student safety—we ensure that our limited security resources are concentrated on the most critical "Level 1" systems vital to the campus's survival.

Selection and Implementation: The OT-Specific Overlay

This is where the RMF intersects with the physical environment. Instead of applying generic IT controls, we choose OT-specific overlays—often based on NIST SP 800-82r3—that are suitable for building automation. This approach prevents "Technical Interference" by ensuring that security controls don't inadvertently cause issues such as boiler failures or fire door malfunctions. It also enables the use of "Compensating Controls," such as leveraging the Digital Twin to detect anomalies when a legacy device cannot support encryption.

Assessment: Verifying the "Sentient" Defense

Assessment within the RMF functions as the validation stage. Instead of simply trusting the firewall's operation, we conduct thorough testing. For a sustainable campus, this also includes confirming that the AI's energy-saving logic does not create security vulnerabilities. We use "Red Teaming" to mimic potential hacker attempts to manipulate the AI's data inputs, aiming to avoid triggers for physical shutdowns. This process guarantees that our "Sentience" serves as a security asset rather than a weakness.

Authorization: The Executive "Buy-In"

Under the RMF, a senior official, known as the Authorizing Official, must formally accept the risks involved in operating a system, establishing Fiduciary Accountability. When the CTO or CFaO signs the "Authority to Operate" (ATO), they affirm that the smart campus technology is secure and resilient enough for public use and students. This shifts cybersecurity from merely a technical task to a high-level risk management decision made by executives.

Continuous Monitoring: The Pulse of the Digital Twin

This is where the RMF truly shines in an AI-driven environment. It shifts monitoring from occasional audits to an ongoing data stream, enabling the CTO and CFaO to move from "Static Security" to "Dynamic Assurance." The AI Sentinel continuously scans the campus for deviations from the "Authorized State." If a vendor changes a configuration or a sensor enters unsafe ranges, the RMF's monitoring system prompts immediate action to ensure a stable and secure campus.

In summary, the RMF is the mechanism that converts technical risk into a **business-grade assurance.** By following this framework, the institution can prove to insurers, grant-funders, and stakeholders that its "Sentient Campus" is not a reckless experiment, but a professionally governed, resilient environment. It ensures that our pursuit of **Sustainability** (efficiency) never comes at the cost of **Stability** (security).

The Intelligence Layer – AI as a Security Sentinel

At a Sentient Campus, there's no need to wait for a technician to detect a vibrating chiller or a security analyst to identify a spike in network traffic. We utilize AI as an automated sentinel to turn cybersecurity from a reactive task into a proactive, self-healing system.

Behavioral Anomaly Detection (BAD): The "Known Good" Baseline

Traditional security relies on identifying "signatures"—recognizing known viruses. However, in OT environments, this method is insufficient because many attacks use legitimate commands to cause physical damage, like instructing a pump to spin until it explodes. The Digital Twin is used to create a detailed "Pattern of Life" for each building. The AI determines the precise power consumption at 2:00 PM on a Tuesday and the specific servers a thermostat communicates with. Any deviation from this baseline, labeled as a "Behavioral Anomaly," is instantly flagged. Instead of questioning "Is this a virus?", we now ask "Is this behavior normal for this' asset?"

Predictive Threat Hunting: Stopping Lateral Movement

The biggest concern in a converged environment is "Lateral Movement"—where a hacker gains access through a low-security device, such as a smart vending machine, and then moves laterally to reach the campus data center. The AI sentinel constantly oversees the "Conduits" specified in our system. When it detects a classroom occupancy sensor trying to "handshake" with the student registration database, it recognizes this as a protocol violation. The AI doesn't just alert; it hunts. By integrating with our software-defined networking (SDN), the sentinel can automatically quarantine a suspicious device into a "Digital Sandbox" before the threat can migrate to mission-critical systems.

Automated Response vs. Human-in-the-Loop

NIST SP 800-82r3 warns that automated responses in OT can have physical consequences. We must balance AI's speed with the wisdom of human oversight. For purely digital threats (e.g., a brute-force login attempt), the AI can autonomously block IP addresses and reset credentials. For anomalies related to mechanical hardware, such as a boiler pressure spike resembling a cyberattack, the AI presents a "Decision Support" dashboard to the Facilities

Officer. It provides the probabilities of an attack versus mechanical failure, enabling the human expert to make the final decision to shut down the system.

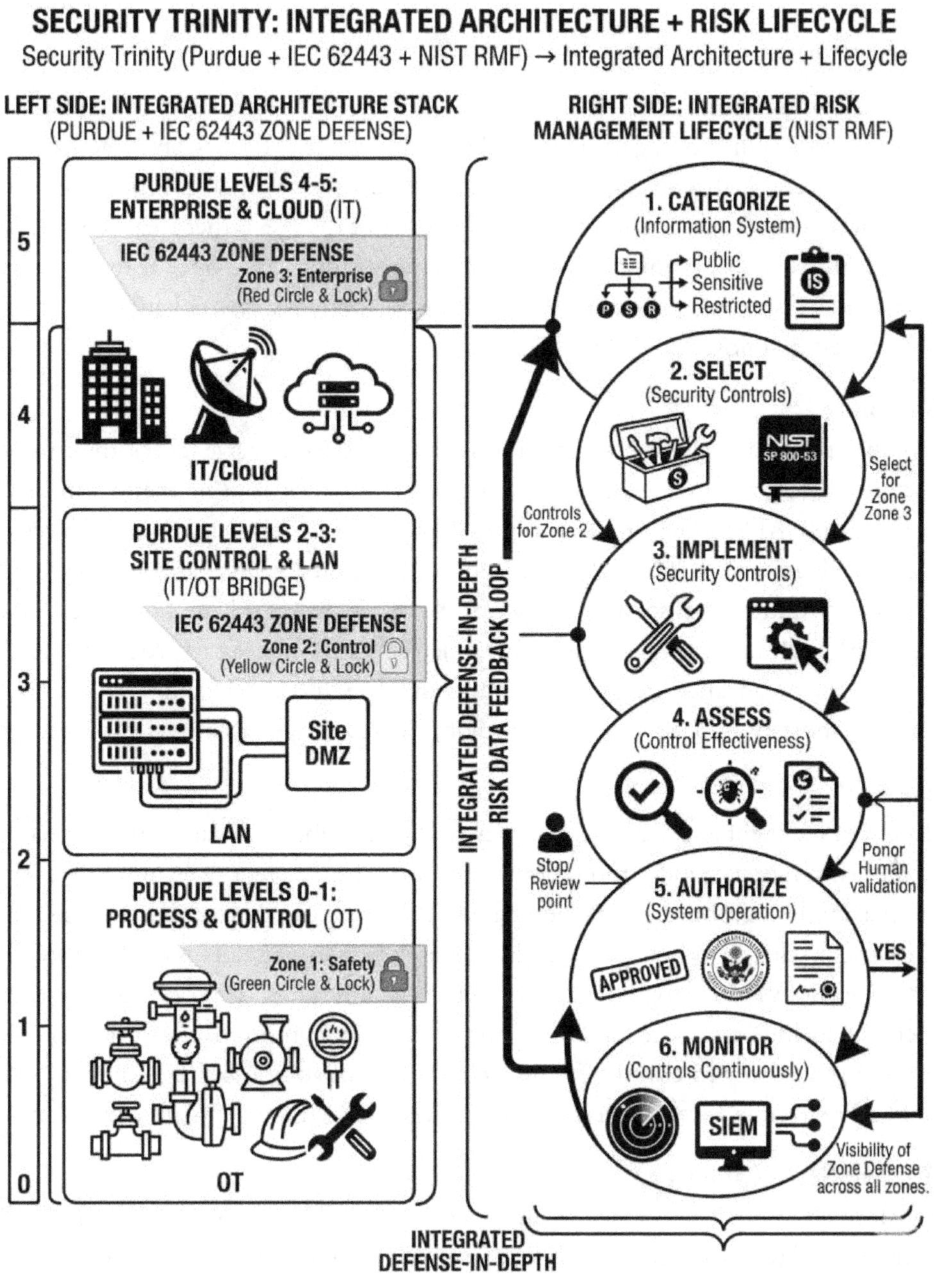

Figure 11. The Security Trinity: Integrated Architecture + Risk Lifecycle

The Feedback Loop of Experience

The ultimate goal of the Intelligence Layer is to ensure the campus gets smarter with every attack. We use our Digital Twin to run "Cyber-Physical Simulations." We can simulate a ransomware attack on the campus power grid in a virtual environment to see how our AI responds, refining its algorithms without ever risking a real-world blackout.

When a real anomaly is detected and resolved, the data is instantly fed back into the central AI model. This ensures that a vulnerability found in a single building on a remote campus is instantly "immunized" against attack, across every other facility in the network. Automated AI monitoring reduces the "Mean Time to Detect" (MTTD) from months to milliseconds. For example, by stopping a breach at the "meaning layer" before it reaches the "physical layer," we avoid the catastrophic costs of equipment replacement and campus-wide shutdowns.

The Efficiency Dividend: Automation as a Force Multiplier

As the campus develops into a cyber-physical ecosystem, automation shifts from being merely convenient to a vital driver of productivity. Transitioning from manual rounds, ad-hoc overrides, and reactive troubleshooting to AI-assisted monitoring, controlled governance, and data-driven operations enables institutions to achieve increasing efficiency gains over time. Automation complements human expertise rather than replaces it, empowering a more streamlined workforce to oversee more assets, respond more quickly, and operate more precisely. This multiplier effect allows teams to focus on higher-value tasks, reduce operational waste, enhance sustainability efforts, and free up resources to invest in innovation and resilience.

Continuous Monitoring.

Continuous monitoring redefines facilities operations from a labor-intensive "check-the-equipment" approach to a precise, exception-based system. Instead of sending technicians for routine manual inspections of boilers, pumps, or air handlers, the AI Sentinel continuously analyzes real-time telemetry data and alerts staff only when performance falls outside expected norms. This transition—acting on actual condition changes rather than fixed schedules—creates a leaner, more focused workforce capable of overseeing a larger, more complex asset portfolio without increasing staff. Each alert corresponds to a real issue, reducing unnecessary truck visits and allowing skilled technicians to concentrate on advanced diagnostics, commissioning, and system tuning. Over time, this workflow enhances productivity, shortens response times, and prolongs asset life by enabling early, precise interventions.

Energy Sovereignty

A secure OT environment enhances safety, efficiency, and predictability. Unauthorized overrides—whether by well-meaning staff, misconfigured vendor tools, or malware—can quietly reduce energy efficiency by pushing systems into less-optimal modes. A robust, NIST-compliant OT architecture prevents these rogue actions and ensures that intentional policies, rather than human error, drive energy performance. This establishes energy sovereignty, meaning the institution, not vendors, rogue devices, or unsecured interfaces, controls and optimizes its energy management. The benefits include a clear reduction in unnecessary consumption, improved reliability of demand-response operations, more precise forecasting for budgets and sustainability, and the ability to redirect savings toward modernization. Predictable systems are easier to optimize, and secure systems are inherently more predictable.

Trust Equity: The Safe Smart Campus

Trust has transformed into a key strategic resource. Federal agencies like NIH, NSF, and the Department of Defense now often demand robust cyber-physical security in environments that handle regulated data, operate specialized labs, or conduct federally funded research. Implementing a NIST-certified, resilient OT architecture demonstrates to these agencies that the institution can protect research integrity, laboratory spaces, and sensitive intellectual property. This "trust equity" acts as a competitive edge—enabling access to high-value grants, speeding up approval processes, and positioning the campus as a dependable partner for public and private research sponsors. Additionally, architectural standards that meet federal requirements also enhance internal governance, operational stability, and institutional reputation, yielding multifaceted returns on investment.

Campus Safety

In today's environment of increased physical and digital uncertainties, confidently claiming that a campus is Resilient by Design offers both an advantage in recruiting students and reassurance. A campus equipped with segmented networks, secure operational technology, reliable access control, and AI-enhanced safety monitoring provides a visibly safer setting for students, staff, and their families. This results in tangible benefits, such as quicker incident response, fewer disruptions in residence halls, more dependable life-safety systems, and trustworthy environmental controls that promote health and well-being. In the competitive higher education sector, resilience acts as a key brand trait, while safety is a core value proposition. Colleges that demonstrate a well-managed, cyber-physically secure campus can stand out in enrollment, retention, and building community trust.

The Unified Command: SOC, NOC, and FOC Convergence

The silos of the past are the vulnerabilities of the present. As campuses evolve into tightly interdependent cyber-physical ecosystems, fragmented response teams can no longer keep pace with the speed, scale, and complexity of modern threats. To eliminate these blind spots, we are establishing a **Unified Convergence Command Center** that merges the Security Operations Center (SOC), Network Operations Center (NOC), and Facilities Operations Center (FOC) into a single operational nervous system. This model creates **shared visibility, synchronized decision-making, and coordinated playbooks for incidents that span** digital and physical domains.

Creating Shared Visibility and Cross-Functional Playbooks

In a converged environment, no single team ever sees the full picture on its own. A tripped sensor may present simultaneously as a cyber-alert in the SOC, a traffic spike in the NOC, and a temperature anomaly in the FOC. In the old model, each of these signals would be investigated separately, with delays, miscommunication, and duplicated effort. In the new unified command model, **all three views appear instantly in a shared "Source of Truth,"** enabling operators to correlate signals, identify root causes, and act decisively. Cross-functional playbooks replace the legacy approach of "IT Playbooks" and "Facilities Procedures." Instead of documenting how each group responds in isolation, new **Integrated Playbooks** outline the coordinated steps SOC, NOC, and FOC teams take together—who assesses what, in what sequence, using which shared tools, and under what communication protocols. This ensures consistency, speed, and clarity across all response actions.

Integrated Incident Response (IIR): A Single Operating Model

The Unified Command enables **Integrated Incident Response (IIR)**—a major evolution beyond traditional computer emergency response or stand-alone facilities escalation. IIR recognizes that cyber events increasingly trigger physical consequences, and physical events often contain digital precursors. In IIR, authority, communication, and escalation paths are **pre-negotiated and documented** across all three domains. If a building must be disconnected from the grid, a lab must be depressurized, or a potentially compromised controller must be isolated, the playbook clearly defines:

- Who has the authority to make that call?
- Who is notified in what order?
- Which systems are isolated and how?
- How decisions are logged, audited, and communicated institutionally

This eliminates the ambiguity that previously led to hesitation, duplication, or unsafe improvisation during emergencies.

Converged Detection and Response

The new unified command model creates a **force multiplier** effect for detection and response. When cyber-physical anomalies surface:

- The **SOC** detects behavioral deviations or attack signals

- The **NOC** identifies unusual network load or routing conditions
- The **FOC** sees real-world impacts such as pressure, flow, voltage, or temperature fluctuations

By combining these perspectives in real time, the campus moves from reactive firefighting to **proactive diagnostics**, thereby shrinking the mean time to identify (MTTI) and the mean time to recover (MTTR). This cohesion is essential as AI-driven OT systems and digital twins increasingly automate parts of the response workflow.

Unified Command as a Governance Mechanism

Beyond operations, the Unified Command serves as a **governance structure** that reinforces consistent standards, agency, and accountability across IT, facilities, and security. It enforces:

- Shared definitions of system states and risk levels
- Co-authored authority frameworks for isolation, shutdown, and recovery
- Standardized semantic models and naming conventions.
- Documentation practices that feed the unified data estate and digital twin
- Auditability and compliance with NIST, IEC 62443, and enterprise governance

This ensures that modernization is not just technical—it is **institutional**, with playbooks taken seriously across every operational and executive tier.

Toward a Cyber-Physical Operations Center of Excellence

As the campus matures, the Unified Command evolves into a **Cyber-Physical Operations Center of Excellence, where AI-assisted dashboards, anomaly-detection** engines, digital twins, and integrated risk registers come together. Staff from all three centers train together, rehearse drills together, and participate in blameless post-incident reviews that strengthen the entire system. The result is a campus where **every operator, every sensor, and every system contributes to a shared situational awareness**, enabling faster decisions, safer outcomes, and a resilient operating environment capable of withstanding and learning from stress.

Engineering Antifragility in the Smart Campus

Antifragility shifts the campus from simply reacting to stress to becoming a system that learns, adapts, and improves through challenges. It's more than a motto; it's a practical strategy. An antifragile campus doesn't just withstand volatility, faults, and near-misses—it leverages them as valuable institutional knowledge, making future operations smarter, safer, and more efficient. In a unified IT/OT university setting, this approach goes beyond securing networks or applying innovative analytics. It necessitates adaptive governance, resilient architectures that can flex without breaking, and staff trained to convert stress into enhanced capabilities.

An antifragile smart campus is structured so that incidents such as a chiller failure, vendor API downtime, or an unexpected increase in occupancy do more than initiate recovery processes. Each event contributes to improving the digital twin, updating operational playbooks, refining segmentation policies, and expanding detection logic throughout the organization. The campus's telemetry data, governance records, and post-incident analyses feed into a dynamic memory system that enhances prediction accuracy, reduces false alarms, shortens response times, and clarifies decision-making roles for IT, Facilities, Safety, and Academic teams.

Risk Transfer, Acceptance, and Insurance

Leveraging structured risk transfer and disciplined acceptance signifies a shift from traditional defensive approaches to an "antifragility" strategy for modern institutions. Unlike conventional systems that aim to resist stress or recover, an antifragile smart campus is built to improve and become stronger through volatility. By combining these risk management methods with AI insights and sustainable infrastructure, the university doesn't merely withstand disruptions—it leverages them as data to enhance its predictive capabilities and resilience. In this view, risk becomes not a threat, but a catalyst that drives the continual growth and self-healing capabilities of an intelligent campus.

Strategic Risk Transfer

In the intricate landscape of smart campus management, institutions cannot and should not shoulder every responsibility alone. Risk transfer involves shifting particular financial or operational risks to partners best suited to handle them. Our "campus-as-a-platform" approach starts with Performance-Based Contracting. Using models such as Energy Savings Performance Contracts (ESPCs), we tie vendor payments to proven results, making our partners responsible for any shortfalls in achieving sustainability objectives.

Additionally, institutions can employ Public-Private Partnerships (P3) to speed up modernization efforts while shifting technical and operational risks related to specific asset classes to specialized organizations. This risk transfer is reinforced through strict Vendor Governance, where institutions incorporate security, interoperability, and audit standards into contracts to hold third-party integrators accountable for maintaining the integrity of the systems they deliver to our "data estate".

Insurance

The cyber-insurance market has changed significantly, shifting from lenient to highly scrutinizing, especially for institutions with complex OT systems. Nowadays, insurers expect more than just confirmation of a firewall; they require proof of a NIST-aligned OT Security Program, which includes segmentation, privileged-access control, continuous monitoring, and OT-aware data management. Demonstrating compliance with the NIST Risk Management Framework (RMF) and implementing automated anomaly detection shows that the institution actively manages and mitigates cyber-physical risks. This positions the university as a low-risk client, potentially resulting in lower premiums, higher coverage limits, and better policy terms.

The Discipline of Risk Acceptance

Innovation on a smart campus requires a calculated appetite for risk. Not every vulnerability can be eliminated, and attempting to do so can lead to institutional paralysis. Instead, mature institutions practice **Disciplined Risk Acceptance**—the formal, documented decision to proceed with a project despite known exposures.

A clear hierarchy organizes this process. CIOs and CFaOs are required to document risks openly via formal exception processes that specify compensating controls and mitigation strategies. Risks that significantly increase our institutional exposure or affect life-safety systems are escalated to enterprise risk management governance and the top leadership, confirming that acceptance is a strategic decision by the Board or Cabinet.

For "Living Labs"—the innovation sandboxes where students and faculty experiment with next-generation AI—we permit more operational experimentation. However, these are always time-limited and documented in a central register, ensuring that accepted risks are actively overseen and remembered.

By clearly identifying where risk is accepted and showing the safeguards in place, the institution proves to insurers that residual exposures are understood, limited, and controlled, rather than accidental or unmanaged. This transparency boosts underwriters' trust and decreases the chance of exclusions or surcharge conditions.

Cyber and Physical "Chaos Engineering"

Resilience and antifragility are skills that require practice. Leaders should go beyond passive oversight and implement controlled stress tests on the campus infrastructure. Conduct these drills quarterly, document the results and corrective measures—using both structured and unstructured data—and incorporate them into the data estate. This approach helps turn each rehearsal into an ingrained part of institutional memory.

Tabletop Exercises

Tabletop exercises simulate realistic cyber-physical crises, enabling key decision-makers to practice coordinated responses in a controlled setting. By systematically walking through potential failures—such as communication breakdowns, sensor outages, vendor errors, or severe weather events—the institution can identify weaknesses in plans, handoff procedures, and authority levels. The outcomes and improved best practices are documented and incorporated into the data systems, ensuring that each rehearsal enhances institutional memory and leads to faster, clearer, and more resilient future responses. Success is not measured by "no incidents," but by how quickly and safely the campus identifies, isolates, and recovers from them.

OT Segmentation Drills

OT segmentation drills intentionally focus on isolating a BAS subnet or critical zone to check that segmentation rules, conduits, operator workflows, and safe-state fallbacks operate as intended. These managed disruptions help determine whether systems remain stable when disconnected from their usual pathways, preventing cyber incidents from causing physical outages. With each drill, antifragility develops as vulnerabilities like misrouted traffic, undocumented dependencies, or unsafe overrides are uncovered and fixed, strengthening the system incrementally.

Vendor Access Exercises

Vendor access exercises assess whether privileged pathways are effectively controlled, reversible, and auditable by intentionally revoking and re-granting third-party credentials. This process verifies time-limited permissions, enforced MFA, session recording, and just-in-time access models. These drills help prevent vendor access from becoming a hidden vulnerability during emergencies or stressful situations. Each cycle enhances trust boundaries and demonstrates that access controls can withstand both accidental errors and malicious actions.

Digital Twin Degradation Tests

Digital twin degradation tests deliberately remove or corrupt data sources—such as occupancy telemetry, energy meters, or HVAC points—to assess how the model performs under incomplete or degraded conditions. These tests verify the model's resilience, ensure fallback mechanisms function properly, and confirm that operators can override AI suggestions when needed. They enhance antifragility by keeping the digital twin effective even with imperfect data, allowing humans to intervene when the system's confidence wanes.

Energy System Scenarios

Energy system scenarios model microgrid islanding, peak price fluctuations, sudden DER outages, and demand-response actions to test the resilience and intelligence of grid-interactive controls. By simulating how the campus electrical ecosystem reacts to shocks—such as load shedding, storage dispatch, and renewable variability—the institution ensures these disruptions enhance system robustness rather than compromise it. These controlled stress tests help improve algorithms, adjust thresholds, and refine recovery methods, making the campus energy system increasingly adaptable.

Lifecycle Decommissioning: Securing the "Exit"

Operational technology typically lasts over 20 years, much longer than the software that runs it. Proper readiness requires knowing when and how to retire these assets. Strict protocols exist for disposing of OT components, as discarded devices like smart-lighting controllers may contain sensitive data, such as Wi-Fi passwords and network diagrams, which must be securely wiped through "digital sanitization" before they leave the facility. For assets that cannot be replaced, such as old boilers and elevators, the CIO/CTO employs "Virtual Patching"—using the AI Sentinel, an AI-powered monitoring and security layer that keeps a close watch on vulnerable or legacy OT systems—creating a digital barrier around these assets until they are decommissioned.

AI Drift Fire Drills

AI drift fire drills introduce harmless anomalies—such as slight sensor deviations, shifts in occupancy patterns, or noise in energy signals—to test the effectiveness of drift detection, rollback procedures, and automated model retraining systems. These drills demonstrate that AI can recognize real-world changes, safely revert to a stable model when necessary, and adapt intelligently without risking operations. Each exercise enhances the AI governance framework, helping models become more precise and robust as conditions evolve.

Readiness Gates and Advancing Only at the Speed of Safe Adoption

Antifragility breaks down when modernization outpaces human and operational preparedness. Leaders must establish Readiness Gates for each project stage. No feature should go live until it passes all gates. This approach ensures rapid progress with controlled risk—advancement that builds resilience instead of increasing fragility.

Human Readiness

Human readiness guarantees that the workforce is prepared to operate, interpret, and troubleshoot new OT/AI-enabled systems safely before deployment. Complete role-based training and field-validated UX ensure technicians can perform tasks confidently without cognitive overload or risky workarounds. An antifragile campus relies on operators who not only understand new tools but also contribute to their improvement by offering feedback that enhances models, workflows, and future change cycles.

Safety and Uptime Readiness

No system development proceeds without documented safe-mode protocols, tested rollback procedures, and formal commissioning approval. This ensures that if a new model, controller, or integration malfunctions under real-world conditions, the environment automatically shifts into a stable, safe state. Uptime preparedness transforms potential disruptions into opportunities for learning, so that each stress event enhances future designs rather than causing widespread campus instability.

Interoperability Readiness

Interoperability readiness verifies that new systems adhere to institutional standards like BACnet/SC and MQTT, align with shared ontologies such as Haystack or Brick, and pass data integrity tests before integration. This process prevents fragile, one-off connections that may degrade with time by ensuring semantic and protocol consistency. It acts as a gatekeeper that guarantees each addition enhances the entire ecosystem, fostering scalable intelligence and avoiding the buildup of technical debt.

Security Readiness

Security readiness guarantees that identity controls (such as SSO/MFA), micro-segmentation, and approved vendor access patterns are established before activating any system. This approach prevents risky shortcuts like open ports, shared credentials, or uncontrolled tunnels from reaching production. By validating security early, the campus becomes antifragile: each new asset enhances visibility and decreases the attack surface rather than enlarging it.

Governance Readiness

Governance readiness mandates that all changes, exceptions, approvals, and risk-acceptance decisions be documented through formal change management processes. This ensures institutional memory, avoids shadow deployments, and promotes accountability in the adoption of new technologies. Incorporating governance as a readiness checkpoint allows the campus to evolve with each iteration—turning operational challenges into improved policies, streamlined processes, and more resilient systems.

Sustainability Readiness

To ensure that sustainability initiatives expand safely and credibly, the institution must progress only after meeting essential readiness milestones. These include audit-grade data on carbon, energy, water, and waste aligned with the GHG Protocol, managed through a unified semantic data estate; commissioning and validation that confirm systems can fail safely and be reversed without impacting uptime; open, interoperable infrastructure compliant with Division 25, with full digital handover; cybersecurity measures like enterprise identity, MFA, and micro-segmentation; AI TRiSM governance with human oversight; circularity and embodied-carbon management via EPDs and material passports; health standards based on equity and IAQ; grid-interactive resilience demonstrated through demand-response and emergency drills; transparent ESG reporting with data ready for audits; workforce readiness roles enabling technicians to operate the new systems; and cross-functional governance (SCSC + Sustainability Council) to enforce decision rights. Collectively, these gates ensure modernization advances at a pace that guarantees reliability, safety, and trust—making sustainability a continual, measurable institutional capability.

A Single Operating Picture

Antifragility acts as a unified executive discipline. Create a dynamic cyber-physical risk register managed by the CTO and CFAO, which includes OT telemetry, vendor evaluations, incident investigations, and sustainability thresholds. Ensure it aligns with decision-making rights—covering identity, segmentation, remote access, and data governance overseen by the CIO—and operational authority—such as safety standards, commissioning, maintenance, and lifecycle planning managed by the CFAO. Present this register, along with the antifragility scorecard, to the Cabinet and Board every quarter.

Secondly, align the register with a unified incident command: the SOC, NOC, and Facilities Operations Center collaborate through shared playbooks, conduct joint tabletop exercises, and coordinate communication strategies to ensure safety and maintain trust.

Establish AI TRiSM as a First-Class Operational System

On an antifragile campus, AI is transparent and well-managed, not a mysterious black box. Consider AI Trust, Risk, and Security Management as a core production service. This approach transforms AI from merely amplifying risks into a key driver of antifragility.

Provenance and Versioning —

In an antifragile and TRiSM-aligned system, provenance and versioning ensure that each AI model can be traced back to its specific data sources, assumptions, governance policies, and regulatory frameworks used in its creation. This lineage functions like a "black box recorder" for AI, enabling teams to understand not only what a model does but also why it responds a certain way under stress. When a model drifts, underperforms, or exhibits unexpected behavior, transparent provenance facilitates quick diagnosis and fixes, turning disruptions into opportunities for improvement. Without this accountability layer, organizations cannot reliably adapt AI systems or learn from failures—both of which are crucial for achieving antifragility.

Continuous Drift and Bias Monitoring

Antifragile systems acknowledge that models will deteriorate in real-world scenarios—patterns in data shift, campus behaviors change, and external factors like weather, schedules, or occupancy can significantly alter input signals. Ongoing drift and bias monitoring function as an early-warning system by using automated checks to identify when a model's predictions diverge from reality or start to impact certain groups negatively. However, automation alone isn't sufficient: human judgment is crucial to ensure alerts lead to informed decisions, contextual understanding, and appropriate corrective measures. This combined approach enables institutions to transform emerging instability into valuable insights, ultimately enhancing the model and its governance over time.

Explainability on Demand

In cyber-physical environments critical to safety, explainability is essential for operations. Providing explanations on demand helps operators, engineers, and decision-makers understand the reasoning, inputs, and constraints behind each AI suggestion. This clarity allows them to verify whether recommendations meet safety standards, regulatory rules, and organizational values. Such transparency fosters trust, reduces hesitation in adopting AI suggestions, and enables quicker responses when adjustments are necessary. In an antifragile setting, explainability also promotes faster learning. When operators understand why a model acted as it did, they can identify opportunities to improve the system through continuous iterations.

Human-in-the-Loop Authority

AI TRiSM requires humans to be the final decision-makers for any actions affecting people, safety, or critical infrastructure. It establishes formal pathways for overrides, appeals, and safety measures, allowing operators to pause, challenge, or revoke AI suggestions when necessary. While AI is proficient at pattern recognition and optimization, humans contribute essential context, ethical judgment, and situational awareness. These override systems enhance resilience by ensuring that stress events—such as cyber anomalies, fluctuating energy demands, sensor malfunctions, or unexpected campus scenarios—prompt human input. This feedback improves models, policies, and operational procedures.

Model Sunsets and Rollbacks

Antifragile systems avoid permanence: models must adapt over time, with outdated versions retired safely and predictably. Processes for model sunset and rollback enable seamless replacement, updating, or reverting of models without operational disruptions or new risks. Rollbacks provide an immediate fallback to a stable model when a new version behaves unpredictably, reducing operational risk and enabling controlled testing. Scheduled sunsets help prevent outdated algorithms from becoming liabilities unnoticed. This lifecycle management fosters a secure environment for continuous innovation—allowing models to evolve under stress, retire properly, and uphold trust within the organization.

Financial Antifragility: Turning Stress Into Capital

Antifragility should enhance the balance sheet, not only the building. The CFaO needs to implement antifragility finance mechanisms that create self-funding learning loops.

Predictive-Maintenance Savings Pool

Set aside the operational cost (OPEX) savings from your modernization efforts in a dedicated, protected budget line to prevent them from being absorbed into the general budget or redirected. These funds are reserved exclusively for ongoing investments in additional controls, modernization, governance, and data quality improvements.

Risk-Adjusted Capital Model

Avoid choosing upgrades solely because of high-efficiency ratings on the equipment label. Instead, prioritize projects that reduce breakdowns (which reduce outages and emergencies and improve reliability) or those that help the campus save or earn money during demand-response events. The DR yield refers to the financial benefit obtained by temporarily lowering energy consumption when the utility offers compensation.

Avoided Debt Ledger

Quantify deferred capital by assessing condition-based renewal and continuous commissioning. Rather than relying on age-based replacements or waiting for failures, the campus employs performance data, fault detection, and ongoing optimization to safely extend asset lifespans. Every case of preventing premature replacement of equipment, such as chillers or control systems, is documented, turning operational data into a strategic financial asset. This approach underscores avoided capital costs and aids reinvestment, helping to safeguard the institution's financial health.

Antifragile Sustainability: From Resilient to Antifragile Performance

A resilient campus can endure stress, but an antifragile campus benefits from it—enhancing its structural intelligence, operational accuracy, and environmental metrics each time it encounters challenges. In an AI-powered smart campus, every disruption serves as input: generating fresh data for the Digital Twin, revealing new patterns for machine learning, updating operation protocols, and strengthening governance safeguards. Instead of merely returning to the original state, the system evolves, becoming more resilient, predictive, and efficient with each cycle.

This transformation occurs because the cyber-physical fabric—sensors, digital twins, OT systems, AI models, and governance—continuously converts real-world anomalies into insight. Every weather event, mechanical disturbance, or environmental deviation becomes a learning moment for both humans and machines.

Smart Indoor Air Quality (IAQ)

When the campus detects an air-quality anomaly—such as poor filtration, elevated CO_2 levels, wildfire smoke infiltration, or pressure imbalance—the antifragile system records all telemetry data into the AI engine. This process enables the system to improve by:

- **Dynamic ventilation thresholds**, based on occupancy and contaminant mix
- **Filtration changeout schedules**, tuned by particulate signatures
- **Pressurization sequences** in labs, residence halls, or high-density spaces
- **Predictive IAQ alerts** tied to weather or wildfire models

The result is a campus whose indoor environment becomes **safer, healthier, and more stable** precisely because it learned from the disruption.

Grid Disturbance → Improved Load Shedding and Storage Dispatch

When the external grid faces instability—such as voltage sags, frequency deviations, or sudden price spikes—a resilient campus might switch to batteries or backup generators. An antifragile campus, however, takes additional steps:

- AI captures telemetry from meters, inverters, batteries, HVAC reactions, and load drops.
- The Digital Twin runs rapid simulations to identify more efficient ways to shed load in future events.
- Storage dispatch algorithms are automatically refined so they discharge at the **optimal moment** and **optimal rate.**

- The microgrid adjusts its predictive models for campus energy flow, occupancy, equipment responsiveness, and local weather patterns.

Each disturbance makes the campus mathematically sharper and operationally more autonomous.

Measuring Antifragile Sustainability

To turn these events from mere stories into lasting institutional strengths, the campus monitors progress through sustainability-focused indicators in the Antifragility Scorecard. These include:

- **Predictive Energy Improvement Rate**
  Measures how much better the system anticipates energy loads, grid disturbances, and optimization opportunities after each stress event.
- **Water-Loss Avoidance**
  Quantifies avoided consumption and damage by comparing leak-detection speed, stormwater response, and predictive modeling before and after each incident.
- **Carbon Reduction Velocity**
  Tracks how quickly the campus reduces emissions as models sharpen and operational efficiency improves.
- **Waste Diversion Accuracy**
  Measures the improvement in recycling and material recovery accuracy as contamination analytics evolve.
- **Sustainability Risk Exposure Index**
  Aggregates climate, water, carbon, IAQ, and grid-risk data to quantify how much risk the campus has shed over time.

These indicators collectively show if the campus is becoming greener, smarter, more efficient, and more adaptable. With each stress test, the campus evolves—its models become more refined, operations more coordinated, governance more accurate, and sustainability outcomes visibly improve. This embodies the essence of antifragile sustainability: a campus where challenges foster growth, disruptions enhance understanding, and each stressor strengthens long-term environmental and institutional resilience.

Human Antifragility: Viewing the Workforce as a Learning System

Building an antifragile campus requires more than just resilient technology; it depends on a workforce that grows stronger through disruption, iteration, and shared learning. In an integrated IT/OT environment, staff serve not only as system operators but also as active sensors, interpreters, and knowledge stewards. Viewing the workforce as a learning system involves creating frameworks that enable teams to adapt during stress, convert near-misses into improvements, and continuously refine practices as the campus becomes more intelligent. By promoting cross-domain learning, rehearsals, frontline feedback, and psychological safety, institutions leverage human expertise as a growing asset. This approach ensures that every event—whether routine or unexpected—enhances the organization's collective resilience and supports confident, safe implementation of the smart-campus model.

Cross-Training and Shadowing

Cross-training enhances resilience by bridging the long-standing gap between digital and physical spheres, enabling teams to become fluent in each other's operations. Network engineers observing the central plant understand the limits of uptime, safety, and seasonal variations; controls technicians learning about identity, segmentation, and access control grasp why cybersecurity is vital for reliability. This mutual learning fosters a common language, minimizes friction, and turns stress events—such as outages or integration issues—into opportunities for shared growth rather than assigning blame across silos.

Scenario Rehearsals

Regular, brief scenario rehearsals strengthen the workforce by making controlled stress routine and developing muscle memory for rare but critical events. Drills involving vendor lockouts, BAS failovers, network segmentation triggers, or safety-exception workflows enable teams to practice coordinated responses without risking operations. Each session uncovers fragile assumptions, undocumented procedures, or unclear authority lines—transforming small failures into opportunities for system-wide improvements. Repeating these exercises helps the campus become more resilient, and operators gain greater confidence in managing actual disruptions.

Voice of the Technician

An antifragile workforce depends on field operators shaping the systems they rely on daily. A standing "Voice of the Technician" forum gives frontline staff direct influence over UX roadmaps, alarm logic, data definitions, and workflow design. Publishing "you said / we did" closes the loop, reinforcing that field expertise is not just welcome—it is the primary input for safe adoption and continuous improvement. This participatory design approach reduces workarounds, improves adoption, and ensures that operational insights strengthen every iteration of the smart-campus ecosystem.

Psychological Safety

Psychological safety serves as the cultural cornerstone of antifragility: individuals need to feel secure in reporting near-misses, unexpected actions, and uncomfortable truths without fearing blame. Promoting near-miss reports, conducting blame-free post-mortems, and rewarding learning foster continuous improvement through structured cycles. When teams prioritize honesty over perfection, they identify issues sooner, innovate with greater confidence, and share insights that strengthen the entire system after each stressor.

Avoiding Fragility Traps

As institutions upgrade their cyber-physical campuses, certain recurring patterns can weaken resilience and lead to long-term operational vulnerabilities. These fragility traps often arise from prioritizing convenience, speed, or short-term gains. Over time, they accumulate technical debt, erode trust, and increase the severity of failures. To create an antifragile campus—one that strengthens through stress, adapts amid change, and maintains human agency—it is vital to avoid these pitfalls. The list below highlights the most common issues that make smart-campus efforts fragile rather than resilient.

Automation Without Agency.

Automation becomes vulnerable when humans cannot see, question, or reverse system actions. AI-controlled systems that operate without transparency, traceability, or the option for operator intervention pose a risk of silent failures—where a mistaken decision spreads through physical systems unnoticed. To stay resilient, every automated process must be explainable, auditable, and interruptible, ensuring that human judgment stays as the ultimate safeguard in safety-critical or mission-critical workflows.

One-Off Integrations

Custom point-to-point integrations may initially solve specific needs, but they often become fragile over time as vendors update their software, change APIs, or exit the market. These fragile points create hidden dependencies that are costly to maintain and can fail without warning. Ensuring reliable interoperability requires standards-based integration methods, shared ontologies, and reusable connectors—avoiding the patchwork approach that can fail under operational stress.

Flat Networks and Permanent Tunnels.

A flat network or "always-on" remote tunnel might seem convenient, but it undermines the boundaries that contain cyber-physical incidents. When OT and IT are connected through unrestricted pathways, a compromised device can spread across buildings and systems. To achieve antifragility, it's essential to implement micro-segmentation, time-limited access, and monitored pathways. These strategies reduce the impact radius, restrict lateral movement, and enhance safety as the network becomes more interconnected.

Pilot Theater

Endless pilots give the false impression of innovation but hinder the institution from developing genuine, repeatable capabilities. Without defined adoption criteria, proper governance, and a clear path to enterprise-wide integration, pilots risk remaining isolated experiments that consume resources without benefiting the campus. Antifragile institutions approach pilots as disciplined experiments, setting specific success criteria—either evolving them into standards or discontinuing them decisively.

Speed Without Readiness

Adding new features into production without adequate operator training, updated procedures, or proven fallback options introduces operational risks. Even dependable technology can be hazardous if the human operators are not prepared. Progressing only at the "speed of safe adoption" guarantees that each modification strengthens rather than weakens the system. Readiness gates protect the workforce, maintain stable operations, and allow the campus to improve consistently even under stress.

Examples of Antifragility

Monsoon Floods

When a major storm overwhelms drainage or surface systems, the campus does more than recover from flooding. AI analyzes every variable—rainfall intensity, sump pump behavior, soil absorption, drain-back pressure, and time-to-overflow—and updates the Digital Twin's hydrological models. This allows the system to:

- Recommend smarter detention strategies
- Optimize pump sequencing to prevent future backflow
- Identify ideal sensor placement for blind spots
- Adjust irrigation windows with real-time soil data

The next storm is not merely endured—it is anticipated and managed with greater accuracy.

HVAC Failure

An antifragile smart campus not only endures a significant HVAC failure but also improves by incorporating the anomaly data into its digital twin and AI systems, boosting future predictive precision. The event ultimately enhances overall system performance, leading to faster detection, smarter automation, and the ability to prevent issues before they occur.

Power Grid Fluctuations

In another example, when a major grid fluctuation impacts campus power, an antifragile system does more than switch to batteries (resilience). Its machine learning algorithms promptly analyze the failure's telemetry, run numerous simulations in the Digital Twin, and automatically update the predictive load-shedding algorithms to address that specific grid issue before it can pose a threat again. As a result, the institution becomes stronger both mathematically and operationally because of the stress.

Cyber-Physical Intrusion

When an attempted unauthorized login targets a building automation controller, a merely resilient campus would block the attempt and continue functioning. An **antifragile campus**, however, uses the event to strengthen its entire identity and access architecture.
The AI Sentinel analyzes the attack vector, correlates identity-misuse patterns, and automatically updates microsegmentation rules, privileged-access thresholds, and anomaly-detection baselines across all OT zones. The Digital Twin logs this behavior as part of a growing "threat ontology," improving the accuracy of future detections. As a result, every future attack—even from unknown sources—becomes easier to detect and isolate because the system learned from the initial stressor.

Antifragility Scenario: Residence Hall Network Outage and Door Access Failure

Late on a rainy Thursday evening, the network segment serving Oak Ridge Residence Hall unexpectedly went down. Students who normally used their phones as their primary credentials suddenly found that the exterior smart lock wouldn't respond—no Wi-Fi, no cloud verification, no mobile credentials. What could have become a safety incident instead became a demonstration of institutional antifragility, because the university had deliberately prepared for this exact failure mode.

Months earlier, during the smart campus planning process, the CTO, Chief Facilities Officer, and Campus Safety had identified "network-dependent access control" as a high-consequence OT risk. In response, they implemented **three resilience layers**:

4. **Local Fail-Secure Mode** – Exterior doors were configured so that, if the network failed, they would fall back to an **on-device credential cache** that stored recently validated student IDs for 24 hours.
5. **Redundant Access Paths** – A battery-powered **Bluetooth Low-Energy offline credential feature** was enabled, allowing phones to communicate directly with the door hardware even without network access.
6. **Operational Playbooks** – Residence Life staff were trained in a simple, pre-staged response: deploy a roving staff member with a digital master credential and set up a temporary check-in station to ensure student safety and calm the situation.

When the outage hit, the system behaved as designed. Students arriving at the hall used the offline Bluetooth fallback, which authenticated them against the cached local list. Those who hadn't opened the door recently—and thus weren't in the cache—were greeted by the RA team at the temporary check-in table, where a staff device with its own offline credential opened the door. No one was locked out. No panic. No unsafe prop-open doors.

The next morning, IT and Facilities conducted a **joint post-incident review**, tracing the root cause to a misconfigured firmware update on a distribution switch. But instead of merely patching the issue, they **improved the ecosystem**:

- They expanded the offline credential cache window from 24 to 72 hours.
- They established automated **pre-deployment switch configuration testing** in a sandbox twin.
- They deployed **edge-based access analytics**, allowing the campus to detect local-lock anomalies without requiring cloud dependency.
- They updated the antifragility scorecard to track "access continuity during network disruption" as a new KPI.

What began as a network failure ended as a capacity-building event: access resilience increased, operational confidence grew, students felt cared for, and the institution's cyber-physical architecture became even more adaptive. In true antifragile fashion, the stressor didn't just test the system—it strengthened it.

A Flooded Mechanical Room

If a water main break floods a mechanical room, a resilient campus would repair the damage and restore operations. An **antifragile campus** goes further: Flood sensors, pump telemetry, and occupancy logs are ingested into the Digital Twin to simulate the hydraulic behavior and spatial vulnerabilities of that specific building. AI models refine their understanding of water intrusion patterns and generate new risk-weighted recommendations—rerouting piping, relocating sensitive electrical panels, adjusting maintenance intervals, or adding predictive moisture sensors in analogous rooms. The next time abnormal moisture is detected anywhere on campus, the system responds faster and with greater precision because it "learned" from the incident, structurally and operationally.

A Network Outage

When a fiber cut disrupts a major part of the campus network, a resilient system switches to backup pathways. In contrast, an antifragile system goes further: AI evaluates which devices have become isolated, how quickly they entered safe mode, alert delays, and human response issues. It then automatically adjusts routing priorities, enhances edge computing autonomy so operational technology (OT) systems can function safely without upstream connections, and improves communication failover procedures. As a result, during the next outage—caused by construction, weather, or cyber threats—decision-making is faster locally, service disruptions are minimized, and safe-mode behaviors are better maintained across the entire network infrastructure.

The Antifragility Maturity Model

Creating a sustainable smart campus powered by AI involves more than just technical improvements; it requires a fundamental shift in institutions toward systems that not only resist stress but actually improve because of it. The Antifragility Maturity Model offers a straightforward pathway for this development, illustrating how campuses evolve from fragile, siloed operations to a flexible, learning-oriented cyber-physical ecosystem. Each stage of maturity signifies greater alignment in governance, data architecture, OT/IT integration, human readiness, and AI Trust, Risk, and Security Management (TRiSM).

Level 1 — Fragile

Siloed projects, flat networks, vendor-managed mystery boxes, and reactive repairs characterize this stage. Decentralized OT, the absence of governance, shadow systems, and limited interoperability and visibility, characterizes it. It serves as the baseline for measuring antifragility.

Level 2 — Robust

Segmented networks with basic monitoring, some standards, and inconsistent adoption. This stage is defined by the establishment of network segmentation, initial monitoring, protocol adoption, and partial governance, although these are not yet consistently implemented. The environment can handle minor shocks but does not adapt or evolve in response.

Level 3 — Resilient

This level features a governed data estate, daily use of digital twins, cross-domain playbooks, and regular scenario exercises such as segmentation drills and vendor access tests. It is characterized by the presence of governed platforms, digital twins used in daily operations, joint playbooks, and periodic drills. At this stage, the campus can recover smoothly from stress and enhance overall reliability.

Level 4 — Antifragile

This stage embodies the peak of antifragility, characterized by continuous chaos engineering, drift detection, model governance, and feedback loops among data, policies, and operations. It focuses on KPI-driven enhancements and self-sustaining modernization by avoiding debt and achieving predictive OPEX savings. The method incorporates ongoing stress testing, feedback-driven learning, and a consistent upward trend in antifragility.

The Antifragility Scorecard

This model uses the Antifragility Scorecard in the Appendix, focusing on trend-based indicators such as predictive lead time, segmentation integrity, digital twin accuracy, avoided capital debt, and post-stress performance improvements to evaluate institutional maturity and progress. The combination of the maturity model and scorecard creates a comprehensive framework enabling leaders to determine their current state, identify key investments, and drive modernization gradually and sustainably. Each implementation cycle offers an opportunity to enhance resilience and strengthen the campus's ability to learn, adapt, and succeed through change.

The Antifragility Scorecard transforms a philosophical idea into a practical measurement tool. It measures advancements in prediction accuracy, recovery speed, governance discipline, AI reliability, data quality, operator confidence, and financial stability, turning volatility into a source of insight rather than a risk. Each disruption creates new data; each stress event helps improve system architecture, operational safety, automation logic, and institutional trust. This way, the scorecard embodies the key trait of an AI-enabled antifragile campus: one that grows stronger through challenges it faces, not despite them.

The Call to Action for CFaO and CTO

The converged, AI-enabled, sustainable smart campus cannot emerge from technology upgrades alone—it must be architected through deliberate, executive-level leadership. The CFaO and CTO sit at the intersection of convergence around digital infrastructure, operational technology, data governance, and evolving institutional missions. Their call to action is not simply to modernize systems, but to **engineer antifragility** into the campus: a condition in which every disruption becomes a catalyst for improvement, every failure accelerates learning, and every iteration strengthens governance, safety, and operational reliability. This requires a radical shift in mindset—from running an IT utility to shaping a cyber-physical ecosystem that adapts, evolves, and improves under stress.

To lead this transformation, CIOs and CTOs must champion **convergence governance**, ensuring that standards, identity, segmentation, and interoperability are no longer optional but foundational controls protecting the entire institution. They must replace opaque vendor ecosystems and "mystery-box" OT with a transparent, interoperable platform governed by shared ontologies, open protocols, and strong AI TRiSM practices. They must operationalize the antifragility scorecard in the appendix as an executive compass, using its trend-based indicators—semantic integrity, predictive lead time, segmentation health, and capital debt avoided—to measure progress and steer investment. This scorecard becomes the institutional mechanism for ensuring that modernization is not episodic but continuous, not reactive but proactive, not fragile but learning-driven.

Most importantly, CFaOs and CTOs must lead the **human transformation** embedded in the smart-campus operating model. Antifragile systems require antifragile people—operators who are trained, empowered, cross-skilled, and supported through psychologically safe cultures where near-miss reporting, post-mortems, and continuous improvement are celebrated rather than hidden. Technology leaders must build structures that treat the workforce as a learning system: cross-training with Facilities, scenario rehearsals, drills, technician-driven UX design, and human-in-the-loop authority that protects both safety and trust.

The call to action is clear: Design with resilience in mind. Govern to ensure interoperability. Build structures that prioritize safety and train personnel to be adaptable. Implement measurements to foster improvement. Lead the effort to turn the campus into a living system that becomes stronger each time it faces a challenge.

Chapter Summary

Chapter 17 redefines risk management as a fundamental leadership skill for the AI-enabled, cyber-physical campus. It is no longer just a perimeter security measure or an after-the-fact audit. Instead, risk is integrated into the architecture through identity and access controls, network segmentation, governed remote access, data stewardship, and daily operational workflows. The operating model is deliberately bipartisan: the CTO oversees identity, secure connectivity, segmentation, and vendor access, while Facilities (CFAO/FOC) handles safety, uptime, commissioning, and building operations. This collaboration helps bridge the gap where most cyber-physical incidents occur.

A modern risk posture is rooted in Zero Trust principles and ISA/IEC 62443 "zones and conduits," incorporating enterprise IAM with MFA and least privilege. It emphasizes centrally managed, auditable remote access and OT-aware data governance to ensure privacy and maintain institutional trust. The chapter puts this into practice through joint tabletop exercises, integrated incident response, and transparent governance, which together eliminate "shadow OT," reduce cascading failures, and speed up recovery. The chapter also transforms this posture into a board-level narrative by framing OT modernization as resilience infrastructure that protects mission continuity, minimizes downtime and emergency costs, enhances duty of care, and fosters durable trust. Overall, this approach guides operations from fragile to robust, resilient, and ultimately antifragile, enabling continuous learning and improvement from stress.

The chapter emphasizes a practical cultural change: in OT settings, the SRA triad—Safety, Reliability, Availability—should complement the traditional IT CIA triad. Systems need to be fail-safe and predictable, and to keep essential, mission-critical data accessible to maintain a clear "meaning layer." By implementing disciplined governance, shared responsibility, and standards-based architecture, the organization shifts from reactive firefighting to proactive prevention and managed recovery. This creates a campus that not only withstands disruptions but also benefits from them.

Key Calls to Action for the CTO and CFAO

- **Own one risk story—establish a Unified Risk Register.**
  Consolidate digital and physical risks into a unified, jointly managed register; review it with the Smart Campus governance body and present trends to Cabinet or the Board, including antifragility insights scorecard.

- **Set non-negotiable guardrails for OT connectivity.**
  Enforce enterprise IAM (SSO/MFA/least privilege), ISA/IEC 62443 segmentation, time-bound, logged vendor access, and OT data governance as prerequisites for any connected system — no unmanaged tunnels, flat networks, or shadow IT integrations.

- **Institutionalize Integrated Incident Response (IIR).**
  Replace isolated IT and Facilities procedures with collaboratively developed playbooks, pre-approved authority pathways (such as who can isolate a plant or disconnect a building), and a unified SOC–NOC–FOC operational rhythm that includes regular joint drills.

- **Evaluate** Readiness **Spot-Checks:**
  - Do we have **analog fallbacks** for our most critical digital processes?
  - Can our team **operate the building if every computer goes dark?**

  - Does staff know the **"cyber-call"**—who in IT to notify the moment a mechanical system behaves strangely?

- **Practice stress to gain strength.**
  Conduct tabletops, segmentation drills, vendor access exercises, digital twin degradation tests, energy scenario evaluations, and AI drift fire drills; publish the results to the data estate to enhance policies, playbooks, and practices with each rehearsal model.
- **Adopt SRA as a design mandate.**
  Integrate Safety, Reliability, and Availability into architecture reviews, commissioning criteria, and change controls to ensure cyber actions do not pose life-safety or uptime risks — and that safe modes and rollbacks are always in place and proven.
- **Make modernization a resilience investment, not a project.**
  Position OT security and governance as essential for mission continuity and cost containment—highlight benefits like fewer outages, quicker recovery, and reduced emergency expenses. Use this framing to ensure ongoing support for capital and OPEX.
- **Eliminate "shadow OT" at the source.**
  Gate procurement in line with standards such as open protocols, semantic tagging, data ownership, and security baselines. Require documented onboarding to the identity, segmentation, and monitoring systems before any connection occurs.
- **Measure what matters—trend antifragility.**
  Monitor track segmentation integrity, predictive lead time (MTTP), recovery acceleration, digital twin accuracy, OT Zero Trust compliance, and avoided capital debt. Use negative trends as triggers for training, architecture fixes, and model retraining.

Chapter 18
Measuring Progress

Institutions build trust and momentum by tracking operational, sustainability, experience, and resilience outcomes using transparent, mission-aligned performance metrics.

A sustainable smart campus isn't just about deploying a cloud platform, installing IoT sensors, or developing a 3D digital twin. Its success hinges on the institution's capacity to convincingly demonstrate—to its Board of Trustees, community, regulators, and itself—that the campus is safer, more resilient, cost-effective, and sustainable thanks to the combined efforts of IT and Facilities. Measurement is essential in turning technological objectives into concrete evidence, thereby fostering lasting trust within the institution.

Traditionally, modernization efforts tend to break at this point. In the old, mechanistic view of higher education, IT and Facilities operated independently, each with its own metrics. The Chief Technology Officer focused on network uptime, packet latency, server utilization, and cybersecurity. Meanwhile, the CFaO tracked boiler uptime, cost per square foot, energy use intensity (EUI), and deferred maintenance backlog. They operated in different spheres, reported to different committees, and measured different aspects of performance.

A smart campus needs a unified narrative. It relies on a common operational language that demonstrates how digital tools enhance the physical environment's performance. This improved performance supports the academic mission, safeguards research activities, and helps fulfill the institution's sustainability goals.

To manage this integrated environment, the Integrative Architect needs to use a comprehensive measurement approach. In a smart campus, metrics are closely interconnected; for example, an IT network metric like latency directly affects a Facilities metric such as actuator response time, which in turn influences an institutional metric like the carbon footprint. It's essential to evaluate the ecosystem as a whole rather than in isolation.

A useful way to frame this operational governance is simple: **OKRs set the direction; KPIs keep the engine healthy.** You need both, but you must use them differently.

The Engine (KPIs) vs. The Compass (OKRs)

If the smart campus is like a highly advanced, autonomous vehicle, then the Key Performance Indicators (KPIs) are its dashboard gauges. These gauges show the engine temperature, tire pressure, and current speed. They provide ongoing, baseline measurements of operational health.

Objectives and Key Results (OKRs) serve as the GPS navigation system. They set a bold destination for the upcoming quarter or year and outline the specific milestones needed to confirm you're heading in the right direction. When the CTO/CFaO sit down to design the measurement strategy, they must explicitly separate metrics that maintain stability (KPIs) from those that drive transformation (OKRs).

KPIs for Cyber-Physical Engine Health

KPIs in a converged environment need to bridge the IT/OT divide. A dashboard showing only network switch health is useless to a mechanic, and one showing only chiller temperatures is useless to a network security engineer. The shared KPI dashboard should represent the overall health of the entire system.

Operational Resilience and Uptime (The "Do No Harm" Metrics)

"No Harm" metrics serve as the core commitment of any smart campus upgrade: ensuring that new digital features do not compromise the stability, safety, or ongoing functionality of the physical environment they support. In a converged IT/OT campus, even small disruptions can impact research integrity, comfort, safety mechanisms, or energy management, making resilience a critical mission requirement, not just a technical goal. These metrics establish a baseline: modernization improves reliability without introducing new vulnerabilities or endangering operational stability.

- **Mean Time to Resolution (MTTR) for Cyber-Physical Faults**
  How quickly can the combined IT and Facilities team identify, diagnose, and fix a failure affecting both the network and mechanical systems?
- **Predictive Work Order Generation Rate**
  What proportion of maintenance work orders are automatically created by the AI Control Plane, predicting failures, compared to reactive tickets filed by occupants reporting issues? An increasing percentage indicates a shift from reactive troubleshooting to proactive management.
- **Sensor Mesh Health / Data Confidence Score**
  What percentage of campus IoT sensors are actively reporting accurate, calibrated data? Since an AI algorithm's effectiveness depends on data quality, if 15% of sensors are offline, the KPI dashboard should flag the analytics as "low confidence."

Cyber-Physical Security (The Fortress Metrics)

Cyber-Physical Security establishes the minimum protective standards for an institution in a landscape where digital breaches can lead to immediate physical impacts. As IT and OT systems merge, vulnerabilities in building controls, sensors, vendor access channels, and cloud-connected systems pose risks to safety, operational continuity, and trust. These metrics verify that each connected device, network segment, and integration path adheres to strict identity, segmentation, and monitoring criteria—creating a secure perimeter that contains breaches, prevents lateral movement, and shields the physical campus from increasingly advanced cyber-physical threats.

- **OT Microsegmentation Coverage**
  What percentage of legacy Building Automation Systems (BAS) and critical research equipment have been successfully isolated into Zero-Trust network segments?
- **Time-to-Patch for Critical OT Vulnerabilities**
  How quickly can the institution secure a newly discovered vulnerability in a mechanical edge controller without disrupting the building's physical operations?

Sustainability and Efficiency (The Decarbonization Metrics)

Sustainability and Efficiency metrics evaluate how well the campus minimizes energy waste, reduces carbon emissions, and converts its built environment into a high-performance, resource-efficient ecosystem. These metrics verify that each smart-campus upgrade makes a meaningful contribution to decarbonization efforts, covering real-time energy improvements and lifecycle-based asset choices. They transform sustainability from a once-a-year report into a daily operational practice—where data, automation, and smart systems consistently lower emissions while enhancing comfort, reliability, and financial management.

- **Occupancy-Adjusted Energy Use Intensity (EUI)**
  Traditional EUI measures energy per square foot. The smart campus KPI measures energy against real-time human utilization. Is the building efficiently shedding load when the spatial analytics indicate it is only 20% occupied?
- **Automated Load Shedding Success Rate**
  When the microgrid AI autonomously curtails HVAC load during a peak utility pricing event, how often does it succeed without triggering human thermal comfort complaints?

Human Factors and Operator Experience (The Cultural Metrics)

Cultural metrics evaluate whether modernization genuinely empowers those who operate, maintain, and engage with the smart campus daily. These indicators track cognitive load, training effectiveness, workflow ease, trust in analytics, and the overall human experience in increasingly digital settings. An upward trend in these metrics indicates that the technology is not just implemented but also understood and embraced, leading to improved performance and greater confidence among operational teams.

- **Nuisance Alarm Reduction Rate**
  A direct measure of "alarm fatigue." Are we successfully using analytics to suppress redundant, non-actionable alerts so operators only see critical signals?
- **Mobile Workflow Adoption Rate**
  What percentage of field mechanics are closing out their work orders, accessing digital twins, and viewing equipment histories via their mobile devices while standing at the physical asset?

Antifragility Metrics

Antifragility must be visible on a dashboard, not only felt in the control room. Leaders can extend current metrics with a dedicated **Antifragility Scorecard**. These KPIs tie directly to outcomes the Board and Cabinet value—reliability, safety, cost stewardship, and trust—while giving operators proof that stress is making the system smarter.

- **Adaptation Speed**
  Time from anomaly informs fleet-wide policy/model update.
  (Target: decreasing trend.)
- **Post-Stress Gain** —
  Improvements in prediction accuracy and in the handling of nuisance information reduce the number of alarms after an incident.
  (Target: increasing trend.)
- **Mean Time to Predict (MTTP)** —
  How far *before* failure does the system signal action, by asset class?
  (Target: increasing lead time.)
- **Semantic Integrity** —
  Drift rate in tags/ontologies affecting digital twins and analytics.
  (Target: decreasing drift.)
- **Capital Debt Avoided** —
  verified asset-life extension and projects deferred due to condition-based insight. (Target: increasing cumulative value.)

The OKRs: Crafting the Compass for Transformation

While KPIs run continuously in the background, OKRs are deployed to force strategic alignment and drive the institution through the phases of transformational change. OKRs require the CTO/CFaO to jointly commit to an ambitious outcome.

The Transition to Predictive Maintenance OKR

The Transition to Predictive Maintenance OKR measures the institution's shift from reactive, failure-driven operations to a proactive, intelligence-driven maintenance model. This metric focuses on how effectively the campus converts sensor telemetry, fault-detection analytics, and digital-twin insights into earlier interventions, fewer emergencies, and longer asset life. It signals whether modernization is truly increasing operational foresight—turning maintenance from "fixing what broke" into "preventing what would have failed."

Example:

Shift the central plant operations culture from reactive firefighting to AI-driven predictive stewardship to protect research continuity.

- **Key Result 1:** Reduce emergency, after-hours HVAC work orders in the science quad by 40% by the end of Q3.
- **Key Result 2:** Successfully train and certify 100% of the central plant operators on the new AI fault-detection dashboard.
- **Key Result 3:** Identify and resolve 50 cascading equipment failures autonomously generated by the AI *before* a physical breakdown occurs.

Bridging the IT/OT Security Gap OKR

Bridging the IT/OT Security Gap: OKR tracks the institution's progress in unifying two historically separate security models into a single, coordinated cyber-physical defense. As building systems, networks, sensors, and enterprise platforms converge, the attack surface expands—and so does the need for shared standards, shared visibility, and shared accountability. This metric measures how effectively the campus establishes common identity controls, segmentation patterns, vendor access governance, and monitoring practices across both digital and mechanical domains. It reflects the institution's readiness to treat IT and OT not as separate risk categories but as a single interconnected security perimeter that safeguards uptime, safety, and institutional trust.

Example:

Secure the campus's vulnerable operational technology without disrupting the physical and academic mission.

- **Key Result 1:** Complete a comprehensive cyber-physical asset inventory of 100% of the campus's legacy controllers and shadow IoT devices.
- **Key Result 2:** Successfully migrate the top 5 most vulnerable legacy buildings into isolated, Zero-Trust network segments by Q4.
- **Key Result 3:** Conduct two joint IT/Facilities tabletop exercises simulating a ransomware attack on the building automation system, achieving a simulated recovery time of under 4 hours.

The Decarbonization OKR

The Decarbonization OKR tracks how effectively the campus converts operational intelligence into measurable reductions in energy use, emissions, and environmental impact. It evaluates whether AI-driven optimization, grid-interactive controls, and real-time carbon visibility are producing verifiable progress toward net-zero commitments. This metric signals whether sustainability has become a governed, data-driven operational practice—not a pledge, not a project, but a continuous institutional capability.

Example:

Leverage the Digital Integration Fabric to aggressively reduce the institution's carbon footprint and utility costs during peak demand.

- **Key Result 1:** Integrate the campus event scheduling software (e.g., 25Live) directly with the BAS via API to automate HVAC setback schedules for all unoccupied lecture halls.
- **Key Result 2:** Achieve a 15% reduction in campus-wide peak electrical demand charges over the summer term using AI-driven thermal load shifting.
- **Key Result 3:** Publish a live, public-facing carbon reduction dashboard to the student body to prove institutional accountability.

The Steering Committee Dashboard

Ultimately, the data collected from KPIs and OKRs must be communicated to the Cabinet and the Board. The Board isn't interested in BACnet IP protocols, MQTT message brokers, or variable frequency drive settings. Instead, they focus on institutional risk, financial management, and the university's academic reputation.

The Smart Campus Steering Committee (SCSC) should utilize these OKRs and KPIs to develop a concise, executive-level narrative. When the CTO/CFaO addresses the Board, they need to present a unified dashboard that clearly demonstrates how technical performance translates into institutional value.

Example:

"By reducing our nuisance alarms by 60% (KPI), we recaptured 1,200 hours of skilled mechanic labor, allowing us to hit our target of transitioning to predictive maintenance (OKR), which ultimately prevented a chiller failure that would have destroyed $5 million in grant-funded research."

When measurement is developed as a collective, quantum-enabled capability, it transforms from a punitive means of micromanaging staff into a powerful tool for fostering institutional trust. It demonstrates that the human expertise of the workforce, enhanced by the intelligence of the smart campus, actively contributes to securing the university's future.

Chapter Summary

Chapter 18 outlines the institutional measurement framework that guarantees smart-campus modernization delivers clear, mission-aligned results. It highlights metrics as the governance tool that transforms IT/OT convergence, AI, and digital twin technologies into measurable operational, financial, sustainability, and experience improvements.

The chapter presents a dual-system approach:

- **KPIs** monitor the ongoing health of the cyber-physical campus—uptime, OT segmentation, predictive work-order rates, sensor reliability, and energy performance.
- **OKRs** drive strategic transformation—advancing predictive maintenance, strengthening cyber-physical security, accelerating decarbonization, and enhancing operator and occupant experience.

Progress is assessed through four key institutional outcome domains: operational resilience, cyber-physical security, sustainability and decarbonization, and human factors/operator experience. These indicators confirm that modernization enhances continuity, minimizes risk, decreases emissions, and promotes frontline adoption.

Model OKRs—focusing on predictive maintenance, IT/OT security alignment, and automated energy and carbon optimization—show how Facilities, IT, and Sustainability leaders track tangible improvements such as fewer emergencies, better segmentation, and lower peak demand.

Measurement is not about proving success;
It is about enabling learning.

Finally, the metrics should be presented in language accessible to executives—emphasizing how modernization protects research, reduces capital debt, enhances resilience, and enriches the campus experience. This method transforms measurement into a strategic asset, encouraging transparency, trust, learning, and continuous improvement across the institution.

Chapter 19

Application Use Case Patterns

Reusable use case patterns accelerate adoption and reduce risk.

A sustainable smart campus becomes real—not in an enterprise architecture diagram, but in a handful of repeatable "case patterns" that translate strategy into operational designs people can run. The goal of this chapter is not to catalog every possible use case (that was inadequately attempted in Chapter 11). It is to provide a small set of proven patterns—each with a common structure—that CTOs and CFaOs can reuse, adapt, and scale.

Each pattern below is intentionally framed the same way:

- **Context**
  (why this facility type matters and what "good" looks like)
- **OT baseline**
  (What must be true in controls, networks, and data before you add sophistication?)
- **Reference architecture**
  (how systems connect—safely—and where analytics and experience layers sit)
- **Governance**
  (who decides, who operates, who secures, who owns outcomes)
- **Metrics that matter**
  (shared CTO + Facilities measures, not vanity dashboards)
- **Lessons learned**
  (what typically goes wrong and how to avoid it)

Pattern 1: Smart Classroom

Context

Smart classrooms are where technology success becomes personal. If the room fails, instruction fails. The north star is not "more technology," but *frictionless teaching and learning*: reliable AV, predictable environmental comfort, simple room controls, and operational visibility that prevents failures before class begins. In a smart campus, the classroom is also a data source: utilization, comfort, and fault signals that help Facilities and IT continuously tune service quality.

OT Baseline

Before adding analytics or experience features, the classroom environment must be operationally stable:

- Room-level HVAC and ventilation controls function reliably and follow schedule/occupancy logic.
- AV and room control systems have standardized configurations and lifecycle support.

- Network pathways are segmented so teaching/learning networks and building controls don't become mutual blast radii.
- A single source of truth exists for room identity (room numbers, schedules, equipment inventory).

Reference Architecture

A classroom pattern works best when it treats "experience" and "operations" as two layers sitting on the same trusted foundation:

- **Edge layer:** room controller, occupancy sensing, CO_2/temperature sensors (where appropriate), AV control, lighting controls.
- **Control layer:** BMS interface for thermal comfort and ventilation; AV management platform for device health; scheduling system for class times.
- **Integration layer:** secure message bus/API gateway that exports *telemetry outward* (room status, faults, IAQ signals) without inviting inbound control risk.
- **Experience layer:** instructor UI (simple "start class" workflow), student support signals (e.g., "room is ready"), and a service view for dispatch.
- **Analytics layer:** reliability analytics (failure prediction), comfort exception detection, and utilization patterns.

The principle is simple: *operators* need the diagnostics; *instructors* need the room to "just work."

Governance

Smart classrooms expose an old gap: IT often owns AV and network, Facilities owns comfort, and academic units own expectations. The governance move is to define a joint service:

- A **classroom service owner** accountable for end-to-end experience (not just equipment).
- A **standard room profile** (Gold/Silver/Bronze) that constrains variability.
- A weekly **"room readiness" operational review**: top recurring faults, time-to-repair, and learning-impact incidents.

Metrics That Matter

- Class-start success rate (no disruption in first 5 minutes)
- Mean time to restore (MTTR) for in-room incidents
- Preventable failure rate (repeat incidents that should have been eliminated)
- Comfort exceptions during occupied class periods
- Instructor satisfaction (brief pulse survey tied to the room, not generic IT surveys)

Lessons Learned

Smart classrooms fail when complexity leaks into the teaching moment. Standardization is the hidden innovation. Limit room-control options, simplify interfaces, and invest in monitoring software that enables condition-based management, catching and addressing problems before they happen—so the classroom is repaired before the first class, not during it.[17]

Pattern 2: Residence Hall: Energy Optimization and Indoor Air Quality

Context

Residence halls run 24/7, have high occupant density, and carry reputational risk: comfort complaints, IAQ concerns, and energy waste all surface quickly and loudly. Here, the smart campus value proposition is tangible: reduce energy intensity while improving comfort and IAQ—especially during shoulder seasons, move-in surges, and extreme weather.

OT Baseline

Residence halls often contain a patchwork of HVAC types, controls vintages, and metering coverage. The baseline requirements:

- Functional zone control with validated sensors (bad sensors destroy trust).
- Metering or submetering strategy that at least supports building-level energy attribution.
- A clear separation between resident networks and building systems; no "convenient" shortcuts.
- A documented setpoint policy—agreed by student life and facilities—so optimization isn't mistaken for indifference.

Reference Architecture

Residence halls benefit from a "comfort and IAQ loop" linked to an "energy loop":

- **Sensing:** occupancy proxy signals, temperature/humidity, CO_2 (selectively), fan status, energy meters.
- **Control:** BMS sequences that support demand-controlled ventilation where appropriate, and sensible setbacks that respect living patterns.
- **Analytics:** anomaly detection for stuck dampers, simultaneous heating/cooling, excessive runtime, and IAQ exceptions.
- **Engagement:** resident communications that clarify what systems do and how issues are reported—because trust is part of IAQ.

A pragmatic approach is to start with fault detection and diagnostics (FDD) and targeted control tuning before attempting aggressive automation.

Governance

Residence halls require a partnership between Facilities, Student Life, Health and Safety, and IT Security. The key is to set explicit guardrails:

- Comfort/IAQ thresholds and escalation paths (who is paged, when).
- Privacy boundaries for occupancy proxies (no creepy inference, no surveillance ambiguity).
- Seasonal commissioning rhythm: pre-heating, pre-cooling, move-in readiness.

Metrics That Matter

- Energy Use Intensity (EUI) per hall and weather-normalized savings
- IAQ exception minutes (CO_2/temperature/humidity out of the agreed band during occupied periods)
- Comfort complaint rate per 100 residents (and time to resolution)
- Preventive vs reactive maintenance ratio for hall HVAC
- Fan energy and runtime anomalies (a leading indicator of waste)

Lessons Learned

If residents do not trust the system, they will override it—space heaters, open windows, improvised fixes—turning optimization into chaos. Pair technical changes with clear communication and a fast response loop. Also: resist campus-wide "one setpoint policy." Start with a few halls, validate outcomes, then scale.

Pattern 3: Central Plant Optimization

Context

The central plant is where financial stewardship and sustainability targets intersect. It is often the single largest controllable lever for decarbonization, cost reduction, and resilience. Unlike classrooms, plant optimization is not primarily a user-experience play; it is a systems-engineering and operational-discipline play—made safer and stronger by data, modeling, and automation.

OT baseline

Central plant success begins with instrumentation integrity and operator confidence:

- Calibrated flow, temperature, pressure, and power measurements.
- Stable controls and sequencing; documented control strategies.
- Clear cybersecurity posture: segmented plant networks, tightly governed remote access, and logging.
- Historian or reliable time-series data collection at appropriate resolution.

Reference Architecture

Central plants are ideal candidates for a staged approach that matures from insight to action:

- **Plant historian / time-series layer:** high-frequency operational data.
- **Optimization engine:** advisory setpoints first (human-in-the-loop), then supervised automation.
- **Integration with enterprise layers:** energy procurement, demand response, emissions signals, work order system for maintenance triggers.
- **Resilience layer:** scenarios for outage response, load shedding, and black-start procedures (modeled, rehearsed, then operationalized).

Digital twin capabilities often begin here—because the plant is a bounded system with strong physics, repeatable loads, and high payoff.

Governance

Optimization can fail if it is perceived as undermining operator expertise. Make governance explicit:

- Operators retain authority; optimization tools provide recommendations with explainability.
- A change control process for sequences and setpoint logic.
- A joint CTO–Facilities review of "automation boundaries" (what can auto-actuate, what must request approval).

Metrics That Matter

- Plant efficiency (kW/ton, boiler efficiency, heat rate)
- Weather-normalized cost per unit of delivered energy
- Peak demand reduction and demand-response performance
- Unplanned downtime and "near miss" events
- Emissions intensity (location-based and, where available, market-based tracking)

Lessons Learned

Central plant optimization is as much governance as it is math. The fastest way to stall is to automate setpoints without trust, visibility, and rollback. Start with advisory optimization, prove savings and stability, and only then move into closed-loop automation.

Pattern 4: Research Lab: Safety, Ventilation, and Compliance

Context

Labs are high-stakes environments: ventilation drives safety, compliance, and enormous energy consumption. Smart lab patterns focus on *risk-managed performance*: maintaining safe air change and pressure relationships while reducing waste through demand-based strategies where feasible. The lab is also where IT/OT convergence must be disciplined; the consequences of control failure are not inconvenience—they are exposure.

OT Baseline

- Verified ventilation performance: sensors, airflow measurement where needed, reliable control of VAV fume hoods, and room supply/exhaust.
- Clear mapping of lab types and hazard levels (not all labs can be treated the same).
- Fail-safe sequences and alarms are tested regularly.
- Network segmentation and strict vendor access control, with audit trails.

Reference Architecture

A safe, smart lab architecture privileges determinism and traceability:

- **Safety control loop:** pressure, airflow, hood sash position, exhaust fan status, alarm escalation.
- **Compliance and reporting layer:** continuous logs for critical parameters; integration with EHS systems.
- **Energy optimization layer:** targeted strategies (e.g., sash management feedback, schedule-aware ventilation adjustments) that never compromise safety thresholds.
- **Operations layer:** proactive maintenance triggers when ventilation performance drifts.

Where digital twin approaches are applied, the first goal is often *verification and simulation of safety scenarios*, not fancy visualization.

Governance

Labs require governance that includes EHS as a first-class owner:

- A lab safety technology steering group (Facilities, EHS, research leadership, IT Security).
- Policy for control changes and validation testing.
- Training and accountability for lab occupants (especially around sash behavior).

Metrics That Matter

- Safety exception events (pressure failures, airflow out-of-range, alarm response time)
- Ventilation energy intensity (normalized to lab type and occupancy)
- Fume hood sash compliance indicators (behavioral, not punitive)
- Deferred maintenance risk score for critical fans and controls

- Audit readiness (completeness of logs and validation records)

Lessons learned

The most common mistake is applying generic "smart building" optimization ideas to labs. Labs are different. Start with safety, integrity, and compliance-grade telemetry; optimize only within clearly defined, validated boundaries.

Pattern 5: Stadium and Arena: Event Operations, Safety, and Peak Loads

Context

Stadiums stress the campus in short bursts: spikes in occupancy, security demands, traffic, concessions, and energy loads. They are also a public-facing symbol of institutional competence. A smart stadium pattern focuses on *event readiness*, *crowd safety*, and *peak-load intelligence*—all while ensuring systems revert cleanly to normal operations afterward.

OT baseline

- Reliable building controls for large-zone HVAC and ventilation transitions.
- Integrated security systems with clear ownership and lawful, policy-aligned use.
- Communications readiness: signage, PA, and emergency alerts tested and rehearsed.
- Network resilience for high-density usage (without bridging into OT networks).

Reference Architecture

- **Event orchestration layer:** schedule triggers that precondition spaces, staff systems, and safety readiness.
- **Crowd safety layer:** integrated situational awareness (entry counts, camera analytics where permitted, queue monitoring), tied to incident management.
- **Energy peak management layer:** pre-cooling strategies, demand limiting, coordination with central plant and utility signals.
- **Post-event reversion:** automated resets and verification checks to avoid lingering faults and waste.

Governance

Stadium governance is operational choreography:

- A defined incident command structure for events (Facilities, Security, IT, Athletics operations).
- Pre-event readiness checklist ownership and sign-off.
- Clear privacy and acceptable-use boundaries for surveillance and analytics.

Metrics That Matter

- Event readiness completion rate (pre-event checks completed on time)
- Incident response time during events
- Peak demand profile and cost per event (energy + staffing impacts)
- Post-event recovery time (return to baseline operations)
- Safety perception measures (patron/staff feedback)

Lessons learned

Events expose integration debt. If systems are stitched together with one-off scripts and unowned workflows, failures will happen publicly. Treat the stadium as a proving ground for operational discipline: runbooks, drills, dashboards that operators actually use, and a post-event "after action" review that feeds continuous improvement.

Case Pattern 6: The Ambient Human Interface (The Smart App)

For the Chief Technology Officer (CTO) and Chief Facilities Officer (CFaO), the Digital Integration Fabric, the unified data ontology, and the API gateways are monumental achievements. However, to a 19-year-old undergraduate or a visiting adjunct professor, middleware is invisible. The end-user does not interact with the architecture; they interact with the physical campus. The "Smart App" is the presentation layer of the smart campus, the digital pane of glass that translates complex IT and OT integrations into an ambient, frictionless human experience.

The Problem: The Friction of Fragmentation

Historically, universities have subjected their communities to digital whiplash. A student navigating a legacy campus must use one app to check the bus schedule, log into a clunky web portal to view their grades, carry a physical plastic card to unlock their dorm, use a separate dining app to check their meal plan balance, and download yet another departmental app to see the weekend's events. This "app sprawl" is the byproduct of institutional silos. When every department procures its own shadow-IT mobile solution, the result is immense cognitive load for the user, a fractured data ecosystem for the institution, and a massive, ungovernable cybersecurity attack surface for the CTO.

The Solution Pattern: The Persona-Driven Smart App

The Smart App pattern abstracts the complexity of the university's backend systems. It replaces departmental app sprawl with a single, highly integrated mobile application powered by the Digital Integration Fabric. By using role-based personas, the app dynamically reconfigures itself based on who is logging in, ensuring that the user sees only the tools and telemetry relevant to their specific relationship with the campus.

Case Example: Northern Arizona University's NAUgo

Northern Arizona University (NAU) provides a gold-standard execution of this pattern through its **NAUgo** application. Built on the Modo Labs platform, NAUgo demonstrates how an institution can aggressively consolidate disparate campus systems into a single, award-winning digital ecosystem that deeply integrates academic, financial, and physical operations.

Key Integrations and Capabilities:

- **Identity as the Anchor (Digital JacksCard):** NAU eliminated the friction of physical plastic IDs by integrating HID mobile access and TouchNet directly into the app. Students add their "Digital JacksCard" to their Apple© Wallet or Android device and use it for contactless entry to residence halls, transit boarding, and point-of-sale dining purchases.
- **Academic and Financial Convergence (LOUIEgo):** Through deep integration with PeopleSoft (branded as "LOUIEgo" within the app), the app pulls academic and financial data out of legacy web portals. Students can view real-time class schedules, grades, financial aid holds, and dining plan balances in a native mobile interface.

- **Physical and Spatial Navigation:** The app taps into the campus's spatial truth. It features real-time transit and shuttle tracking, alongside interactive GIS mapping that guides users not just to buildings but to highly specific physical assets such as EV charging stations, gender-neutral restrooms, and ATMs.
- **Dynamic Personas:** NAUgo does not force a "one-size-fits-all" interface. It dynamically curates the experience using distinct personas: *Flagstaff Campus Student*, *Statewide/Online Student*, *Faculty/Staff*, and *Guest*. It even uses a temporary New Student Orientation persona to guide incoming first-year students through onboarding before "graduating" them to the standard student persona upon matriculation.

Architectural and Operational Integration

To achieve the NAUgo pattern, the CTO must enforce a strict API-first architecture. The mobile app itself houses very little data; rather, it acts as a secure viewing glass into the enterprise data lake. From an IT/OT convergence perspective, the Smart App is where physical telemetry meets user behavior. When the digital ID is used to unlock a physical door (OT), the transaction is authenticated through the enterprise active directory (IT), and the dining points are deducted from the financial system (IT). The app proves that convergence is not just a backend theory, but a frontline reality.

Governance and Change Management

The hardest part of building a unified Smart App is not the coding; it is the institutional politics. Departments are fiercely protective of their "own" apps and communication channels.

- **The Steering Committee as Gatekeeper**
  NAU successfully governs this through a cross-functional steering committee. If a department wants a new "screen" in the app or wishes to send a campus-wide push notification, they must submit a formal Content Request. The committee acts as the gatekeeper, ensuring the app remains a curated, high-value tool rather than becoming cluttered with digital junk mail.
- **Deprecating Shadow Apps**
  The CTO must use the launch of the unified app to decommission legacy, single-use departmental apps forcibly. This requires executive alignment and a commitment to migrating the displaced department's functionality into the new, centralized platform.

Metrics and Antifragile Outcomes

A successful Smart App provides immediate, measurable ROI and fundamentally shifts how the institution interacts with its community.

- **Daily Active Users (DAU) as a Proxy for Trust**
  NAUgo boasts roughly 25,000 unique daily users. High DAU proves that the app has successfully embedded itself into the necessary daily routines of the campus.
- **Cost Avoidance and Sustainability**
  By transitioning to the Digital JacksCard, the institution drastically reduces the recurring operational costs, plastic waste, and administrative overhead associated with printing and replacing thousands of physical ID cards every semester.
- **Crisis Agility**
  Because the app is the trusted, centralized communication hub, the institution possesses an immediate, high-success-rate channel (push notifications) for emergency life-safety alerts, weather closures, and dynamic operational shifts.

The Smart App proves to the campus community that the institution respects their time and their digital expectations, serving as the ultimate, visible capstone of the smart campus transformation.

A Closing Principle: Standardize the Loop, Not the Building

Across these six use case patterns, a clear pattern emerges: establish the operational cycle, standardize how integration is handled, and oversee the service—focusing on the service itself, not just the technology. This approach makes scaling less about copying specific projects and more about building a reusable capability: a combined IT and Facilities strength for secure connectivity, reliable data, and measurable results.

Chapter 20
The Campus Digital Twin

Implementing digital twins to connect spatial, static, and live data with AI into a predictive intelligence layer enhances decision-making and operational performance.

A campus digital twin should be viewed as an evolving capability rather than just a tool that the university acquires. It signifies a shift where the built environment—traditionally managed through fragmented systems, local expertise, and periodic reports—becomes a clear, measurable, and continuously improving enterprise asset. The digital twin is a decision-making system that acts as the campus's operational mirror: a dynamic model that integrates the university's assets, expected performance, and real-time operations. When implemented correctly, it not only reveals performance metrics but also enhances performance by closing the gap between observation and action, allowing the university to operate with greater foresight, confidence, and stewardship.

A digital twin is not just a data visualization project;
It is a real-time campus operating and decision system.

The justification of the digital twin is simple: Universities constantly struggle to reduce energy consumption, improve reliability, manage deferred maintenance, and enhance user experience, all while operating under limited budgets and staffing. The campus estate is particularly complex, comprising hundreds of buildings from various construction periods, featuring incompatible control systems, inconsistent documentation, and a disconnect between IT systems that manage information and OT systems that oversee physical operations.

A digital twin offers a new organizational approach by connecting these separate elements into a cohesive operational narrative—integrating spatial layouts, asset data, and real-time activities within a shared framework, all governed by common standards and used by campus maintenance teams.

Historical Foundations

The definition of a digital twin distinguishes it from traditional 3D models or static simulations through its dynamism and real-time connectivity. While a standard simulation might model how a building *could* perform under hypothetical conditions, a digital twin captures its current performance.[18] It is a living digital model that pulls in data from sensors, equipment, and building systems to provide a dynamic view of the facility's internal operations. For a campus, which often comprises a heterogeneous mix of historic structures and modern installations, the digital twin acts as a "single source of truth," centralizing information that is typically scattered across fragmented databases, emails, and paper-based maintenance logs.[19]

Exhibit 17. Evolution of Campus Asset Management

Management Stage	Primary Tools	Data Characteristics	Operational Mode
Traditional FM	Paper logs, spreadsheets, legacy CMMS	Static, siloed, often outdated [20]	Reactive (fix when broken) [11]
BIM-Centric	3D models (Revit), CAD	High geometric fidelity, static post-construction [6]	Design and Construction focused [21]
Descriptive Digital Twin	Integrated BIM-GIS-IoT platforms	Real-time, dynamic, bidirectional [22]	Proactive/Predictive [5]
Cognitive Twin	AI/ML modules, Autonomous agents	Interpretive, learning, prescriptive [23]	Autonomous/Prescriptive /Condition-Based[17]

Communicating the Why

Deploying a digital twin on a campus serves multiple purposes, including improving operational efficiency, enhancing sustainability, and enriching occupant experience. Digital twins offer real-time virtual models of physical assets, allowing facility managers to monitor systems remotely, perform diagnostics, and detect inefficiencies early, before they cause failures. This transition from reactive to predictive maintenance is especially important in academic settings, where building operations are crucial for fostering a productive environment for learning and research.[24] Additionally, digital twins deliver prioritized insights and extensive data to manage facilities more efficiently, supporting informed decisions throughout the entire asset lifecycle.

Strategic Operational Platform

A digital twin encompasses infrastructure, cybersecurity, vendor ecosystems, finance, operations, sustainability reporting, and human workflows. It shapes how issues are identified, how priorities are set, and how success is measured. In a university context, the twin should not be viewed merely as "a better dashboard." Instead, it should be seen as a strategic operational platform that boosts mission fulfillment: a smarter path to sustainability, a more disciplined approach to asset management, and a more reliable campus experience. When leadership communicates this consistently, patiently, and clearly, the twin can earn the legitimacy necessary to become a core element of the campus's operational framework.

From Descriptive to Cognitive: A Defining Digital Twin Maturity Shift

Most institutions start their digital twin journey with a descriptive model. At this stage, the twin mainly acts as a mirror of reality, combining spatial models, asset records, and live telemetry into a single view of the campus. It answers basic questions like 'What exists?' 'Where is it?' and 'What's happening now? replacing fragmented dashboards, drawings, and spreadsheets with a unified source of operational truth. This alone is a significant advancement, boosting visibility, speeding diagnoses, and reducing institutional 'blind hours.' However, despite its benefits, the descriptive twin remains primarily observational—showing conditions but not yet capable of reasoning about them.

A descriptive digital twin can be seen as an intelligent map rather than an autonomous intelligence system. It offers context and links various elements—connecting rooms to sensors, assets to alarms, and buildings to energy meters—but depends heavily on human input to turn insights into actions. Operators determine what is important, why it is important, and the next steps. Optimization is performed manually, and learning occurs informally. The twin itself does not evolve automatically; it stays static unless explicitly updated. While this stage stabilizes operations at a basic level, it does not yet enable transformation.

The Cognitive Digital Twin: From Representation to Reasoning

The cognitive digital twin marks a significant qualitative change rather than just an incremental update. It transitions from being a passive display of campus data to an active reasoning entity. Instead of merely showing telemetry, it now interprets it. Through the integration of AI models, historical data, operational rules, and institutional priorities, the cognitive twin can answer complex questions such as: Why is this occurring? What might happen next? What intervention would lead to the best result?

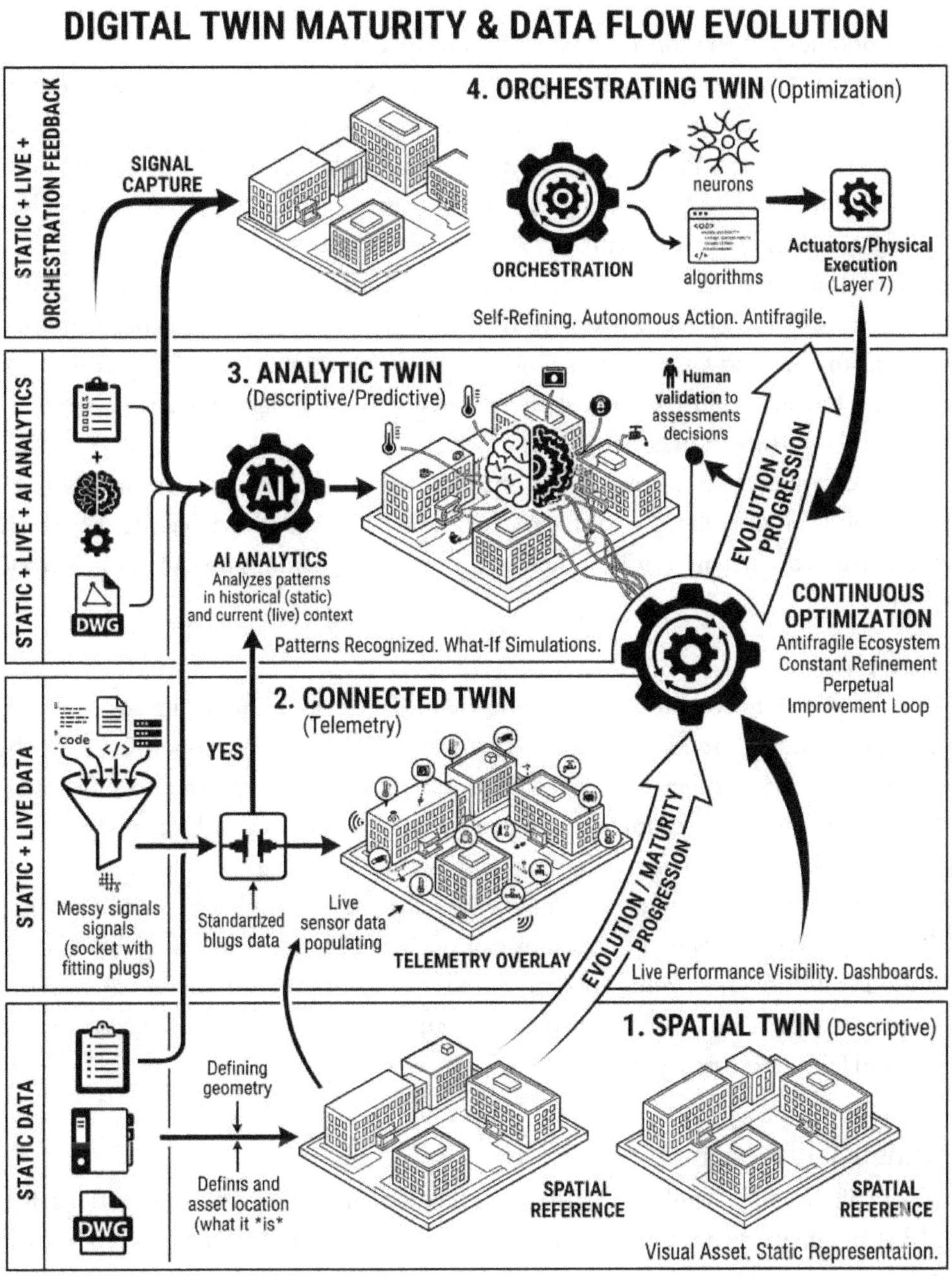

Figure 12. Digital Twin Maturity & Data Flow Evolution

This transition is characterized by the twins' shift from simple correlation to understanding causality and making predictions. Instead of just notifying staff about an inefficient chiller, the cognitive twin pinpoints the likely root cause, assesses failure risks, measures energy and carbon impacts, and suggests a prioritized response. It encompasses seasonality, occupancy trends, tariff schemes, and progress in asset health. Essentially, the twin functions as an institutional memory and analytical brain, constantly learning from results and improving its suggestions.

Importantly, a cognitive twin complements human judgment rather than replacing it. Human authorization still governs decisions, but these are now supported by detailed, prioritized, and contextual evidence. This enables the organization to move from reactive troubleshooting to proactive, condition-based operations, resolving issues before occupants notice them or they become failures.

Prescriptive and Autonomous Operation as a Maturity Threshold

The shift from descriptive to cognitive twins also unlocks access to prescriptive and eventually selective autonomous functions, which is a key sign of maturity for the smart campus. In prescriptive mode, the twin doesn't just forecast results but also suggests specific actions, provides confidence levels, analyzes trade-offs, and offers rollback options. For instance, it might advise pre-cooling a building overnight to reduce peak demand charges, or modify ventilation sequences based on predicted occupancy and air-quality trends.

As trust, governance, and data quality improve, some low-risk decisions can be automated, with the twin functioning within strict boundaries. These are not full "autopilot" modes but targeted optimizations—like load shifting, setpoint adjustments, and fault suppression—where the institution explicitly permits machine intervention. Human oversight remains essential, particularly for safety-critical or dignity-related systems, but operational responsibility is gradually shifting from humans to platforms.

Achieving this stage is more about institutional preparedness than technical achievement. Developing autonomous capabilities depends on disciplined governance, reliable data, well-defined decision rights, and a workforce that understands and trusts the system. Without these core elements, pursuing autonomy can introduce risk instead of resilience.

Why This Transition Matters

The shift from descriptive to cognitive digital twins defines whether a smart campus remains a **visualization project** or becomes an **operating model**. Descriptive twins improve awareness; cognitive twins change behavior. Descriptive twins shorten response time; cognitive twins reduce the need for response altogether. Most importantly, only cognitive and prescriptive twins enable the campus to scale intelligence without scaling labor—an essential requirement in an era of workforce scarcity, deferred maintenance, and rising operational complexity.

In maturity terms, this transition marks the point at which the campus stops merely *seeing itself* and starts **learning from itself**. The digital twin becomes not just a mirror of the institution, but a mechanism through which the institution continuously improves—turning volatility, stress, and constraint into sources of insight. This is the moment when the digital twin ceases to be a project deliverable and becomes a **core institutional capability**, embedded in daily operations and long-term strategy alike.

Measurable Financial Impact

The integration of digital twins significantly affects the financial profile of campus operations. Research suggests that combining BIM and digital twins can enable lifecycle cost savings of 20% to 40% through enhanced data integration and predictive maintenance.6 In mega-projects and large-scale campus developments, these technologies reduce design conflicts by 30% and rework by 25% during the construction phase alone, while providing a 15% total project cost saving through real-time sensor feedback and early problem recognition.6 Moreover, the digital twin's value extends to energy management and space utilization, helping managers identify ways to reduce energy consumption and potentially lower real estate costs by improving space efficiency.1

Digital Twin Architecture

The latest advancements in campus digital twins involve integrating different technical fields into a unified architecture. This integration goes beyond visual combination, focusing on semantic modeling of assets, spaces, and people so that AI can understand these relationships. Central to these systems is the combined use of BIM, GIS, and IoT.

The Synergy of BIM, GIS, and IoT

BIM serves as the structural foundation of the digital twin, providing high-fidelity 3D geometry and metadata for building components such as HVAC systems, plumbing, and electrical grids. However, BIM alone lacks the broader spatial context required to manage a campus. GIS integration addresses this by providing geographic context, allowing stakeholders to visualize assets within the physical campus layout and connect them to external infrastructure such as utilities and transportation networks. The live operational layer is provided by IoT integration, where unique API endpoints receive data streams from sensors and meters, mapping them directly to the corresponding elements in the 3D model.

Standard Ontologies for Smart-Campus Digital Twins

As universities update their operational technology systems, the main obstacle to scalability isn't a data shortage but a lack of shared understanding across different systems. Standard ontologies address this issue by establishing a unified language for buildings, equipment, rooms, sensors, and their interrelations. Three key standards—Brick Schema, RealEstateCore, and Digital Twins Definition Language (DTDL)—have become central to this emerging data framework.

Brick Schema: The Technical Equipment Ontology

Brick Schema offers a precise, engineering-grade vocabulary to describe mechanical and electrical systems—such as HVAC components, sensors, meters, points, and control relationships. It effectively captures the technical foundation of the built environment. Brick's detailed modeling of equipment and telemetry enables clear mappings among BAS systems, controllers, and real-time operational data.

RealEstateCore (REC):
The Organizational and Spatial Ontology

RealEstateCore complements Brick by capturing the business and spatial aspects of campus operations—including spaces, leases, zones, functions, and usage. It links building systems to real estate concepts such as occupancy, usage types, and cost centers. This forms the semantic connection essential for information analytics, capital planning, and space optimization.

DTDL:
The Cloud-Scale, Interoperability-Driven Modeling Layer

Microsoft's Digital Twins Definition Language (DTDL) provides a flexible, cloud-ready modeling framework that harmonizes multiple ontologies into a unified graph. It enables scalable digital-twin representations that integrate BACnet data, IoT telemetry, space models, and enterprise systems into a coherent "campus knowledge graph."

Why Ontological Standards are Important

For CTOs, CFaOs, and campus leaders, these ontologies go beyond technical interest; they act as strategic tools that transform fragmented building data into meaningful institutional insights.

The field is moving rapidly toward **ontology convergence**, with harmonization efforts between Brick and RealEstateCore eliminating the need for institutions to choose between standards. This convergence allows campuses to use "the best of both": Brick for technical equipment and RealEstateCore for spatial and organizational context. The result is a shared semantic foundation that enables AI systems to understand not only *what* telemetry is coming in, but *why it matters* and *what action it should trigger*. This semantic alignment is essential for:

- Accurate fault detection and diagnostics
- Consistent interoperability across vendors
- Scalable digital-twin deployments
- Reliable, cross-campus optimization and automation
- Future AI enablement, including predictive and agentic capabilities

Selecting between DTDL (Digital Twin Definition Language), RealEstateCore (REC), and Brick Schema isn't necessarily about choosing one over the other, but about focusing on the layer of the digital twin stack you want to emphasize. In many modern smart campus projects, these three often work in tandem. DTDL provides the modeling language (syntax), while RealEstateCore or Brick offers the domain-specific vocabulary.

When these harmonized ontologies are in place, institutions can:

- Reduce integration costs and vendor lock-in
- Deploy digital twins faster and at lower risk
- Support enterprise-grade sustainability reporting
- Enable measurable ROI from AI-driven optimization
- Build a future-proof architecture for decades of modernization

Standard ontologies are the data governance equivalent of adopting a common language. Once established, every modernization effort—from energy optimization to predictive maintenance to digital-twin operations—moves faster, costs less, and scales reliably across the entire campus portfolio.

The Types of Data that Feed the Digital Twin

The Digital twin must be fed more than just abstract data; it needs to incorporate the three distinct types of data that influence the built environment. The first is spatial data: maps, floor plans, geospatial data, and models that connect insights to specific locations.

The second type is static asset data, which includes information such as manuals, warranties, maintenance histories, commissioning reports, and the essential relationships between systems that provide the institution with information about its equipment identity, purpose, and performance history.

Third is operational data: live signals from meters, sensors, and building management systems that represent real-time conditions—such as temperatures, flows, energy consumption, occupancy, air quality, alarms, and various performance indicators.

The digital twin becomes significantly more effective when these data are incorporated into a managed "data estate," allowing the organization to go beyond visualization towards diagnosis, prediction, and optimization.

Standards-Based Approach

All smart campus initiatives encounter a common challenge: the campus infrastructure is heterogeneous. Buildings differ in age, control systems, renovation history, documentation quality, and local team practices. Consequently, the digital twin should be built for seamless integration rather than a single configuration. Using open protocols and standards-based approaches is crucial, as they guarantee that the twin can be expanded and adapted over time.

Similarly, API-first integration is essential, not just a luxury, allowing the twin to connect smoothly with existing systems like work order management, energy platforms, identity management, procurement, safety, communications, and analytics. Without this strategy, the digital twin could become an expensive, fragile custom tool. When managed correctly, the twin evolves into a scalable platform that expands through replication rather than reinvention.

Driving Organizational Change and Acceptance

Even with a sleek design, digital twin programs often face challenges unrelated to technology. The built environment is managed by people, not dashboards—many of whom have deep craft skills working under imperfect conditions. A twin provides a new level of transparency, revealing faults earlier, uncovering long-standing inefficiencies that were hidden, and sometimes exposing legacy constraints or delayed investments. This transparency may be seen as useful or as judgment. Leaders must ensure it is understood as assistance.

Organizational Change Management

At this stage, a university's change strategy is vital. The most effective way to implement it is to see it as widespread individual change throughout the organization. The **ADKAR (Awareness, Desire, Knowledge, Ability, and Reinforcement)** model is particularly useful here, as it urges leaders to do more than announce the change; they must also focus on sustaining adoption over time.

Awareness should be built through a narrative that links the twin to urgent institutional concerns such as outage costs, increasing deferred maintenance, risks associated with under-instrumented buildings, and the credibility needed to meet sustainability goals. Desire is stimulated by early involvement of facilities, energy managers, safety stakeholders, and IT partners, making the twin feel co-created rather than externally imposed.

Knowledge must be role-specific and grounded in actual campus systems, not generic vendor training. Ability is developed when the twin becomes integrated into daily tasks—prioritizing issues, scheduling maintenance, monitoring performance, and managing controls. Reinforcement, the most difficult step, requires leaders to sustain focus after launch by celebrating achievements, publicly tracking progress, adjusting strategies as needed, and creating feedback channels for frontline teams to shape the system's evolution.

Scaling Impactful Pilot Projects

A crucial leadership decision is to roll out digital twins gradually instead of all at once. Although they provide extensive insights, and institutions may feel compelled to implement them widely to justify expenses, campuses do not become smarter merely through declarations. Instead, they develop this intelligence by creating a strong foundation and progressively scaling effective practices.

The strategic approach begins by choosing a few ready and impactful projects. These initial projects serve as test cases in which the institution refines device standards, naming conventions, cybersecurity protocols, and integration techniques, and addresses practical concerns such as control precision and sensor calibration. Taking these steps helps establish the campus's credibility, paving the way for scaling. The most effective strategy is simple: earning early trust is more beneficial than trying to cover everything broadly with untested solutions. The initial aim is not to develop an ideal campus model but to create a trustworthy baseline.

A digital twin that provides consistent, useful diagnostics begins to influence staff behavior. It can, for instance, identify zones that are over-cooled due to a stuck damper, detect abnormal power use by equipment such as pumps, or show that lights remain on overnight in empty spaces because of scheduling errors. The value of the twin lies in delivering insights in a language that operations teams understand—fewer surprises, clearer causes, and faster resolutions.

Taking Operations Back to Building Design

Once visibility is established, the next phase involves **analysis and refinement.** This step evolves the digital twin from a simple alert mechanism to a tool that detects patterns and performance issues beyond mere faults. It allows the institution to compare actual operations with the original design, expected results, and internal standards. This helps differentiate between outright failure and ongoing underperformance. Such distinctions uncover much hidden value, as many campuses are not failing outright but are gradually losing efficiency—examples include control sequences drifting, setpoints being overridden for convenience, systems operating outside optimal ranges, or ventilation and lighting loads mismatching actual space use.

At this stage, insights need to be put into action; if analytics remain limited to reports or dashboards, they won't impact results. The real value of the digital twin lies in its ability to help prioritize actions—showing which problems need urgent focus, which investments deliver rapid gains, and which buildings have the greatest potential for performance enhancements.

Leverage Insights into Action

The third stage—**transform and expand**—signifies when a campus begins to view the twin as an operational tool rather than merely an analytical platform. At this stage, the focus shifts from merely deriving insights to implementing ongoing actions: HVAC strategies based on occupancy that cut energy consumption in unoccupied rooms without sacrificing comfort, automated processes turning diagnostic data into maintenance tickets with appropriate context, and more sophisticated energy optimization that accounts for time-of-use rates, peak demand, and grid carbon intensity.

In later stages, this involves smart load management, coordinating flexible loads and distributed energy resources like storage, solar, and EV charging to reduce costs, enhance resilience, and support decarbonization. Nevertheless, caution is essential in this phase. Automation shouldn't be rushed; it must be founded on trust. Systems that automate too early without establishing credibility often encounter resistance and risks. In contrast, those that automate after demonstrating reliability can be expanded and scaled effectively.

Expanding Learned Skills, Standards, and Conventions

Throughout these phases, the data estate acts as the central foundation impacting success. While most universities have data, fewer have well-managed and governed data. Creating a digital twin demands more than simple storage — it requires proper organization. Asset identifiers must be consistent across systems. Equipment relationships should be accurately mapped. Telemetry data needs standardization, including uniform naming conventions, units, and metadata. Spatial data must be precisely anchored to enable the detection and localization of faults. Lineage information should be clear enough so teams can trust the data they see. Furthermore, access controls must be strong enough to support responsible data use in both operational and academic contexts. Without these practices, a digital twin is just a collection of loosely connected visuals. With them, it becomes a trustworthy, accurate replica of operational reality.

Prioritizing Calculable ROI

Empirical studies on large-scale construction projects have validated the financial benefits of digital twin constructs. The integration of real-time sensor data with high-fidelity 3D models has been shown to reduce time overruns by an average of 20%. Furthermore, the transition from reactive to predictive maintenance strategies allowed for the identification of potential equipment failures weeks in advance, preventing costly unplanned downtime.[25] In academic building implementations, AI-driven digital twins for energy forecasting have enabled facility managers to identify hidden inefficiencies in MEP (mechanical, electrical, and plumbing) systems, leading to more sustainable operations.[26]

In this context, institutional leaders are anxious to leverage digital twins. Indeed, they can estimate avoided costs and projected energy savings using physics models and equipment data. But while these estimates are useful and often accurate in the right direction, they are not precise guarantees. Actual outcomes can be affected by many variables: weather changes, tariff policies, occupant habits, deferred maintenance, and operational overrides — factors that no model can fully account for. The best strategy is to treat twin-derived savings as a guide for prioritization rather than a strict promise, and to verify results through measurement and verification. This maintains credibility and allows for continuous improvement without turning the twin into an unreliable promise.

Updating Legacy Facilities

Legacy buildings demand special attention since they form the backbone of most campuses. Many structures constructed before 2000 lack sensor density, modern controls, and secure network infrastructure, often featuring limited panel capacity, cramped mechanical rooms, and proprietary or analog controls. Documentation can also be incomplete. Effective upgrades depend on pragmatic strategies: employing modular and wireless sensors where appropriate, phased electrical updates, standardized secure network closets, carefully sequenced work to minimize disruptions, and recognizing that the new systems' reliability depends on the existing controls and instrumentation. Retrofitting these older buildings is more than a technical challenge; it offers an opportunity to prolong their lifespan, reduce emissions, and improve learning and working environments without starting from scratch.

Keep the Focus on People

Implementing a digital twin is fundamentally a leadership challenge in enhancing institutional intelligence. It begins with establishing a precise baseline, moves on to turning insights into actionable steps, and evolves by converting validated insights into continuous actions. Achieving scale depends on interoperability, strong governance, and replication. Success relies on change leadership that respects frontline teams' expertise, builds trust, and incorporates the twin into daily campus operations and accountability structures. The universities that succeed will be those that intentionally deploy technology thoughtfully—developing a credible, trustworthy, and well-governed digital twin that can serve as a durable institutional asset.

Implementation Strategies and Best Practices

Implementing a campus-wide digital twin is a complex undertaking that requires a structured approach to technical, organizational, and financial variables. Industry experts advocate for a multi-stage roadmap that begins long before the final digital model is rendered. Building a digital twin is a transformative, multi-phase journey that evolves from basic data collection to sophisticated, autonomous orchestration. Below is an expanded detail on each of the six steps, incorporating best practices and risk mitigation strategies to ensure a successful implementation.

Exhibit 18. Digital Twin Implementation Roadmap

Timeline	Phase	Key Activities and Deliverables
Stage 1	Assessment and Strategy	Evaluate current BIM/IT capabilities; identify ROI-driven value drivers; assemble the cross-functional team.
Stage 2	Technology Foundation	Assess hardware needs for IoT sensors; upgrade site connectivity; define hardware deployment for drones/scanners.
Stage 3	Data Foundation	Establish data standards and naming conventions; implement a data governance plan defining ownership and access.
Stage 4	Living Twin Activation	Connect live telemetry streams to the 3D model to enable bidirectional communication between physical and virtual assets.
Stage 5	Testing and Validation	Measure results against initial KPIs; document lessons learned; run pilot "what-if" scenarios.
Stage 6	Expansion and Scaling	Deploy advanced AI modules; expand capabilities to additional buildings or campus infrastructure.

Stage 1: Assessment and Strategy

This initial stage anchors your Digital Twin initiative project in a mission-aligned vision.

Tasks

Evaluate your current BIM/OT capabilities, static data repositories, and overall institutional readiness. Leverage existing or assemble a cross-functional "Smart Campus Steering Council" that is tied to IT/OT governance structures and includes IT, facilities, academics, and sustainability leadership to ensure diverse perspectives are represented.

Best Practices

Conduct a comprehensive assessment of existing OT infrastructure (IoT sensors, utility meters, BACNet), OT systems (BMS, lighting controls, etc.), and capabilities (e.g., work management systems and processes). Define clear ROI-driven value drivers (e.g., energy savings, decarbonization, campus safety, and health), potential funding sources, and secure executive sponsorship.

Risks to Avoid

Sidestepping the "Why." Starting without a documented vision leads to disjointed technology experiments rather than a unified institutional transformation.

Stage 2: Technology Foundation

This stage builds the "digital nervous system" required for data flow.

Tasks

Assess hardware needs for IoT sensors, modernize campus network connectivity (e.g., Wi-Fi 6 or 5G), and strengthen cybersecurity.

Best Practices

Prioritize modular, adaptive, and intelligent architecture to avoid vendor lock-in. Ensure secure, high-speed, low-latency networks are in place to support integrations.

Risks to Avoid

Ignoring legacy infrastructure. Most campus systems weren't designed to be intelligent; they plan for technical orchestration between analog building controls and new digital platforms.

Stage 3: Data Foundation

Data is the lifeblood of the digital twin; this phase focuses on establishing a "Data Estate".

Tasks

Establish data standards and naming conventions, and implement a data governance plan that defines ownership, quality, and access.

Categorize data into spatial (3D layouts), static (asset specs), and live (real-time telemetry) streams.

Best Practices

Implement a "sources of truth" by breaking down departmental silos through API-driven data integration. Embed data privacy, security, and ethical design principles from the start.

Risks to Avoid

Siloed Data. Data trapped in proprietary systems prevents the creation of a comprehensive virtual replica, leading to "partial truths" in decision-making.

Stage 4: Living Twin Activation

At this stage, the model shifts from a static map to a dynamic "Living Twin".

Tasks

Connect live telemetry streams to the 3D model and activate bidirectional communication. Use the twin to visualize and analyze physical assets, systems, and processes in real-time.

Best Practices

Implement automated fault detection and diagnostics to move from reactive to proactive maintenance. Use AI to simulate "what-if" scenarios, such as modeling the impact of a power outage.

Risks to Avoid

Technical Debt. Unmanaged IoT endpoints can create hidden backdoors; cybersecurity must be woven into the design through zero-trust architectures.

Step 5: Testing and Validation

Benchmarking results measure success against the initial Key Performance Indicators (KPIs).

Tasks

Run pilot "what-if" scenarios and document lessons learned. Measure energy reduction, occupant comfort, and equipment reliability against established baselines.

Best Practices

Adopt a "build-measure-learn" loop. Use the digital twin to validate plans and prepare for emergencies with high precision.

Risks to Avoid

Misinterpreting Savings. Predicted "avoided costs" are data-driven estimates; acknowledge that external variables such as weather and occupant behavior can influence actual financial outcomes.

Step 6: Expansion and Scaling

The final stage moves beyond pilot buildings to the entire campus ecosystem.

Tasks

Deploy advanced AI modules and expand capabilities to additional buildings or campus infrastructure. Evolve toward an "Autonomous Digital Twin" capable of independent actions.

Best Practices

Position the campus as a "Living Lab" for research, inviting faculty and students to co-innovate on the platform. Continue to monitor, report, and update the master plan to reflect new opportunities as they arise.

Risks to Avoid

"Boiling the Ocean." Do not scale too rapidly; expansion should prioritize "low-hanging fruit" or buildings with the highest priority needs first to maintain momentum.

Additional Implementation Considerations

New and Renovation Construction

One of the most consistently cited best practices is to initiate the digital twin during the planning and design phase of a project, rather than waiting until post-construction. Early incorporation enables more accurate modeling and improved data collection during the construction phase. For property owners, the impact of every dollar spent early in the lifecycle is an order of magnitude higher than a dollar spent five years later as a late adopter. This approach prevents the "information loss" that typically occurs during the handover from contractors to facility managers.

Investing in the "smartification" of documents during the design phase is another critical best practice. This involves removing the information from static PDFs and CAD drawings and recreating it with relevant metadata to ensure it is queryable within the digital twin environment.[27]

A structured, automated facility handover process can save significant time during the operations phase; studies indicate that some organizations spend up to 2 years organizing documentation after an insufficient handover. By digitizing this information into a "digital backbone," organizations ensure that maintenance technicians have immediate access to the specs and maintenance history they need to troubleshoot effectively.

The "Human Factor"

Despite the technological advancements, a "mega-study" of LLM-powered digital twins revealed that they are currently more reliable for capturing relative differences between participants than for making exact individual-level predictions. [28]The study found a modest correlation (approx. 0.2) between human and "twin" answers in survey settings, suggesting that while AI can simulate general behaviors, it cannot yet fully replace human judgment in complex, high-stakes individual scenarios. This reinforces the best practice that digital twins should be used to *amplify* the efforts of human facility teams rather than *replace* them.

Engage Stakeholders Early and Often

The success of a digital twin implementation is often more dependent on organizational alignment than technical prowess. A major hurdle is resistance to change, often expressed through the acceptance of the sub-optimal status quo. To overcome this, it is essential to involve major stakeholders early, including operations supervisors who must transition from paper-based processes to digital interfaces. Demonstrating "early wins" is a key strategy for building confidence and securing long-term buy-in from leadership.[29]

Defining a clear ROI is vital for justifying the initial investment, which can be high due to the costs of IoT infrastructure and professional services.[30] The business case should be itemized across several value categories:

- **Cost Savings:** Reduced downtime, improved energy efficiency, and lower maintenance costs through proactive interventions.
- **Efficiency Gains:** Faster troubleshooting and service response times for maintenance crews.
- **Risk Mitigation:** Improved safety through emergency scenario simulation and better compliance reporting.
- **Sustainability:** Enhanced tracking of carbon footprints and progress toward environmental targets.

AI and Digital Twins

Artificial Intelligence is the catalyst that transforms a digital twin from a passive monitoring tool into an intelligent, autonomous entity. By integrating machine learning (ML) and Generative AI based on RAGs/LLMs with a robust data estate behind the digital twin, campuses can move towards anomaly detection and condition-based predictive management.

Anomaly Detection and Condition-Based/Predictive Maintenance

In the predictive domain, AI models analyze historical and real-time telemetry to forecast future system states and identify deviations from expected behavior. For example, Long Short-Term Memory (LSTM) neural networks—a type of recurrent AI neural network—are particularly effective at identifying temporal patterns in building data. These models can be trained on years of historical energy consumption and weather data to generate highly accurate energy forecasts, allowing facility managers to optimize HVAC schedules based on upcoming climate conditions and occupancy patterns.

Anomaly detection relies on statistical and unsupervised learning methods to identify when a system is behaving "abnormally". This is more sophisticated than simple threshold alerts (e.g., alert if the temperature is> 80°F). AI-driven anomaly detection uses a "dynamic baseline," which continuously learns what is normal for a specific system under given load conditions. If the AI identifies a temperature spike in a server room as atypical for the current cooling cycle, it can trigger an alert or an automated response before a physical failure occurs. Also, in renewable energy systems, such as campus microgrids, AI-powered digital twins have achieved fault-prediction accuracy for inverters and batteries exceeding 90%.

Generative AI and the Conversational Interface

The most recent innovation in the field is the integration of Generative AI (GenAI) to bridge the "field knowledge gap". As a significant portion of the specialized maintenance workforce approaches retirement, organizations risk losing decades of institutional knowledge. GenAI-powered "Copilots" address this by providing a natural language interface grounded in the building's specific data and documentation.

Willow Copilot, for example, allows technicians to interact with the digital twin using natural language. A technician can ask, "How do I fix the high-pressure alert on the chiller?" and the AI will pull relevant troubleshooting steps from vectorized O&M manuals and the asset's specific maintenance history in the Knowledge Graph. This use of Retrieval-Augmented Generation (RAG) ensures that the AI's responses are technically accurate and specific to the customer's actual facility. Furthermore, GenAI is being used to automate the digitization process itself, extracting structured knowledge from legacy blueprints and diagrams to build the digital twin's "brain" at a speed that traditional manual entry cannot match.

Exhibit 19. AI Applications in Campus Digital Twins

AI Methodology	Primary Application in DT	Operational Benefit
Supervised Learning	Asset segmentation and classification	Automated population of Knowledge Graphs with correct asset tags.
LSTM (Deep Learning)	Temporal pattern prediction (Energy, HVAC)	High-accuracy forecasting for demand-side management and energy load balancing.
Generative AI (RAGs/LLMs)	Conversational interfaces (Copilots), automated digitization	Bridging the workforce knowledge gap and accelerating twin onboarding.
Reinforcement Learning	Autonomous system control and optimization	Dynamic setpoint adjustments and valve control without manual intervention.
Unsupervised Learning	Anomaly detection and clustering	Identification of subtle system drifts and hidden inefficiencies in telemetry streams.

Incorporating Data to Enable AI: Engineering Requirements

For AI to provide reliable insights, the data infrastructure supporting the digital twin must meet rigorous engineering standards. This process, often called "data prep," is the most critical factor in the success of AI-driven projects.

Data Hygiene and Quality Gates

The principle of "garbage in, garbage out" is particularly relevant for digital twins. Inaccurate or incomplete data can lead an AI model to produce unreliable predictions or miss a critical failure signal. To ensure high standards, data must pass through "quality gates" along the path from the sensor to the dashboard.

- **Sensor Gate**
  Verifies that sensors are alive and calibrated, ensuring that readings fall within physically possible limits (e.g., no negative flow in a pipe).
- **Gateway Gate**
  Checks for timeliness and drops duplicate or delayed messages that might distort the time-series analysis.

- **Data Platform Gate**
  Validates that the schema (units, naming conventions) matches the system requirements before the data is committed to the main database.
- **Digital Twin Engine Gate**
  Uses physics-based rules to check for consistency across the system (e.g., the temperature must increase across a heating coil when the valve is open).

Each critical dataset should have a human owner—such as a field engineer for sensor quality and a data scientist for model quality—to ensure accountability if data health degrades. Furthermore, the digital twin itself can serve as a guard: by comparing its predictions with real-world outcomes, the system can identify "sick" sensors that are drifting over time.

Normalization, Transformation, and Feature Engineering

Data gathered from a campus is often heterogeneous in its scale and format. To make it suitable for ML training, it must undergo normalization and transformation.

- **Min-Max Normalization**
  Rescales data values to a range between 0 and 1. This ensures that features with large numerical ranges (such as energy consumption in kWh) do not drown out features with smaller ranges (such as temperature) during model training.
- **Z-score Standardization**
  Adjusts data to have a mean of zero and a standard deviation of one, which is helpful for models that assume a Gaussian distribution.
- **Feature Engineering**
  This is the process of creating new variables from raw data to help the AI identify patterns. For a campus digital twin, this might include "Day Type" (workday vs. holiday) or "Building Occupancy Percentage" as features that help the AI understand that high energy use is normal during exam weeks but abnormal during winter break.

Data Architecture

The sheer volume of raw data generated by a campus digital twin can reach multiple terabytes, necessitating scalable storage solutions. Cloud-native data lakes, such as those built on Azure Data Lake Storage Gen2, provide a centralized repository for storing diverse data types—including telemetry, images, and documents—without the rigid constraints of a traditional data warehouse. These environments are optimized for the intensive compute workloads required for AI model training.[31]

However, not all data should be sent to the cloud. For latency-sensitive use cases, such as autonomous valve control or AR/VR interfaces, hosting the digital twin and its processing at the edge is best practice. High-sensitivity AR/VR applications require a round-trip time (RTT) latency of less than 5 ms, which can only be achieved if the data is processed near its physical source. A hybrid architecture—where large datasets remain at the edge for low-latency tasks while AI/ML services are accessed in the public cloud for complex analysis—offers the best balance of performance and cost.

The Future Roadmap: Prescriptive and Autonomous Buildings

The ultimate trajectory for campus digital twin technology is the evolution toward "prescriptive" and "autonomous" systems. This maturity level represents the "Cognitive Digital Twin," where the system not only monitors and simulates but also interprets data, learns from it, and makes decisions independently.[32]

Exhibit 20. Maturity Levels of Digital Twin Evolution

Maturity Level	Characteristic	Key Technology
Level 1: Descriptive	Passively mirrors current conditions.	BIM + IoT Dashboards.
Level 2: Predictive	Forecasts future states and failures.	Machine Learning (LSTM), Anomaly detection.
Level 3: Prescriptive	Suggests optimal actions to operators.	Simulation-based decision support, Opti-algorithms.
Level 4: Autonomous	Executes actions without human intervention.	Multi-agent systems (MAS). Reinforcement Learning,

The transition from Level 3 to Level 4 involves the deployment of "multi-agent systems" (MAS), where autonomous software agents work collaboratively with both each other and human operators. These agents communicate their roles and states through a digital twin "service layer," allowing for a gradual, modular transition from classical automation to full autonomy.[33] For example, in a campus microgrid, one agent might manage EV charging while another manages solar storage; the digital twin serves as the shared medium that allows these agents to coordinate their actions to balance the grid's overall load.[34]

The value of a digital twin is realized only through decisions and actions taken.
—never by dashboards alone.

Key Strategies

Digital twins are rapidly evolving from static visual representations to a true, AI-enabled operating system for the built world—an operational AI platform that continuously connects data, skills, and agents to drive decisions, execute work, and deliver outcomes at campus and portfolio scale. In the near future, digital twins will be extensively integrated into campus systems and embedded throughout enterprise workflows, facilitating quicker and more confident decision-making at scale.

1. Speed to value is now the real maturity test

The maturity of digital twins is gauged by the speed at which they deliver operational value, rather than by technical complexity or model accuracy. Owners and operators face pressure to reduce energy consumption, improve reliability, and manage expanding portfolios with leaner teams. As a result, projects that require months or years to show results are no longer practical.

The leading digital twin platforms deliver results within days or weeks by offering use cases as ready-to-deploy capabilities rather than custom projects. Fast delivery of value indicates more than just technical preparedness; it shows the platform's understanding of actual operational workflows, limitations, and priorities. When organizations see quick results, their confidence increases, adoption speeds up, and they become more willing to expand.

2. Champions determine success

Technology alone doesn't lead transformation—people are essential. The success of implementing a digital twin depends on an internal advocate responsible for the results, not merely the setup. This advocate connects strategy with execution, ensuring the platform's features support real-world workflows.

Champions are essential for promoting adoption, transforming workflows, and maintaining change after the initial implementation. Without champions, digital twins might be seen as mere innovation trials instead of integral operational tools. When supported by strong champions, digital twins integrate into everyday decision-making, shaping team priorities, actions, and success metrics across a portfolio.

3. Trust and transparency matter more than perfect models

In operational settings, trust outweighs precision. A model that is perfectly accurate but not understood or trusted by users won't influence decisions. Operators require clear information on the system's knowledge, its limitations, and how it produces recommendations.

Transparency fosters trust by establishing realistic expectations and clarifying limitations. When users understand the confidence level behind insights or decisions, they are more inclined to act on them. Digital twins that prioritize explainability and accountability, rather than striving for perfect theoretical accuracy, promote consistent, large-scale use and ultimately provide greater value than high-fidelity models that go unused.

4. Dashboards are giving way to decision engines

Dashboards were created for interpretation, but now the operational environment requires action as building data volume and diversity increase, relying on humans to analyze dashboards and decide the next steps becomes unsustainable.

Modern digital twins are evolving from simple reporting tools to active decision-making engines that analyze data, incorporate intelligence, and suggest clear, prioritized actions. AI copilots help reduce cognitive effort by highlighting important information, explaining its significance, and recommending next steps. This transformation enables operators to act more quickly and make better decisions without being overwhelmed by data, fundamentally changing the way work is performed.

5. Digital Twins are becoming the operating system of the built world

Digital twins are transforming from isolated visual models into the central connecting layer that integrates systems, data, and workflows within the built environment. Instead of acting solely as standalone representations, they now serve as the operational backbone, enabling coordination of work orders, energy strategies, maintenance, capital planning, and performance management.

Chapter Summary

Research and lived experiences show that the campus digital twin represents more than just a slight improvement to facility management software; it signifies a fundamental shift in how the built environment is managed. By consolidating data from BIM, GIS, and IoT into a semantic knowledge graph, organizations can achieve a new level of operational intelligence that was once unattainable. The use of AI—from predictive machine learning models to GenAI-driven conversational tools—addresses essential workforce knowledge gaps and supports proactive energy management.

However, the realization of these benefits is contingent upon adherence to strict data engineering best practices. The creation of a Data Estate with rigorous quality gates is essential for ensuring that AI-driven insights are trusted and actionable. Platforms like Willow have set the benchmark for this "state-of-the-art" approach, demonstrating composability and extensibility built on standardized ontologies, secure integrations, and cloud-native architectures, and proving they can deliver measurable, significant lifecycle cost savings.

For campus administrators and IT leaders, success relies on early adoption, strong data governance, and a strategic vision that considers the digital twin as the crucial foundation for building an autonomous, sustainable campus of the future. Instead of advocating a "big-bang" approach, the Manifesto describes the digital twin as an evolving layer for control and decision-making, one that develops in tandem with the institution's data systems and governance frameworks. It emphasizes the need to integrate the twin into daily operations—such as work management, sustainability reporting, emergency response, and scenario planning—ensuring that insights consistently translate into action.

The chapter emphasizes that the digital twin's value depends on human actions, decisions, defined decision rights, trained operators, and feedback mechanisms that verify results and enhance performance over time. When implemented effectively, the campus digital twin becomes a sustainable asset for the institution—constantly minimizing blind spots, boosting antifragility, and transforming the built environment into a dynamic source of operational intelligence.

Chapter 21

Bringing It All Together

Integrated governance, digital architecture, data strategy, and leadership
Create a stable, future-ready operating model for the AI-enabled sustainable smart campus.

An AI-enabled, sustainable smart campus isn't a simple product or a one-time program to implement. Instead, it's an institutional capability that you develop—an operational approach that connects mission, location, data, and people into a cohesive, learning system. When effectively established, it transforms how a university manages resources, supports its community, and responds to uncertainty. It reflects the institution's character, evident in building performance, student experience quality, digital service reliability, and the credibility of sustainability efforts.

Throughout this manifesto, a consistent theme has emerged: the smart campus is mainly about strategy and governance, with technology as an enabler. Sensors, AI, digital twins, and advanced networks are important. Still, their significance is maximized when they support specific goals such as reducing carbon emissions, promoting healthier environments, ensuring community safety, enhancing operational resilience, and providing more equitable access to opportunities.

Core Commitments That Make the Vision Real

Bringing it all together requires commitments that are both practical and strategic:

- Leadership and collaboration must become structural, not episodic—engaging students, faculty, staff, and external partners in ways that shape priorities and sustain momentum.
- Sustainability must be embedded across planning and operations, supported by goals, metrics, and accountability rather than slogans.
- Student experience must remain central: the most sophisticated campus is a failure if it does not support learning, well-being, belonging, and accessibility.
- Infrastructure modernization must be pursued with cybersecurity and interoperability as first principles.
- Community engagement must extend beyond campus borders, so the institution learns from others and contributes to regional and global solutions.
- Innovation and research must be operationalized through pilots and living labs that mature into scalable patterns.
- Digital transformation must be framed as stewardship—deploying IoT, AI, analytics, and digital twins in service of mission outcomes, not technology for its own sake.

Higher education holds a unique opportunity. Universities, as institutions, possess the moral authority, research expertise, and convening influence necessary to credibly lead on issues such as climate, ethics, and technology-driven social progress. AI-powered sustainable smart campuses serve as a prominent example of this leadership—they effectively turn values into tangible realities.

The task ahead isn't just about modernizing buildings or adopting new platforms. It's about creating campuses that learn—reducing their environmental footprint while enhancing the user experience, building trust while expanding capabilities, and addressing complexity with wisdom rather than fragility.

The most effective way to foresee the future is by shaping it. In higher education, this involves designing campuses that reward collaboration, foster a legacy of learning and discovery, and promote environmental responsibility, social equity, and ethical use of technology—enhancing sustainability, economic strength, and fair influence for future generations.

A New Institutional Operating Model

From Silos to Shared Outcomes

Historically, campuses have been managed in separate domains. IT handled networks, identity, and enterprise systems. Facilities managed buildings, energy, maintenance, and capital projects. Sustainability teams focused on climate commitments, reporting, and behavioral initiatives. While each domain delivered value, it also faced limitations imposed by the others.

The AI-driven sustainable smart campus promotes a collaborative approach: shared outcomes, infrastructure, and governance. This requires CTOs and CFaOs to shift from occasional coordination to continuous operational partnership. The "convergence" mentioned throughout this manifesto is not just a theoretical concept; it reflects the everyday management of large-scale cyber-physical systems. The campus data becomes a collective asset. The digital twin serves as a shared tool. Additionally, risk assessment, metrics, and investment decisions are all shared responsibilities.

This convergence is not only about efficiency. It is about integrity. In a campus where systems influence safety, air quality, learning continuity, and energy performance, the institution must be able to explain—not just internally but publicly—how decisions are made, what tradeoffs are being optimized, and how technology remains aligned with values.

A New Normal and Persistent State

Furthermore, smart campus initiatives are often misinterpreted as short-term projects—limited-time programs, technology rollouts, or pilot collections that end once systems are implemented or dashboards are activated. This perspective is not only incomplete but also actively harmful. An AI-enabled sustainable smart campus should be seen as an ongoing institutional capability—integrated into the university's standard procedures, governance, and daily decision-making. Similar to cybersecurity, financial management, or academic quality assurance, the smart campus is not a finished product. It is an ongoing operation for the institution.

When viewed as a permanent capability rather than a one-time project, the smart campus becomes integral to the university's ongoing planning, construction, operations, assessment, and continuous improvement. Governance structures, capital planning, procurement standards, cybersecurity, facilities management, sustainability reports, and academic collaborations are all guided by a unified, institutional operating model. Data is managed as a long-term asset. Digital twins are kept as continuously updated sources of operational facts. AI systems are regularly monitored, audited, and improved as part of routine operations. Crucially, roles, responsibilities, and decision-making authority are formalized to ensure progress is not reliant on individual champions or temporary funding. Still, it endures through leadership changes, budget cycles, and technological shifts.

Redefining the smart campus in this way signifies a fundamental shift—from episodic updates to ongoing institutional learning and development. Instead of viewing it as a one-time upgrade, the campus evolves into a living system that continuously senses, analyzes, adapts, and improves itself. Success is now judged not by the number of connected devices or platforms, but by the consistent delivery of intelligence, sustainability, resilience, and positive human experiences as core outcomes of the university's operations. From this perspective, the smart campus isn't a separate initiative but a core capability integral to the institution's mission.

Evolving Business Models: Financing the Smart Campus

As sustainability demands grow and digital expectations increase, financing strategies need to adapt. Traditional capital planning typically supports large, isolated projects like renovations, chiller replacements, or new buildings. However, smart campus features operate differently. They generate ongoing value via optimization, continuous commissioning, and learning loops. Their costs also continue through operating budgets, vendor management, cybersecurity, and workforce development. The institutions that lead will increasingly adopt financing approaches that match this reality.

Performance Investment

Performance-based investments will become more common, tying vendor compensation or internal funding triggers to measurable outcomes such as energy reduction, uptime, comfort compliance, or service response times. Public-private partnerships can accelerate modernization when carefully structured—especially when the institution wants to de-risk new capabilities while preserving governance, interoperability, and long-term control. Impact-aligned investment strategies will expand the conversation beyond narrow ROI toward mission outcomes such as student retention, health and well-being, resilience, and public trust.

Lifecycle Costing and Total-Cost-of-Ownership

Critically, lifecycle costing and total-cost-of-ownership thinking become non-negotiable. In a converged environment, "cheap now" often becomes expensive later—especially when proprietary integration patterns create lock-in, when systems cannot export clean data, or when security obligations exceed what was assumed at procurement. Smart campus investments must be framed as *long-duration institutional infrastructure*, not short-term technology projects.

Planning as a Foresight Discipline: Building What Can Adapt

The future of the sustainable smart campus remains uncertain. It will be influenced by factors such as climate volatility, regulatory changes, changing student expectations, new AI capabilities, and evolving social norms concerning privacy and equity. As a result, foresight becomes an essential planning discipline rather than a luxury.

Foresight planning isn't about predicting the future with complete accuracy. Instead, it's about creating an institution capable of responding intelligently to complexity. This involves engaging in scenario planning, maintaining a flexible structure for change, and developing governance that can adapt smoothly without chaos. In this approach, planners aren't just implementing systems; they're orchestrating new relationships among people, places, data, and goals.

That is why systems thinking is crucial in this work. The campus functions as an ecology: energy systems influence comfort; comfort impacts learning; learning affects retention; retention influences revenue; revenue determines the capacity to invest; investment drives modernization; modernization impacts emissions and resilience. In a smart campus, these connections can be quantified—and thus managed—only if the institution perceives them as interconnected components rather than separate departmental objectives.

A Smart Campus Becomes a Living Laboratory and a Lasting Legacy

A sustainable campus provides enduring benefits that go beyond immediate environmental effects. By implementing energy-efficient systems, renewable resources, and data-informed operations, campuses lower their carbon footprint and operating expenses while fostering healthier, more resilient spaces. Equally important, they promote a culture of sustainability that students carry into their future careers and communities.

The AI-enabled sustainable smart campus serves as a living laboratory, providing genuine research and learning opportunities. Students and faculty can experiment with ideas in real-world settings, assess results, and refine their approaches. This transforms the institution from merely a place to learn about sustainability into a space where sustainability is actively practiced, tested, and continuously enhanced.

This is the legacy: resilience, responsibility, and forward-thinking—an institution that remains credible and competitive because it can adapt to what the world requires next.

Roles of Participants, Benefactors, and Stakeholders

Along with the Chief Technology Officer (CTO), Chief Facilities Officer (CFaO), and Chief Sustainability Officer (CSO), many other campus leaders are essential in developing sustainable smart campuses. If these initiatives are not improved, they will impede institutions in the decades ahead. To preserve its global leadership, American academic institutions must take the initiative. Additionally, the CEO must highlight how AI can transform the physical world, altering how individuals collaborate, innovate, learn, and interact with their environments, devices, and energy. Maintaining the current state signals an unsustainable future for our institutions, communities, regions, and the planet.

Exhibit 21. Strategic Role Matrix

Role	Core Responsibilities	Skills and Competencies
President / Chief Executive Officer	Champion the long-term vision for an AI-enabled sustainable campus; secure resources and board support; model an innovation-first culture.	Vision casting, strategic storytelling, coalition building, and high-level resource orchestration.
Provost / Chief Academic Officer	Align curricula, research agendas, and faculty incentives with smart-tech and sustainability goals; embed "living-lab" pedagogy.	Academic planning, cross-college diplomacy, change management, and accreditation savvy.
Chief Financial Officer (CFaO)	Fund smart-campus portfolios; structure ROI models, green-bond/ESG financing, and long-range capital strategies.	Strategic budgeting, capital-markets fluency, risk analytics, and incentive-based allocation.
IT and Facilities Management Leaders	Implement and maintain IoT, BAS, and digital-twin platforms; retrain staff; sustain operations during transformation.	OT/IT convergence, program management, workforce development, vendor negotiation.
Trades Workers	Share tacit knowledge of building systems; operate and maintain evolving smart infrastructure.	Craft expertise, digital-literacy upskilling, collaborative problem-solving, safety compliance.

Role	Core Responsibilities	Skills and Competencies
Faculty and Staff	Leverage smart tools to enhance teaching, research, and admin workflows; provide user-experience feedback.	Pedagogical innovation, process redesign, data-informed decision-making, and a continuous-learning mindset.
Students	Co-design, pilot, and evaluate smart-campus solutions; voice priorities through student governance.	Digital fluency, peer engagement, feedback articulation, and advocacy.
External Partners	Advise, co-develop, and promote smart-campus projects; share best practices and training resources.	Solution architecture, partnership brokering, knowledge transfer, and industry insight.
Facilities Manager	Oversee building operations, space strategy, maintenance, and compliance in a smart-enabled environment.	Organizational leadership, building-systems literacy, budget control, and KPI tracking.
Energy and Sustainability Specialist	Deploy efficiency and renewable projects; monitor performance; advance environmental stewardship.	Energy analytics, green-building standards, carbon accounting, stakeholder education.
Smart Campus Coordinator	Orchestrate deployment and daily operation of smart systems; manage vendor SLAs; liaise across departments.	Project coordination, IoT familiarity, communication excellence, and issue triage.
Director of Integration	Align disparate systems; lead user training and change adoption from the job site to the C-suite.	Systems thinking, stakeholder translation, onboarding design, and continuous improvement.
IoT Systems Engineer	Design, secure, and optimize the campus IoT backbone; ensure interoperability among devices and data flow.	Network architecture, edge computing, protocol mastery (MQTT, BACnet), and troubleshooting.
Data Analyst	Transform smart-system data into actionable insights for operations, sustainability, and student success.	Statistical modeling, BI tools, data storytelling, cross-domain interpretation.
Security Manager	Protect cyber-physical systems and data; embed zero-trust architecture; manage physical security.	Cybersecurity frameworks, risk assessment, incident response, and crisis leadership.
Sustainability Program Coordinator	Run campus-wide sustainability campaigns, events, and reporting; cultivate culture change.	Program logistics, stakeholder outreach, behavior-change tactics, and metrics tracking.
Building Automation Specialist	Configure and fine-tune BAS (HVAC, lighting, etc.) for efficiency and comfort.	Controls programming, BACnet/LonWorks expertise, commissioning, and diagnostic analytics.
Renewable Energy Consultant	Advise on solar, wind, and geothermal integration; perform audits; craft optimization roadmaps.	Renewable-tech know-how, financial modeling, policy literacy, and feasibility assessment.
Environmental Health and Safety Officer	Ensure EH&S compliance, implement health and safety programs, and monitor IAQ and hazards.	Regulatory mastery, inspection and audit skills, safety-training delivery, and incident investigation.
Strategic Planning Coordinator	Embed smart-campus priorities into institutional strategic plans; track progress and adjust roadmaps.	Strategic facilitation, scenario planning, KPI alignment, stakeholder diplomacy.
Director of Campus Planning and Development	Lead physical-growth strategy, integrating smart infrastructure and sustainable design.	Master planning, BIM/digital-twin literacy, capital-project leadership, regulatory navigation.
Director of Energy Management	Manage campus energy portfolio, renewables adoption, and smart EMS deployment.	Energy engineering, demand-response strategy, utility negotiation, MandV analytics.
Director of Environmental Health and Safety	Oversee compliance, risk mitigation, and safe implementation of sustainability initiatives.	Comprehensive EH&S knowledge, risk analytics, emergency preparedness, and training programs.

Role	Core Responsibilities	Skills and Competencies
Director of Transportation and Parking	Design low-carbon mobility solutions—EV charging, bike-sharing, transit partnerships, and smart parking.	TDM expertise, GIS analytics, partnership management, and user-experience design.
Director of Student Affairs	Engage students in sustainability culture through programs, events, and living-lab opportunities.	Student-engagement strategy, co-curricular design, equity lens, outcomes assessment.
Director of Communications	Craft and disseminate the smart-campus narrative; manage crisis comms; visualize impact metrics.	Strategic communications, brand storytelling, data visualization, and multichannel engagement.

These contributors and stakeholders—collaborating with the CTO, CFaO, and CSO—transform executive intent into operational reality, propelling the institution up the Building Management Hierarchy of Needs toward an AI-enabled, resilient, and student-centered smart campus.

Future Trends in Smart Campus Innovation

We are entering a new era where the campus is seen not merely as a collection of buildings and systems, but as a dynamic, continuously monitored environment that learns and adapts.

That future won't come from a single breakthrough. Instead, it will emerge through convergence: intelligent materials, ubiquitous sensing, edge and cloud computing, semantic data layers, and more advanced AI systems—each purposefully governed and aligned with institutional values.

The next campus platform is not an "app",—
it is a responsive, composed environment.

Today, many smart-campus efforts still resemble "projects": a new analytics dashboard, a lighting retrofit, a pilot sensor network, a digital twin for one facility. In the future state, those elements become *a platform*—a modular, adaptive environment that connects physical operations, digital services, and human experience into a single operational fabric. The strategic shift is subtle but profound:

- From systems to ecosystems
- From monitoring to prediction
- From prediction to coordinated action
- From coordinated action to institutional learning (what worked, what didn't, and why—at campus scale)

This shift relies on creating an intuitive, persona-centered platform, because a campus that is technically 'smart" but difficult to use will not meet expectations. Design thinking becomes infrastructure: it's not just about engineering technology but also about building trust, encouraging adoption, and shaping daily behaviors.

Very soon, all mechanical rooms will feature 'standard' AI capabilities that autonomously optimize operations—diagnosing issues, scheduling maintenance, and adjusting in real-time without human input. The completion criteria for construction projects will evolve from merely "systems commissioned" to "AI operational and learning," fundamentally transforming how CIOs and CFaOs approach IT and building handovers.

Intelligent materials and energy-responsive infrastructure will reshape what "efficiency" means

The campus of the near future will not rely solely on "smarter control systems." It will increasingly rely on *smarter physics*—materials and surfaces that reduce energy demand before software ever has to optimize it.

Electrochromic (dynamic) windows are a strong example: instead of treating sunlight and heat gain as problems to be corrected by HVAC, the building envelope becomes responsive. Department of Energy reporting has shown substantial potential reductions in cooling loads and meaningful annual energy savings in cooling-dominated contexts, alongside peak-demand reduction benefits that matter deeply for institutional energy costs and resilience strategies.[35]

Layer in phase-change materials, advanced insulation (including aerogels), reflective or adaptive roofing, and longer-life infrastructure materials, and you get a new planning logic: *reduce demand structurally, then optimize what remains digitally.* This is exciting because it turns sustainability from a perpetual "behavioral campaign" into an engineered outcome—while still preserving the human-centered mission.

Digital twins will evolve from "models" into predictive engines—and then into campus operating systems.

Most campuses still treat digital twins as visualization or asset context—valuable but limited. As operational technologies progress, digital twins will transition from simple efficiency tools to vital components of institutional capacity. The incorporation of "Active Control" and Operational AI marks the beginning of a new era in which buildings transform from static structures into smart, responsive entities that support campus objectives. Future advancements will enhance efficiency and effectiveness by integrating with smarter infrastructure and robotics, empowering trade experts to maintain campuses that are adaptable to diverse needs, promote better living and health, and prioritize learning and discovery. This environment can predict needs, minimize waste, ensure safety, and enhance user experience in real time. Ultimately, the future campus will resemble a unified, responsive ecosystem rather than a set of isolated services.

When a digital twin evolves from a descriptive or diagnostic tool into a **predictive system**, it crosses a critical threshold. It stops answering *"What is happening?"* or even *"Why did this happen?"* and begins answering *"What will happen next—and what should we do about it?"* At that point, the twin becomes more than a model of the campus; it becomes an **operational intelligence layer** for the institution. This transition reshapes planning, operations, and optimization—and ultimately redefines how universities steward their physical environments in service of mission, sustainability, and resilience.

Planning Shifts from Static Forecasts to Scenario-Rich Exploration

In a predictive twin environment, planning is no longer a linear exercise constrained by historical averages and static assumptions. Instead, it becomes **scenario-rich, probabilistic, and evidence-driven.** Rather than producing single forecasts or master plans that assume stability, leaders can explore *multiple plausible futures* before committing capital or policy.

- **Decarbonization pathways** can be modeled against real operating constraints—weather volatility, grid carbon intensity, occupancy change, and utility pricing—revealing which investments produce durable reductions rather than symbolic gains.
- **Resilience strategies** can be stress-tested under compound disruptions: extreme heat coupled with grid instability, flooding combined with staffing shortages, or cyber-physical incidents during peak academic periods.

- **Space reallocation** decisions can be simulated based on actual usage patterns, hybrid schedules, and future enrollment scenarios—reducing the risk of overbuilding or misallocating scarce capital.
- **Emergency response strategies** can be rehearsed virtually, allowing leaders to test decision rights, communications, and system behaviors before lives or research are at stake.

In this future state, planning moves from *opinions and projections* to **evidence-backed choices**, enabling governance bodies to fund fewer initiatives—but with far greater confidence in their long-term value.

Operations Become Anticipatory Instead of Reactive

Perhaps the most profound shift occurs in day-to-day operations. When the twin becomes predictive, the campus no longer waits for failures, complaints, or alarms to act. Instead, the digital twin continuously looks for **early signals of drift, degradation, or emerging risk**, such as:

- HVAC systems gradually losing efficiency long before comfort complaints appear
- Ventilation patterns that will lead to poor IAQ under forecasted occupancy or weather conditions
- Electrical or mechanical assets trending toward failure weeks or months in advance
- Safety or access patterns that indicate elevated risk before incidents occur

This enables a shift from:

- **Break–fix → condition-based maintenance**
- **After-the-fact reporting → forward-looking risk management**
- **Heroic response → quiet prevention**

Critically, this does not replace human expertise. Instead, it **augments scarce technical talent**, allowing facilities and operations professionals to focus on high-value decisions rather than constant firefighting. The predictive twin becomes a diagnostic partner—surfacing insights, prioritizing work, and reducing cognitive load—while humans retain authority and accountability. Over time, this shift materially reduces downtime, emergency spend, deferred maintenance growth, and staff burnout.

Optimization Becomes Multi-Objective and Systemic

Traditional campus optimization efforts are siloed by necessity. Energy teams optimize for cost. Facilities teams optimize for uptime. Academic units optimize schedules. Safety teams optimize for risk reduction. Sustainability teams optimize for emissions. Each goal is rational—but often in conflict with the others. A predictive digital twin changes this equation. Because it understands **relationships and tradeoffs across systems**, optimization becomes **multi-objective and coordinated**, rather than competitive:

- Energy efficiency can be optimized *without* degrading indoor air quality.
- Comfort and learning conditions can be balanced dynamically against real-time energy prices and grid carbon intensity.
- Academic schedules can inform building readiness automatically, rather than relying on static timetables.
- Safety, access, and occupancy insights can be aligned without creating surveillance overreach, governed by clear data policy.

Instead of asking, "Which metric matters most?" leaders can ask, "What is the best combined outcome for mission, people, and sustainability—under today's conditions?" This is a foundational shift from **local optimization to institutional optimization**.

4. Governance Evolves from Policy Oversight to Operational Stewardship

As the digital twin becomes predictive, governance itself must evolve. Boards and executive leaders are no longer simply approving projects or receiving lagging metrics. They are stewarding a **living system**—one that continuously learns, adapts, and improves. This requires:

- Clear decision rights over when predictive insight becomes automated action
- Ethical and privacy-forward controls over sensing, analytics, and AI
- Transparent metrics that measure outcomes rather than activity
- Shared accountability across IT, Facilities, Sustainability, and Academic leadership

In this model, governance is not a brake on innovation—it is the enabler of trust at scale.

The Campus Becomes a Unified, Responsive Ecosystem

Ultimately, when the digital twin becomes predictive, the campus itself changes character.

It no longer behaves like a collection of independent services—IT here, Facilities there, sustainability somewhere else. Instead, it functions as a **unified, responsive ecosystem**:

- One that anticipates needs rather than reacts to failures
- One that minimizes waste quietly and continuously
- One that embeds safety, health, and comfort into daily operations
- One that adapts in real time to people, learning, and environment

This marks the shift from smart components to a truly sentient campus—specifically, not as a speculative idea but through a disciplined, manageable, and mission-oriented approach. For campus leaders and planners, this has significant practical implications: the digital twin evolves from being merely a mirror into a collaborative planning and design tool, a fiscal management tool, and ultimately into a partner for coordination and action—integrating facilities, IT/OT operations, sustainability, and user experience into a unified campus operating system.

> ***IoT 2.0 and "Edge AI" will make responsiveness real—***
> ***and make privacy and antifragility easier to achieve.***

Edge AI Architecture Enhances Speed and Resiliency

Another key trend is not a flashy device but a shift in where intelligence is processed. **Edge AI** brings analytics and decision-making nearer to the data source — in buildings, controllers, gateways, and local compute nodes. This shift matters because it lowers latency, decreases bandwidth needs, and enhances privacy by minimizing the movement of raw data.[36]

For the AI-enabled sustainable smart campus, edge intelligence changes the feasible use cases:

- **Real-time IAQ and ventilation control** that adapts minute-by-minute to actual occupancy and outdoor conditions
- **Local anomaly detection** for equipment and safety conditions
- **Resilient operations during connectivity degradation** (because "cloud-only" control is fragile in emergencies)
- **More defensible privacy architectures**, where only summarized or policy-filtered information is exported

Edge AI will become a fundamental capability for campuses aiming to scale responsibly, particularly as sensor density grows and the demand for real-time responsiveness increases.

Synthetic Data Bridges Gaps and Enhances Operational AI Models

As campuses gather increasing data on space use, learning activities, movement patterns, and building efficiency, concerns about privacy and ethics grow. In this context, synthetic data transcends being merely a technical tool; it becomes a key component of governance.

> *Synthetic data and federated models will unlock innovation without forcing a privacy tradeoff.*

Synthetic datasets enable institutions to train and evaluate models even when real data is incomplete, sensitive, or legally restricted. Federated learning techniques facilitate cross-campus benchmarking and collective model enhancement without aggregating raw data centrally. This approach offers the strategic advantage of speeding up innovation, minimizing risk, and supporting regional collaboration without breaching trust. In the future, "privacy-by-design" will primarily involve reducing raw data transfer, establishing secure training processes, and ensuring transparent accountability for model behavior.

Evolving at the Speed of Change

A key characteristic of AI-enabled Sustainable Smart campus environments is that objectives are constantly shifting. Policies need updating. New technologies emerge frequently. Student expectations continue to evolve. New risks constantly arise. The optimal solution for a smart ecosystem application today might be outdated or inappropriate by next year—or even next semester.

> *All objectives are non-stationary; therefore, governance cannot be either.*

This core aspect of non-stationary objectives is that the model's behavior must consistently align with evolving institutional goals and constraints. Achieving this demands deliberate lifecycle management, including feedback loops, ongoing monitoring, revalidation, retraining, and alignment with policies. It also emphasizes a fundamental leadership principle: IT/OT is not just a one-time purchase but an ongoing institutional capability that requires continuous stewardship and planned investment.

The future campus will also feel different because *interfaces will change*:

- **Mixed reality** will make learning spaces and facilities work visible: students see contextual overlays; technicians see live equipment state and guided repair workflows; emergency response teams see situational maps that update in real time.
- **Embodied AI and robotics** will move beyond novelty into operational value: logistics, waste handling, cleaning, inspection, and mobility support—especially when paired with edge AI for safety and reliability.
- **Agentic copilots** for facilities and IT/OT operations will reduce cognitive load: summarizing alarms, recommending next actions, auto-generating work orders, and learning from resolution outcomes.

This is where the excitement truly takes shape: you start to envision a campus where operational excellence and student experience harmonize instead of conflict. A campus can be more sustainable exactly because it adapts more effectively.

All these trends lead to a key strategic shift: the data estate is transitioning from merely a backend repository to a vital decision-making engine. The most advanced campuses will build:

- integrated spatial + operational + experience data layers
- semantic/ontology structures to make cross-system data interpretable
- model lifecycle pipelines that treat AI as a managed asset
- governance that is agile, ethical, transparent, and participatory

In that future state, the campus doesn't just "report" sustainability outcomes—it *continuously discovers new leverage points* for decarbonization, resilience, and experience improvement.

A Future Worth Building, If Built With Purpose

The idea of an AI-enabled, sustainable smart campus should invite imagination rather than anxiety. At its best, it promises campuses that are healthier, more resilient, more humane, and more aligned with their academic mission. Yet inspiration alone is never enough. Vision only becomes reality when institutions pair ambition with discipline—and deliberately avoid two persistent and costly traps.

The first is **trend-chasing without architecture**: the accumulation of disconnected "smart" point solutions that solve local problems but fragment the institution as a whole. These investments may deliver short-term wins, but without a shared architecture they increase complexity, lock in vendors, and erode long-term adaptability.

The second trap is **automation without governance**: scaling sensing, analytics, and AI capabilities faster than trust, accountability, and oversight can keep pace. When intelligence outstrips governance, institutions risk privacy violations, cyber-physical exposure, cultural backlash, and loss of legitimacy—ultimately slowing or reversing progress rather than accelerating it.

Leading institutions will resist both traps by focusing not on technology adoption, but on **architecting for adaptability**. They will treat the campus as a living, evolving, and responsive learning system—one that is deliberately designed to sense, learn, and improve over time. In this model, buildings are no longer static containers for activity, and digital platforms are no longer isolated utilities. Instead, physical infrastructure, digital systems, data, and human workflows are integrated into a coherent, mission-aligned whole.

This approach places equal weight on **human-centered design and ethical governance** as on sensors, automation, and AI. It recognizes that trust is infrastructure; that privacy and transparency are not constraints but enablers; and that intelligence only creates value when it is understandable, governable, and actionable by people. By embedding clear decision rights, shared standards, and disciplined execution into the operating model, institutions ensure that innovation scales safely and sustainably.

When this integration is achieved, the future stops being a collection of vision statements, pilot projects, or "smart" initiatives competing for attention. It becomes a lived reality. The campus becomes safer—not just through surveillance, but through predictive insight and resilient design. Health improves—not through isolated upgrades, but through environments that continuously adapt to occupants. Carbon emissions decline—not through slogans, but through operational intelligence embedded in daily decisions. Most importantly, the institution becomes **anti-fragile**—able not only to withstand disruption, but to learn from it and emerge stronger.

A future worth building is not defined by how advanced the technology appears, but by how well the campus serves its people and purpose. When built with intention, governance, and care, the AI-enabled sustainable smart campus becomes what it was always meant to be: a trustworthy, adaptive foundation for teaching, learning, research, and community—designed not just to keep up with change, but to shape it responsibly.

Epilogue

Our names will not be written on the stones,
but in them and in the hearts of those who pass through these halls.

For centuries, university campuses have been characterized by their physical presence—brick walls, ivy, and concrete structures. They served as static spaces for the evolving quest for knowledge. However, with the advent of the AI-enabled Operational Technology era, this traditional view has fundamentally changed.

The classrooms, laboratories, halls, and buildings we create and oversee today are more than passive shells for learning activities; they are active, intelligent participants. Thanks to the merging of Information Technology, Operational Technology, and Artificial Intelligence, the campus has come alive. Collectively, the campus now functions as a highly responsive, interconnected cyber-physical system that can sense its surroundings, forecast issues, and adapt to safeguard its inhabitants.

Beyond the Mechanistic Campus

To realize the vision of an AI-enabled Sustainable Smart Campus, higher education leaders need to move beyond traditional, linear management approaches. For many years, campuses operated in isolated silos: Facilities handled physical infrastructure, IT handled the wiring, and the departments rarely collaborated.

The AI-Enabled Sustainable Smart Campus requires a new leadership style. It needs a perspective that recognizes the complex, almost quantum-like interconnectedness of our operations. In this changing environment, updating a digital algorithm can instantly affect a research lab's thermal conditions; a cyber weakness in a mechanical chiller can threaten the entire network; and the psychological safety of a frontline mechanic directly affects the outcome of a million-dollar software project. In short, managing parts separately is no longer effective. Instead, leaders must oversee the entire system.

The True Measure of the Smart Campus

In this book, we've examined the complex architecture needed for the future: Zero-Trust networks, the Digital Integration Fabric, semantic ontologies, and predictive AI control planes. We've detailed how to leverage this technology to decarbonize institutions, safeguard research efforts, and shift operational resilience from fragile to antifragile.

The true essence of a smart campus is that it is fundamentally a human endeavor
that enhances the dignity, abilities, and purpose
of the people who live and work there.

Nevertheless, even with all the technological complexities, the true essence of a smart campus is fundamentally human. A campus isn't considered smart simply because it has 10,000 IoT sensors or a perfect AI-enabled Digital Twin. It is truly "smart" when its technology enhances the dignity, abilities, and purpose of the people who live and work there.

A digital dashboard doesn't measure your true success. Instead, it's evident in the veteran mechanic mentoring a young digital native with a mobile copilot, or the trust between a network engineer and a plant operator working together to defend against cyber-physical threats. It's also reflected in a student thriving in an inclusive, responsive environment, unaware of the complex algorithms silently working behind the scenes to support their success.

The Call to the Integrative Architects

To the Chief Technology Officers, Chief Facilities Officers, Cabinet members, and Board members: the responsibility is yours. Shifting to the AI-Enabled Sustainable Smart Campus will be the most culturally complex, politically sensitive, and technologically demanding change your institution has faced. This task cannot be delegated to just one department or vendor. Instead, it requires an Integrative Architect—a courageous leader willing to break down long-standing tribal silos, curious enough to bridge the mechanical and digital realms, and compassionate enough to support your workforce through this major modernization.

View this transformation not merely as an infrastructure project but as your legacy, establishing the physical and digital foundation for a century of discovery and human progress. Lead courageously, govern with integrity, and design a campus that drives its evolving mission.

Appendices

The Smart Campus Operating Model

This is a summary of the key decisions, governance structures, standards, and safeguards related to the planning, implementation, and evolution of the Smart Campus.

Purpose
Establish a repeatable, secure, interoperable operating model that enables IT/OT convergence, governed data use, and scalable smart campus outcomes (cost, carbon, reliability, experience).

Decision Rights (Who Decides What)

- **CTO (Accountable):**
  - Enterprise architecture for smart campus platforms (integration, cloud/edge patterns, identity, APIs)
  - Cybersecurity baseline for IT/OT and connected devices (minimum controls, segmentation, monitoring)
  - Institutional Data Estate design (data domains, access model, tooling standards)
- **CFAO (Accountable):**
  - Building/utility priorities, asset lifecycle plans, operational workflows, OT system readiness
  - Maintenance strategy alignment (predictive vs. reactive) and field adoption
- **Chief Data Officer / Data Governance Lead (Accountable):**
  - Data governance policies (quality, stewardship, cataloging, retention)
  - Data access approvals for research/learning/analytics use cases (in coordination with privacy)
- **CISO / Security Lead (Accountable):**
  - Threat modeling, incident response, security monitoring requirements, risk acceptance decisions
  - Vendor security review and continuous compliance expectations
- **Chief Sustainability Officer (Accountable):**
  - Carbon/accountability framework, reporting requirements, sustainability KPI definitions
  - Alignment to climate commitments and external disclosures
- **Provost / Academic Leadership (Consulted):**
  - Academic use cases, learning analytics boundaries, research enablement requirements
- **Procurement / Legal (Accountable for contracting terms):**
  - Data ownership, interoperability clauses, security SLAs, audit rights, exit/portability provisions

Governance (How We Operate)

- **Smart Campus Governance Council (monthly):** CTO + Facilities + CISO + CDO + Sustainability + Provost rep + Procurement
  - Approves portfolio priorities, standards updates, and scaling decisions
- **Architecture and Standards Review Board (biweekly):** enterprise architects + OT engineers + security + data
  - Enforces reference architecture, integration patterns, and device onboarding rules
- **Use-Case Portfolio Management (quarterly):**
  - Ranks initiatives by value, feasibility, risk, and scalability; prevents duplicative pilots

Standards (What Must Be True to Scale)

- **Interoperability:** open protocols where feasible, published APIs, common semantic models, vendor commitments to integration
- **Data:** shared naming/metadata conventions, data cataloging, master data alignment for assets/spaces, quality thresholds
- **Security:** network segmentation for OT, identity-based access, secure device onboarding, patching/firmware policy, logging/monitoring
- **Reliability:** uptime targets, fail-safe modes for critical building systems, backup, and recovery expectations
- **Sustainability Measurement:** consistent definitions for carbon/energy/water KPIs; auditable reporting lineage

Guardrails (Non-Negotiables)

- **No device or system connects without** approval for onboarding, segmentation, and monitoring.
- **No "one-off integrations"** that bypass the reference architecture or create unowned technical debt.
- **No vendor lock-in by accident:** data portability, documented interfaces, and exit provisions required.
- **Privacy and ethics by design:** approved purposes, role-based access, retention limits, and transparency.
- **Operational continuity first:** OT safety and reliability requirements override convenience features.

Implementation Roadmap Stages

Exhibit 22. Smart Campus Implementation Stages

Stage	Primary Objective	Primary Owners	Key Dependencies	Measurable Outcomes (Examples)
Stage 1: Connect and Monitor	Establish visibility and control foundations across priority buildings/systems; create baseline performance and governance.	CTO (platform + security), Facilities (OT readiness), CISO (controls), CDO (governance)	Asset inventory; OT network segmentation plan; device onboarding process; integration platform selection; baseline KPI definitions	Coverage: % of priority buildings connected; Data quality: % assets/points mapped to standard naming; Reliability: reduced critical alarms/unknown states; Security: % devices onboarded with approved controls; Baseline: energy/comfort/space utilization baselines established
Stage 2: Analyze and Improve	Convert telemetry into operational insight and targeted optimization; deliver repeatable wins.	Facilities (operations), CTO/CDO (analytics + data estate), Sustainability (KPI tracking)	Data pipelines stabilized; analytics tooling; trained operational workflows; change management; "use-case factory" process	Energy: measurable reduction in EUI / peak demand; Maintenance: % shift from reactive to planned work; Space: improved utilization in targeted zones; Experience: improved comfort tickets/time-to-resolve; Governance: % initiatives passing architecture/security review without exceptions
Stage 3: Transform and Expand	Scale automation and digital twins; institutionalize continuous optimization and scenario planning	CTO (platform at scale), Facilities (operational transformation), CISO (resilience), Provost (living-lab enablement)	Mature data estate; interoperable integrations across major systems; robust IAM; vendor performance governance; funding model for lifecycle	Digital twin: operational twin live for priority district/buildings; Predictive: % critical assets under predictive maintenance; Carbon: measurable emissions reduction with auditable lineage; Resilience: improved incident response times / simulated scenario readiness; Academic enablement: # approved research/teaching use cases leveraging governed campus data
Stage 4: Continuous Optimization	Embed smart campus as standard operating practice; optimize continuously, refresh lifecycle, and expand innovation.	CTO + Facilities (shared), Sustainability (accountability), CDO (data stewardship)	Lifecycle renewal plan; metrics governance; continuous commissioning; operating budget for platform sustainment	Operational excellence: sustained KPI improvement year-over-year; Lifecycle: reduced deferred maintenance growth rate; Trust: high compliance to data/security standards; Innovation: faster time-to-deploy new use cases with lower marginal cost

CTO and CFaO Responsibilities

Exhibit 23. Technology vs Facilities vs Shared Responsibilities

Domain	CTO Core Knowledge and Oversight	Chief Facilities Officer Core Knowledge and Oversight	Shared responsibility (must be jointly governed)
Primary mission	Digital services, cyber risk, academic/admin enablement, user experience	Safe, reliable, compliant operation of the built environment; building and utility uptime	Translate mission into a single, agreed "campus reliability + experience" agenda (students/faculty outcomes and safe operations)
Operating "platform"	Apps, data platforms, networks, identity, cloud	Buildings, central plants, utilities, trades, contractors	Treat buildings as a cyber-physical platform: operational uptime requires secure digital enablement; digital services depend on physical continuity (power/cooling/space)
Signature enterprise systems	ERP/SIS/CRM/HRIS/LMS, IAM/SSO, ITSM, SIEM/SOAR, data lake/warehouse	BMS/BAS, EMS, SCADA (as applicable), CMMS/EAM, metering, lighting controls, fire alarm, access control, video, conveyance systems	Systems integration strategy (APIs, historians, time-series, CMMS-BMS workflows) and "system-of-systems" architecture decisions
OT/BMS scope	Needs working fluency with OT realities and constraints	Owns OT operations and vendor ecosystem	OT governance charter: what counts as OT, what is in scope, and who owns lifecycle decisions across buildings/central plant/campus-wide platforms
Standards bodies (typical)	NIST, ISO/IEC, IETF/IEEE, ITIL/COBIT; privacy regs	ASHRAE, NFPA, ICC, NEC, OEM, and AHJ code interpretations	A unified standards registry: cybersecurity + safety codes + interoperability requirements, aligned to procurement and project delivery
Key OT protocols and integration	Must understand BACnet/IP, Modbus TCP, MQTT, OPC UA (if present), and SNMP patterns and security implications	Deep expertise in BACnet (IP/MS/TP), Modbus (RTU/TCP), LonWorks, KNX, DALI, proprietary buses	Protocol selection guidance, gateway strategy, tagging/metadata conventions, and secure integration patterns (including vendor access methods)
Network vs plant emphasis	LAN/WAN, Wi-Fi, DNS/DHCP, segmentation, firewalls, identity	Central plant operations, HVAC, hydronics, steam/chilled water, boilers/chillers	Network design that respects operational constraints: segmentation, deterministic performance needs, RF realities, and maintenance windows
Wiring and distribution	Low-voltage structured cabling, fiber, PoE, telecom rooms, edge compute	High-voltage distribution, switchgear, generators, ATS, plus control wiring	Power + network interdependence: UPS strategy for OT/edge, generator-backed comm rooms, and outage coordination that protects both IT and life-safety

Domain	CTO Core Knowledge and Oversight	Chief Facilities Officer Core Knowledge and Oversight	Shared responsibility (must be jointly governed)
Cybersecurity	Security policy, monitoring, incident response, third-party risk, and identity	Physical/operational realities; safety-first operations; vendor practices	OT security program (e.g., segmentation, MFA for remote access, logging, vulnerability management) tuned to uptime + safety constraints
Safety and compliance	Data/privacy compliance; cyber policy	Life safety (fire alarm/egress), OSHA/EHS, inspections, code compliance	Joint risk governance for cyber-physical safety: how cybersecurity controls interact with life-safety and emergency procedures
Change cadence	Fast cycles: patches/releases; hardware refresh 3–7 yrs	Slow cycles: assets last decades; controls mid-cycle modernization	Portfolio planning that bridges cadences: align "IT refresh rhythm" with "capital renewal rhythm," and define when to modernize controls vs. full system replacement
Lifecycle time horizons	Short: software weeks–months; endpoints/servers 3–7 years; cloud contracts 1–3 years	Long: buildings 30–70+ years; major MEP 15–30; controls 10–20; roofs/envelopes 20–40	A shared lifecycle map and funding model: design for "upgradeability" (modular controls, open protocols) while managing long-lived assets and shorter-lived digital layers
Performance metrics	Availability, security posture, user satisfaction, cost-to-serve	IAQ, comfort, EUI, service calls, uptime, asset life, backlog	A single scorecard that links digital + physical KPIs (e.g., IAQ/comfort tied to controls reliability, energy tied to analytics, uptime tied to both OT and network resiliency)
Procurement lens	Licensing/subscriptions, data clauses, security requirements, APIs	Capital projects, service contracts, commissioning specs, and maintainability	Joint procurement standards: security-by-design, interoperability, data ownership, commissioning requirements, and "no black-box" clauses where feasible
Vendor ecosystem	SaaS/cloud/MSPs/telecom/cyber	A/E, GC, MEP, controls contractors, OEMs, commissioning agents, ESCOs	Unified vendor governance: remote access rules, incident notification, data sharing, commissioning acceptance, and performance-based contracting
Workforce and skills	Network/cloud/security/data engineers, architects, service desk	Trades, operators, BAS techs, facilities engineers, planners	Joint training plan: OT cybersecurity basics for facilities; building systems literacy for IT; shared "operator UX" and adoption focus for new platforms
Documentation and sources of truth	CMDB, architecture standards, data dictionaries	As-builts, O&M manuals, sequences, TAB/Cx reports, one-lines, point lists	A harmonized "digital backbone" for documentation: asset registry + system diagrams + point metadata + change history, maintained as part of operations.

Domain	CTO Core Knowledge and Oversight	Chief Facilities Officer Core Knowledge and Oversight	Shared responsibility (must be jointly governed)
Data estate and governance	Data architecture, access controls, retention, analytics, privacy	Point accuracy, sensor calibration, operational context, maintenance history	Shared data governance: naming/tagging standards, data quality SLAs, access policies, privacy boundaries, and ethical transparency for pervasive sensing
Digital twin strategy	Platforms, integration, security, analytics, and user experience	Ground truth, spatial/static accuracy, operational workflows	Joint digital twin roadmap: what problems to solve first, what data is required, how insights become work orders, and how value is measured and sustained
Incident response	Cyber IR playbooks, SOC escalation	Operational response, safety protocols, vendor dispatch	A combined cyber-physical incident playbook: who declares, who isolates networks, how fail-safe modes work, and how to restore operations safely
Common failure modes	Shadow IT, misconfiguration, credential sprawl, vendor lock-in	Controls drift, calibration failure, deferred maintenance, and undocumented changes	Shared controls to prevent drift: change control, commissioning discipline, configuration baselines, and periodic "re-commissioning + security validation" cycles

Note: The shared-responsibility column is the real bridge: it's where universities either build a coherent smart-campus operating model—or end up with duplicated platforms, unsafe workarounds, and modernization fatigue.

Smart Campus Reference Architecture

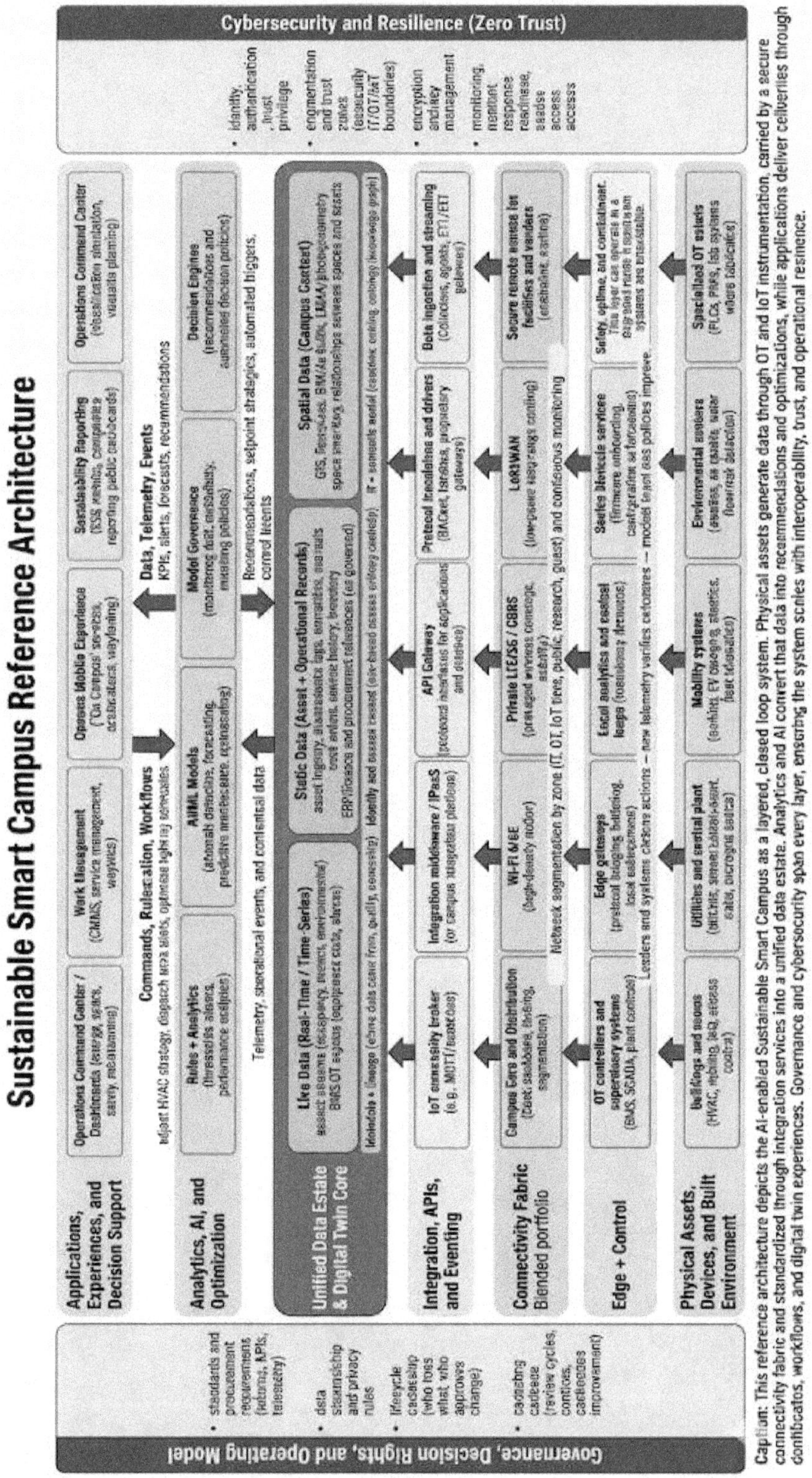

Figure 13. Smart Campus Reference Architecture

Cyber-Physical Asset Audit Form

This is a sample form that could be digitized rather than collected on paper, ensuring data flows directly into the Data Estate for us across various platforms. Of course, institutions should augment their data collection to meet their specific needs and nomenclature.

Section A: General Asset Info

- **Asset Name/ID:** (e.g., AHU-04-North)
- **Physical Location:** (Building, Floor, Room #, or x, y, z coordinates)
- **Asset Category:** [] HVAC [[]Lighting [ ] Security/CCTV [ ] Power/Utility [ ] Network Gear [ ] Other
- **Manufacturer/Model:** ______________________

Section B: The "Cyber" (IT Layer)

- **Connection Type:** [ ] Hardwired Ethernet [ ] Wi-Fi [ ] Zigbee/Z-Wave [ ] Cellular [ ] Proprietary Serial
- **IP/MAC Address:** ______________________
- **Firmware Version:** ______________________
- **Security Status:** [ ] Default Password Changed? [ ] Encrypted Comm? [ ] Patchable?
- **VLAN Assignment:** ______________________

Section C: The "Physical" (Facilities Layer)

- **Primary Function:** (e.g., VAV Box controlling Zone 202)
- **Power Source:** (e.g., Panel LP-1, Circuit 12)
- **Upstream Dependencies:** (e.g., Fed by Chiller-01)
- **Condition Rating:** [ ] New [ ] Functional [ ] End-of-Life [ ] Critical Failure Risk

Section D: Infrastructure and Edge Readiness

- **Local Signal Strength:** (dBm for Wi-Fi/Cellular at asset location)
- **Edge Compute Availability:** [ ] Local Gateway Nearby? [ ] Rack Space Available?
- **Fiber/Copper Uplink Capacity:** [ ] Available [ ] Limited [ ] Saturated

Collect and store the data in electronic format for use across the Data Estate.

Sample Governance Charter

Purpose and intent

The Smart Campus Steering Committee (SCSC) exists to turn the institution's smart campus vision into coordinated decisions, disciplined execution, and responsible outcomes. Its goal is not to add bureaucracy; it is to establish a reliable process for making trade-offs across silos and managing risk as campus systems become more interconnected and data-driven. This charter defines scope, authority, decision-making procedures, reporting schedule, and escalation processes as recommended governance deliverables.

Mandate and scope

The SCSC governs the smart campus as an institutional program spanning OT modernization, data integration, cybersecurity guardrails, user-facing services, and sustainability performance enablement. It oversees the smart campus portfolio from pilot selection through scaling decisions. It also ensures that work aligns with institutional principles: safety, privacy, cybersecurity, interoperability, accessibility, and long-term financial stewardship.

Authority and decision rights

The SCSC is empowered to approve the smart campus roadmap and sequencing, including "pilot-to-scale" gates and decommissioning of duplicative approaches. It validates enterprise standards—interoperability requirements, security baselines, and data governance expectations—in coordination with existing IT/security/data governance bodies. It resolves cross-silo conflicts and escalates unresolved decisions to the Executive Sponsor and, when appropriate, to Cabinet. It approves KPI definitions and the reporting cadence for the shared scorecard, ensuring that outcomes remain mission-aligned and auditable.

Membership and roles

The committee is co-chaired by the CTO and the senior Facilities leader, reflecting shared ownership of outcomes for digital and physical infrastructure. A top-level Executive Sponsor (President, Provost, COO, or designee) ensures authority for cross-institution decisions. Standing members include the CISO (risk and monitoring), the CDO or data governance lead (stewardship and ethical use), Sustainability leadership (performance accountability), Finance (lifecycle funding logic), Procurement/Legal (contract guardrails and data rights), and an Academic representative (teaching/research enablement). Communications/change management representation is included to ensure decisions translate into adoption and trust.

Operating cadence

The SCSC meets monthly to review portfolio status, approve scaling decisions, manage escalations, and review scorecard progress. A biweekly Architecture and Security Review—reporting into the SCSC—functions as a deployment gate to ensure identity, segmentation, remote access, logging/monitoring, and data handling meet institutional standards before systems connect. The Executive Sponsor receives quarterly updates designed for Cabinet and Board-level reporting, emphasizing outcomes, risk posture, and major decisions.

Decision process and escalation

The committee is consensus-driven by default, but it is designed to decide. When consensus cannot be reached, co-chairs elevate to the Executive Sponsor. Decisions are documented with a rationale, risks, dependencies, affected stakeholders, and success measures. Exceptions to standards are permitted only when time-bound, risk-assessed, and explicitly approved.

Required artifacts and deliverables

Within the first 90–120 days, the SCSC produces and maintains: (1) a mission-aligned vision statement endorsed by the Executive Sponsor; (2) a RACI model clarifying decision rights across IT/OT, data governance, and cybersecurity; (3) a shared scorecard with KPI definitions, baselines, and reporting cadence; and (4) a stakeholder engagement and communication plan that institutionalizes transparency and feedback loops.

Sample Governance Decision Rights, Operating Model, and Cadence

1. Purpose

This charter establishes the governance architecture for designing, deploying, operating, and continuously improving an AI-enabled Sustainable Smart Campus (SSC). It defines the decision rights, operating model, and standing cadence through which the institution aligns technology, the built environment, data, and ethical commitments to mission outcomes—including sustainability performance, operational excellence, safety, resilience, and an improved campus experience. The SSC is not a discrete initiative; it is an institutional capability. Accordingly, governance is designed to be durable across leadership transitions, adaptable to evolving technologies, and disciplined enough to protect safety, privacy, and institutional trust.

2. Scope

This charter governs the planning and operation of all smart-campus capabilities that sense, connect, analyze, or automate campus operations and experiences, including but not limited to:

- Building and infrastructure systems (BMS/BAS, HVAC, lighting, metering, access control, surveillance, life-safety integrations, energy plant systems, microgrid components).
- Digital networks that enable OT/IoT connectivity (wired, wireless, edge, segmentation, identity).
- IoT devices, sensors, actuators, controllers, gateways, and related firmware/software.
- The institutional "data estate" used for analytics, reporting, AI/ML, digital twin, and decision support.
- AI-enabled decision systems that influence operations, resource allocation, safety, student support, or other institutional outcomes.
- Cross-system integrations among OT, enterprise IT, academic systems, and campus service platforms.

3. Guiding Principles

Governance decisions and operating practices shall be anchored in the following principles:

1. **Mission Alignment:** Smart-campus investments and policies will prioritize measurable contribution to institutional mission and strategic priorities.
2. **Safety and Continuity:** Life-safety, critical operations, and occupant well-being are non-negotiable; availability is treated as a primary design constraint in OT environments.
3. **Security by Design:** Cybersecurity controls are embedded from inception, not appended after deployment; OT/IoT security shall be governed with risk-based discipline appropriate to cyber-physical systems.
4. **Privacy, Consent, and Dignity:** Data practices shall be transparent, minimal, purpose-bound, and aligned with institutional commitments to human dignity and rights.
5. **Ethical AI:** AI systems must be explainable, auditable, and governed with equity safeguards, bias mitigation, and mechanisms for human oversight and redress.
6. **Interoperability and Portability:** Technology choices will prioritize open standards, well-documented APIs, and design patterns that reduce vendor lock-in and preserve future flexibility.

7. **Evidence and Outcomes:** Governance will prioritize decision-making grounded in trusted data, baseline metrics, and demonstrable outcomes—not novelty, vendor momentum, or anecdote.
8. **Shared Stewardship:** The SSC is a joint responsibility across IT, Facilities, Sustainability, Academic/Student domains, Risk, and Finance, requiring coordinated planning and accountability.

4. Governance Structure

The SSC governance model consists of four interconnected layers: Executive Sponsorship, Steering Council, Domain Councils, and Operational Teams.

4.1 Executive Sponsors

Role: Provide strategic mandate, approve risk posture and investment envelope, and ensure sustained institutional alignment.
Typical members: President/Chancellor (or designee), EVP/CFaO, Provost (or designee), CTO, Chief Facilities Officer, Chief Sustainability Officer, General Counsel (as needed), Chief Risk/Compliance (as needed).
Responsibilities:

- Approve institutional outcomes and guardrails for SSC (privacy, ethics, risk tolerance).
- Authorize funding envelopes for multi-year SSC programs.
- Receive quarterly outcomes and risk briefings from the Steering Council.

4.2 Smart Campus Steering Council (SCSC)

Role: Primary cross-functional decision body overseeing the SSC portfolio, integration priorities, and institutional tradeoffs.
Membership (standing):

- CTO (co-chair)
- Chief Facilities Officer (co-chair)
- CISO (or equivalent security leader)
- Chief Sustainability Officer (or designee)
- Finance/Budget leader (or designee)
- Campus Safety leader (or designee)
- Institutional Data Governance lead
- Academic/Student Affairs representative (rotating or standing)
- Procurement leader (as needed)
- Legal/Privacy representative (as needed)
- Program Executive / Smart Campus Director (non-voting facilitator)

Mandate and Authority: The SCSC is empowered to:

- Set portfolio priorities and sequencing across IT/OT modernization, data estate, digital twin, and AI-enabled operations.
- Resolve cross-domain conflicts (e.g., security vs. operational continuity; analytics ambition vs. privacy constraints).
- Approve enterprise integration standards and the "approved patterns" required for scale.

- Confirm baseline metrics and target outcomes for SSC programs.
- Escalate investment or risk decisions to Executive Sponsors when thresholds are exceeded.

4.3 Domain Councils

Domain Councils translate strategy into enforceable standards, review high-impact decisions within their remit, and ensure consistency across projects.

A) Data Governance Council (DGC)

Purpose: Govern the institutional data estate supporting OT/IT convergence.

Key responsibilities:

- Define data classification, retention, access, and sharing rules.
- Establish authoritative sources of truth and reconciliation rules across systems.
- Set data quality standards for "decision-grade" analytics.
- Approve data-sharing agreements and privacy-impact review requirements.

B) Cybersecurity and Resilience Council (CRC)

Purpose: Govern cyber risk for infrastructure, with OT-appropriate controls and resilience discipline.

Key responsibilities:

- Approve OT/IoT network segmentation standards and device onboarding security controls.
- Set vulnerability management expectations for constrained and legacy devices.
- Define incident response playbooks spanning IT and Facilities operations.
- Establish risk acceptance thresholds and exception processes.

C) Architecture Review Board (ARB)

Purpose: Ensure the campus evolves as an integrated, scalable system—not a set of proprietary islands.

Key responsibilities:

- Approve integration patterns, API requirements, canonical data models, and interoperability standards.
- Review major platform selections for lock-in risk, portability, and lifecycle viability.
- Maintain reference architectures for OT/IT convergence, digital twin feeds, and analytics pipelines.

D) AI Ethics and Oversight Board (AIEB)

Purpose: Ensure responsible and equitable use of AI in campus operations and services.

Key responsibilities:

- Review high-impact AI use cases (those affecting people, safety, access, services, or resource allocation).
- Require documentation of model intent, training data lineage, performance, and limitations.
- Set monitoring requirements for drift, bias, and unintended consequences.
- Define transparency, explainability, and redress mechanisms for affected stakeholders.

4.4 Operational Teams

Operational Teams execute approved work, manage production systems, and sustain day-to-day performance.

- OT/IoT Operations Team: device lifecycle management, connectivity, commissioning, monitoring, maintenance coordination.
- Security Operations (SOC) + OT Security Function: detection, response, vulnerability coordination, and incident handling.
- Data Platform / Integration Team: pipelines, middleware, data quality, observability, access provisioning.
- Facilities Operations and Engineering: plant and building operations, maintenance planning, setpoint strategies, occupant comfort, service continuity.
- Program Management Office (PMO) / Smart Campus Program Office: portfolio tracking, benefits realization, governance documentation, communications.

5. Decision Rights and Thresholds

Governance depends on explicit decision boundaries. The following decisions require review and approval as defined below.

5.1 Reserved for Executive Sponsors

- Institutional privacy commitments and AI ethical guardrails with campus-wide impact.
- Multi-year capital commitments above defined thresholds (to be set by the CFAO).
- Risk acceptance that materially elevates institutional exposure (as defined by Chief Risk/CISO).
- Programs that materially change campus safety posture, emergency response protocols, or life-safety integrations.

5.2 Reserved for the Steering Council

- SSC roadmap approval and sequencing across domains.
- Approval of enterprise smart-campus platforms (BMS consolidation strategy, digital twin platform selection, enterprise IoT management tooling, analytics, and AI foundations).
- Cross-domain tradeoffs that affect multiple divisions or create ongoing operational costs.
- Approval of SSC baseline metrics, targets, and public reporting approach.

5.3 Reserved for Domain Councils

- Data-sharing and integration policies; data access standards; major data model decisions (DGC).
- OT/IoT segmentation standards; security exception approvals; incident response protocols (CRC).
- Integration patterns; vendor platform architecture reviews; interoperability standards (ARB).
- High-impact AI model approvals; transparency/redress requirements; ongoing monitoring policies (AIEB).

5.4 Delegated to Operational Teams

- Routine maintenance, patching schedules within approved constraints, and operational configuration that does not alter policy boundaries.

- Implementation of approved projects, following reference architectures and security/data requirements.
- Minor changes below established thresholds (to be documented and auditable).

6. Operating Model: The Three Pipelines

Governance is operationalized through three continuous pipelines:

6.1 Design-to-Deploy (Built Environment Release Discipline)

All SSC changes that affect OT/IoT systems must follow a controlled path: design → review → commissioning → release → monitoring → post-implementation validation.

Minimum requirements:

- Named system owner and accountable operator
- Commissioning plan and rollback plan
- Security and privacy review as applicable
- Change window aligned with academic calendar and occupancy realities
- Post-change validation against defined performance indicators (energy, comfort, uptime, safety)

6.2 Data-to-Decision (Decision-Grade Data Discipline)

All dashboards, analytics, and models used for SSC decisions must be supported by:

- Authoritative data sources and reconciliation rules
- Data quality scoring and observability
- Access controls aligned to least privilege and purpose
- Documentation sufficient for audit and institutional transparency

6.3 Policy-to-Behavior (Trust Discipline)

All surveillance, access, sensor, and AI-enabled services with human impact must include:

- Plain-language transparency to affected communities
- Purpose limitation and data minimization
- Consent boundaries where applicable and lawful
- Defined mechanisms for questions, challenges, and redress
- Periodic review for equity impact and unintended consequences

7. Cadence and Meeting Rhythm

To ensure governance is both real and sustainable, the following standing cadence is adopted:

- **Weekly:** OT/IoT Operations and Security Huddle (30–45 minutes) Focus: exceptions, vulnerabilities, device onboarding, operational incidents, and upcoming maintenance windows.
- **Biweekly:** Built Environment Release / Change Review (60 minutes)Focus: approve production changes that affect BMS, controls, access systems, digital twin feeds, or automated actions.
- **Monthly:** Smart Campus Steering Council (90 minutes)Focus: portfolio progress, risk decisions, cross-domain conflicts, outcomes dashboard review, sequencing updates.
- **Quarterly:** Executive Sponsors Review (60–90 minutes)Focus: outcomes against targets, risk posture, major investment decisions, public accountability metrics.

8. Metrics, Reporting, and Accountability

The governance system will maintain a defined baseline and target outcomes across at least four categories:

1. **Sustainability Outcomes:** energy use intensity, emissions metrics, water intensity, waste diversion.
2. **Operational Outcomes:** uptime of critical systems, maintenance performance, comfort metrics, and reliability indicators.
3. **Security and Risk Outcomes:** vulnerability exposure, incident metrics, segmentation compliance, and audit results.
4. **Experience Outcomes:** service performance indicators, space effectiveness measures, satisfaction/sentiment where appropriate.

The Steering Council is responsible for ensuring these metrics remain decision-grade, well-governed, and used to guide investment and improvement.

9. Policy and Procurement Alignment

All procurement and contracting for SSC-related systems must align with governance requirements, including:

- interoperability and data portability expectations
- security requirements (including device lifecycle and patch pathways)
- privacy, retention, and data ownership clauses
- service-level expectations tied to operational continuity
- vendor disclosure requirements for AI components and model behavior, where applicable

Procurement and Legal shall partner with the ARB, CRC, and AIEB to ensure contract language enforces institutional governance commitments.

10. Exceptions, Risk Acceptance, and Auditability

Exceptions to standards may be granted only through documented risk acceptance aligned to defined thresholds. All exceptions must:

- identify scope, duration, compensating controls, and the responsible owner
- include an expiration date and remediation plan
- be recorded in a governance-accessible register
- be reviewed on a standing cadence until retired

11. Charter Review and Evolution

This charter shall be reviewed annually by the Steering Council and presented to Executive Sponsors for reaffirmation or revision. Revisions may be triggered sooner by:

- major cyber incidents or risk posture changes
- regulatory changes affecting privacy or safety systems
- significant technology shifts (e.g., BMS consolidation, digital twin platform adoption, AI governance expansion)
- material changes to institutional strategy or climate commitments

12. Formal Adoption

This charter becomes effective upon endorsement by the Executive Sponsors and shall be communicated institution-wide as the governing operating model for smart-campus transformation.

Signed / Endorsed by:

- President/Chancellor (or designee)
- EVP/Provost (or designee)
- EVP/COO (or designee)
- Chief Technology Officer
- Chief Facilities Officer
- Chief Sustainability Officer
- Chief Information Security Officer (or equivalent)

ADKAR Matrix for IT/OT Convergence

To implement the joint change management mandate effectively, the Chief Information Officer (CIO) and the CTO can utilize this concise matrix as a practical field guide. It aligns the five stages of the ADKAR change model with specific stakeholder groups essential for developing the AI-Enabled Sustainable Smart Campus. Instead of a generic communication plan, this matrix clearly outlines the role-specific deliverables needed to transition each group from isolated silos to a cohesive cyber-physical operating model.

Exhibit 24. ADKAR Phases and Smart Campus Deliverables

ADKAR Phase	Facilities Operators and Mechanics	IT Security and Network Engineering	Sustainability and Energy Management	Campus Leadership (Cabinet and Deans)
Awareness *(The "Why")*	Deliverable: "Voice of the Technician" kickoff.	Deliverable: OT constraints briefing.	Deliverable: Data visibility charter.	Deliverable: Cyber-physical risk and ROI briefing.
	Translate modernization into local reality: "This system will reduce 2:00 a.m. nuisance alarms and provide mobile access to equipment manuals."	Shift the IT mindset: establish that physical life-safety and environmental continuity absolutely supersede standard IT patching and reboot schedules.	Demonstrate how breaking down data silos and normalizing BACnet telemetry will finally provide the granular proof needed to achieve decarbonization goals.	Frame convergence not as an IT project, but as institutional risk mitigation, operational antifragility, and a capital cost-avoidance strategy.
Desire *(The "Willingness")*	Deliverable: Co-design workshops.	Deliverable: Joint operating rhythm.	Deliverable: Living Lab alignment.	Deliverable: Integrated master planning.
	Operators sit with UI developers to map their daily physical workflows directly onto the new dashboard interface, ensuring the tools address real field pain points.	Establish shared SLAs with Facilities. IT sees that convergence gives them the authority to secure the "Shadow IoT" devices that previously kept them up at night.	Partner with faculty to tie the new sensor mesh directly to academic research grants and student applied-learning opportunities.	Leadership actively grafts the digital infrastructure strategy into the 10-Year Physical Master Plan, securing baseline funding.
Knowledge *(The "How")*	Deliverable: Scenario-based field training.	Deliverable: ICS/OT architecture training.	Deliverable: Digital Twin and analytics training.	Deliverable: AI governance education.
	Kinetic, hands-on training for new	Network engineers learn the physics of	Sustainability teams are trained to query	The Cabinet is educated on the AI Governance

ADKAR Phase	Facilities Operators and Mechanics	IT Security and Network Engineering	Sustainability and Energy Management	Campus Leadership (Cabinet and Deans)
	semantic tags, mobile CMMS tools, and understanding the logic behind AI-generated fault alerts.	the central plant, industrial control protocols, and the mechanics of OT microsegmentation.	the normalized data lake and use the simulation engine for load-shedding models.	Charter, data privacy boundaries, and the shift from CapEx to OpEx funding models.
Ability *(The "Action")*	Deliverable: The Sandbox and Lighthouse Pilot.	Deliverable: Implementing Zero-Trust for OT.	Deliverable: Executing algorithmic control.	Deliverable: Activating the SCSC.
	Operators confidently execute real work orders using the new tools in a safe, controlled pilot building. They test AI overrides without fear of breaking the system.	IT successfully deploys segmented networks, implements secure vendor-remote-access gateways, and implements behavioral anomaly detection.	The team successfully runs an automated Demand Response event, shedding HVAC load during peak pricing using the new AI control plane.	The Smart Campus Steering Committee actively evaluates and approves/rejects new vendor purchases against established interoperability standards.
Reinforcement *(The "Sustainment")*	Deliverable: Transparent UX Backlog and Public Praise.	Deliverable: Joint incident post-mortems.	Deliverable: Institutional ESG reporting.	Deliverable: The Pre-Procurement Gate.
	Operators see their feedback turn into software updates ("Because you asked..."). Leaders publicly celebrate the cross-functional teams that resolve complex cyber-physical issues.	When a failure occurs, IT and Facilities conduct blameless reviews together to refine the incident response playbook.	The campus publishes live, transparent carbon-reduction scorecards directly from the smart campus platform, proving ROI.	Leadership holds the line. They formally refuse to fund any new construction or software that bypasses the unified digital architecture.

ADKAR for the Smart Campus Checklist

To implement the human aspect of IT/OT convergence, the CTO/CFaO can utilize this quick-reference checklist. It applies the ADKAR methodology into clear, yes-or-no questions that leadership needs to answer before starting any smart campus project.

1. Awareness (The Institutional "Why")

- [ ] **Mission Alignment:** Have we explicitly communicated how this technology protects the university's core mission (e.g., research continuity, student safety) rather than just framing it as an IT upgrade?
- [ ] **Risk Honesty:** Have we clearly articulated the risks of the status quo (e.g., deferred maintenance, siloed cyber vulnerabilities, rising utility costs) to the frontline staff?
- [ ] **Unified Front:** Are the CTO and CFaO delivering the same message to both the IT network engineers and the central plant mechanics?

2. Desire (The Willingness to Change)

- [ ] **Co-Creation Validation:** Have we actively included frontline operators—our "Innovators" and "Early Adopters"—in the design and selection of the new dashboards and tools?
- [ ] **Pain-Point Resolution:** Does the new system explicitly solve a known daily frustration for the workforce (e.g., reducing 2:00 AM nuisance alarms)?
- [ ] **Psychological Safety:** Do operators trust that the introduction of AI and automation is meant to elevate their roles from reactive firefighting to proactive stewardship, rather than replace them?

3. Knowledge (The Capability to Change)

- [ ] **Role-Specific Enablement:** Have we abandoned generic vendor tutorials in favor of targeted, role-based training (e.g., teaching network engineers about HVAC thermodynamics, and teaching mechanics about network microsegmentation)?
- [ ] **Scenario Rehearsals:** Have we conducted joint IT/Facilities tabletop exercises to practice responding to cyber-physical crises (e.g., a ransomware lock-out of the building automation system)?
- [ ] **Embedded Guidance:** Are digital job aids and contextual checklists built directly into the new mobile apps and CMMS interfaces to reduce cognitive load during a crisis?

4. Ability (The Proven Execution)

- [ ] **The Lighthouse Pilot:** Have we designated a single, controlled "sandbox" building where staff can physically test the new digital twin and AI fault detection without the fear of causing a campus-wide outage?
- [ ] **Human-in-the-Loop Verification:** Can operators easily and safely override the AI control plane and return systems to manual control if physical safety or research integrity demands it?
- [ ] **Readiness Gates:** Have we formally agreed *not* to scale the technology to the broader campus until the operators in the pilot building demonstrate confident, frictionless capability?

5. Reinforcement (The Institutionalization)

- [ ] **The Pre-Procurement Gate:** Is the Smart Campus Steering Committee actively rejecting departmental requests to purchase shadow-IT systems that bypass the new digital integration fabric?

- [ ] **Shift in KPIs:** Are we officially measuring and rewarding the workforce for acting on *predictive* AI insights rather than just logging their response times to *reactive* emergency failures?
- [ ] **The Iterative Loop:** Do we have a transparent, public process for taking operator feedback on software bugs or UI flaws and visibly deploying those corrections?

Guide to RACI

RACI is a framework for assigning responsibilities that helps clarify roles in complex tasks by explicitly stating who is **Responsible** for performing the work, who bears **Accountability** for the results, who needs to be **Consulted** before decisions, and who should be **Informed** of progress or outcomes. Its roots go back to mid-20th-century project and operations management, especially in large-scale engineering, defense, construction, and government initiatives where work was distributed among numerous teams and contractors.

- **Responsible**
  The role(s) that perform the work. In smart campus contexts, this often includes Facilities operators, integrators, or IT platform teams. Multiple roles may be responsible.
- **Accountable**
  The single role that owns the outcome and makes final decisions. This is non-negotiable. For example, Facilities may be accountable for safe operation, while the CTO is accountable for identity and cybersecurity governance.
- **Consulted**
  Roles whose input is required before decisions are made—risk, safety, sustainability, legal, or academic stakeholders.
- **Informed**
  Roles that must be kept aware of outcomes but do not influence decisions directly

Historically, RACI (and related variants like ARCI or RAMs—Responsibility Assignment Matrices) aimed to reduce duplication, address gaps in ownership, and resolve authority conflicts in environments where organizational charts alone did not clearly define how work was actually coordinated. Over time, RACI has become a standard tool in project management, IT service management, and process engineering because it offers a simple yet effective way to align accountability with execution—particularly in matrix organizations where multiple functions collaborate to achieve a single goal.

RACI is important to the planning, implementation, and operation of the Smart Campus because IT/OT convergence often fails due to unclear accountability rather than technical complexity. When multiple groups touch the same systems—identity, networks, controls, data, safety—ambiguity becomes risk. RACI (Responsible, Accountable, Consulted, Informed) provides a simple but powerful structure to eliminate that ambiguity.

Where RACI Is Most Critical

- OT identity and vendor access
- Network segmentation and change control
- Digital twin data ownership
- Sustainability reporting and ESG metrics
- Incident response and escalation
- Capital project digital handover

What Good RACI Enables

- Faster decisions with fewer exceptions

- Reduced "shadow OT" and workarounds
- Clear escalation during incidents
- Board-ready accountability
- Trust between IT and Facilities

In short, RACI turns shared ambition into **shared execution.** It is the quiet architecture that allows the smart campus to operate as a coherent, governed system rather than a collection of well-intentioned efforts.

How to Use RACI Effectively

- Apply RACI **at the service level,** not just the project level
- Publish RACIs as part of the **Bilingual Service Catalog**
- Revisit annually as capabilities mature (descriptive → cognitive → prescriptive)
- Treat deviations as **risk exceptions,** not informal workarounds

RACI Examples

Microgrid / DER Dispatch and Demand Response

The "New Energy Edge" makes the campus a VPP; dispatch/security is shared across IT/OT with carbon-aware orchestration.

R	Facilities/Energy Operations – operational dispatch and safe operating envelopes
A	CTO – secure integration to platforms and markets; telemetry governance
C	Sustainability – carbon-aware load shaping; Risk/Safety – resilience impacts
I	Vendors – controls, interconnection, performance

Incident Response for Cyber-Physical Events

A cyber-physical IR playbook with cross-training and table-tops; shutting down digital assets has physical consequences that must be governed.

R	Risk/Safety + SOC/NOC/FOC unified playbook – initiate and coordinate human safety with cyber containment.
A	CTO – converged IR playbook; balance containment with life-safety/continuity
C	Facilities – safe modes, failover; Campus Police, where appropriate
I	Vendors, Data Governance – platform forensics, evidence

Digital Twin Data Ownership and Model Updates

OT data is an institutional asset governed with clear purpose, controls, and accountability; twin trust depends on semantics and commissioning truth at the source.

R	**Data Governance** – ontology, lineage, retention, access roles; update cadence
A	**CTO** – platform stewardship and integration guardrails
C	**Facilities, Sustainability** – point accuracy, commissioning status, emissions use cases
I	**Vendors** – contractually required "digital handover" payloads

Network Segmentation and Change Control (ISA/IEC 62443 "Zones and Conduits")

Segmentation limits the blast radius and prevents IT incidents from becoming building incidents; the CTO owns the design standard, with Facilities consulted to maintain continuity.

R	Network Architecture (IT/OT Platform Team) – design/enforce segmentation and conduits
A	CTO – campus segmentation standard and exception process
C	Facilities, Risk/Safety – operational impacts, safe states, outage windows
I	Vendors, Data Governance – integration pathways, monitoring

Exhibit 25. Compact RACI Grid

Governance Domain / Decision Area	R – Responsible	A – Accountable	C – Consulted	I – Informed
OT Identity and Vendor Access (IAM, MFA, least privilege)	IT / Security Engineering	CTO	Facilities, Risk/Safety, Vendors	Data Governance
Network Segmentation and Change Control (Zones and Conduits)	Network / OT Platform Team	CTO	Facilities, Risk/Safety	Vendors, Data Governance
Remote Access (Vendors and Staff)	IT Security	CTO	Facilities, Vendors	Risk/Safety
Digital Twin Data Ownership and Model Updates	Data Governance	CTO	Facilities, Sustainability	Vendors
Cyber-Physical Incident Response	Risk/Safety (Unified Command)	CTO	Facilities, Security Ops	Vendors, Data Governance
Capital Project Digital Handover (Div. 25)	Facilities + Vendors	CTO	Data Governance, Sustainability	Risk/Safety
Microgrid / DER Dispatch and Demand Response	Facilities / Energy Ops	CTO	Sustainability, Risk/Safety	Vendors
IAQ Policy and CO_2-Aware Ventilation	Facilities	Sustainability	Risk/Safety, Data Governance	CTO
Space Optimization (Occupancy Analytics)	Data Governance	CFAO	Sustainability, CTO	Risk/Safety
ESG / Sustainability Reporting (Scopes 1–3)	Sustainability	Data Governance	Facilities, CTO	Vendors
Material Passports and Scope 3 Supply Chain	Sustainability	Data Governance / Procurement	Facilities, CTO	Vendors

How to Read This Grid

- **Accountable is singular**
  Every row has exactly one "A." When accountability is shared, decision-making slows and risk increases.
- **Responsible can be plural**
  Execution often spans teams, but accountability does not.
- **Consulted ≠ Veto**
  Consultation ensures safety, uptime, privacy, and sustainability are honored—without paralyzing execution.
- **Informed enables trust**
  Transparency reduces surprise escalations and reinforces the legitimacy of governance.

ASHRAE and the Smart Campus

For university leaders, incorporating ASHRAE (American Society of Heating, Refrigerating and Air-Conditioning Engineers) standards has evolved from simply ensuring mechanical compliance to becoming a key element of the "Sentient Campus' strategy. As detailed in The Smart Campus Manifesto, the integration of Information Technology (IT) and Operational Technology (OT) demands strict adherence to these standards to support goals related to environmental sustainability, cost efficiency, and occupant health. By leveraging smart building innovations—such as Digital Twins and AI-powered predictive maintenance—that align with ASHRAE's energy efficiency and indoor air quality guidelines, institutions can turn passive systems into dynamic, data-driven ecosystems. These ecosystems can adapt in real-time to the needs of students and staff, significantly lowering the campus's carbon footprint. The following table aligns key ASHRAE standards with the strategic pillars of a modern, AI-enabled smart campus.

To foster an antifragile smart campus, leadership needs to see these standards not merely as separate engineering tasks but as part of the university's Technical Governance framework. Requiring all OT systems—such as HVAC and lighting—to use ASHRAE-compliant protocols, such as BACnet (Standard 135), ensures that physical assets are "AI-ready." This approach enables the campus to sense, learn, and react cohesively, functioning as a unified, intelligent system.

Exhibit 26. ASHRAE Standards & Smart Campus Integration

ASHRAE Standard	Focus Area	Smart Campus Linkage	Strategic Value for Leadership
90.1	Energy Standard for Buildings	Digital Twins & Energy Modeling: Provides the baseline for AI cognitive engines to optimize energy loads in real-time.	Ensures long-term fiscal sustainability and compliance with state/federal carbon mandates.
55	Thermal Environmental Conditions	Sentient Occupancy Sensing: Integrates with IoT sensors to adjust HVAC based on actual human density rather than fixed schedules.	Enhances student success and faculty retention by maximizing cognitive comfort and well-being.
62.1 / 62.2	Ventilation & Indoor Air Quality (IAQ)	Health & Wellness Dashboards: Enable real-time monitoring and automated response to CO_2 and pollutant levels.	Mitigates institutional risk related to public health and enhances the "Healthy Campus" brand.
135	BACnet Protocol	System Interoperability: The "language" of the smart campus allows IT/OT convergence and data democratization.	Prevents vendor lock-in and creates a unified "Institutional Data Estate" for campus-wide analytics.
189.1	High-Performance Green Buildings	Holistic Sustainability: Integrates site sustainability, water use, and resource efficiency into the Digital-Physical Ecosystem.	Demonstrates institutional commitment to ESG (Environmental, Social, and Governance) goals.

ASHRAE Standard	Focus Area	Smart Campus Linkage	Strategic Value for Leadership
211	Commercial Building Energy Audits	Data-Driven Capital Planning: Uses historical IoT data to automate audits and prioritize deferred maintenance.	Optimizes capital expenditure (CapEx) by identifying high-ROI infrastructure upgrades.
221	Field HVAC Performance	Predictive Maintenance: Uses AI to detect performance drift from ASHRAE benchmarks before equipment failure occurs.	Reduces operational expenditure (OpEx) and prevents disruptive emergency repairs.

Case Study: Northern Arizona University

A case study of a sustainable smart campus master plan.

Executive Summary

Northern Arizona University (NAU) has emerged as a national leader in imagining how a public research university can operate—**not just more efficiently, but more intelligently, sustainably, and resiliently.** Through an ambitious, board-approved AI-Enabled Sustainable Smart Campus strategy, NAU has transformed aging infrastructure, siloed operations, and constrained resources into a coordinated, data-driven campus ecosystem.

At the heart of this transformation is the convergence of **IT and Facilities (OT)**, enabled by a campus-scale **digital twin**, institution-wide governance, and applied artificial intelligence. The results include measurable energy savings, avoided facilities failures, improved safety, enhanced student and faculty experiences, and a fundamentally new operating model—one that NAU describes as **antifragile**: capable of learning from stress, improving through disruption, and getting stronger over time.

The Context: A Complex Campus Under Pressure

NAU is a large regional public university serving more than **28,000 students**, with a flagship mountain campus in Flagstaff, Arizona—**over 600 acres, 6.8 million square feet, and 100+ buildings**, many dating back more than a century. The campus sits at **an elevation of 7,000 feet**, adjacent to the Coconino National Forest. It faces unique climate and infrastructure challenges, including exposure to extreme weather, aging facilities, and limited access to skilled trades labor.

By the late 2010s, several pressures converged:

- Rising energy and utility costs
- Deferred maintenance across an aging building portfolio
- Difficulty recruiting and retaining facilities trades staff
- Increasing dependence on external partners for critical building systems
- Growing expectations for sustainability, safety, and student experience
- A need to compete for students amid demographic decline

Incremental fixes were no longer sufficient. The existing operating model—reactive, siloed, and reliant on institutional memory—had reached a breaking point.

Early Smart Initiatives

With the appointment of a new CIO in 2016 to oversee the consolidation of IT services across the institution and to promote a modern IT service vision, early smart campus initiatives were integrated into a new IT strategic plan. Much work remained to establish governance and infrastructure capable of supporting advanced technology initiatives and the CIO's vision for "smart" campus capabilities.

Foundational Infrastructure Initiatives

- Establish IT and Data Governance

- Improve wireless infrastructure (WiFi 6, LoRaWAN, CBRS)
- Improve cybersecurity program (policies, firewalls, virtual private network)
- Improve Internet bandwidth and resilience
- Move key applications to the SaaS model (LMS)
- Adopt a cloud computing strategy and partner (Microsoft©)
- Implement technology standards (Hyflex classrooms)
- Adopt and apply the ITSM program (Service Now)
- Identify redundant and shadow technology (email, web servers, etc.)
- Reduce technical debt and update aged systems. (WebTMA)

Early Smart Campus Initiatives

- Condition-based maintenance with remote servicing of classrooms
- Modern collaboration and integrated communication platform (Microsoft© Teams)
- Smartphone application with digital identity (NAUgo)
- Reimagined learning ecosystem (Canvas™ LMS, Smart Sparrow™ adaptive learning , TopHat™, Packback™, iClicker™, etc.)
- Smart trash/recycle receptacles.
- Space management and dynamic reservation
- Federated digital signage
- Administrative decision support dashboards (Enrollment, retention, financial)
- Student success monitoring (Salesforce CRM)
- Smart parking enforcement
- Bus, vehicle, and asset tracking
- Campus safety and surveillance.
- Improve/expand door-access and
- Delivery robots (NAU was 2nd to implement Starship™)
- IoT development and telemetry integration (occupancy, meters)

It would take several years to build substantial infrastructure. Still, by the time the COVID pandemic was declared in March of 2020, the infrastructure and capabilities were sufficient to shift the campus to remote work and learning overnight. Progress in updating infrastructure, improving applications, and deploying smart projects continued throughout the pandemic. The early smart initiatives that came from the IT strategic plan had demonstrated value in creating efficiencies and experiences that supported institutional strategic priorities.

From Technology Projects to Institutional Strategy

While IT-led "smart" initiatives had existed at NAU since 2016, the turning point came in 2022, when the university incorporated smart campus features directly into its 10-year Comprehensive Sustainable Smart Campus

Master Plan. Approved by the Arizona Board of Regents in 2023, the plan redefined smart technologies as essential institutional infrastructure rather than optional upgrades.[37]

The master planning vision was clear: a technology-enabled (AI, Cloud Computing), sustainable smart campus that fosters meaningful and "sticky" living and learning experiences for students, while improving efficiencies in human resources, energy consumption, facilities management, space utilization, resilience, and enterprise risk reduction. Several key themes guided the plan:

- **Student-Centered Design:** Creating inclusive, collaborative, and wellness-oriented spaces that support learning beyond the classroom.
- **Resource Optimization:** Prioritizing renovation, adaptive reuse, and deferred maintenance reduction over physical expansion.
- **Sustainability & Carbon Neutrality:** Integrating the Climate Action Plan and committing to carbon neutrality for Scope 1 and 2 emissions by 2030.
- **Indigenous Planning:** Expanding the Native American Cultural Center, integrating Indigenous knowledge and design, and honoring the land and its communities.
- **Smart Campus Evolution:** Leveraging technology to enhance safety, accessibility, operations, and the learning environment.
- **Statewide Access:** Strengthening NAU's presence across Arizona through high-demand programs and potential branch campuses.

The master plan was not just an extension of the existing IT strategic plan. The comprehensive plan covered all areas of authority and university functions. Ownership of the master plan was held by the Chief Operating Officer, with the CIO, Chief Facilities Officer, Chief Sustainability Officer, and members of Data Governance sharing responsibility under a governance model that prioritized transparency, participation, and measurable results.

The Digital Twin: One Pane of Glass for the Built World

A key enabling capability was NAU's deployment of the Willow digital twin, making NAU the first higher education institution in the U.S. to implement Willow and launch a campus-wide, real-time digital twin across multiple building management systems toward full automation of building controls. The twin unified previously disconnected data silos, including:

- Building Automation and HVAC systems
- Energy and utility metering
- Work order and maintenance systems
- BIM and GIS spatial data
- IoT telemetry and sensor data
- Occupancy and scheduling proxies
- Real-time grid emissions data

By establishing a comprehensive data estate, common ontology, and real-time building telemetry, NAU shifted facilities management from experience-based intuition to evidence-based decision-making—without removing humans from the loop.

Within six months, Willow-generated insights helped identify more than $ 200,000 in avoidable operational costs and dramatically improved the speed and confidence of facilities' responses. These early results would prove to be just the tip of the iceberg.

From Reactive to Predictive—and Preventive

The operational impact was evident. During one incident, digital twin data uncovered unusual water usage in a building thought to be leak-free. Despite initial skepticism from experienced trades staff, the data proved accurate twice, leading to the discovery and repair of several hidden leaks. In another case, digital twin alerts identified conditions that could have caused a repeat of a previous flooding event that had shut down an academic building and disrupted research. This time, disaster was averted.

These moments proved transformational—not just technically, but culturally. Over time, trust shifted from anecdote to analytics, and the campus moved toward condition-based and predictive maintenance at scale.

Energy savings became evident almost immediately, especially during extended campus shutdowns, when building energy use approached normal levels. With smart controls and real-time insights, NAU finally aligned occupancy with energy consumption.

People First: Change, Trust, and Adoption

NAU leaders were honest that the biggest risk was not technology failure but adoption. Facilities staff initially feared losing jobs; IT was concerned about OT cybersecurity risks; leadership worried about growing too quickly. To address this, NAU emphasized:

- Phased implementation aligned with funding realities
- Early proof points and "low-hanging fruit."
- Embedded IT project management within Facilities
- Intentional human-in-the-loop design for work control and set-points
- Ongoing training, retraining, and psychological safety
- Transparent governance through IT and Data Councils

As adoption increased, so did confidence. Facilities teams started to "drink the Kool-Aid," using insights to enhance uptime, prioritize tasks, and evaluate vendor performance. Trust between IT and Facilities grew significantly over time.

Outcomes That Matter to Leadership

Today, NAU leadership tracks smart campus outcomes through institutional KPIs that roll up into university-level priorities:

- Energy reduction and decarbonization progress
- Improved confidence in sustainability and carbon reporting
- Avoided downtime from flooding and mechanical failure
- Improved classroom uptime and faculty satisfaction
- Enhanced campus safety through smart access and surveillance
- Improved operator efficiency with fewer staff
- Reduced cyber and vendor risk in OT environments

Critically, NAU now oversees its campus as a unified system instead of just a group of buildings.

Objectives and Key Results

Objectives and Key Results (OKRs) are essential to Northern Arizona University's Sustainable and Smart Campus transformation by turning long-term vision into tangible, campus-wide results. As NAU shifted from standalone technology projects to a board-approved, operational plan, leadership adopted OKRs to align investments across facilities, IT, sustainability, mobility, and academic operations with the university's strategic goals. This framework prioritizes outcomes over activities—focusing on carbon reduction, student and employee experience, Indigenous engagement, operational resilience, and fiscal responsibility—while offering a clear structure for governance, accountability, and ongoing learning. Overall, these Objectives and Key Results define its strategic intent and ensure that its smart campus initiatives support the goals of NAU 2025: Elevating Excellence.

Exhibit 27. NAU's Objectives and Key Result Areas

Objective	Key Results (KRs)
1. Achieve a carbon-neutral, resilient campus by 2030	• Eliminate Scope 1 & 2 emissions by 2030 • Reduce campus Energy Use Intensity annually • Convert district heating to low-temp hot water • Implement a geothermal or biomass heating solution• • Complete deep energy retrofits in high-priority buildings
2. Transform campus mobility to a pedestrian-first, low-carbon system	• Prioritize pedestrians, bicycles, and transit over private vehicles • Standardize and extend the Pedway as the campus mobility spine • Add safe east–west multimodal connections • Reduce vehicle miles traveled and parking demand through TDM
3. Reduce transportation and fleet emissions	• Electrify passenger fleet vehicles to define its strategic intent and ensure the use of electric vehicles by 2030 • Electrify bus and shuttle fleet by 2035 • Install EV charging infrastructure campus-wide
4. Embed Indigenous identity, lifeways, and ways of knowing across campus	• Expand the Native American Cultural Center (NACC) • Establish medicine gardens, ceremonial, and teaching spaces • Create a Tribal Sovereignty / Walk of Nations path • Integrate Indigenous language, art, and storytelling throughout campus
5. Operate the campus as an intelligent, data-driven system	• Implement a campus digital twin for planning and operations • Deploy real-time energy and water dashboards • Integrate IoT sensors for occupancy and predictive maintenance • Use analytics to prevent failures and reduce downtime
6. Improve safety, access, and experience through smart campus technologies	• Expand Wi-Fi across Pedway, open spaces, and parking • Implement smart signage at gateways and key corridors • Standardize smart ingress/egress and security systems • Enable future-ready paths for robotics and autonomous delivery
7. Optimize space utilization and shift from ownership to sharing	• Increase classroom utilization to ~30 hours per week • Achieve ~80% average seat fill across instructional spaces • Implement centralized scheduling for shared spaces • Consolidate and right-size academic districts
8. Professionalize campus operations and governance	• Establish a Building Manager Coordinator role • Standardize building manager training and responsibilities • Improve transparency and accountability for facilities performance • Align policies with sustainability and collaboration goals

Sustainable & Smart Campus – Core KPIs

While OKRs identified strategic intent, the university established KPIs as measures that directly align with NAU 2025 – Elevating Excellence priorities, such as student success, sustainable stewardship, Indigenous engagement, workforce effectiveness, and operational excellence. The metrics emphasize outcomes over projects—highlighting the transition from technology implementation to overall institutional performance management. Each KPI can be summarized for executive reports or examined via the digital twin and dashboard platforms.

Exhibit 28. NAU Sample Smart Campus Master Plan KPIs

KPI Category	Key Performance Indicator (KPI)	Description / Measurement Focus
Sustainability & Climate	Scope 1 & 2 Emissions	Annual greenhouse gas emissions from onsite energy, fleet, and purchased electricity.
	Energy Use Intensity (EUI)	Campus and building-level energy use per square foot
	Carbon Neutrality Progress	Milestones toward 2030 Scope 1 & 2 carbon neutrality
	Energy Use Reduction	Reduce the total electrical energy used.
	Water Use Reduction	Total potable water use and reclaimed water utilization
	Waste Diversion Rate	Percentage of waste diverted from landfill (recycling, composting)
Mobility & Transportation	Mode Share	Percentage of trips by walking, biking, transit vs. single-occupancy vehicles
	Vehicle Miles Traveled (VMT)	Annual VMT by campus commuters and fleet
	Parking Demand	Utilization of perimeter vs. interior parking facilities
	Fleet Electrification	Percentage of passenger, facilities, and transit fleet electrified.
Facilities & Infrastructure	Deferred Maintenance Backlog	Total deferred maintenance value and annual reduction rate
	Facilities Downtime Avoided	Number of incidents avoided through preventive or predictive maintenance
	Utility System Modernization	Progress toward low-temperature hot water and plant interconnection
Smart Campus & Technology	Asset Coverage in Digital Twin	Percentage of campus buildings and utilities represented
	Predictive Maintenance Adoption	Share of work orders generated condition-based vs. reactive.
	Real-Time Monitoring Coverage	Percentage of buildings with live energy/water data
	Network Availability	Wi-Fi and network coverage and uptime in academic and outdoor spaces
Space Utilization	Classroom Utilization	Average scheduled hours per classroom per week
	Seat Fill Rate	Average classroom occupancy percentage
	Space Sharing Adoption	Percentage of centrally scheduled classrooms and shared spaces
Student, Faculty & Staff Experience	Classroom Uptime	Percentage of instructional spaces available as scheduled
	Facilities Response Time	Average time to resolve facilities service requests
	Campus Safety Indicators	Safety incidents influenced by lighting, access, or smart systems.
Indigenous Planning & Equity	Indigenous Spaces Delivered	Number of Indigenous-designed or culturally significant spaces implemented
	Indigenous Engagement	Frequency of formal engagement sessions with Indigenous communities
	Native Language & Art Presence	Count of integrated signage, art, and storytelling installations

KPI Category	Key Performance Indicator (KPI)	Description / Measurement Focus
Governance & Operations	Cross-Functional Governance Participation	Representation across IT, Facilities, Sustainability, and Indigenous Affairs
	Building Manager Training Coverage	Percentage of buildings with trained, supported managers
	KPI Reporting Maturity	Dashboards and transparency of institutional KPI reporting

Antifragility as a Design Principle

Perhaps NAU's most distinctive contribution is its embrace of **antifragility**—the idea that campuses should not merely resist failure but **learn from it and improve**. Each incident, anomaly, or inefficiency feeds back into the data estate, sharpening insight and accelerating response.

With AI increasingly layered atop this foundation, NAU is now positioned to:

- Scale smart services campus-wide
- Modernize its enterprise data architecture
- Enable generative AI across facilities and sustainability
- Support new student services, auxiliaries, and revenue models

Today, NAU can manage buildings at scale, verify outcomes, coordinate IT/OT decisions, and respond in near real time—something that was not possible under the previous operating model.

Lessons for Peer Institutions

NAU's journey offers clear guidance for others:

1. Start with **people and governance**, not technology
2. Treat smart campus initiatives as **institutional strategy**, not projects
3. Identify and measure **KPIs and OKRs from day one**
4. Focus on **low-risk, high-value early wins**
5. Build trust in data through transparency and proof
6. Keep humans intentionally in the loop
7. Let AI and digital twins **augment expertise**, not replace it

Northern Arizona University's AI-Enabled Sustainable Smart Campus is not just a vision slide—it is a working reality. Through disciplined governance, bold leadership, and a relentless focus on results, NAU has transformed constraints into opportunities and complexity into an advantage. In doing so, NAU is not only building a smarter campus but also reshaping how universities operate in an era of uncertainty, scarcity, and rapid change.

Case Study: The Willow Digital Twin

A Review of Willow Inc.'s Digital Twin Platform
in the AI-Enabled, Sustainable Smart Campus Ecosystem

Today's higher education campuses face a complex mix of operational challenges, financial constraints, and environmental concerns. Universities are balancing large portfolios of aging infrastructure with cutting-edge research facilities, making a unified, smart management system essential rather than optional. The old approach of reactive maintenance and isolated data systems is giving way to the digital twin model, which offers a more integrated and proactive management solution. [7]

Willow Inc. has become a leading provider in this space, delivering an AI-powered platform that unifies various operational technology (OT) systems into a single, intelligent "built world" layer. More than just a "digital twin," the Willow platform serves as an operational AI platform grounded in data, skills, and agents, enabling swift, seamless execution of use cases. By democratizing the effort required to achieve outcomes, Willow allows organizations to move beyond custom integrations and harness value at scale.[1]

Willow's capabilities are powered by combining spatial geometry, static asset histories, and real-time telemetry into a single, reliable source of truth that forms the basis of operational AI capabilities, skills, and prescriptive actions. This integration enhances predictive analytics and supports autonomous controls, ultimately increasing ROI. Additionally, built on the robust and scalable Microsoft© Azure platform and featuring the advanced Willow Object Model (WOM) semantic ontology, Willow enables universities to leverage their existing talent and knowledge. This transforms their physical estates into dynamic, data-driven environments. [4]

Deep Dive: The Willow Digital Twin Ecosystem

Willow is a prominent player in the digital twin sector, focusing on the "built world"—including large-scale campuses, airports, and commercial real estate portfolios. The Willow platform is distinguished by its comprehensive use of Operational AI and its commitment to standardized, secure data architectures.

Architecture and Data Ingestion

The Willow platform is built on Microsoft© Azure, leveraging its PaaS (Platform as a Service) capabilities for global scalability and enterprise-grade security. The platform functions as a "single intelligent model" that synthesizes three primary segments of data:

- **Spatial Data:** Digital representations of the physical environment, typically sourced from BIM and GIS.
- **Static Data:** Documentation such as blueprints, O&M manuals, and warranty details.
- **Live Data:** Real-time telemetry from BMS, CMMS, and IoT devices.

Willow currently processes over 10 million telemetry points in real-time, integrating with over 75 built-world systems. This integration allows for a unified view of systems that traditionally operate in isolation, such as HVAC, fire safety, and access control.[2]

The Willow Knowledge Graph and DTDL

At the heart of Willow's platform is the Knowledge Graph, built on a rich ontology using DTDL (Digital Twin Definition Language). This ontology provides a structured representation of every asset and its relationship to the surrounding space and systems. For example, a "Pump" isn't just a data point; it is modeled as an asset that belongs to a specific "Piping System," which serves a specific "Zone," which is located on a specific "Floor" of a specific "Building". This semantic depth is what allows Willow's AI to perform complex diagnostics—such as identifying that a temperature imbalance on the 4th floor is caused by a malfunctioning valve three floors below.[5]

Operational AI and the Insights Spectrum

Willow employs a sophisticated "Insights Spectrum" that guides facility teams from reactive to prescriptive operations.

1. **Identify Faults + Diagnostics:** The platform turns raw telemetry into actionable insights, identifying equipment that is failing or running outside of its design spec.
2. **Dynamic Baselines:** Instead of fixed alerts, Willow uses historical data to learn "normal" behavior. If a moisture detector triggers, the system evaluates the reading against past patterns to determine the severity.
3. **Occupancy-based Optimization:** By combining access control data (badge swipes) with people counting sensors and Wi-Fi logs, the system builds a reliable picture of space utilization. This allows the twin to cross-validate presence and reduce energy waste in unoccupied zones.
4. **Carbon/Cost-based Optimization:** The digital twin leverages real-time grid carbon intensity data to generate insights that help facility teams shape energy usage to minimize environmental impact and costs.

Exhibit 29. Willow's AI Design Patterns and Use Cases

Design Pattern	Application in Willow	User Impact
GenAI Subgraph Generation	Digitizing single-line diagrams and reports.	Accelerates building onboarding from weeks to hours.
Decision Intelligence	Impact scoring and KPI prioritization (e.g., Avoidable Cost).	Helps managers focus on the most financially significant issues first.
Intelligent Automation	Autonomous valve closing or setpoint adjustment.	Reduces the need for manual intervention in critical events.
Knowledge Discovery	Conversational chat with O&M manuals via Copilot	Empowers technicians with decades of specialized knowledge in real-time.
Perception (Vision/Signals)	Interpreting foot traffic and presence data.	Optimizes space utilization and improves safety monitoring.

Security and Governance

Security is a cornerstone of Willow's approach, especially given the sensitivity of building access and occupancy data. The platform is ISO 27001 and SOC 2 Type 2 certified and follows zero-trust principles. Most services are deployed in dedicated instances within the customer's single-tenant environment, ensuring data isolation. Willow differentiates between "Operational Data" (equipment telemetry used for product training) and "Personally Identifiable Information" (PII), ensuring that any data used for product improvement is strictly controlled and never shared across customer instances. Prompts and responses for GenAI features like Willow Copilot remain within the customer's selected Azure region, maintaining strict data residency.

Research Outcomes and Empirical Case Studies

The implementation of campus digital twins across various global institutions has yielded a wealth of empirical data on the technology's effectiveness and current limitations.

Academic Institutions and Smart Campus Results

At institutions like the National University of Singapore (NUS) and Cornell, research focuses on human-model interaction and the gap between "marketing rhetoric" and the "technical realities" of digital twin platforms. Meanwhile, the University of Melbourne has successfully deployed digital twins for structural health monitoring of infrastructure and stormwater capacity analysis. These projects highlight that the value of the digital twin extends beyond the building envelope to the campus's civil infrastructure and environmental resilience.

One standout application is the use of digital twins for drone management at HKUST. The platform serves as a "hyper-interactive 3D map" of the campus airspace, acting as a smart air-traffic control system for drone swarms. By feeding real-time data into the digital twin, the system orchestrates drone movements to prevent collisions and enable complex maneuvers safely. This represents a "state-of-the-art" use of digital twins for the "low-altitude economy" on large campuses.

Architecture

The Willow platform aims to address the fragmentation within the "built world.' Typically, buildings and campuses consist of separate systems—such as HVAC, lighting, security, and elevators—that generate large volumes of telemetry data but seldom communicate with one another. The platform functions as an Independent Data Layer (IDL), standardizing these diverse data streams into a single virtual model of the physical environment. This framework integrates three main data types: spatial, static, and live data.

Spatial data offers the critical geometric context. Using Building Information Modeling (BIM) and 3D laser scanning, the platform crafts a digital model of the campus's physical layout, including the relationships between rooms, levels, and infrastructure networks. Static data enriches this model with the "technical history" of assets—detailing equipment specs, warranties, maintenance records, and asset classifications. Live data completes the system by streaming real-time telemetry from IoT sensors, meters, and building management systems (BMS), monitoring environmental variables like temperature, CO_2 levels, occupancy, and energy flow. By combining these layers, the platform provides facilities teams with "situational awareness," enabling them not only to identify when an asset is failing but also to pinpoint its exact location, service zone, and past performance in a unified interface.

Exhibit 30. Data Components, Sources, and Utility

Data Component	Sources and Technical Origins	Primary Operational Utility
Spatial Layer	BIM (Revit), Terrestrial Laser Scanning (TLS), GIS	Visualizes physical adjacencies and system topologies for rapid navigation.
Static Layer	CMMS records, warranties, O&M manuals, asset registers	Provides the foundational "identity" and specifications of every campus asset.
Live Layer	IoT sensors, BMS (BACnet/Modbus), smart meters, Wi-Fi	Delivers real-time operational status and environmental telemetry.

By integrating these layers, the platform extends beyond basic monitoring capabilities. It enables a shift from reactive maintenance—where issues are fixed only after they occur—to a proactive and predictive approach. This is made possible through a "Knowledge Graph" that illustrates the connections among all twins within the environment.1 In a campus context, this means the platform recognizes that a particular Air Handling Unit (AHU) serves a specific set of Variable Air Volume (VAV) boxes, which control the temperature in designated classrooms. Such relational understanding is what enables the platform to detect "Avoidable Costs," such as systems operating in empty rooms or equipment operating outside their intended parameters.

The Willow Object Model (WOM) and Semantic Ontology

The core intelligence of a digital twin depends on its underlying ontology. An ontology provides a standardized vocabulary and rules for organizing data and defining relationships between entities. In the built environment, Willow uses the Willow Object Model (WOM), a highly comprehensive ontology for buildings and infrastructure. WOM is not a standalone system but builds on the open-source RealEstateCore (REC) ontology, which incorporates aspects of the Brick Schema for detailed technical system descriptions.

The WOM uses the Digital Twin Definition Language (DTDL) to describe entities. This choice is strategic because DTDL is a machine-readable language based on JSON-LD that enables the precise definition of properties, telemetry points, commands, and relationships. Within the WOM, a "twin" is an instance of a specific model. For example, a campus may have hundreds of twins based on a single "Chiller" model, each inheriting the same set of standard telemetry points, such as SupplyWaterTemperature and PowerConsumption.

Exhibit 31. Ontology Layers

Ontology Layer	Composition and Scale	Strategic Function
Model Layer	3,700+ DTDL Models	Standardizes the definition of assets, spaces, and systems.
Property Layer	1,400+ Properties	Defines static characteristics (e.g., manufacturer and serial number).
Relationship Layer	120+ Relationship Types	Establishes the functional and spatial links between twins.

The relationships defined in the WOM are directional and hierarchical, creating a dense web of dependencies known as "graph density". Key relationship types include *isPartOf,* which defines physical containment (e.g., a room is part of a building). *Isfedby*, which captures the flow of substances or energy (e.g., a VAV box is fed by an AHU) Another critical relationship is *CapabilityOf,* which links sensors or actuators to the assets they monitor or control. This semantic structure enables the platform to perform impact assessments. If a main water valve fails, the platform can traverse the graph to identify every asset and space that will lose water supply, enabling facilities managers to communicate outages precisely to affected departments rather than guessing the scope of the impact.

Furthermore, the WOM facilitates "semantic normalization." In legacy campus environments, data points are often named inconsistently across buildings or vendors (e.g., "RTU-1_Temp" vs. "Unit01_DischargeAirTemp"). The WOM maps these disparate tags to a common language, enabling the platform's analytical "Skills" and AI models to be deployed across the entire campus without manual, building-specific reconfiguration. This scalability is essential for higher education institutions that often manage hundreds of buildings of varying ages and technological sophistication.

Technical Foundation: Built on Microsoft© Azure

The platform's ability to manage millions of telemetry points across global portfolios is made possible by its foundation on the Microsoft© Azure cloud ecosystem. This partnership provides the enterprise-grade security, scalability, and integration capabilities required for large-scale institutional deployments. The architecture is primarily centered on three core Azure services: Azure Digital Twins (ADT), Azure IoT Hub, and Azure IoT Edge.

Azure Digital Twins (ADT) serves as the orchestration engine, hosting the "twin graph" where the relationships defined in the WOM are instantiated. ADT maintains the current state of the environment, processing incoming telemetry to update twin properties in real-time. Because ADT is built on DTDL, it provides the flexibility to model complex interactions between people, places, and things. To drive updates to these twins, the platform utilizes Azure Functions, which act as serverless compute resources to translate incoming raw data into the standardized DTDL format.[14]

Exhibit 32. Microsoft© Azure Components

Azure Component	Role in Platform Architecture	Impact on Campus Operations
Azure Digital Twins	Twin graph orchestration and state management.	Provides the digital twin's real-time "brain".
Azure IoT Hub	Secure ingestion and device management.	Ensures encrypted, bidirectional communication with millions of devices.
Azure IoT Edge	Localized processing and protocol translation.	Enables offline operability and reduces latency for critical systems.
Azure Data Explorer	Time-series data historization.	Powers long-term trend analysis and predictive ML models.

Connectivity between the campus and the cloud is managed through Azure IoT Hub and Azure IoT Edge. IoT Hub provides a secure, bidirectional communication path for millions of connected devices.14 For the higher education sector, where network security is paramount, Azure IoT Hub integrates with Azure Active Directory (AAD) to provide robust access control and identity management.14 Azure IoT Edge further enhances the architecture by enabling cloud-native workloads—such as AI models or data filtering logic—to run locally on campus hardware.24 This "edge intelligence" is particularly useful for reducing bandwidth costs and ensuring that critical building safety systems continue to function even during periods of internet connectivity loss.

For historical analysis and forecasting, the platform uses Azure Data Explorer (ADX). While ADT manages the current state of the twins, ADX stores every telemetry point, creating a massive repository of time-series data.1 This historical record enables the platform's machine learning models to identify patterns over months or years, such as the gradual degradation of a chiller's efficiency or seasonal patterns in energy usage that can be used for more accurate utility budgeting.

OT and Smart Campus Integration: Device Incorporation

A significant challenge for smart campuses is the convergence of Information Technology (IT) and Operational Technology (OT). OT systems, which govern physical operations like HVAC and security, often use legacy protocols such as BACnet and Modbus that were never designed for cloud connectivity. Willow addresses this "protocol fragmentation" by acting as a bridge between the physical and digital worlds.

The Role of Gateways and Protocol Translation

Most modern smart building systems utilize BACnet (Building Automation and Control Networks), which holds a global market share of over 77%. BACnet is preferred for HVAC and lighting because of its object-oriented approach, which aligns well with the digital twin concept. However, critical industrial-grade equipment such as boilers and meters often utilizes Modbus, a simpler protocol that relies on register-address mapping.

The platform incorporates these devices using IoT Gateways (such as the USR-M300) or Azure IoT Edge devices. These gateways perform "translation," taking raw BACnet or Modbus packets and converting them into JSON payloads that the cloud platform can ingest. This allows the platform to integrate not just modern sensors but also legacy equipment that may be decades old—an essential capability for universities with historical building stock.[29]

Exhibit 33. Protocol Features of BACnet and Modbus

Protocol Feature	BACnet	Modbus
Architecture	Object-Oriented / Peer-to-Peer	Register-Based / Master-Slave
Security	Supports BACnet/SC (Secure Connect)	Traditionally unencrypted; requires VPN/Gateways
Interoperability	High standard for BMS	Specialized; common in energy and industrial devices

The Shared Data Estate and IoT Ecosystem

Beyond mechanical systems, the platform incorporates a wide range of IoT devices to enhance campus intelligence. This includes Wi-Fi access points, which can provide real-time occupancy data based on device connection counts, and environmental sensors for CO2, humidity, and particulate matter. By overlaying occupancy data with HVAC settings, the platform creates a "shared data estate" that enables traditionally siloed systems to collaborate.

For example, when a university classroom is empty, as indicated by Wi-Fi or scheduling systems, the platform can automatically signal the BMS to adjust the HVAC setpoints to an energy-saving state. This integration of "people, purpose, and environment" is the hallmark of a true smart campus. The platform's ability to process over 10 million telemetry points in real-time ensures that these optimizations occur with sub-second latency, maintaining occupant comfort while maximizing efficiency.

Operational AI: Generative, Predictive, and Autonomous

The intelligence of the Willow platform is manifested through its Operational AI layer, which employs a spectrum of techniques from traditional machine learning to cutting-edge generative models. This AI is not a separate module but is "woven into the fabric" of the platform, utilizing the Knowledge Graph as its foundational data source.

Willow Copilot and GenAI

Willow Copilot is an AI-powered natural language interface designed to ground large language models (LLMs) in the institution's specific data. This prevents the "hallucinations" common in generic AI models by ensuring that every answer is derived from the campus's live telemetry or static documentation. Copilot enables technicians to ask complex questions like, "Which VAV box is responsible for the overheating in Room 402?" or "Summarize the maintenance requirements for the backup generator". By using Retrieval Augmented Generation (RAG), Copilot can search through thousands of uploaded manuals and maintenance logs to provide instant troubleshooting guidance, significantly reducing the "time to repair".[22]

Machine Learning and Predictive Analytics

The platform utilizes supervised and unsupervised machine learning (ML) models for anomaly detection and forecasting.

Anomaly Detection

By analyzing telemetry streams, the system identifies deviations from normal operating patterns that may indicate a developing fault. For instance, a "Discharge Fan Speed Requires Tuning" skill can detect excessive variations in fan speed that suggest a PID loop is poorly calibrated, identifying a problem before it leads to equipment failure.

Energy Forecasting

ML models integrate historical usage data with external weather forecasts and occupancy schedules to predict future energy demand. This enables campuses to participate in "demand response" programs or "dynamic load shedding," where they reduce consumption during peak times to save money and reduce their carbon footprint.

Active Control: The "Closed Loop"

One of the most advanced features of the platform is "Active Control," which closes the loop between insight and action. While standard digital twins might only "alert" a human operator, Active Control allows the platform to adjust system setpoints based on real-time data autonomously. In a campus environment, this might involve automatically closing a water valve if a major leak is detected or scaling back HVAC intensity in a building when occupancy drops below a certain threshold. This capability is critical for achieving the high levels of energy efficiency required by modern sustainability mandates.[30]

The Microsoft Strategic Alliances

The Willow platform does not exist in a vacuum; its effectiveness is amplified by a robust ecosystem of strategic partners that cover technology, implementation, and long-term facility management. The partnership with Microsoft© is the cornerstone of Willow's go-to-market strategy. Beyond the technical foundation of Azure, Willow is integrated with Microsoft© Dynamics 365 for field service management and the Microsoft© Cloud for Sustainability. This integration allows for a seamless workflow where an AI-generated insight in the digital twin can automatically trigger a work order in Dynamics 365, dispatching a technician with the exact parts and tools needed for the job. Willow's recognition as a finalist for the 2025 Microsoft© Education Partner of the Year Award underscores its leadership in the higher education vertical.[23]

Implementation Case Studies in Higher Education

The theoretical benefits of the digital twin platform are best illustrated through real-world implementations at major universities. These case studies highlight the challenges of legacy silos and the measurable impact of the platform's deployment.

Northern Arizona University (NAU)

Northern Arizona University's (NAU) "Smart Campus Master Plan" is a 10-year strategy to modernize and enhance the campus by utilizing smart technology and integrating OT data across a 600-acre area with over 129 mixed-use buildings. Before implementation, the university's sustainability officer spent months manually collecting reports from various IoT devices, flow meters, and BMS systems that could not communicate with each other.

Implementation Details

The university converged over 11 data silos—including HVAC, utility billing, CO_2 grid emissions, and occupancy sensors—into a single Knowledge Graph. They utilized "Willow Activate Packs" to rapidly connect these systems and surface insights. By correlating classroom schedules with real-time occupancy and HVAC data, NAU found that many academic offices and classrooms were occupied only 50% of the time despite being fully conditioned.

Operational Outcomes

NAU realized over **$200,000 in avoidable costs** within the first six months. They automated HVAC setbacks across hundreds of rooms, reducing energy waste and advancing their goal of carbon neutrality by 2030. Furthermore, the platform solved a critical workforce issue: new technicians now become effective within their first week by using the digital twin to learn the campus's complex systems. This process previously took decades of "on-the-job" experience.

Georgia Southern University

Georgia Southern University is a large public research institution in Georgia, serving more than 26,000 students across its three campuses in Statesboro, Savannah (Armstrong Campus), and Hinesville. It is among the state's largest universities and a flagship institution of the University System of Georgia.[9]

The Statesboro Campus, the university's largest and oldest, spans more than 900 acres. The campus features academic buildings, housing, green spaces, lakes, indoor and outdoor recreational spaces, and research facilities, all within a traditional residential setting. The Armstrong Campus in Savannah occupies 268 acres in a suburban setting on the city's south side.

Across its three locations, Georgia Southern maintains hundreds of buildings, including extensive academic complexes, student housing, research facilities, libraries, recreational centers, and administrative structures, as documented in its multi-campus building inventory system. Collectively, the university's physical footprint spans more than 1,200 acres statewide, reflecting its scale, research ambitions, and role as a major regional economic engine.

Georgia Southern University implemented the Willow Digital Twin platform to manage more than 100 buildings, with a focus on energy efficiency and maintenance automation. The university unified 11 data repositories, including HVAC, energy meters, utility bills, and its work-ticket system. The university used the platform's "Active Control" to dynamically adjust HVAC setpoints based on classroom schedules.

Within just nine months, the university saved approximately $1 million in operational expenses. The platform's built-in diagnostics reduced recurring maintenance issues by verifying whether a closed work order actually resolved the underlying fault. This efficiency allowed the university to reinvest funds into the "student experience" rather than waste them on reactive maintenance and ghost energy consumption.

Other Willow Case Studies

While higher education use cases are compelling, valuable insights can also be gained from commercial spaces and campuses. Those interested in a digital twin should consider visiting industry leaders and viewing properties whose applications overlap with higher education. The Willow customer base includes major real estate and technology firms that serve as both partners and clients. For example:

- **CBRE and BNP Paribas Real Estate**
  These partners use the platform to manage large-scale portfolios, providing "Portfolio-wide Intelligence" to help maximize asset value.
- **Jacobs and Siemens**
  These engineering firms integrate Willow's technology into the design and construction phases of projects, ensuring that a "digital thread" is maintained from the very first BIM model through to building occupancy.
- **Walmart**
  Willow has proven the scalability of its platform across Walmart's vast real estate portfolio.
- **Dallas Fort Worth International Airport**
  DFW is one of the busiest airports in the world. They have extensively implemented Willow throughout the airport to manage complex aviation infrastructure and enhance the experience for travelers and airlines. [32]

Conclusion

Transforming a university campus into a "smart" environment isn't a one-time event but an ongoing process of developing institutional intelligence. The Willow digital twin platform offers the key technical and semantic infrastructure to support this continuous evolution. By connecting outdated OT protocols with modern cloud-based AI, the platform helps institutions realize the hidden potential in their existing building data. The achievements of early adopters such as Northern Arizona University and Georgia Southern University show that the benefits are real: millions saved in operational costs, a notable reduction in carbon footprints, and a more resilient, well-prepared workforce.

As the built environment continues to evolve, digital twins are moving beyond simple efficiency tools to become foundational elements of institutional capacity. The introduction of *Active Control* and Operational AI marks the start of a new era—one in which buildings transition from static infrastructure into intelligent, responsive systems that actively advance campus goals. Looking ahead, continued integration with smarter infrastructure, automation, and robotics will further amplify efficiency and effectiveness. These advances will empower skilled trades and facilities professionals to steward campuses that are adaptable to diverse needs, promote health and well-being, and place learning and discovery at the center of the physical environment.

What began as static visual models is evolving into a truly AI-enabled operating system for the built world—an operational AI platform that continuously connects data, expertise, and intelligent agents to inform decisions, execute work, and deliver outcomes at building, campus, and enterprise scale. Willow exemplifies this shift. More than a traditional "digital twin," it functions as an integrated operational AI platform that enables rapid, seamless execution of use cases while significantly reducing the effort required to achieve measurable results.

ROI is not just an ideal—it's proven through the Willow Digital Twin across various industries. In higher education, digital twins generate millions in annual OPEX savings, cut downtime by over 20%, decrease deferred maintenance growth by 40%, and reduce energy and carbon emissions via load optimization. It also lowers capital expenses by identifying building use characteristics and enhancing space utilization. These results are possible because Willow replaces guesswork with governed, real-time, cross-domain insights. In today's climate, where institutions must "do more with less," Willow isn't merely a tech upgrade or integration of building management systems; it's a foundational operational and financial model for the future of higher education sustainability, campus operations, and strategic planning.

For higher education leaders, the question is no longer whether to adopt digital twin technology but which one, and how quickly they can develop the necessary "data estate" to support a sustainable and efficient future campus.

By investing in strategic partnerships with organizations like Willow and Microsoft© and employing standardized, advanced ontologies and capabilities, universities are positioned to lead in developing the most intelligent AI-enabled Sustainable Smart Campuses ever dreamed of.

Case Study End Notes

1. Willow Digital Twin, https://willowinc.com/willow-digital-twin
2. Partner Innovation series: smarter buildings with Azure Digital Twins, https://partner.microsoft.com/he-il/blog/article/partner-innovation-series-smarter-buildings-with-azure-digital-twins
3. Joshua Ridley - Council on Tall Buildings and Urban Habitat, https://verticalurbanism.org/people-profile/joshua-ridley/
4. WillowInc/opendigitaltwins-building: WillowTwin open .- GitHub, https://github.com/WillowInc/opendigitaltwins-building
5. Willow Knowledge Graph, https://willowinc.com/knowledge-graph
6. Building a better tomorrow: Key trends driving ... - Willow, https://willowinc.com/wp-content/uploads/2025/04/Willow_Trends-Whitepaper_Design_Final.pdf
7. Digital Twins Are Changing University Campus Operations - GovTech, https://www.govtech.com/education/higher-ed/digital-twins-are-changing-university-campus-operations
8. A Playbook for Unifying IT and OT: Why Now Is the Moment to Unite IT and OT Behind a Shared Data Estate - Willow, https://willowinc.com/a-playbook-for-unifying-it-and-ot-why-now-is-the-moment-to-unite-it-and-ot-behind-a-shared-data-estate/2654
9. Georgia Southern University's Digital Transformation With Willow ..., https://willowinc.com/georgia-southern-universitys-digital-transformation-with-willow/2806
10. IBcon Commercial Real Estate Conference and Expo 2024 Tampa - Realcomm, https://www.realcomm.com/ibcon-2024/program/schedule/
11. Willow Platform, https://willowinc.com/willow-platform
12. A Framework for Scalable Digital Twin Deployment in Smart Campus Building Facility Management - ResearchGate,
https://www.researchgate.net/publication/398720230_A_Framework_for_Scalable_Digital_Twin_Deployment_in_Smart_Campus_Building_Facility_Management
13. A Dynamic Digital Twin Framework for Sustainable Facility Management in a Smart Campus: A Case Study of Chiang Mai University. https://www.mdpi.com/2227-7080/13/10/439
14. Announcing Azure Digital Twins: Create digital replicas of spaces and infrastructure using cloud, AI, and IoT | Microsoft© Azure Blog, https://azure.microsoft.com/en-us/blog/announcing-azure-digital-twins-create-digital-replicas-of-spaces-and-infrastructure-using-cloud-ai-and-iot/
15. Digital Twins for a Physical World | Autodesk University, https://www.autodesk.com/autodesk-university/class/Digital-Twins-Physical-World-2019
16. Willow Skills and Insights, https://willowinc.com/willow-skills-and-insights/3002
17. Introduction to the RealEstateCore Ontology | English | ABB BuildingPro, https://docs.buildings.ability.abb/collection/eliona-english/academy/introduction-to-ontologies/introduction-to-the-realestatecore-ontology
18. RealEstateCore Full - doc, https://doc.realestatecore.io/3.3/full
19. Connecting IoT Hub Data to Azure Digital Twins, https://nestdigital.com/uncategorized/connecting-iot-hub-data-to-azure-digital-twins/

20. Azure Digital Twins | Microsoft© Learn, https://learn.microsoft.com/en-us/azure/digital-twins/overview

21. Willow's Approach to AI - Willow, https://willowinc.com/willows-approach-to-ai/2697

22. Northern Arizona University Puts Innovation Into Action with Willow ..., https://willowinc.com/northern-arizona-university-puts-innovation-into-action-with-willow/2015

23. Willow recognized as a finalist for 2025 Microsoft© Education Partner of the Year Award, https://willow-inc.com/willow-recognized-as-a-finalist-for-2025-microsoft-education-partner-of-the-year-award/2944

24. Use Azure IoT Edge as a gateway for downstream devices, https://docs.azure.cn/en-us/iot-edge/iot-edge-as-gateway?view=iotedge-1.5

25. IoT Edge | Cloud Intelligence - Microsoft© Azure, https://azure.microsoft.com/en-us/products/iot-edge

26. Azure IoT Edge runtime and architecture explained - Microsoft© Learn, https://learn.microsoft.com/en-us/azure/iot-edge/iot-edge-runtime

27. Modbus vs. BACnet: A Deep Dive into Building and Industrial Communication Protocols, https://www.andivi.com/modbus-vs-bacnet-a-deep-dive-into-building-and-industrial-communication-protocols/

28. A Solution to Protocol Compatibility Challenges in BACnet-to-Modbus TCP Conversion via IoT Gateway - PUSR, https://www.pusr.com/blog/A-Solution-to-Protocol-Compatibility-Challenges-in-BACnet-to-Modbus-TCP-Conversion-via-IoT-Gateway

29. Using Gateways for BACnet Integration - Contemporary Controls, https://ccontrols.com/enews/2023/1123story3.htm

30. Willow: The new way to manage buildings, https://willowinc.com/

31. Building Energy and Operations - Willow, https://willowinc.com/activate/building-energy-operations

32. Presenting the First-Ever AI 75: Meet the Most Innovative Leaders in Artificial Intelligence in Dallas-Fort Worth | Groove Jones, https://www.groovejones.com/the-most-innovative-leaders-in-artificial-intelligence-in-dallas-fort-worth

33. 2021 Building Tech 50 List - BuiltWorlds, https://builtworlds.com/insights/2021-building-tech-50-list/

34. Digital Twin ROI: Breaking Down the 20-30% Cost Reductions with Real Numbers - Simio, https://www.simio.com/digital-twin-roi-breaking-down-the-20-30-cost-reductions-with-real-numbers/

35. Digital Twins in Higher Education Estates: A Practical Guide - Twinview, https://www.twinview.com/insights/digital-twins-in-higher-education-estates-a-practical-guide

36. Digital Twins Could Make Facilities Management in Higher Ed More Efficient | EdTech Magazine, https://edtechmagazine.com/higher/article/2025/12/digital-twins-could-make-facilities-management-higher-ed-more-efficient

37. Element Six. (2026, February 16). Digital twins that drive decisions: Speed to value, adoption, & operational AI with Rick Szcodronski [Video]. YouTube. https://www.youtube.com/watch?v=LQQKfp6Yd2M

Top-10 OT Risks Register

This risk register is organized to support the board's oversight of **cyber-physical risk** by aligning each item to NIST CSF 2.0 outcomes across **Govern, Identify, Protect, Detect, Respond, and Recover**[38].

Exhibit 34. Risk Statements: Likelihood/Impact, Gaps, and Actions

Risk statement (CSF lens)	Likelihood / Impact	Current controls (typical)	Gaps (common)	Next actions (prioritized)
Unmanaged OT asset exposure: The institution cannot secure, monitor, or recover what it cannot inventory (IDENTIFY).	H / H	Partial BAS inventories; vendor lists; CMMS asset records	Incomplete OT/IoT inventory; unclear ownership; unknown network paths and dependencies	Create authoritative OT/IoT inventory (controllers, gateways, sensors, vendor touchpoints) mapped to owner, criticality, and segment; institute onboarding/offboarding process for devices and vendors
Identity and privileged access weakness: Shared accounts, weak authentication, or over-privileged access enable unauthorized control changes and limit accountability (GOVERN/PROTECT).	H / H	Some MFA on IT systems, local OT accounts, vendor credentials	OT not integrated to enterprise IAM; limited least privilege; missing break-glass; weak credential lifecycle	Require enterprise IAM where feasible, MFA for privileged access, role-based entitlements, unique accounts; implement break-glass with time-bound access + mandatory review (aligns to baseline practices emphasized by CISA CPGs) (CISA)
Remote access compromise (vendor + staff): Always-on tunnels or unmanaged remote tools provide a high-impact pathway for intrusion (PROTECT/DETECT).	H / H	VPN used inconsistently; some logging; vendor "standard" tunnels	No centralized governance; no time-bound access; incomplete session logging/recording	Standardize secure remote access: approved jump paths, MFA, time-bound access windows, logging/session auditing; eliminate unmanaged tunnels (CISA remote access guidance) (CISA)
Network flatness and weak segmentation: A compromise in one zone can laterally move into BAS core, utilities plant, security subsystems, or research-critical environments (PROTECT).	M/H / H	VLANs in some areas, ad hoc firewall rules	No consistent "zones and conduits" model; limited enforcement; exceptions unmanaged	Establish OT segmentation baseline (OT separated from IT by default; restricted zones for BAS core/utilities/security/labs; tightly controlled conduits); document exception process and risk acceptance (OT tailored controls are emphasized by NIST OT guidance) [39]
Change enablement failure: Unmanaged changes (patches, configuration, integrations) destabilize operations or safety sequences, causing outages and loss of	M/H / H	Informal maintenance windows:	No OT-aware change policy; limited rollback plans;	Implement OT-aware change enablement: pre-change risk review, maintenance windows, rollback

Risk statement (CSF lens)	Likelihood / Impact	Current controls (typical)	Gaps (common)	Next actions (prioritized)
trust (GOVERN/PROTECT/RECOVER).		vendor change notes	weak configuration baselines	plans, "safe state" validation, and post-change verification; track configuration baselines for critical systems
Monitoring blind spots in OT: The institution lacks detection of abnormal behavior and cannot triage incidents without disrupting operations (DETECT/RESPOND).	M/H / H	IT SOC monitoring IT logs; limited BAS alerting	OT telemetry not integrated; limited anomaly detection; unclear escalation paths.	Extend monitoring to include OT signals (SIEM integration or equivalent), OT-aware anomaly detection, and joint escalation procedures; develop OT playbooks + tabletop exercises (NIST OT guidance highlights OT's unique needs) (NIST Computer Security Resource Center)
Patch/vulnerability management mismatch: Legacy OT, constrained maintenance windows, and vendor dependencies create exploitable exposure or unsafe updates (IDENTIFY/PROTECT).	H / M/H	IT patching tools for endpoints; vendor advisories	No OT vulnerability intake process; weak compensating controls; no "patch vs mitigate" governance	Establish OT vulnerability management: asset criticality, maintenance windows, vendor coordination, and compensating controls (segmentation, allowlisting, monitoring) when patching is not feasible; document risk acceptance.
Third-party / supply chain dependency risk: Integrators and vendors introduce insecure defaults, weak support practices, or opaque changes; offboarding leaves lingering access (GOVERN/PROTECT/RESPOND).	M/H / H	Contract language varies; ad hoc vendor oversight	Inconsistent security requirements, unclear evidence requirements, and weak offboarding	Standardize OT procurement/security clauses: secure remote access, logging, vulnerability disclosure, patch commitments, evidence and response timelines; enforce vendor offboarding checklists; align with baseline "secure by design/demand" expectations where practical (U.S. Department of War)
Privacy and trust failure from sensor/data fusion: OT data becomes sensitive when combined with identity/location/behavior; perceived surveillance erodes legitimacy (GOVERN).	M / H	General privacy policies; inconsistent signage/notice	No explicit OT data purpose/limits; unclear access boundaries; retention creep	Define OT data governance: purpose limitation, role-based access, aggregation/de-identification defaults, retention rules, transparent communications; create review/appeal mechanisms (CSF 2.0 elevates "Govern" to align risk with enterprise oversight) (NIST Publications)

Risk statement (CSF lens)	Likelihood / Impact	Current controls (typical)	Gaps (common)	Next actions (prioritized)
Resilience and recovery gaps for cyber-physical incidents: Backups, restore procedures, and fallback modes are untested; recovery is slow and improvisational (RECOVER/RESPOND).	M / H	IT backup and restore procedures	OT configs not backed up; restore not tested; unclear priorities for critical loads and systems.	Implement OT resilience plan: configuration backups for critical OT, tested restores, defined fallback modes, prioritized systems/loads, and rehearsed recovery playbooks; integrate into emergency operations planning (OT guidance stresses reliability/safety context) (NIST Computer Security Resource Center)

Implementation Roadmap

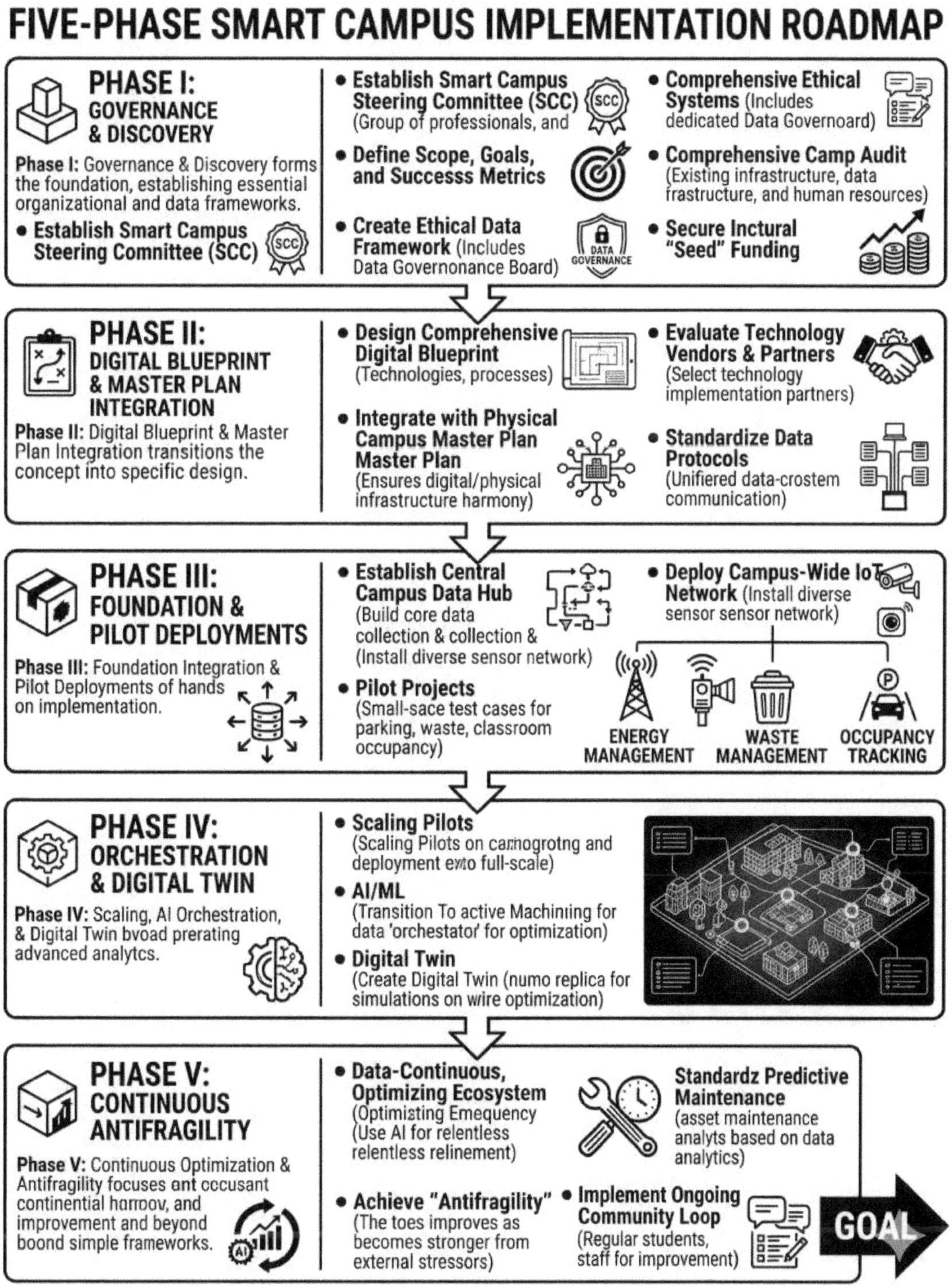

Figure 14. Smart Campus Implementation Roadmap

Readiness Assessment Framework

Before an institution can successfully navigate the complexities of IT/OT convergence and AI implementation, leadership must establish an accurate baseline of its current capabilities. A smart campus initiative built on a fractured foundation will rapidly magnify existing operational and cultural flaws.

This Readiness Assessment Framework is designed for the CTO, Chief Facilities Officer (CFaO), and the Smart Campus Steering Committee (SCSC) to evaluate the institution's starting position jointly. It moves beyond simple technical checklists to evaluate the holistic, cyber-physical, and human environments required for sustainable transformation.

How to Use This Framework

This assessment should not be done in isolation. Instead, it needs to be scored collaboratively during a joint session with leaders from IT, Facilities, the Sustainability Office, Student Services, Research, and Academic Affairs, among others.

Score each criterion on the following maturity scale:

- **Level 1 (Siloed and Reactive)**
  Legacy operations; IT and OT are entirely separate; culture is defensive.
- **Level 2 (Developing)**
  Ad-hoc integration; isolated pilots exist; standards are drafted but not enforced.
- **Level 3 (Integrated)**
  Standardized architecture; strong cross-functional governance; proactive enablement.
- **Level 4 (Antifragile)**
  Campus-wide orchestration; self-healing systems; culture of continuous co-creation.

Continued on next page.

Domain 1: Architectural and Data Readiness

Evaluates the physical network, device visibility, and the semantic foundation necessary to feed an AI control plane.

Exhibit 35. Architectural and Data Readiness Assessment Rubric

Assessment Criteria	Level 1: Siloed and Reactive	Level 2: Developing and Stabilizing	Level 3: Integrated and Governed	Level 4: Antifragile	Level Score (1–4)
OT Asset Visibility	"Shadow OT" is rampant. IT and Facilities maintain separate, conflicting inventories. Asset ownership and lifecycle responsibility are unclear.	A consolidated asset inventory exists, but it is manually maintained and periodically outdated. New OT assets are often added through projects without consistent onboarding or review.	A centralized OT asset inventory is actively maintained and shared across IT, Facilities, and Security. New assets must be registered through governed processes tied to procurement and commissioning.	100% of OT assets are dynamically mapped, continuously updated, and monitored. Asset visibility, ownership, and lifecycle governance are enforced through the Smart Campus Steering Committee (SCSC).	_____
Network Infrastructure	OT relies on flat, legacy networks or unmanaged daisy-chained connections. Segmentation is minimal or absent, increasing cyber and operational risk.	Basic segmentation between IT and OT exists, often firewall-based. Network capacity and reliability vary across buildings, limiting scalability and real-time data use.	OT networks are purpose-built, segmented, and engineered for reliability and security. Wireless, wired, and edge connectivity are planned as a portfolio aligned to operational criticality.	A high-bandwidth, low-latency fabric—combining Wi-Fi 7, LoRa, private 5G, edge compute, and resilient backbone networks—blankets the physical plant and safely supports real-time control, analytics, and resilience.	_____
Data Estate & Ontology	Data is limited, scattered, and system-bound. Buildings use proprietary, vendor-specific naming conventions (e.g., "AHU2_Z4_T"). There is no shared data estate—only isolated data exhaust. Data meaning is implicit, inconsistent, and difficult to reuse beyond the originating system. Analytics are largely manual,	Some operational data is centralized, but the data estate is incomplete and inconsistently governed. Partial tagging or semantic efforts exist, often varying by vendor or project. Data is used primarily for dashboards and reporting rather than decision-making. Retrofitting semantics is costly and error-prone, limiting confidence in analytics and cross-domain use.	A defined institutional data estate exists and is treated as a shared asset. A common semantic model (e.g., Brick Schema or Haystack) is enforced for new systems, and legacy data is progressively normalized. Data is actively leveraged for predictive maintenance, energy optimization, space analytics, and operational decision support across IT, Facilities, and Sustainability.	The data estate functions as institutional operational truth. A unified, enterprise semantic standard is strictly enforced across all spatial, asset, and telemetry data. Data is continuously validated, machine-readable, and interoperable, and it is systematically leveraged by AI models, digital twins, automation workflows, and strategic planning processes. Insights generated from the data estate directly improve operations, resilience, sustainability outcomes, and institutional learning over time.	_____

Assessment Criteria	Level 1: Siloed and Reactive	Level 2: Developing and Stabilizing	Level 3: Integrated and Governed	Level 4: Antifragile	Level Score (1–4)
Integration Architecture	Point-to-point custom code connects systems. Integrations are brittle, undocumented, and break easily when systems change or vendors rotate.	Some shared integration tools exist, but many systems still rely on custom scripts or one-off connectors. Reuse is limited, and technical debt accumulates.	A standardized integration platform (APIs, event brokers, message queues) supports secure, reusable system connections. New integrations follow defined patterns and governance.	A secure Digital Integration Fabric—using API gateways, event streaming, and message brokers—delivers scalable, resilient orchestration of all OT and enterprise telemetry with built-in monitoring and governance.	_____

Domain 2: Cybersecurity and Risk Readiness Rubric

Evaluates the ability to expand connectivity without increasing institutional vulnerability.

Exhibit 36. Cybersecurity and Risk Readiness Rubric

Assessment Criteria	Level 1: Siloed and Reactive	Level 2: Hardened but Fragmented	Level 3: Governed and Resilient	Level 4: Antifragile	Level Score (1–4)
Network Segmentation	OT systems share flat networks with enterprise IT or the public internet. A compromise in one system can cascade across campus operations.	Basic IT/OT separation exists, often using perimeter firewalls. Internal OT segmentation is inconsistent and varies by building or vendor.	OT environments are segmented using zone and conduit logic aligned with operational criticality. Segmentation rules are documented, monitored, and enforced.	Micro-segmentation is continuously validated. Segmentation gaps discovered during incidents permanently improve architecture and reduce future blast radius.	_____
Identity & Remote Access	Shared vendor credentials and always-on tunnels are common. Access is convenient but unaudited and difficult to revoke.	Identity controls exist but are unevenly applied. MFA and role-based access are used inconsistently for OT and vendors.	Privileged access is identity-based, time-bound, logged, and reviewed. Vendor access follows least-privilege principles and formal approval workflows.	Remote access patterns are continuously tested, audited, and refined. Access failures and near-misses strengthen controls rather than eroding trust.	_____
Incident Response	IT and Facilities respond independently. Cyber incidents are treated as IT problems; operational implications are discovered too late.	Informal coordination exists, but processes are untested. Response depends on personal relationships rather than institutional playbooks.	Joint IT/OT incident response playbooks are documented and exercised through tabletop drills. Safety, uptime, and cyber risk are managed together.	Incidents and near-incidents are treated as learning events. Post-incident reviews improve controls, playbooks, and training across domains.	_____
Risk Governance	Cyber-physical risk is not visible at the executive level. Risk acceptance is implicit and undocumented.	Cyber risk is discussed, but OT risk is poorly understood outside Facilities. Reporting is inconsistent.	OT risk is explicitly included in enterprise risk management. Decision rights and escalation paths are clear and enforced.	Risk governance continuously evolves. Metrics show that the institution becomes safer and more predictable after stress and disruption.	_____

Domain 3: Operational and Workflow Readiness Rubric

Evaluates the translation of data into actionable, operator-centric outcomes.

Exhibit 37. Operational Assessment Readiness Rubric

Assessment Criteria	Level 1: Reactive and Firefighting	Level 2: Preventive and Stabilizing	Level 3: Predictive and Optimized	Level 4: Antifragile	Level Score (1–4)
Maintenance Posture	Maintenance is primarily break-fix. Failures are addressed after a disruption occurs, often at the expense of overtime and emergency spending.	Preventive maintenance is formalized, but mostly calendar-based. Condition data is reviewed manually and inconsistently.	Predictive maintenance is operational using telemetry and analytics. Interventions occur before failure with measurable reductions in downtime and cost.	Failures continuously shorten diagnostic time and improve prediction accuracy. Maintenance maturity increases after each disruption.	_____
Alarm and Event Management	Operators experience alarm fatigue. Alerts lack context and overwhelm staff, leading to critical signals being missed.	Alarm rationalization is underway, but many alerts still lack prioritization or actionable guidance.	Alerts are contextual, prioritized, and tied directly to workflows. Operators trust alerts because they consistently lead to correct action.	Alert effectiveness improves over time as false positives are eliminated and root causes are permanently addressed.	_____
Workflow Integration	Insights live in dashboards disconnected from work execution. Data rarely translates into action.	There is partial integration between analytics tools and CMMS or dispatch systems. Manual handoffs are still common.	Analytics, CMMS, mobile tools, and dispatch are integrated. Insights automatically generate prioritized, actionable work.	Workflows continuously adapt. Operational intelligence flows seamlessly from sensing to action to learning.	_____
Operational Metrics	Effort and heroics rather than outcomes measure success. Performance trends are unclear.	Basic KPIs exist, but are lagging and retrospective.	KPIs focus on reliability, efficiency, and predictive outcomes. Data supports decision-making.	Metrics demonstrate improvement after stress. Volatility becomes a source of insight, not instability.	_____

Domain 4: Cultural and Change Enablement Readiness

Evaluates the psychological safety, cross-functional trust, and human adaptability of the workforce.

Exhibit 38. Cultural Assessment Readiness Rubric

Assessment Criteria	Level 1: Resistant and Adversarial	Level 2: Cooperative but Fragile	Level 3: Collaborative and Enabled	Level 4: Antifragile	Level Score (1–4)
Cross-Functional Trust	IT and Facilities operate with competing priorities and mutual skepticism. Change is seen as an imposed risk.	Collaboration exists but depends on individuals rather than structures. Misalignment resurfaces under pressure.	Trust is institutionalized through shared goals, metrics, and governance. Teams plan and solve problems together.	Trust strengthens under stress. Cross-functional collaboration improves after disruption rather than degrading.	_____

Assessment Criteria	Level 1: Resistant and Adversarial	Level 2: Cooperative but Fragile	Level 3: Collaborative and Enabled	Level 4: Antifragile	Level Score (1–4)
Change Management	Training is one-time and tool-focused. Adoption is assumed rather than managed.	Change efforts are project-specific and inconsistent. Early adopters succeed; others lag.	Role-based change enablement is embedded in programs. Adoption is measured and reinforced over time.	Continuous change is normalized. The organization becomes better at change itself with each initiative.	_____
Operator Experience	Tools increase cognitive load. Operators develop workarounds or disengage.	UX improves, but is not consistently prioritized. Feedback is informal and slow to act on.	Operator experience is designed intentionally. Feedback loops are formal, visible, and acted upon.	Operators are co-designers. Systems measurably reduce cognitive load and burnout over time.	_____
Learning Culture	Failures trigger blame and avoidance. Lessons are lost.	Some lessons are captured, but rarely institutionalized.	Learning is structured through post-incident reviews and improvement cycles.	The institution reliably improves after failure. Learning velocity is a strategic advantage.	_____

Domain 5: Governance and Strategic Alignment

Evaluates the executive commitment to treating the smart campus as a permanent institutional capability.

Exhibit 39. Governance Readiness Rubric

Assessment Criteria	Level 1: Fragmented and Project-Driven	Level 2: Coordinated but Inconsistent	Level 3: Governed and Portfolio-Based	Level 4: Antifragile	Level Score (1–4)
Decision Rights	Authority is unclear. Projects proceed without cross-domain review. Exceptions are common and undocumented.	Governance bodies exist but lack enforcement power. Decision rights vary by project.	Decision rights are clearly defined, documented, and enforced across IT, Facilities, and Sustainability.	Decision frameworks are stable yet adaptive. Governance improves continuously through measured outcomes.	_____
Procurement & Standards	Smart systems are procured independently. Standards are optional or bypassed under schedule pressure.	Standards exist but are weakly enforced. Exceptions accumulate technical debt.	Procurement enforces interoperability, cybersecurity, and data standards. Exceptions are rare and time-bound.	Standards evolve based on operational learning. Vendor behavior improves over time through governance.	_____
Strategic Alignment	Smart campus efforts are disconnected from institutional strategy and capital planning.	Strategy exists but is inconsistently translated into execution.	Smart campus investments are managed as a strategic portfolio aligned to mission outcomes.	Strategy, funding, execution, and measurement reinforce one another. The institution scales intentionally and safely.	_____
Executive Visibility	Leaders view smart campus work as a technical detail rather than an institutional capability.	Reporting exists but focuses on activity, not outcomes.	Executive reporting focuses on outcomes tied to risk, cost, sustainability, and experience.	Governance conversations center on learning, improvement, and long-term institutional strength.	_____

Interpreting the Readiness Assessment Results

1. **Majority Level 1 (High Risk)**
   The institution is not ready for advanced smart technologies, AI, or large automation projects. Immediate focus must be redirected to Stage 1: Discovery, Baseline Assessment, and establishing the Smart Campus Steering Committee.
2. **Majority Level 2 (Transitioning)**
   The foundation is being laid, but scaling now will result in platform chaos. Focus on standardizing the data ontology, enforcing microsegmentation, and executing a single, highly governed "Lighthouse" pilot to prove value.
3. **Majority Level 3 (Ready to Scale)**
   The institution has strong governance and secure architecture and is strengthening its resilience. Leadership should confidently advance the deployment of smart initiatives, the AI control plane, digital twins, and deeper algorithmic optimization across an integrated IT/OT portfolio.
4. **Majority Level 4 (Industry Leader)**
   The campus is operating as an antifragile, self-healing ecosystem. Focus shifts to continuous algorithmic auditing, kinetic integration (robotics/AVs), and the use of the campus as a premier academic "Living Lab."

Smart Campus Readiness Scorecard

Smart Campus Readiness is the institution's capacity to govern, operate, and continuously improve a secure, interoperable cyber-physical environment that advances mission outcomes, sustainability, resilience, and human experience.

This scorecard is based on the Smart Campus Readiness Assessment Framework as defined in the *Manifesto for the AI-Enabled Sustainable Antifragile Smart Campus.* Taking the time to complete it provides several key advantages to institutional planners:

1. Enables shared operational truth and understanding while fostering collaboration.
2. Prevents technology overreach, emphasizes human aspects
3. Surface governance and cultural blockers early
4. Creates a defensible baseline for boards, auditors, and funders
5. Helps align actions directly with phased execution and strategic goals

How to Use This Workbook

1. Score collaboratively – this is not a solo exercise
2. Evidence beats opinion – note concrete examples
3. Disagreement is a signal – surface it, do not smooth it over
4. If between levels → score the lower level
5. The final output is about the overall profile, not a single number, and the priority actions that need to be taken.

Scoring Key (Used Throughout)

Refer to the five domains of readiness rubrics for explanations of each criterion and level. In general, the scoring levels indicate the following:

Exhibit 40. General Meaning of Readiness Levels

Score Level	Manifesto-Aligned Meaning
1 Siloed, Reactive, and Heroic	Fragmented IT and Facilities operations with limited visibility, unmanaged or "shadow" IT/OT, inconsistent standards, and reactive response to failures. Risk is absorbed through heroics rather than governance or architecture.
2 Stabilizing, Developing, Identifying Risk	Cross-functional collaboration has begun, and visibility is improving. Some governance and standards exist, but they are inconsistently applied. The institution is reducing fragility, containing risk, and stabilizing operations — yet remains dependent on point solutions and manual coordination.

Score Level	Manifesto-Aligned Meaning
3 Integrated, Governed, and Resilient	IT and OT are formally integrated under shared governance and architectural standards. Shadow IT/OT is eliminated. Systems are observable and recover reliably from disruption. Improvements are deliberate and governed, but learning is still episodic rather than systemic.
4 Carefully Architected, and Antifragile	The campus is intentionally designed as a governed cyber-physical system. Enterprise standards, data semantics, security, and workflows are continuously enforced. Disruptions generate insight that improves future performance. Learning, optimization, and adaptation are continuous, measurable, and institutionalized.

Workbook

Institution: ______________________________

Assessment Date(s): ______________________________

Assessment Coordinator/ Facilitator: ______________________________

Participants (Roles):

☐ IT / Security ______________________________

☐ Facilities / Operations ______________________________

☐ Sustainability ______________________________

☐ Budget & Finance ______________________________

☐ Academics / Research ______________________________

☐ Campus Safety ______________________________

☐ Others (list):______________________________

____

____

____

Additional Notes

Domain 1: Architecture, Data, and Integration Readiness

1.1 OT Asset Visibility

Score: ☐ 1 ☐ 2 ☐ 3 ☐ 4

Evidence/examples observed:

- ______________________________
- ______________________________

1.2 Network Infrastructure

Score: ☐ 1 ☐ 2 ☐ 3 ☐ 4

Evidence/examples observed:

- ______________________________
- ______________________________

1.3 Data Estate, Ontology & Semantics

Score: ☐ 1 ☐ 2 ☐ 3 ☐ 4

Evidence/examples observed:

- ______________________________
- ______________________________

1.4 Integration Architecture

Score: ☐ 1 ☐ 2 ☐ 3 ☐ 4

Evidence/examples observed:

- ______________________________
- ______________________________

Domain 1 Summary

Domain Score (1–4): _______ (sum / 4)

Identified Strengths:

Opportunities For Improvement:

Action Plan:

Domain 2: Cybersecurity & Cyber-Physical Risk Readiness

2.1 Network Segmentation

Score: ☐ 1 ☐ 2 ☐ 3 ☐ 4

Evidence/Examples Observed:

- ____
- ____

2.2 Identity & Remote Access (Staff + Vendors)

Score: ☐ 1 ☐ 2 ☐ 3 ☐ 4

Evidence/Examples Observed:

- ____
- ____

2.3 Incident Response (IT + OT)

Score: ☐ 1 ☐ 2 ☐ 3 ☐ 4

Evidence / Examples Observed:

- ____
- ____

2.4 Risk Governance & Executive Visibility

Score: ☐ 1 ☐ 2 ☐ 3 ☐ 4

Evidence/Examples Observed:

- ____
- ____

Domain 2 Summary

Domain Score (1–4): _______ (sum of scores/4)

Identified Strengths:

Most Significant Exposure/Risk/Issues:

Action Plan:

Domain 3: Operational & Workflow Readiness

3.1 Maintenance Posture

Score: ☐ 1 ☐ 2 ☐ 3 ☐ 4

Evidence/Examples Observed:

- ______________________________
- ______________________________

3.2 Alarm & Event Management

Score: ☐ 1 ☐ 2 ☐ 3 ☐ 4

Evidence/Examples Observed:

- ______________________________
- ______________________________

3.3 Workflow Integration (Data → Action)

Score: ☐ 1 ☐ 2 ☐ 3 ☐ 4

Evidence/Examples Observed:

- ______________________________
- ______________________________

3.4 Operational Metrics & Feedback

Score: ☐ 1 ☐ 2 ☐ 3 ☐ 4

Evidence/Examples Observed:

- ______________________________
- ______________________________

Domain 3 Summary

Domain Score (1–4): _______ (sum of scores/4)

Identified Strengths:

Operational pain points surfaced:

Action Plan:

Domain 4: Culture & Change Enablement Readiness

4.1 IT–Facilities Trust & Collaboration

Score: ☐ 1 ☐ 2 ☐ 3 ☐ 4

Evidence/Examples Observed:

- ______________________________
- ______________________________

4.2 Change Management Discipline

Score: ☐ 1 ☐ 2 ☐ 3 ☐ 4

Evidence/Examples Observed:

- ______________________________
- ______________________________

4.3 Operator Experience & Adoption

Score: ☐ 1 ☐ 2 ☐ 3 ☐ 4

Evidence/Examples Observed:

- ______________________________
- ______________________________

4.4 Learning from Failure

Score: ☐ 1 ☐ 2 ☐ 3 ☐ 4

Evidence/Examples Observed:

- ______________________________
- ______________________________

Domain 4 Summary

Domain Score (1–4): _______(sum of scores/4)

Cultural Strengths:

Cultural Constraints and Issues:

Action plan:

Domain 5: Governance & Strategic Alignment Readiness

5.1 Decision Rights & Accountability

Score: ☐ 1 ☐ 2 ☐ 3 ☐ 4

Evidence/Examples Observed:

- ____
- ____

5.2 Procurement & Standards Enforcement

Score: ☐ 1 ☐ 2 ☐ 3 ☐ 4

Evidence/Examples Observed:

- ____
- ____

5.3 Strategic & Capital Alignment

Score: ☐ 1 ☐ 2 ☐ 3 ☐ 4

Evidence/Examples Observed:

- ____
- ____

5.4 Executive Visibility & Outcomes

Score: ☐ 1 ☐ 2 ☐ 3 ☐ 4

Evidence/Examples Observed:

- ____
- ____

Domain 5 Summary

Domain Score (1–4): _______ (sum of scores / 4)

Governance Current Strengths/Abilities:

Overall Assessment Implications/ Summary

Action Plan:

Summary of Domain Scores

Record the domain assessment scores below. The score should be the average of the scores.

Domain	Score
Architecture & Data	____
Cybersecurity & Risk	____
Operations	____
Culture & Change	____
Governance	____

Key Insight

The lowest-scoring domain represents the primary inhibitor and the maximum safe speed of the Smart Campus program, and likely the area of initial action for greater readiness.

Top 3 Readiness Gaps (Priority Order)

1.
2.
3.

Immediate Priority Actions (Next 90 Days)

1.
2.
3.

Longer Term Action Plan:

Facilitator Notes

Recommended Action Level Priorities

Phase Gate / Level I:
Governance and Discovery — Establishing Operational Truth

Objective: Prove the campus is governable before it becomes programmable.

This initial gate validates that the institution has the leadership structures, decision rights, and visibility required to initiate smart-campus work responsibly. The focus is on discovery, not deployment. Asset inventory is treated as the establishment of **operational truth**, exposing shadow OT, unmanaged vendors, undocumented dependencies, and latent cyber-physical risk.

Phase Gate / Level 1 is achieved when the institution can demonstrate:

- A formal Smart Campus Steering Committee (or equivalent) with authority
- Clear, documented IT, Facilities, Security, and Data decision rights
- A consolidated view of OT assets, networks, and vendor access
- Agreed on guiding principles for standards, risk tolerance, and scope control

Phase Gate /Level II:
Planning and the Digital Blueprint — Architecture Before Integration

Objective: Prove that we know the *why* and the *how*.

This phase gate ensures the institution has translated vision into a coherent, shared design. Architecture is treated as a leadership discipline rather than a technical artifact. The Digital Blueprint establishes the future-state reference models for IT/OT architecture, cybersecurity, data, integration patterns, and lifecycle alignment.

Phase Gate / Level 2 is achieved when the institution can demonstrate:

- Approved reference architectures for IT/OT, security, and data estate
- Defined semantic and interoperability standards
- Documented integration, segmentation, and remote-access patterns
- Alignment between smart-campus priorities and capital/sustainability plans

Phase Gate / Level III:
Foundation Integration — Stabilize and Integrate

Advanced complex initiatives do not scale beyond pilots until this gate is met.

Objective: Prove the campus is stable, observable, and secure enough to trust.
This gate validates that foundational integration has been executed according to the blueprint and that operations are reliable under normal conditions. The emphasis is on stability, observability, and control, not advanced intelligence. Data begins flowing into the governed data estate, identity and access controls are enforced, and fragile point-to-point integrations are eliminated.

Phase Gate / Level 3 is achieved when the institution can demonstrate:

- Secure, segmented IT/OT connectivity as designed
- Enterprise identity and vendor access controls in force
- Reliable ingestion of core telemetry into the data estate
- Reduced operational surprises and improved recovery predictability

Phase Gate / Level IV: Scaling and Expanding — From Pilots to Platforms

Objective: Prove value before multiplying complexity.
This level ensures successful, governed patterns to scale across buildings, systems, and domains. Scaling is evidence-driven and portfolio-managed, not opportunistic. The institution demonstrates that smart-campus capabilities are repeatable, interoperable, and supported by shared workflows and governance.

Phase Gate / Level 4 is passed when the institution can demonstrate:

- Repeatable deployment patterns validated across multiple assets
- Consistent data, integration, and security behavior at scale
- Measurable improvements in efficiency, resilience, sustainability, or experience
- Reduced dependence on proprietary or bespoke solutions

Phase Gate / Level V: Continuous Optimization — From Resilience to Antifragility

Objective: Prove that the campus learns and improves because of stress.
This final gate represents the Manifesto's target state. The campus operates as an **antifragile cyber-physical system**—one that not only resists disruption, but systematically improves because of it. Optimization, commissioning, and learning are continuous, measurable, and institutionalized across technology, operations, and culture.

Phase Gate / Level 5 is sustained when the institution can demonstrate:

- Continuous commissioning and optimization loops
- Predictive and prescriptive insights embedded into daily operations
- Documented learning from failures, near-misses, and stress events
- Ongoing gains in performance, sustainability, and human experience

Smart Campus Protocols and Standards

This appendix is meant to be practical rather than comprehensive. It includes (1) a standards matrix organized by campus domain, (2) a preferred standards approach highlighting what to standardize and what to tolerate, and (3) procurement-ready clauses that you can insert into RFPs and renewal agreements. The goal is straightforward: to ensure interoperability, cybersecurity, and data reuse are enforceable across capital projects, enabling the campus to function as a platform-oriented environment rather than relying on gateways.

Standards and Protocols Matrix (by Domain)

Legend:

- **Preferred** = strongly recommended standard for new deployments
- **Accepted** = common legacy standard you may need to support
- **Governance Note** = what you must specify to avoid one-off integrations

Exhibit 41. Protocols by Domain

Domain	Preferred protocols /standards	Common legacy /accepted	Governance note
BAS / HVAC controls	BACnet/IP, BACnet/SC (secure IP connectivity) (BACnet International)	BACnet MS/TP, proprietary vendor variants	Require secure remote access, segmentation alignment, and a semantic tagging standard (Haystack/Brick) so points are reusable across buildings (project-haystack.org)
Utilities plant / industrial systems	OPC UA (where available), structured telemetry interfaces (dragos.com)	Modbus RTU/TCP, vendor-native protocols	Treat plant as high-consequence: enforce zones/conduits segmentation and OT-aware controls per NIST guidance (NIST Computer Security Resource Center)
Metering (electric/thermal/water/gas)	IP metering + publish/subscribe (MQTT), where applicable; standardized data export patterns	Modbus, pulse outputs, vendor gateways	Define canonical data model (units, intervals, asset IDs). Meter data must map to space/asset master data.
Lighting controls	IP-native lighting controls; integration via BACnet/IP where feasible	KNX, DALI/LonWorks (common in legacy estates)	Require a documented integration path and consistent naming/tagging; avoid "lighting as a parallel platform."
IoT sensors (IAQ, leaks, occupancy, etc.)	MQTT pub/sub; standardized onboarding and topic conventions	Vendor clouds with proprietary APIs	Declare topic taxonomy, data retention rules, and security onboarding (certs/keys). MQTT gives decoupling and reuse

Domain	Preferred protocols /standards	Common legacy /accepted	Governance note
Video/security systems	Standards vary; require documented APIs + event export	Proprietary VMS analytics	Treat as a high-trust domain: require privacy governance, role-based access, auditing, and defined escalation workflows.
EV charging	OCPP for charger ↔ central system interoperability (Open Charge Alliance)	Vendor-locked charger ecosystems	Require OCPP support and the ability to integrate with campus energy/load management systems.
Demand response / DER coordination	OpenADR for DR and grid/DER signaling (openadr.org)	Utility-specific DR programs	Require OpenADR compatibility if participating in DR/DER programs to reduce lock-in and enable automation.
Semantic interoperability	Project Haystack (tagging + web services) (project-haystack.org) and/or Brick Schema (ontology + relationships) (docs.brickschema.org)	Ad hoc point naming; vendor naming	Pick a standard and enforce it in commissioning; semantics are what make digital twins and analytics portable.
Cybersecurity architecture	IEC 62443 zones and conduits (segmentation model) (gca.isa.org)	Flat VLANs, ad hoc rules	Make zones/conduits part of design deliverables and as-builts—not "tribal knowledge."
OT security controls guidance.	NIST SP 800-82r3 OT security guidance (NIST Publications)	Generic IT-only controls	Require OT-tailored controls (maintenance windows, compensating controls) and joint response playbooks.
Enterprise cyber risk language	NIST CSF 2.0 (Govern/Identify/Protect/Detect/Respond/Recover) (NIST Publications)	Unstructured risk registers	Use CSF terms for board reporting, policy, and accountability

"Preferred Standards" Posture: What to Standardize vs. What to Tolerate

A campus reference architecture becomes governable when it makes **three kinds of declarations:**

1) "Preferred for new builds and major retrofits."

- **BACnet/SC** (where feasible) and BACnet/IP for BAS modernization.
- **MQTT** for large-scale IoT telemetry (sensors, edge gateways)
- **Haystack and/or Brick** for semantic tagging and relationships
- **IEC 62443 segmentation** as the reference model for zones and conduits
- **OCPP** for EV charging interoperability
- **OpenADR** for DR/DER signaling where grid programs apply

2) "Accepted / must interoperate with."

- **Modbus** (plant equipment and meters), and legacy control protocols (KNX/LonWorks), where present

- Proprietary vendor systems **only if** they expose documented interfaces and comply with campus governance (security + data + audit)

3) "Discouraged unless exception approved."

- Always-on vendor tunnels
- Flat networks for OT
- One-off integration gateways that become single points of failure
- Deployments that cannot support semantic tagging/export and are "dashboard-only."

Procurement-Ready Language (by Domain)

Below are clauses you can adapt. They are written so they map directly to CTO decision rights: identity, segmentation, remote access, data governance, and interoperability.

1) BAS / Controls RFP Language (BACnet + secure connectivity + semantics)

Interoperability

- "Systems must support BACnet/IP and provide a documented integration interface. Where feasible for IP deployments, BACnet/SC support is required for secure communications between BAS devices."

Semantic standard

- "All points and equipment must be delivered with campus-required semantic tagging (Project Haystack and/or Brick) and a complete point list that maps tags/relationships to equipment, spaces, and systems."

Data export

- "The vendor must support export of telemetry, alarms, and equipment metadata to the campus data platform using documented interfaces. Point naming conventions, units, sampling intervals, and alarm metadata must conform to campus standards."

2) Metering RFP Language (data quality + canonical model)

- "Meters must provide time-series data export with documented protocols and include units, timestamps, and interval definitions. Vendor must provide a mapping file that links meters to campus asset IDs and space/building IDs."
- "Data must be available for integration into EMIS and analytics platforms without screen-scraping or manual report extraction."

3) IoT Sensor and Edge Gateway Language (MQTT + topic governance)

- "Sensors/gateways must support publish/subscribe telemetry (MQTT preferred) and adhere to campus topic conventions, device identity requirements, and secure onboarding controls."
- "Vendor must provide a device inventory file including serials, firmware versions, network requirements, and ownership metadata."

4) EV Charging Language (OCPP + energy integration)

- "Charging stations must support OCPP and enable interoperability between charge points and the campus central management platform."
- "Charging platform must expose APIs/events for load management, pricing/policy integration, and sustainability reporting."

5) Demand Response / DER Language (OpenADR readiness)

- "If participating in demand response or DER programs, the solution must support OpenADR signaling or provide a standards-based method of receiving DR/DER events and executing automated response strategies."

6) OT Cybersecurity and Segmentation Language (IEC 62443 + NIST OT guidance)

- "Solution must be deployable within a segmented architecture consistent with IEC 62443 zones and conduits. Vendor must provide network diagrams, required ports/protocols, and recommended zone placement."
- "Security controls must align with OT operational constraints (uptime, safety, maintenance windows) consistent with NIST SP 800-82 OT security guidance."

7) Remote Access Language (time-bound, audited, governed)

- "Remote access must be time-bound, centrally governed, and logged. Always-on tunnels are prohibited without explicit exception approval and documented compensating controls."
- "Privileged sessions must be auditable (logs; session recording where feasible). Vendor must support incident evidence requirements and response timelines."

8) Identity and Access Language (least privilege + role alignment)

- "All privileged access must support MFA and least privilege. Access must be role-based and tied to named individuals (no shared admin accounts). Break-glass procedures must be defined, time-bound, logged, and reviewed."

9) Monitoring / Incident Response Language (OT-aware response)

- "System must generate security-relevant logs and OT operational events suitable for integration into the campus monitoring ecosystem. Vendor must support OT-aware incident response playbooks and participate in tabletop exercises upon request."
- "Escalation paths must preserve safety and continuity; response procedures must be compatible with OT operational constraints."

10) Governance and Board Reporting Language (CSF alignment)

- "Vendor shall provide evidence and documentation required to support institutional risk reporting aligned to NIST CSF 2.0 functions (Govern/Identify/Protect/Detect/Respond/Recover)."

The "Minimum Standards Bundle" (what to require campus-wide)

If you want one compact standard set to anchor the book (and to govern modernization), a pragmatic bundle looks like this:

1. **BAS:** BACnet/IP + BACnet/SC for secure IP BAS where feasible
2. **IoT telemetry:** MQTT for sensor/event publish-subscribe patterns
3. **Semantics:** Haystack and/or Brick for tagging and relationships
4. **Segmentation:** IEC 62443 zones/conduits architecture as the shared model
5. **OT security controls:** NIST SP 800-82 as the OT-aware guidance baseline

6. **Enterprise risk language:** NIST CSF 2.0 for governance and board reporting
7. **Mobility/energy:** OCPP and OpenADR, where relevant

The Antifragility Scorecard

A measurement framework for a campus that grows stronger because of shocks and stress.

The antifragile campus demands proof—not aspiration—that every shock, anomaly, and disruption increases the institution's intelligence, stability, and operational maturity. This scorecard converts that principle into measurable indicators that CTOs, CFaOs, Facilities, IT, and the Smart Campus Steering Committee can evaluate quarterly. It reflects the manifesto's emphasis that antifragility arises from governed data, continuous AI monitoring, learning loops, secure IT/OT architectures, and disciplined change-enablement practices.

Exhibit 42. The Antifragility Scorecard

Domain	Metric	What It Measures	Why It Indicates Antifragility	Target Trend
Predictive Intelligence	Mean Time to Predict (MTTP)	How far before a failure does the system detect an anomaly across BAS, utilities, labs, and critical OT domains?	Higher MTTP means the campus is learning from prior stress and forecasting further ahead.	Increasing
	Post-Stress Predictive Gain	Improvement in detection accuracy after each incident, failure, or vendor outage.	Shows that the system is incorporating stress events into improved models and digital-twin logic.	Increasing
Operational Resilience	Recovery Acceleration Rate	Reduction in time required to recover from cyber-physical incidents or equipment faults.	Faster recovery over time means the institution is learning from disruptions and institutionalizing fixes.	Decreasing recovery time
	Safe-Mode Continuity Score	Frequency and effectiveness of safe fallback modes during OT/IT disruptions.	Demonstrates that systems are designed to fail safely and improve safe-state logic based on real incidents.	Improving
Cyber-Physical Security	Segmentation Integrity Index	% of OT assets in fully compliant ISA/IEC 62443 segmentation zones and conduits.	Strong segmentation reduces blast radius and is refined as attacks/tests reveal gaps.	Increasing
	Adaptive Threat-Response Time	Time from anomaly → containment across IT/OT networks.	A shrinking delta indicates the security layer is learning from prior events.	Decreasing
AI and Automation Governance	Model Drift Detection Rate	How quickly drift or degradation is identified and corrected across AI models.	Improved drift detection signals that the AI governance system itself is learning.	Increasing speed and accuracy
	Explainability Compliance	% of AI decisions with traceable, reviewable logic and human-in-the-loop approval.	Ensures AI strengthens trust under stress rather than automating fragility.	Approaching 100%
Data Estate and Digital Twin Quality	Semantic Integrity Score	Accuracy and consistency of tagging, naming, and metadata across BAS, IoT, GIS, BIM, and CMMS.	Reducing "semantic rot" means data quality improves after operational stress reveals inconsistencies.	Increasing
	Telemetry Reliability Rate	Percentage of critical sensors with healthy, validated signals.	High reliability indicates strong data governance and adaptive maintenance cycles.	Increasing

Domain	Metric	What it Measures	Why It Indicates Antifragility	Target Trend
Human and Organizational Learning	Post-Incident Integration Rate	% of incidents with documented lessons learned integrated into workflows, policies, or automation logic within 60 days.	Captures whether the campus actually learns from disruption.	Approaching 100%
	Operator Confidence Index	Survey-based measure of technician trust in systems, tools, and workflows.	Human confidence improves as systems become more predictable.	Increasing
Financial Antifragility	Capital Debt Avoided	Verified dollars saved by extending asset life, avoiding emergency replacements, and preventing failures.	A financially antifragile campus converts stress into capital efficiency.	Increasing
	Predictive OPEX Reinvestment Ratio	Percent of savings reinvested into modernization, data quality, security, and training.	Reinvestment closes the loop and compounds antifragility.	Increasing
Sustainability	**Operation KPIs**			
	Energy	Automated DR Yield; Dynamic Carbon Intensity Alignment; Predictive Energy Improvement Rate	Energy KPIs strengthen antifragility by ensuring the campus can sense, forecast, and autonomously adapt to grid, weather, and load volatility, reducing operational vulnerability and improving stability.	Improving +/-
	Water	Water-Loss Avoidance; Irrigation Optimization Index.	Water KPIs increase antifragility by detecting leaks, inefficiencies, and stress patterns early, enabling faster corrective action and protecting critical operations from supply or system disruptions.	Improving +/-
	Waste	Diversion Accuracy; Food-Waste Reduction Rate; Material Reuse Yield	Waste KPIs enhance antifragility by revealing inefficiencies and resource flows that, when optimized, reduce operational strain and convert disruptions into opportunities for circular reuse and resilience.	Improving +/-
	IAQ/Health	IAQ Health Index; Predictive Comfort Yield.	IAQ and health KPIs make the campus antifragile by enabling real-time environmental adaptation that protects well-being, prevents disruptions, and turns stress events into improvements in comfort and safety.	Improving +/-
	Mobility	Electrified Fleet Utilization; Commuter Emissions Trend.	Mobility KPIs support antifragility by providing visibility into transportation patterns and emissions, allowing the campus to adapt to changing behaviors, reduce operational risk, and maintain continuity during disruptions.	Improving +/-
	Financial/ESG	Avoided Debt; Avoided Carbon; ESG Disclosure Completeness	Financial and ESG KPIs build antifragility by converting operational stress into measurable savings, reduced carbon liability, and stronger institutional credibility, reinforcing long-term resilience.	Improving +/-

How to Use the Antifragility Scorecard

- For Quarterly **Governance Review (CTO/CFaO Co-Chairs)**
  The Smart Campus Steering Committee reviews trends rather than single-point values. Antifragility is shown in **directionality**—systems that get better after each shock.
- **To Obtain Cross-Functional Feedback**
  Facilities, IT, Security, Risk, Academic Affairs, and Sustainability leaders co-interpret results to ensure improvements reflect the institutional mission, not just technical metrics.
- **Operationalizing Governance**
  Each indicator maps to the decision rights outlined for CTOs (identity, segmentation, data governance) or CFaOs (operational performance, commissioning, life cycle planning).
- **Inform Continuous Improvement Loops**
  Each negative trend signals a need for:
    - New/more training
    - Vendor accountability
    - Architecture refinements
    - Policy updates
    - Semantic corrections
    - Model retraining
    - Resilience drills

The key is not to seek or expect perfection—it is improving trend lines that demonstrate the campus is learning from and adapting through adversity.

Glossary

Actuator
A device that executes a control action in the physical world (e.g., opening a valve, changing damper position, dimming lights) based on a command from a controller or platform. *See also: Sensor, Control Loop, BAS, BMS.*

ADKAR (Prosci)
A change management model describing how individuals adopt change through *Awareness, Desire, Knowledge, Ability,* and *Reinforcement. See also: Change Management, Training, Reinforcement, Governance Cadence.*

AI (Artificial Intelligence)
Computational methods that perform tasks like prediction, classification, optimization, and language interaction are used to improve campus operations, learning, sustainability, and safety. *See also: Machine Learning, LLM, Predictive Analytics, Automation.*

AI Ethics
Principles and practices that ensure AI is used responsibly—fairness, transparency, privacy, accountability, and human impact. *See also: Algorithmic Accountability, Bias, Consent, Data Governance.*

AI-Native Architecture
Data and platform design optimized for continuous AI use—supporting training pipelines, real-time inference, monitoring, and retraining. *See also: Data Estate, Model Lifecycle Management, MLOps, Feedback Loop.*

AI Sentinel
An AI-driven monitoring and protection layer that continuously watches over vulnerable or legacy OT systems—especially those that cannot be patched, upgraded, or replaced due to age, operational constraints, or capital limitations.

AI TRiSM (Trust, Risk, and Security Management)
A governance framework ensuring AI systems are monitored, explainable, auditable, secure, and aligned with ethical and operational expectations.
See also: AI Ethics, Algorithmic Accountability, Model Drift.

Air Handling Unit (AHU)
A central HVAC component that conditions and circulates air (heating, cooling, filtration, ventilation). *See also: HVAC, IAQ, VAV, FDD.*

Algorithmic Accountability
A governance discipline that assigns responsibility for algorithm behavior and outcomes through documentation, audits, explainability, and remediation. *See also: AI Ethics, Model Drift, Audit, Transparency.*

Algorithmic Energy Management
AI-driven optimization of HVAC, lighting, microgrids, and building systems using occupancy, weather, peak pricing, and carbon-intensity data. *See also: Demand Response, GEB, Microgrid, Sustainability.*

Anomaly Detection
Analytics methods that identify unusual behavior (e.g., abnormal energy use, unexpected equipment patterns) require investigation. *See also: FDD, Predictive Maintenance, Monitoring, Baselining.*

Antifragility
Antifragility is the characteristic of a system that improves through disruption. In the smart campus context, it describes an operating state in which cyber-physical systems, data, and governance learn from stress, failure, and variability—becoming more resilient, efficient, and capable over time rather than merely returning to baseline. *See also: Resilience, Sentient Campus.*

API (Application Programming Interface)
A standardized method for software systems to exchange data and commands. *See also: Integration Layer, Interoperability, Data Fabric, Middleware.*

Asset Registry
A structured inventory of campus assets (equipment, meters, controllers, spaces), including identifiers, locations, attributes, and maintenance context. *See also: CMMS, BIM, Digital Twin, Master Data.*

Audit (Compliance / Controls)
A formal review of processes, access, data handling, model behavior, or security controls to verify adherence to standards and regulations. *See also: Compliance, FERPA, GDPR, NIST CSF, ISO 27001.*

Automation
The ability for systems to execute actions with limited human intervention (e.g., adjusting HVAC based on occupancy). *See also: Control Loop, Orchestration, Digital Twin, Edge Computing.*

Avoided Debt Ledger
A disciplined record of capital expenses was prevented through condition-based renewal and continuous commissioning, demonstrating how OT intelligence extends asset life and reduces future debt obligations.

BACnet
A common building automation protocol enabling communication among building devices and controllers. *See also: Modbus, LonWorks, Interoperability, BAS.*

Baseline (Performance Baseline)
A "starting point" measurement used to compare improvement over time (energy, IAQ, uptime, response times). *See also: KPIs, Measurement and Verification, EUI, Benchmarking.*

Behavioral Data
Data describing how people use systems and spaces (app usage, occupancy patterns, mobility flows). *See also: Privacy, Consent, Digital Well-Being, Equity.*

Benchmarking
Comparing performance against peers, historical results, or design intent to identify gaps and opportunities. *See also: Baseline, Analyze Per Design, EUI, OKRs.*

BIM (Building Information Modeling)
A digital model of a facility's physical and functional characteristics; increasingly linked to operations and digital twins. *See also: Digital Twin, Asset Registry, Spatial Data.*

Bias (Algorithmic Bias)
Systematic unfairness in a model's outcomes—often affecting specific groups due to data imbalance or design choices. *See also: AI Ethics, Equity Impact Assessment, Algorithmic Accountability.*

BMS (Building Management System)
A supervisory platform that monitors and controls building systems (often used interchangeably with BAS). *See also: BAS, OT, HVAC, FDD.*

BAS (Building Automation System)
Controls and monitoring for building functions (HVAC, lighting, sometimes power, and security). *See also: BMS, Controller, Sensor, Actuator.*

BOMA / BOMI
The Building Owners and Managers Association (BOMA) International and the associated institute, the Building Owners

and Managers Institute (BOMI), are trade associations and credential providers in the commercial real estate industry. See https://boma.org/

Bridge Problem
The structural friction between IT's CIA risk model (Confidentiality, Integrity, Availability) and Facilities' SAS model (Safety, Availability, Sustainability) creates misalignment in governance, security, and operational priorities.
See also: IT/OT Convergence, Cyber-Physical Risk, Governance.

Campus Digital Nervous System
The integrated connectivity, identity, telemetry, and data exchange layers enable real-time sensing, coordination, decision-making, and automation across the campus. *See also: Integration Layer, Network Segmentation, Data Estate.*

Campus Digital Twin
A digital representation of buildings/campus systems combining static and real-time data to visualize, analyze, simulate, and optimize. *See also: Digital Twin, BIM, Data Estate, Scenario Planning.*

Campus Intelligence Layer
The analytics, AI models, digital twins, and data pipelines that transform raw telemetry into actionable insights, predictions, and optimizations.
See also: Analytics Layer, Data Estate, Operational AI.

Campus Sensorium
The full network of sensors, meters, building systems, IoT devices, occupancy data, IAQ monitors, and telemetry streams provides real-time visibility into the campus environment.
See also: IoT, Live Data, Edge Computing, Data Estate.

Carbon Accounting
Measurement and reporting of greenhouse gas emissions (often by scopes), supporting planning and public accountability.
See also: Sustainability Reporting, TCFD, SASB, Decarbonization.

Change Management
Methods to move people and processes to a new operating model—communications, training, stakeholder engagement, reinforcement. *See also: ADKAR, Human Factors, Operating Model, Governance.*

Chief Data Officer (CDO)
A senior role responsible for data strategy, governance, quality, and value realization. *See also: Data Governance, Data Stewardship, Data Estate.*

Chief Facilities Officer (CFaO)
The executive leader is responsible for the stewardship, reliability, and evolution of the physical campus environment. In an AI-enabled smart campus, the CFaO serves as a co-architect of the cyber-physical institution—overseeing Operational Technology, asset lifecycles, sustainability performance, life-safety systems, and occupant experience—while partnering with the CTO to ensure secure IT/OT convergence, data-driven operations, and long-term institutional resilience. *See also: Operational Operating Model; Sustainability; Antifragility; Chief Technology Officer (CTO)Technology (OT).*

Chief Information Officer (CIO)
The executive leader is responsible for the institution's digital strategy, information systems, data governance, and cybersecurity. In an AI-enabled smart campus, the CIO acts as an integrative architect—overseeing enterprise platforms, identity, networks, and data estates—while partnering with the Chief Facilities Officer to govern IT/OT convergence, enable digital-physical intelligence, and support the campus as a resilient institutional operating model. *See also: Information Technology (IT); Operational Technology (OT); IT/OT Convergence; Institutional Operating Model; Data Estate; Cyber-Physical Campus; Digital-Physical Ecosystem; Chief Facilities Officer (CFaO).*

Chief Sustainability Officer (CSO)
A senior role responsible for sustainability strategy, targets, reporting, and cross-campus alignment. *See also: Carbon Accounting, Sustainability Reporting, Decarbonization, Governance.*

CMMS (Computerized Maintenance Management System)
A system of record for work orders, assets, preventive maintenance schedules, and maintenance history. *See also: Asset Registry, Predictive Maintenance, Work Order.*

COBie (Construction Operations Building Information Exchange)
A structured data standard that defines how essential building information is captured and exchanged from design and construction into operations and maintenance. COBie focuses on asset-level data—such as spaces, systems, equipment, warranties, and maintenance requirements. In smart campus architectures, COBie serves as a critical bridge between BIM and operational systems, ensuring that buildings transition from projects to managed institutional assets without loss of information. *See also: BIM, CMMS.*

Cognitive Campus
A cyber-physical environment in which buildings, infrastructure, and services continuously sense, learn, adapt, and respond using governed AI, OT telemetry, and integrated data platforms.
See also: Digital Twin, Operational AI, Campus Sensorium, AI-Enabled Sustainable Smart Campus.

Commissioning (Cx)
Verification that building systems perform as intended. Continuous commissioning extends this into ongoing optimization using analytics. *See also: Continuous Commissioning, FDD, Measurement and Verification.*

Common Data Model
A shared structure and meaning for data across systems to support interoperability and analytics. *See also: Ontology, Semantic Layer, Interoperability, Data Normalization.*

Compliance
Meeting legal, regulatory, and policy obligations (privacy, safety, accessibility, environmental reporting). *See also: FERPA, GDPR, CCPA, Environmental Compliance.*

Condition-Based Maintenance (CBM)
Maintenance triggered by real-time asset condition (vibration, pressure, acoustics, load) rather than scheduled intervals, enabling early detection and reduced downtime. *See also: Predictive Maintenance, IoT Sensors, Fault Detection and Diagnostics.*

Consequence-Based Prioritization
A capital and maintenance prioritization method based on mission impact, safety, resilience, carbon outcomes, and operational dependencies rather than asset age alone. *See also: Risk Management, Capital Planning, Predictive Maintenance.*

Continuous Commissioning
An ongoing process of monitoring and tuning systems to maintain performance over time—often supported by FDD and dashboards. *See also: Commissioning, FDD, KPIs.*

Control Loop
The feedback mechanism that measures a condition and adjusts a system to meet a target (e.g., temperature control). *See also: Sensor, Actuator, Controller, BAS.*

Controller (OT Controller)
A device or software component that receives sensor inputs and issues commands to actuators. *See also: PLC, BAS, Control Loop, SCADA.*

Convergence Governance
The cross-functional governance model aligning IT, Facilities, Safety, Sustainability, and Finance through shared decision

rights and standards for cyber-physical systems.
See also: SCSC, Decision Rights, Operating Model.

Cyber-Physical Campus
A campus environment where digital systems (networks, software, data, and AI) are inseparably integrated with physical infrastructure (buildings, utilities, safety systems, and mobility) to operate as a single, coordinated system. In a cyber-physical campus, actions in the digital domain directly sense, influence, and optimize physical conditions—requiring unified governance, secure architecture, and shared accountability to ensure safety, reliability, and trust. ***See also: Digital-Physical Ecosystem; IT/OT Convergence; Operational Technology (OT); Institutional Operating Model; Secure IT/OT Architecture; Digital Twin; Cyber-Physical Risk.***

Cyber-Physical Risk
Risk arising when cyber compromise affects physical systems (safety, operations, infrastructure). *See also: OT Security, Segmentation, Incident Response, Resilience.*

Data Estate
The total set of campus data assets, pipelines, storage, analytics, governance, and security—treated as an enterprise capability. *See also: Data Governance, Data Fabric, Data Stewardship, Digital Twin.*

Data Fabric
An architectural approach that unifies data access and movement across systems via metadata, APIs, and integration services. *See also: Integration Layer, Middleware, Data Normalization.*

Data Governance
The policies, roles, and decision rights that ensure data quality, compliance, security, and appropriate use. *See also: Data Stewardship, Privacy, Access Control, Audit.*

Data Normalization
The process of standardizing data formats, units, timestamps, naming conventions, and definitions across systems. *See also: Common Data Model, Interoperability, Data Quality.*

Data Quality
The accuracy, completeness, consistency, timeliness, and reliability of data are critical for trustworthy analytics and automation. *See also: Data Governance, Monitoring, Baselining.*

Data Stewardship
Assigned responsibility for data definitions, quality, and lifecycle management across domains. *See also: Data Governance, Master Data, Metadata.*

Decarbonization
Reducing greenhouse gas emissions through efficiency, electrification, renewables, and operational optimization. *See also: Carbon Accounting, Microgrid, DER, Energy Management System.*

Demand Response
Reducing or shifting electricity use during peak periods in response to price signals or grid events. *See also: Load Shedding, Peak Demand, DER, Energy Storage.*

Design Thinking (Human-Centered Design)
A problem-solving approach grounded in empathy, user journeys, prototyping, and iterative improvement. *See also: Persona, User Experience, Equity, Adoption.*

Digital Divide
Unequal access to devices, connectivity, and digital skills can create inequities in learning and campus services. *See also: Equity, Accessibility, Digital Literacy.*

Digital-Physical Ecosystem
An integrated environment in which digital platforms, data, and intelligence are tightly coupled with physical infrastructure, assets, and human activity—enabling real-time sensing, coordinated control, and adaptive decision-making across the campus. In the Smart Campus, the digital-physical ecosystem replaces siloed systems with a unified, cyber-physical fabric that supports reliability, sustainability, safety, and continuous improvement. See also: *Cyber-Physical Campus; IT/OT Convergence; Institutional Operating Model; Digital Twin; Institutional Data Estate; Operational Technology (OT); Antifragility.*

Digital Twin
A digital representation of a physical asset or system that combines static and real-time data for insight, simulation, and optimization. *See also: BIM, Campus Digital Twin, FDD, Scenario Planning.*

Digital Well-Being
The health and social impacts of digital environments on students and staff (privacy, stress, accessibility, agency). *See also: Behavioral Data, Equity, Consent, Surveillance.*

Distributed Energy Resources (DER)
On-site or local energy assets—solar, storage, controllable loads—used for efficiency, cost control, and resilience. *See also: Microgrid, Energy Storage, Smart Grid.*

Demand Response Yield (DR Yield)
The financial benefit you get when you temporarily reduce energy use because the utility is paying you to do so.

Edge AI
Edge AI refers to artificial intelligence models and inference that run directly on local or near-device hardware—such as IoT gateways, building controllers, or onsite servers—rather than relying solely on cloud processing. In the smart-campus context, Edge AI enables real-time decision-making for functions that require low latency, privacy preservation, or resilience during network disruption (e.g., IAQ control, anomaly detection, autonomous valves, AR/VR interactions). *See also: Generative AI, Operational AI, Machine Learning, Digital Twins.*

Edge Computing
Processing and analytics near devices/buildings to reduce latency and improve reliability. *See also: IoT 2.0*, **Latency, Resilience, Cloud.**

Energy Management System (EMS)
A platform that monitors and optimizes energy use across buildings/campus, often integrating meters, BAS, tariffs, and analytics. *See also: EUI, Demand Response, DER, Carbon Accounting.*

Energy Storage (Batteries / Thermal Storage)
Systems that store energy for later use—supporting load shifting, resilience, and renewables integration. *See also: DER, Microgrid, Demand Response.*

Environmental Compliance
Adherence to environmental standards and reporting requirements (emissions, waste, water, hazardous materials). *See also: Sustainability Reporting, SASB, TCFD.*

ESG Operating Model
A governance-driven institutional framework that integrates environmental performance, social outcomes, and accountable decision-making into the daily operation, capital planning, and digital architecture of the campus—ensuring sustainability is durable, measurable, and mission-aligned.

Equity Impact Assessment
A structured evaluation of how a technology or policy may affect different groups—intended to prevent unintended harm. *See also: AI Ethics, Accessibility, Digital Divide.*

Explainability
The ability to understand and communicate why an AI system produced an outcome. *See also: Algorithmic Accountability, Transparency, Model Monitoring.*

Fault Detection and Diagnostics (FDD)
Analytics that detect equipment faults, identify causes, and prioritize corrective actions. *See also: Anomaly Detection, Continuous Commissioning, Predictive Maintenance.*

Feedback Loop (Operational / Model)
Structured input (from users, sensors, logs) is used to improve systems and models over time. *See also: Non-Stationary Objectives, MLOps, Continuous Improvement.*

FERPA
U.S. law governing privacy of student education records. *See also: Privacy, Consent, Data Governance.*

Federated Governance
A governance model that balances centralized standards with distributed ownership across units. *See also: Data Governance, Portfolio Intake, Decision Rights.*

Fleet Electrification
Transitioning campus vehicles to electric, requiring charging infrastructure and grid/energy planning. *See also: DER, Peak Demand, Energy Storage.*

Fitwel
A research-based building certification system developed by the U.S. Centers for Disease Control and Prevention (CDC) and the General Services Administration (GSA) that focuses on improving occupant health, well-being, and productivity through evidence-based design and operational strategies. Fitwel emphasizes practical, scalable interventions—such as walkability, safety, access to healthy amenities, and inclusive design—that can be implemented across existing and new facilities.

Fragility Debt
The operational risk that accumulates when new digital capabilities sit atop aging, inconsistent, or unstable OT systems—leading to brittleness and failure. *See also: Deferred Maintenance, Controls Modernization, Cyber-Physical Risk.*

Generative AI
Generative AI (GenAI) refers to artificial intelligence systems that create new content—such as text, images, code, simulations, or narratives—by learning patterns from large datasets. Within the Smart Campus context, Generative AI primarily augments human work by synthesizing information, accelerating documentation and analysis, and enabling natural-language interaction with institutional systems, rather than directly controlling physical infrastructure. *See also: Operational AI, Machine Learning, Digital Twin.*

Governance Cadence
A predictable rhythm of decision forums and reviews (steering committee, intake gates, risk reviews, KPI reviews). *See also: Operating Model, Portfolio Intake, Risk Management.*

Grid-Interactive Efficient Building (GEB)
A building that responds dynamically to utility grid conditions through on-site generation, storage, controllable loads, and automated demand-response strategies. *See also: Virtual Power Plant, Demand Response, Energy Optimization.*

HVAC (Heating, Ventilation, and Air Conditioning)
Systems that provide thermal comfort and air quality. *See also: IAQ, AHU, VAV, BAS.*

Infrastructure as a Service (IaaS)
A cloud-computing model in which a provider delivers on-demand virtualized computing resources—such as servers, storage, and networking—over the internet. Institutions retain control over operating systems, applications, and data while avoiding

the capital expense and maintenance burden of physical hardware. IaaS forms the foundational layer on which higher-level cloud services operate. *See also: PaaS, Saas.*

Indoor Air Quality (IAQ)
refers to the real-time environmental conditions of the air inside campus buildings, measured continuously using sensors that monitor factors such as carbon dioxide (CO_2), particulate matter (PM2.5/PM10), volatile organic compounds (VOCs), humidity, temperature, and airflow/pressurization. IAQ is a core dimension of **health, comfort, cognitive performance, and safety,** and serves as a key sustainability and risk management indicator across classrooms, residences, laboratories, and high-density spaces. *See also: HVAC, Sensors, Wellness, Compliance.*

Identity and Access Management (IAM)
Processes and systems that manage user identities, authentication, and authorization—especially critical when extending access into OT environments. *See also: MFA, Least Privilege, Zero Trust.*

Incident Response
The playbook and operational practice for detecting, containing, and recovering from cybersecurity or operational incidents. *See also: Resilience, Monitoring, OT Security.*

Institutional Operating Model
A structured framework that defines how an institution translates mission and strategy into coordinated action—aligning governance, decision rights, processes, data, technology, and culture to deliver consistent outcomes over time. In the Smart Campus context, the institutional operating model integrates digital and physical systems (IT and OT) to enable reliable operations, sustainability, resilience, and continuous learning. *See also: AI-Enabled Sustainable Smart Campus; IT/OT Convergence; Governance; Decision Rights; Digital Twin; Institutional Data Estate; Antifragility.*

Integration Layer (Middleware)
Technology that connects systems through APIs, message brokers, and data pipelines. *See also: Interoperability, Data Fabric, API.*

Interoperability
The ability of systems and devices to exchange data and work together reliably using standards, models, and integration patterns. *See also: Open Standards, Data Normalization, API.*

Integrative Architect
An executive role—typically embodied by the CIO or CTO—is responsible for unifying digital and physical systems into a coherent institutional operating model. The Integrative Architect designs and governs IT/OT convergence, establishes architectural guardrails, aligns data and platforms, and ensures that technology modernization advances mission outcomes, resilience, sustainability, and the human experience rather than being fragmented into projects. *See also: Chief Information Officer (CIO); Chief Technology Officer (CTO); Chief Facilities Officer (CFaO).*

IoT (Internet of Things)
Networked sensors and devices that collect data and/or perform actions in physical environments. *See also: Edge Computing, Telemetry, OT.*

IoT 2.0
A next-generation IoT posture with embedded edge intelligence, autonomous response, and improved scale management. *See also: Edge AI, Automation, Resilience.*

ISO 27001 (Information Security Management Systems)
An international standard that defines requirements for establishing, implementing, maintaining, and continually improving an Information Security Management System (ISMS). ISO 27001 is a risk-based framework that extends beyond traditional IT systems to encompass cyber-physical infrastructure, operational technology, digital twins, and institutional data estates—ensuring that connectivity, automation, and AI-enabled operations scale in alignment with governance, trust, and accountability. *See also: NIST CSF, Compliance, Audit.*

ISO 41001 (Facility Management Systems)
An international standard that sets requirements for establishing, implementing, maintaining, and improving a facility management system. ISO 41001 emphasizes alignment between facility operations and organizational strategy, enabling campuses to manage the built environment as a strategic asset through consistent governance, lifecycle planning, service performance, and continual improvement.

ISO 50001 (Energy Management Systems)
An international standard that provides a structured framework for managing and improving energy performance. ISO 50001 enables institutions to establish energy baselines, set measurable objectives, monitor performance, and implement continuous improvement processes—treating energy as a managed system rather than a utility expense, and supporting measurable decarbonization and cost control.

IT (Information Technology)
The domain is responsible for digital systems that manage data, applications, networks, identity, and computing services supporting the institution's academic, administrative, and research missions. In the Smart Campus, IT extends beyond traditional enterprise systems to serve as the digital backbone, enabling secure connectivity, data governance, analytics, and integration with Operational Technology (OT). *See also: Operational Technology (OT); IT/OT Convergence; Institutional Data Estate; Cyber-Physical Campus; Secure IT/OT Architecture; Digital-Physical Ecosystem; Institutional Operating Model.*

Key Performance Indicators (KPIs)
Operational measures that track health and performance (uptime, EUI, response time, satisfaction). *See also: OKRs, Dashboard, Baseline.*

LEED (Leadership in Energy and Environmental Design)
A globally recognized green building certification system that evaluates building and campus performance across categories such as energy efficiency, water use, materials, site design, and indoor environmental quality. Within smart campus architectures, LEED functions increasingly as an ***operational reference point****, informing data models, digital twins, and performance benchmarks rather than serving solely as a design-phase certification.*

Lifecycle Costing
Financial analysis that evaluates total cost across acquisition, operations, maintenance, upgrades, and end-of-life. *See also: TCO, Capital Planning, Operating Budget.*

Living Laboratory / Living System (Campus as Research and Education Platform)
A governance and data-sharing model that safely opens campus systems, telemetry, and digital twins for teaching, research, and innovation partnerships. *See also: Academic Engagement, Data Governance, AI Ethics.*

Load Shedding
Reducing energy use by turning down or cycling non-critical loads during peak demand or grid stress. *See also: Demand Response, Peak Demand, Energy Storage.*

LonWorks
A building automation protocol found in legacy systems, especially in older campuses. *See also: BACnet, Modbus, Interoperability.*

LLM (Large Language Model)
A language-based AI model used for conversational interfaces, summarization, and knowledge assistance; it requires governance in campus settings. *See also: Non-Stationary Objectives, AI Ethics, Policy-Aware Alignment.*

Machine Learning (ML)
A subset of artificial intelligence that uses data-driven models to identify patterns, make predictions, and improve performance over time without explicit reprogramming. On a smart campus, machine learning enables capabilities such as predictive maintenance, energy optimization, space utilization analytics, and early risk detection by learning from operational and environmental data streams. *See also: Artificial Intelligence (AI); Digital Twin; Data Estate.*

Master Data
Authoritative reference data for core entities (buildings, rooms, assets, departments), essential for integration consistency. *See also: Data Stewardship, Common Data Model, Asset Registry.*

Measurement and Verification (MandV)
Methods for confirming performance improvements and savings (energy, emissions, operations), often using accepted protocols. *See also: Baseline, Commissioning, EUI.*

Metadata
Data about data—definitions, lineage, ownership, quality indicators, and access rules. *See also: Data Governance, Data Fabric, Model Cards.*

Microgrid
A local energy network that can operate connected to or independent of the utility grid, improving resilience and enabling renewables integration. *See also: DER, Energy Storage, Resilience.*

Minimum Viable Convergence (MVC)
A defensible baseline for IT/OT integration and controls (inventory, segmentation, remote access control, data export) that enables early value safely. *See also: Segmentation, Remote Access, Data Export, Zero Trust.*

Monitoring-Based Commissioning (MBCx)
Ongoing commissioning using live telemetry, analytics, and FDD tools to keep systems tuned, detect drift, and sustain performance over time and seasons. *See also: Commissioning, Baseline, Optimization, Digital Twin.*

Model Drift
Degradation of an AI model's performance over time due to changes in data patterns, context, or policies. *See also: Non-Stationary Objectives, Model Monitoring, MLOps.*

Model Lifecycle Management
The governance and operations required to deploy, monitor, update, retrain, and retire AI models responsibly. *See also: MLOps, Algorithmic Accountability, Audit.*

Model Monitoring
Ongoing measurement of AI performance, bias, reliability, and safety in production. *See also: Model Drift, Feedback Loop, Algorithmic Accountability.*

Modbus
A common industrial/OT protocol used in building and plant systems. *See also: BACnet, SCADA, PLC.*

Net-Zero (Campus Context)
A target state where campus emissions are reduced and balanced to achieve a defined neutrality boundary (often Scope 1 and 2, sometimes Scope 3). *See also: Decarbonization, Carbon Accounting, Sustainability Reporting.*

Network Segmentation
Separating networks into zones with controlled pathways to reduce cyber risk and contain incidents. *See also: Zero Trust, OT Security, and Least Privilege.*

New Edge (The OT Digital Frontier)
The emerging boundary where legacy physical systems—BAS, utilities, meters, access control—become cloud-connected, sensor-rich, AI-enabled, and tied to enterprise IT. *See also: Edge Computing, OT Modernization, Cyber-Physical Interfaces.*

NG911 (Next Generation 911)
IP-based emergency communications enabling voice, text, photos, and improved location routing—often integrated with campus alerting. *See also: Emergency Notification, Resilience, Safety Systems.*

Non-Stationary Objectives
The concept that "correct" AI behavior changes over time as policies, contexts, and expectations evolve—requiring continuous oversight and adaptation. *See also: Model Drift, Model Lifecycle Management, Governance.*

Occupancy Analytics
Using sensors and data to understand how spaces are used and to optimize HVAC, scheduling, and user experience. *See also: Space Utilization, IAQ, Energy Optimization.*

OKRs (Objectives and Key Results)
A strategic framework linking aspirational objectives to measurable outcomes over a defined period. *See also: KPIs, Portfolio Intake, Strategic Alignment.*

Ontology
A structured "meaning model" that enables data from many systems to be interpreted consistently (e.g., knowledge graphs). *See also: Common Data Model, Interoperability, Knowledge Graph.*

OODA (Observe–Orient–Decide–Act)
The OODA Loop is a decision-making framework that describes how organizations sense their environment, interpret signals, choose actions, and execute responses through continuous feedback. In The Smart Campus Manifesto, OODA is extended beyond a human-only decision cycle to a cyber-physical, data-driven loop—where sensors observe, analytics and AI orient, governance-informed systems decide, and operations act, with outcomes continuously feeding back into learning. The OODA model enables campuses not just to respond faster, but to learn through disruption, steadily increasing institutional intelligence, resilience, and antifragility over time.

Operating Model
How people, process, governance, and technology work together day-to-day (roles, decision rights, workflows). *See also: Governance Cadence, RACI, Change Management.*

Operational AI
Artificial intelligence applied directly to the real-time management, optimization, and resilience of physical and cyber-physical systems. In a smart campus, Operational AI analyzes live and historical telemetry from buildings, utilities, space, safety, and infrastructure to enable predictive maintenance, adaptive energy control, automated fault response, and scenario-based decision support—shifting operations from reactive control to continuous, learning-based optimization. ***See also:***
Artificial Intelligence (AI); Machine Learning (ML); Digital Twin; Operational Technology (OT); Cyber-Physical Campus; Predictive Maintenance; Antifragility; Data Estate.

Operational Drift
The gradual divergence of system behavior from design intent due to overrides, degraded sensors, undocumented changes, or sequence failures.
See also: MBCx, FDD, Stability, and Reliability Metrics.

Operator Experience (OX)
The usability, cognitive load, clarity, and navigability of systems used daily by Facilities operators are primary determinants of successful modernization.
See also: UX, Workflow Integration, Change Management.

Operational Truth
A validated, evidence-based representation of how systems actually operate—derived from telemetry, commissioning, and field verification rather than documentation.
See also: Inventory, Baseline, Readiness Assessment.

OT (Operational Technology)
Hardware/software that monitors and controls physical systems (BAS, meters, plant controls, access systems). *See also: IT/OT Convergence, SCADA, PLC, BMS.*

OT Security
Cybersecurity practices for operational environments where safety and uptime are critical (segmentation, monitoring, and access control). *See also: IT, Network Segmentation, Zero Trust, Remote Access Control.*

Platform as a Service (PaaS)
A cloud model that provides a managed application-development platform—including operating systems, databases, middleware, and deployment tools—without requiring organizations to manage the underlying infrastructure. PaaS accelerates development, standardizes environments, and supports modern application architectures used across smart-campus systems. *See also: IaaS, Saas).*

Peak Demand
The maximum electricity demand over a billing interval, often driving significant costs and grid impact. *See also: Demand Response, Load Shedding, Energy Storage.*

Penetration Testing
Controlled security testing intended to identify vulnerabilities before adversaries do. *See also: OT Security, Audit, Vulnerability Management.*

Persona (Persona-Centric Design)
A representative user profile used to design services and systems around real needs and journeys (student, faculty, trades, visitors). *See also: Design Thinking, User Experience, Accessibility.*

Pilot (Proof of Value)
A limited deployment designed to validate value, feasibility, risk controls, and adoption before scaling. *See also: Living Lab, Scale, Portfolio Intake.*

PLC (Programmable Logic Controller)
An industrial controller used to automate processes in plants, utilities, and infrastructure. *See also: SCADA, Controller, OT.*

Policy-Aware Alignment
Ensuring AI behavior reflects current institutional policies, privacy rules, and ethical boundaries. *See also: Non-Stationary Objectives, AI Ethics, Governance.*

Portfolio Intake
A structured process for evaluating and prioritizing use cases and projects based on value, risk, readiness, and capacity. *See also: Governance Cadence, OKRs, Risk Management.*

Predictive Analytics
Methods that forecast outcomes (failures, demand, occupancy, risk) using historical and real-time data. *See also: AI, Machine Learning, Predictive Maintenance.*

Predictive Maintenance
Maintenance triggered by risk signals from analytics rather than a fixed schedule, reducing downtime and cost. *See also: FDD, CMMS, Reliability.*

Privacy-by-Design
Building privacy protections into systems from the start (data minimization, consent, access controls, transparency). *See also: Consent, Data Governance, FERPA, GDPR.*

Privileged Access Management (PAM)
Controls for high-risk accounts (admin/vendor) including MFA, time-limited access, and session recording. *See also: Remote Access Control, IAM, Least Privilege.*

Procurement (Strategic Procurement)
Acquiring technologies and services in a way that supports interoperability, security, and long-term value—often using standards and performance clauses. *See also: Vendor Management, Open Standards, Lifecycle Costing.*

Psychological Safety (for Convergence Teams)
A team environment where IT, Facilities, and Security staff can speak openly about risks, constraints, and uncertainties without fear of blame—critical for converged operations. *See also: Change Management, Workforce Enablement, Human Factors.*

Resilience
The ability to withstand disruptions (climate events, outages, cyber incidents) and recover quickly. *See also: Antifragility, Disaster Recovery, Microgrid, Incident Response.*

Remote Access Control (OT Context)
Secure methods for staff and vendors to access OT environments—MFA, PAM, segmentation, and session logging. *See also: OT Security, Zero Trust, Vendor Management.*

RACI
A role clarity tool defining who is Responsible, Accountable, Consulted, and Informed for key decisions and processes. *See also: Operating Model, Governance, Decision Rights.*

Responsibility Mirage
Responsibility Mirage refers to the false assumption that migrating systems or services to cloud or SaaS platforms transfers full security and risk accountability to the vendor. The mirage arises when teams mistake infrastructure security for application and configuration security—resulting in misconfigured storage, over-privileged accounts, weak vendor access pathways, and data exposure that is ultimately the institution's responsibility to prevent.

Risk Management (Smart Campus Context)
An integrated approach to identifying and mitigating risks across social, economic, environmental, governance, and technology/data domains. *See also: Five-Domain Framework, Cyber-Physical Risk, Compliance.*

Safety–Availability–Sustainability (SAS Triad)
The operational and risk framework that guides Facilities organizations: protecting life safety, ensuring uptime, and enabling long-term environmental stewardship. *See also: CIA Triad, Risk Domains, Facilities Leadership.*

SASB
A sustainability reporting framework focusing on financially material metrics by sector. *See also: TCFD, Sustainability Reporting, Environmental Compliance.*

SCADA
Supervisory control systems used to monitor and control industrial processes (central plant, utilities). *See also: PLC, OT, Cyber-Physical Risk.*

Scenario Planning
Exploring plausible futures to guide strategy and resilience, often using simulations and digital twins. *See also: Digital Twin, Foresight, Resilience Mapping.*

Segmentation (Network Segmentation)
Creating network zones and controlled pathways to reduce risk and constrain lateral movement. *See also: OT Security, Zero Trust, Minimum Viable Convergence.*

Semantic Rot
The slow degradation of data accuracy or meaning as equipment changes, spaces are repurposed, or metadata standards drift, undermining analytics, and digital twins. *See also: Ontology, Data Estate, Normalization, Digital Twin Accuracy.*

Sensor
A device that measures conditions (temperature, humidity, CO_2, occupancy, vibration, power) and reports data to systems. *See also: Actuator, Telemetry, IoT.*

Sentient Campus
A Sentient Campus is an advanced smart campus operating state in which tightly integrated digital and physical systems continuously sense, learn, and adapt. Using real-time operational data, artificial intelligence, and governed automation, the campus functions as a living, cyber-physical system that anticipates needs, responds intelligently to change, and improves over time. Unlike a reactive or merely resilient campus, a Sentient Campus is antifragile—it converts disruption, stress, and variability into insight that strengthens safety, sustainability, reliability, and the human experience. *See also: Smart Campus, Antifragility.*

Service Desk
The operational function that receives, routes, and resolves technology service requests and incidents. *See also: User Experience, Adoption, KPIs.*

Shadow OT
Shadow OT refers to operational technology systems—such as sensors, controllers, smart devices, or vendor-managed platforms—that are deployed without enterprise visibility, governance, or security controls, creating untracked cyber-physical risk to campus operations. *See also: Cyber-Physical Campus, IT/OT Convergence, Zero Trust Architecture, Operational Risk*

Shadow AI
Unapproved or ungoverned use of AI models or tools circumventing enterprise governance, posing privacy, security, and safety risks—analogous to Shadow OT. *See also: Shadow OT, AI Governance, TRiSM.*

Smart Building
A building that integrates sensing, controls, analytics, and automation to optimize energy, comfort, safety, and maintenance. *See also: BAS, FDD, Digital Twin.*

Smart Campus
A campus-wide approach that integrates buildings, infrastructure, digital services, and governance through shared data and user-centered design. *See also: IT/OT Convergence, Data Estate, Sustainability.*

Smart Grid
A digitally enabled electricity grid that supports real-time monitoring, distributed energy, and demand flexibility. *See also: Demand Response, DER, Microgrid.*

Soft Landings
A structured post-construction process ensuring that designers, engineers, and contractors remain engaged to stabilize systems, support early operations, and train staff. *See also: Commissioning, Operator Experience, Change Management.*

Software as a Service (SaaS)
A cloud delivery model where fully prepared, ready-to-use applications are offered over the internet via a subscription. The provider handles all infrastructure, updates, and security, allowing institutions to access the service via a web browser or API. SaaS is widely used for enterprise systems, learning platforms, work management tools, and specialized OT-integrated applications. *See also: IaaS, PaaS.*

Space Utilization
Measurement and optimization of how rooms and buildings are used (occupancy rates, scheduling efficiency). *See also: Occupancy Analytics, Scheduling, Campus Experience.*

Software Bill of Materials (SBOM)
A formal and organized inventory that details all components, libraries, modules, and dependencies—whether

open-source or proprietary—that comprise a software application. Often called a "list of ingredients," it offers transparency within the software supply chain, aiding in the management of security risks, licensing issues, and compliance.

Sustainability
The disciplined, long-term stewardship of institutional resources—physical, financial, environmental, and human—such that present needs are met without compromising future capacity. In the smart campus context, sustainability is an ***operational property****, expressed through continuous measurement, standards-based design, governed data, and adaptive systems that reduce waste, extend asset life, support health and equity, and strengthen institutional resilience over time, See also: Decarbonization, Circularity, Governance.*

Sustainability Reporting
Public and internal reporting of performance against sustainability commitments, targets, and risk exposures. *See also: Carbon Accounting, SASB, TCFD.*

Semantic Layer
The semantic layer standardizes meaning. It enforces consistent tags, units, hierarchies, and naming conventions across the enterprise. By establishing this shared vocabulary, the institution can scale analytics and AI across dozens of buildings without rewriting the logic for each individual "vendor dialect."

Synthetic Data
Artificially generated data that mimics real patterns is used to train/test models while protecting privacy. *See also: Privacy, Model Training, Federated Learning.*

TCFD
A climate-risk disclosure framework emphasizing governance, strategy, risk management, and metrics/targets. *See also: Sustainability Reporting, Climate Resilience, Risk Management.*

TCO (Total Cost of Ownership)
Full lifecycle cost of a system, including implementation, integration, operations, maintenance, and upgrades. *See also: Lifecycle Costing, Vendor Management, Operating Budget.*

Telemetry
Continuous operational data streams from systems and devices (time-series signals). *See also: IoT, Monitoring, Data Estate.*

Training (Capability Building)
The structured development of skills and confidence is needed to adopt new systems and workflows. *See also: ADKAR, Change Management, Digital Literacy.*

Transparency (Institutional / Algorithmic)
Clear communication about how systems work, what data is collected, and how decisions are made. *See also: Consent, Explainability, Trust.*

Trust (Institutional Trust)
The confidence stakeholders place in the institution's technology and governance—built through transparency, fairness, and reliability. *See also: AI Ethics, Privacy, Governance.*

'Turn Down'
Flow turndown describes the relationship between the maximum and minimum flow rates that a system, valve, or meter can reliably handle or control. It is defined and applied in two primary methods, depending on the equipment type. In airflow systems such as fans and VAV units, turndown is expressed as a percentage. In valves and flow meters, turndown is conveyed as a ratio of the maximum flow to the minimum controllable flow installed in the system. For example, a 10-to-1 turndown on a 1,000-ton chiller means it can still operate effectively at 100 tons.

V2V (Vehicle-to-Vehicle)
Cars communicate with each other about speed, location, braking, and hazards.

V2I (Vehicle-to-Infrastructure)
Vehicles exchanging data with road infrastructure (traffic lights, sensors, signs).

V2P (Vehicle-to-Pedestrian)
Alerts and communication between vehicles, pedestrians, or cyclists via mobile devices or sensors.

V2G (Vehicle-to-Grid)
Energy exchange between electric vehicles and the power grid (charging, discharging, load balancing).

V2N (Vehicle-to-Network)
Connectivity to cloud services for navigation, traffic, or software updates.

V2X
V2X enables vehicles to act as intelligent, connected nodes within a broader transportation and energy ecosystem. *See also: V2V, V2I, V2G, V2N.*

Vendor Management
The ongoing management of vendor performance, security, support, SLAs, and roadmap alignment. *See also: Procurement, Vendor Lock-In, Remote Access Control.*

Vendor Lock-In
Dependence on proprietary platforms or contracts that reduce flexibility and increase switching costs. *See also: Open Standards, Interoperability, Procurement.*

VAV (Variable **Air Volume)**
An HVAC strategy that adjusts airflow by zone based on need, improving comfort and efficiency. *See also: HVAC, AHU, IAQ.*

Virtual Power Plant (VPP)
Aggregation of many distributed energy assets to act as a coordinated resource for grid services. *See also: DER, Demand Response, Energy Storage.*

Vulnerability Management
Processes for identifying, prioritizing, and remediating security weaknesses in systems. *See also: OT Security, Penetration Testing, and Patch Management.*

WELL Building Standard
A performance-based building standard focused on human health and well-being within the built environment, addressing factors such as air quality, lighting, thermal comfort, acoustics, safety, and accessibility. In smart campuses, WELL principles are operationalized through continuous sensing, analytics, and governance—embedding health, equity, and occupant experience into daily operations rather than discrete wellness programs.

Workforce Scarcity (Facilities Domain)
The structural shortage of skilled trades and hybrid digital-mechanical technicians is caused by retirements, complexity, and competition for talent.
See also: Knowledge Capture, Training Pathways, Digital Twin.

Work Order
A formal maintenance task record (reactive, preventive, predictive) tracking labor, materials, and resolution. *See also: CMMS, KPIs, Predictive Maintenance.*

Zero-Day Attack
A cyber-physical intrusion in which an adversary exploits a software, firmware, or hardware vulnerability that is unknown to the vendor and has no available patch or mitigation. Because developers have "zero days" to respond, attackers gain a unique advantage—using the flaw to bypass defenses, execute malicious code, steal data, or disrupt operations before the organization or security community is even aware the weakness exists. *See also: Zero-Day Vulnerability, Exploit, Cyber-Physical Risk, Threat Detection.*

Zero Trust Architecture
A security model that assumes no implicit trust—requiring verification, least privilege, segmentation, and continuous monitoring. *See also: IAM, MFA, OT Security, Segmentation.*

Bibliography

Almawave. (n.d.). *Digital twin technology: Definition, benefits, and industry impact.* https://www.almawave.com/digital-twin-technology/

Aramark. (n.d.). *What is digital twin technology for facilities management?* https://www.aramark.com/our-services/facilities-management-services/k-12-facilities-resource-center/what-is-digital-twin-technology-for-facilities-management-

Asian Development Bank. (2024). *Digital twin framework: A practical guide.* https://www.adb.org/sites/default/files/publication/1051191/digital-twin-framework-guide.pdf

Aaron. (2024, April 10). *Creating a shared vision.* Leader Navigation. https://www.leadernavigation.com/shared-vision/

Adamkó, A. (2018). Building a smart university using innovative technology and architecture. In *Smart universities: Concepts, systems, and technologies* (Vol. 4, pp. 161–188).

Adamko, N., et al. (2021). Toward a smart and sustainable campus future vision. In *[Springer book/volume].* Springer. https://link.springer.com/chapter/10.1007/978-3-030-63567-1_20

Afanasiev, M., and Lysenkova, M. (2019). How university acts in the development of "smart cities." In *SHS Web of Conferences* (Vol. 71, Article 05011). EDP Sciences.

Akila. (2023, March 16). *The five dimensions of digital twins.* https://www.akila3d.com/blog/insights/five-dimensions-digital-twins/

Alade, K. T., Lawal, A. F., and Akinyele, D. (2017, April). *Smart materials and technologies for next generation energy-efficient buildings.* IEEE Smart Grid. https://smartgrid.ieee.org/bulletins/april-2017/smart-materials-and-technologies-for-next-generation-energy-efficient-buildings

Anand, V. (2022). *Smart sustainable campus: Challenges and approaches using edge-based system design.* Capgemini Engineering. https://www.capgemini.com/insights/expert-perspectives/smart-sustainable-campus/

Andersen, G., and MoldStud Research Team. (2024, January 24). *Creating sustainable campus infrastructure: Role of directors of engineering.* MoldStud. https://moldstud.com/articles/p-creating-sustainable-campus-infrastructure-role-of-directors-of-engineering

Anthology. (2022, November 10). *How universities can meet the growing expectations of students.* THE Campus. https://www.timeshighereducation.com/campus/how-universities-can-meet-growing-expectations-students

Bell, M. (2024, March 18). *What top challenges facilities leaders are facing in 2024.* EAB. https://eab.com/resources/blog/strategy-blog/top-challenges-facilities-leaders-are-facing-2024/

Bhatia, S. (2018). Sustainable smart universities for smart cities. *Journal of Economics, Management and Trade, 21*(12), 1–?.

Bonderud, D. (2022, December 14). *What is a smart campus and what technologies support one?* EdTech Magazine. https://edtechmagazine.com/higher/article/2022/12/what-smart-campus-and-what-technologies-support-one-perfcon

BrainBox AI. (2024, January 4). *10 ways to leverage AI in facilities management.* https://brainboxai.com/en/articles/10-ways-to-leverage-ai-in-facilities-management

Brick Schema. (n.d.). *Brick Schema and RealEstateCore announce a major integration.* https://www.realestatecore.io/brickrec/

Capgemini. (2024). *Smart sustainable campus.* https://www.capgemini.com/insights/expert-perspectives/smart-sustainable-campus/

Caputo, P., and Moretti, N. (2025). Digital twins from building to urban areas: An open opportunity to energy, environmental, economic and social benefits. *Applied Sciences, 15*(19), 10795. https://doi.org/10.3390/app151910795

Center of Excellence for Clean Energy. (n.d.). *Sustainability specialist.* Green Buildings Career Map. https://greenbuildingscareermap.org/jobs/sustainability-specialist

Chameleon. (2024, May 15). *Good data, good twins: A simple guide to data quality for digital twins.* Medium. https://medium.com/@chameleon13/good-data-good-twins-a-simple-guide-to-data-quality-for-digital-twins-8fa6fa3d1bd0

Chenoweth, H. (2019, June 27). *University facilities leaders tackle their toughest challenges.* Higher Ed Facilities Forum. https://info.higheredfacilitiesforum.com/blog/university-facilities-leaders-tackle-their-toughest-challenges

Christiansen, B. (2024, February 23). *The roles and responsibilities of a facilities manager.* Limble CMMS. https://limblecmms.com/blog/facilities-manager-roles-and-responsibilities/

Cox, H. (2015). *A model for creating a campus sustainability plan.* Society for College and University Planning. https://www.scup.org/resource/a-model-for-creating-a-campus-sustainability-plan/

Creasey, T. (2023, December 19). *The Prosci ADKAR model: Why it works.* Prosci. https://www.prosci.com/blog/why-the-adkar-model-works

Davis, C., and Scott, L. (2024). *Lessons in sustainability: Three universities creating green campuses.* Facilities Management Advisor. https://facilitiesmanagementadvisor.blr.com/green-building/lessons-in-sustainability-three-universities-creating-green-campuses/

Deloitte. (n.d.). *Smart campus: The next generation connected campus* [PDF]. https://www2.deloitte.com/content/dam/Deloitte/us/Documents/strategy/the-next-generation-connected-campus-deloitte.pdf

Element Six. (2026, February 16). Digital twins that drive decisions: Speed to value, adoption, & operational AI with Rick Szcodronski [Video]. YouTube. https://www.youtube.com/watch?v=LQQKfp6Yd2M

Evans, N. (2024, January 3). *5 best practices for digital twin implementation.* CTO. https://www.CTO.com/article/1270042/5-best-practices-for-digital-twin-implementation.html

Equinix. (2023, January 24). *Data management best practices for a new era of digital twins.* Equinix Blog. https://blog.equinix.com/blog/2023/01/24/data-management-best-practices-for-a-new-era-of-digital-twins/

Farhodi, F. (2024, March 25). *Learn about the transformative potential of smart materials in building design and construction.* NeuroJect. https://neuroject.com/smart-materials-in-buildings/

Fischer, M. (2021). *5 facilities trends every higher ed leader should be watching.* EAB. https://eab.com/resources/research-report/5-facilities-trends-leaders-should-watch/

FMX. (2024). *The essential guide to developing a comprehensive campus master plan.* https://www.gofmx.com/blog/campus-master-plan/

Froehlich, A. (2025, May 13). *The real value of digital twins in smart buildings. Buildings.* https://www.buildings.com/articles/real-value-digital-twins-smart-buildings

Froehlich, A. (2025, June 26). *Retrofitting aging buildings requires strategic tech integration.* TechTarget. https://www.techtarget.com/retrofitting-aging-buildings-requires-strategic-tech-integration

Future Today Strategy Group. (2025). *2025 tech trends report* (18th ed.). https://ftsg.com/wp-content/uploads/2025/03/FTSG_2025_TR_FINALGeomaus. (n.d.). *Digital twin technology for facility management.* https://geomaus.com/digital-twin-technology-facility-management/

Gupta, D. (2024, November 11). *A guide on smart campus technology for higher education.* Appinventiv. https://appinventiv.com/blog/smart-campus-technology-for-higher-education

Hall, A. (2024). *The power of shared vision: Building successful teams.* Aaron Hall Blog. https://aaron-hall.com/the-power-of-shared-vision-building-successful-teams/

Hartman, T. (2023, June 15). *The future of learning: How smart campus technology is changing higher education.* Hartman Executive Advisors. https://hartmanadvisors.com/the-future-of-learning-how-smart-campus-technology-is-changing-higher-education-2/

Hong Kong University of Science and Technology. (2024). Navigating the skies: HKUST's game-changing digital twin. https://hkust.edu.hk/news/navigating-skies-hkusts-game-changing-digital-twin

Hong Kong University of Science and Technology. (2025). *HKUST launches "Sustainable Smart Campus as a Living Lab" inspiring a culture for sustainability innovation.* https://ssc.ust.hk/

Hopping, D. (2024, April 29). *Tracking toward carbon neutrality: Digitalization's role in sustainable higher education campuses. Forbes.* https://www.forbes.com/sites/siemens-smart-infrastructure/2024/04/29/tracking-toward-carbon-neutrality-digitalizations-role-in-sustainable-higher-education-campuses/

Hopping, D. (2024, August 5). *Digitalization in sustainable higher education campuses. Sustainability Magazine.* https://sustainabilitymag.com/articles/digitalisation-in-sustainable-higher-education-campuses

Ideta. (2024, September 8). *Smart campus solutions: Enhancing college life with AI technology.* https://www.ideta.io/blog-posts-english/smart-campus-solutions-enhancing-college-life-with-ai-technology

Interview Guy Editorial Team. (2025). *Building automation specialist job description* [Updated for 2025]. Interview Guy. https://interviewguy.com/building-automation-specialist-job-description/

iScano. (n.d.). *6-month digital twin roadmap for construction: Laser scanning and LiDAR best practices.* https://iscano.com/laser-scanning-lidar-best-practices/digital-twin-implementation-roadmap/

IT Executives Council. (2023, October 23). *8 essential questions every CTO must address before initiating digital transformation.* https://itexecutivescouncil.org/8-essential-questions-every-CTO-must-address-before-initiating-digital-transformation/

Izourane, F. Z., Ardchir, S., Ounacer, S., and Azzouazi, M. (2024). Smart campus based on AI and IoT in the era of Industry 5.0: Challenges and opportunities. In *Industry 5.0 and emerging technologies: Transformation through technology and innovations* (pp. 39–?). Springer. https://link.springer.com/chapter/10.1007/978-3-031-70996-8_3

Jones, M. (2022). *What is a smart campus, and what are the benefits for college students and faculty?* Coxblue. https://www.coxblue.com/what-is-a-smart-campus-and-the-benefits-to-college-students-and-faculty/

Kellton. (n.d.). *AI data quality: Ensuring robust data hygiene layer for better AI accuracy.* https://www.kellton.com/kellton-tech-blog/ai-data-quality-data-hygiene-for-better-ai-accuracy

Khan, M. A., and Almusharraf, A. (2025). AI-driven digital twins for predictive maintenance. *Proceedings of the 2025 European Conference on Computing in Construction (EC3).* https://ec-3.org/publications/conferences/EC32025/papers/EC32025_369.pdf

Kieler, A. (2022, March 7). *The top issues facilities leaders need to be thinking about.* Higher Ed Facilities Forum. https://info.higheredfacilitiesforum.com/blog/the-biggest-issues-facilities-leaders-need-to-consider-right-now

Lamb, T. (2024, September 30). *Smart campuses: 6 trends shaping the future of higher education.* Performance Services. https://www.performanceservices.com/resources/smart-campus-6-trends-shaping-the-future-of-higher-education/

Malki, D. (2024). Smart and sustainable city experience on smart campus: A case study from Hassan Ist University. *Asian Journal of Research in Computer Science, 17*(5), 43–?.

Manning, S. (2024, February 21). *University facilities leaders discuss planning and sustainability challenges.* Higher Ed Facilities Forum. https://info.higheredfacilitiesforum.com/blog/university-facilities-leaders-discuss-planning-sustainability-challenges

Maxsenti, M. (2022, January 19). *How technology is shaping modern educational buildings.* Genea. https://www.getgenea.com/blog/the-power-of-the-smart-campus/

Microsoft. (n.d.). *Environmental sustainability.* Microsoft© Corporate Social Responsibility. https://www.microsoft.com/en-us/corporate-responsibility/sustainability

Microsoft. (2021, February). *RealEstateCore, a smart building ontology for digital twins, is now available.* Azure IoT Blog. https://techcommunity.microsoft.com/blog/iotblog/realestatecore-a-smart-building-ontology-for-digital-twins-is-now-available/1914794

Microsoft. (2024). *Adopting DTDL-based industry ontologies: Azure Digital Twins.* https://learn.microsoft.com/en-us/azure/digital-twins/concepts-ontologies-adopt

Missouri State University. (n.d.). *Sustainability coordinator* [Job description]. https://www.missouristate.edu/Human/JobDescriptions/1311.htm

Mousa, M., and Al-Sadoon, A. (2025). AI-enhanced IoT systems for predictive maintenance and affordability optimization in smart microgrids: A digital twin approach. *arXiv.* https://arxiv.org/html/2511.12175v1

Nagda, V. (n.d.). *IoT engineer job roles and responsibilities.* Instrumentation Tools. https://instrumentationtools.com/iot-engineer-job-roles/

National Association of Energy Service Companies. (2024, August 14). *Oberlin: Campus energy and resource senior manager* [Job posting]. https://members.naesco.org/jobs/Details/oberlin-campus-energy-resource-senior-manager-

National University of Singapore. (n.d.). *Digital twin models: A comparative approach to global methods.* Asia Research Institute. https://ari.nus.edu.sg/grant/digital-twin-models-a-comparative-approach-to-global-methods/

Orenstein, D. E., Troupin, D., Segal, E., Holzer, J. M., and Hakima-Koniak, G. (2019). Integrating ecological objectives in university campus strategic and spatial planning: A case study. *International Journal of Sustainability in Higher Education, 20*(2), 190–?.

Palter, J. (2024, May 8). *Enhancing campus safety with smart security systems.* Real Time Networks. https://www.realtimenetworks.com/blog/enhancing-campus-safety-with-smart-security-systems

Pearce, K. (n.d.). Four best practices to go beyond digital twins. *Manufacturing Today.* https://manufacturing-today.com/news/four-best-practices-to-go-beyond-digital-twins-by-kyle-pearce/

Pinnacle Infotech. (n.d.). *Enhancing facility management with digital twin technology: A guide.* https://pinnacle-infotech.com/enhancing-facility-management-with-digital-twin-technology/

Polin, K., Yigitcanlar, T., Limb, M., and Washington, T. (2023). The making of smart campus: A review and conceptual framework. *Buildings, 13*(4), 89.

Pratt, M. (2024, September 23). *15 digital transformation questions every CTO must answer.* CTO. https://www.CTO.com/article/652979/10-digital-transformation-questions-every-CTO-must-answer.html

Prosci. (n.d.). *The Prosci ADKAR® model.* https://www.prosci.com/methodology/adkar

ProptechOS. (n.d.). *RealEstateCore: The universal language for smart buildings.* https://proptechos.com/realestatecore/

Qnex Campus. (2024). *5 smart campus solutions you must know for sustainable campus.* https://qnexcampus.com/5-smart-campus-solutions-you-must-know-for-sustainable-campus/

Rao, M., Neetha, S., Rao, S., Sneha, M., Kotian, S., and Rao, N. (2018). An IoT based smart campus system. *International Journal of Scientific and Engineering Research, 9*(4), 146–151. https://www.ijser.org/researchpaper/An-IoT-Based-Smart-Campus-System.pdf

Rutherford, J. (2023, August 16). *The design of hybrid teaching environments: 10 questions answered.* EDUCAUSE Review. https://er.educause.edu/articles/2023/8/the-design-of-hybrid-teaching-environments-10-questions-answered

Sand Technologies. (n.d.). *5 steps to prepare your data for AI.* https://www.sandtech.com/insight/5-steps-to-prepare-your-data-for-ai/

Sharma, R., and Gupta, S. (2024). Integration of Building Information Modeling (BIM) and digital twins for lifecycle cost optimization in mega-projects. *ResearchGate.* https://www.researchgate.net/publication/396874076_Integration_of_Building_Information_Modelling_BIM_and_Digital_Twins_for_Life cycle_Cost_Optimisation_in_Mega_Projects

Sharma, H., and Kanwal, N. (2024). Video surveillance in smart cities: Current status, challenges and future directions. *Multimedia Tools and Applications.* https://doi.org/10.1007/s11042-024-19696-6

Simio. (n.d.). *Master digital twin creation: Practical guide for beginners.* https://www.simio.com/master-digital-twin-creation-practical-guide-for-beginners/

Sims, T. (2018, January 21). *8 (new) energy efficient materials architects should know.* ArchDaily. https://www.archdaily.com/886414/8-new-energy-efficient-materials-architects-should-know

Singh, A. (2024). AI powered digital twins for predictive maintenance and operational optimization of renewable energy systems. *ResearchGate.* https://www.researchgate.net/publication/394169381_AI_Powered_Digital_Twins_for_Predictive_Maintenance_and_Operational_Optimization_of_Renewable_Energy_Syste ms

Singh Kanhaiya, K. S. (2024, June 27). *Revitalizing change: A guide for project managers on the "R" of ADKAR.* Institute of Project Management. https://instituteprojectmanagement.com/blog/a-guide-for-project-managers-on-the-r-of-adkar/

Sinopoli, J. (2016). *Advanced technology for smart buildings.* Artech House.

Stanford University. (2024). *The 10 best ways to make your school campus more sustainable.* MAHB Blog. https://mahb.stanford.edu/blog/the-10-best-ways-to-make-your-school-campus-more-sustainable/

Switch Automation. (2019). *Digital facilities management: Leveraging the power of people, data and technology.* https://www.switchautomation.com/

Teachmint. (2024, April 16). *What is a smart classroom: Complete overview.* https://blog.teachmint.com/smart-classroom-overview-tmx/

University of Melbourne. (n.d.). *Digital twin developments projects.* Centre for Spatial Data Infrastructures and Land Administration. https://eng.unimelb.edu.au/csdila/projects/digital-twin

Valks, B., Arkesteijn, M. H., Koutamanis, A., and den Heijer, A. C. (2021). Towards a smart campus: Supporting campus decisions with Internet of Things applications. *Building Research and Information, 49*(1), 1–20.

Verhoef, L., and Bossert, M. (2019). *The university campus as a living lab for sustainability: A practitioner's guide and handbook.* Delft University of Technology.

VIATechnik. (n.d.). What *is a digital twin? And why should facilities teams care?* https://www.viatechnik.com/insights/blog/what-is-a-digital-twin-and-why-should-facilities-teams-care/

Villegas, A., & Wiek, H. (2026, April 10). *Analyst note: The state of sustainable investing in the private markets.* PitchBook Data, Inc.

Wagner, C., and Enshaei, R. (2024). From automation to autonomy: A digital twin framework for transparent agent and human collaboration in industrial multi-agent systems. *Systems, 14*(1), 76. https://doi.org/10.3390/systems14010076

Weagle, S. (2019, April 1). *From safe campus to smart campus.* Campus Security Today. https://campussecuritytoday.com/articles/2019/04/01/from-safe-campus-to-smart-campus.aspx

Willow. (n.d.). *The new way to manage buildings.* https://willowinc.com/

Willow Inc. (n.d.). Data security with Willow in the era of AI innovation. https://willowinc.com/data-security-with-willow-in-the-era-of-ai-innovation/3052

Willow Inc. (n.d.). From data to decisions: A journey across the insights spectrum. https://willowinc.com/from-data-to-decisions-a-journey-across-the-insights-spectrum/2869

Willow Inc. (n.d.). Willow platform. https://willowinc.com/willow-platform

Willow Inc. (n.d.). Willow's approach to AI. https://willowinc.com/willows-approach-to-ai/2697

Zito, P. (2021, April 20). *The ultimate guide to building automation careers.* Smart Buildings Academy. https://guides.smartbuildingsacademy.com/the-ultimate-guide-to-building-automation-careers

List of Figures and Exhibits

Figures

Exhibits

Index

K

L

M

N

O

P

R

S

T

U

V

W

Z

End Notes

[1] See the video "The Knowledge Navigator" (circa 1985) at https://youtu.be/umJsITGzXdo?si=rau2L2TsKdzQMBts

[2] Antifragility refers to the property of a system that does not merely withstand stress but improves because of it. In the smart campus context, it describes an institutional environment in which cyber-physical infrastructure, governance practices, and operational culture adapt, learn, and strengthen in the face of volatility, failures, or disruption. Rather than returning to baseline after a shock, an antifragile campus becomes measurably more predictable, more efficient, and more resilient over time—transforming stressors into sources of intelligence and long-term capability.

3 Polycrisis describes a state where multiple crises—such as economic, demographic, geopolitical, technological, environmental, and cultural—are intertwined, amplifying each other's effects and creating combined impacts that exceed their individual contributions. The term gained popularity through historian Adam Tooze's work, which highlighted the complex, mutually reinforcing disruptions shaping the 21st century. Earlier research in systemic risk and complexity science also contributed to the development of this field. In higher education and the smart-campus ecosystem, polycrisis reflects how pressures such as enrollment declines, financial instability, climate risks, cybersecurity threats, and rapid technological change do not occur in isolation—they interact, intensify, and exacerbate one another. Recognizing this interconnectedness helps institutional leaders understand that building resilient, even antifragile, systems requires addressing the entire network of dependencies, not just isolated issues. Tooze's foundational, widely cited essay explaining his use of the term and its intellectual lineage can be found at https://adamtooze.com/2022/10/29/chartbook-165-polycrisis-thinking-on-the-tight-rope/

4 The ISA Global Cybersecurity Alliance (ISAGCA) is a global consortium working to secure critical infrastructure. For more information, see https://gca.isa.org

5 The National Institute of Standards and Technology (NIST) provides guidance on how to secure operational technology (OT) while addressing its unique performance, reliability, and safety requirements. OT encompasses a broad range of programmable systems and devices that interact with the physical environment (or manage devices that interact with the physical environment). These systems and devices detect or cause a direct change through the monitoring and/or control of devices, processes, and events. Examples include industrial control systems, building automation systems, transportation systems, physical access control systems, physical environment monitoring systems, and physical environment measurement systems. The document provides an overview of OT and typical system topologies, identifies common threats and vulnerabilities to these systems, and provides recommended security countermeasures to mitigate the associated risks. See https://csrc.nist.gov/pubs/sp/800/82/r3/final?utm_source=chatgpt.com

6 The Cybersecurity and Infrastructure Security Agency publishes information to assist organizations in making decisions for designing, implementing, and managing OT environments to ensure they are safe and secure, as well as enable business continuity for critical services. For more information, see https://www.cisa.gov/resources-tools/resources/principles-operational-technology-cyber-security

7 The National Institute of Standards and Technology Cybersecurity Framework is available at https://nvlpubs.nist.gov/nistpubs/CSWP/NIST.CSWP.29.pdf

8 MQTT is a Client Server publish/subscribe messaging transport protocol. It is light weight, open, simple, and designed to be easy to implement. These characteristics make it ideal for use in many situations, including constrained environments such as for communication in Machine to Machine (M2M) and Internet of Things (IoT) contexts where a small code footprint is required and/or network bandwidth is at a premium for more information see https://docs.oasis-open.org/mqtt/mqtt/v5.0/mqtt-v5.0.html?utm_source=chatgpt.com

[9] For more information about ASU's Syllabot and other initiatives see https://www.linkedin.com/pulse/from-static-syllabus-living-learning-systems-lev-gonick-8yopc/

10 For more info on IT Service Management standards visit https://www.opentext.com/what-is/itsm

[iii] For more information about ASHRAE, its standards, and numerous resources see https://www.ashrae.org/

12 First published in 1962, Everett M. Rogers' Diffusion of Innovations. The theory explains how, why, and at what rate new ideas and technology spread through a population. He mapped adoption to a standard bell curve (a normal distribution), which, mathematically, yields the exact percentages you are using. For more information, see Rogers, E. M. (2003). Diffusion of Innovations (5th ed.). Free Press.

13 Nmap ("Network Mapper") is a free, open-source tool used for network discovery and security audits. It is widely utilized by system and network administrators for tasks such as network inventory management, scheduling service upgrades, and monitoring hosts or services uptime. See https://nmap.org/

14 OODA (Observe–Orient–Decide–Act) is a continuous decision loop that enables organizations to sense conditions, interpret context, choose responses, and act quickly—learning and adapting faster than the environment changes. It is a well-established methodology. For more information, see https://miro.com/strategic-planning/what-is-ooda-loop/

15 For more information on sourcing ADKAR training, see https://events.educause.edu/educause-prosci-change-management-programs

16 Based at the University of California San Diego, ALERTCalifornia is a public safety program working to understand natural disasters and determine short and long-term impacts on people and the environment to inform management decisions. For more information, see https://alertcalifornia.org/

17 Examples of AV media control/room management software include Crestron Fusion, Extron GVE, Q-Sys QSC Biamp AMX

18 *See Digital Twin Technology: Definition, Benefits, and Industry Impact* - Almawave, https://www.almawave.com/digital-twin-technology/

19 See *What Is a Digital Twin? And Why Should Facilities Teams Care?* - VIATechnik, https://www.viatechnik.com/insights/blog/what-is-a-digital-twin-and-why-should-facilities-teams-care/

20 See *From BIM to digital twin. IOT data integration in asset management platform*, at https://www.researchgate.net/publication/387451435_From_BIM_to_digital_twin_IOT_data_integration_in_asset_management_platform

21 Integration of Building Information Modeling (BIM) and Digital Twins for Lifecycle Cost Optimization in Mega Projects.,
https://www.researchgate.net/publication/396874076_Integration_of_Building_Information_Modelling_BIM_and_Digital_Twins_for_Lifecycle_Cost_Optimisation_in_Mega_Projects

22 Digital Twins from Building to Urban Areas: An Open Opportunity to Energy, Environmental, Economic and Social Benefits - MDPI, https://www.mdpi.com/2076-3417/15/19/10795

23 Reference 6-Month Digital Twin Roadmap for Construction at, https://iscano.com/laser-scanning-lidar-best-practices/digital-twin-implementation-roadmap/

24 The Digital Continuum: The Digital Transformation Roadmap - Simio, https://www.simio.com/de/whitepapers/the-digital-continuum-the-digital-transformation-roadmap/

25 From BIM to digital twin. IOT data integration in asset management platform, https://www.researchgate.net/publication/387451435_From_BIM_to_digital_twin_IOT_data_integration_in_asset_management_platform

26 See *Digital Twin AI: Opportunities and Challenges from Large Language Models to World Models,* at https://arxiv.org/html/2601.01321v1

27 Reference *5 Best Practices for Digital Twin Implementation* - WGI Engineering, https://wginc.com/5-best-practices-for-digital-twin-implementation/

28 See *A Mega-Study Of Digital Twins Reveals Strengths, Weaknesses And Opportunities For Further Improvement*, Stanford SCALE Initiative, https://scale.stanford.edu/ai/repository/mega-study-digital-twins-reveals-strengths-weaknesses-and-opportunities-further

29 Reference *AI Data Preparation Guide for Quality and Optimal Inference*, Tredence, https://www.tredence.com/blog/ai-data-preparation

30 For more information about the Hong Kong University of Science and Technology's Digital Twin, see https://digitaltwin.hkust.edu.hk/

31 Digital Bricks designs cloud-native data lakes that provide a single, flexible repository for storing vast amounts of raw, structured, and unstructured data. For more information about data lakes, see https://www.digitalbricks.ai/build-innovate/data-lakes

32 Reference, *Digital Twins for a Physical World*, Autodesk University, https://www.autodesk.com/autodesk-university/class/Digital-Twins-Physical-World-2019

33 Reference, *Digital Twins in Higher Education Estates: A Practical Guide*, Twinview, https://www.twinview.com/insights/digital-twins-in-higher-education-estates-a-practical-guide

34 Reference, *Modbus vs. BACnet: A Deep Dive into Building and Industrial Communication Protocols*, https://www.andivi.com/modbus-vs-bacnet-a-deep-dive-into-building-and-industrial-communication-protocols/

35 See the Department of Energy's publication, *Better Windows, Better Outcomes: How Electrochromics Improve Health, Productivity, and Efficiency* at https://www.energy.gov/sites/default/files/2023-12/bto-electrochromic-window-report-121123.pdf

36 Reference, , *Edge AI: A Comprehensive Survey of Technologies, Applications, and Challenges, Shankar, at* https://www.researchgate.net/publication/385370366_Edge_AI_A_Comprehensive_Survey_of_Technologies_Applications_and_Challenges

37 The NAU Sustainable Smart Campus Master Plan is available at https://in.nau.edu/wp-content/uploads/sites/139/2023/11/NAU-Smart-Campus-Master-Plan_11202023_DLR-Group-Part-1.pdf and https://in.nau.edu/wp-content/uploads/sites/139/2023/11/NAU-Smart-Campus-Master-Plan_11202023_DLR-Group-Part-2.pdf

38 See https://nvlpubs.nist.gov/nistpubs/CSWP/NIST.CSWP.29.pdf for more information about the NIST CSF 2.0 framework for risk.

39 See NIST Computer Security Resource Center at https://csrc.nist.gov/pubs/sp/800/82/r3/final?utm_source=chatgpt.com